Pennsylvania Weather

Ben Gelber

A 400-Year CHRONICLE of HISTORIC EVENTS

Pennsylvania Weather

Ben Gelber

A 400-Year CHRONICLE *of* HISTORIC EVENTS

press

Book Design & Production:
Columbus Publishing Lab
www.ColumbusPublishingLab.com

All rights reserved.
This book, or parts thereof, may not be
reproduced in any form without permission.

Library of Congress Cataloging-in Publication Data
Pennsylvania Weather / Ben Gelber p. cm.
Includes bibliographical references and index.

Paperback ISBN: 978-1-63337-467-6
E-Book ISBN: 978-1-63337-468-3

Copyright © 2025 by Benjamin D. Gelber
LCCN: 2020924273

Printed in the United States of America
1 3 5 7 9 10 8 6 4 2

Material adapted from Benjamin D. Gelber,
The Pennsylvania Weather Book.
New Brunswick, N.J.: Rutgers University Press, 2002

Cover photos: Ben Gelber, fall photos credit
POCONOMOUNTAINS.COM with permission.

Dedication

*To my parents, Norman and Judith Gelber,
who inspired and nurtured my enthusiasm
for all things weather and journalism,
and my loving family.*

Contents

Figures ix

Tables xv

Preface and Acknowledgments xvii

Introduction 1

Chapter One: Pennsylvania Landscape and Climate 3

Chapter Two: Chroniclers of Pennsylvania Weather 25

Chapter Three: Seasons on Parade 45

Chapter Four: Winter Storms and Out-of-Season Snowfalls 65

Chapter Five: Cold Waves 173

Chapter Six: Rainstorms and Floods 207

Chapter Seven: Thunderstorms, Tornadoes and Whirlwinds 267

Chapter Eight: Heat Waves and Droughts 361

Chapter Nine: Tropical Weather Systems 389

Chapter Ten: Climate Trends 435

Appendices and Climate Tables

    Monthly Weather Extremes 463

    Annual Weather Extremes 466

    Seasonal Averages and Extremes 466

    Monthly Averages and Extremes 467

    Climate Extremes for Selected Cities 467

    Pennsylvania Weather Stations 474

Bibliography 477

Index 487

About the Author 507

Figures

Fig. 1.1 Pennsylvania counties.

Fig. 1.2 Major Pennsylvania cities.

Fig 1.3 Physiographic regions of Pennsylvania.

Fig. 1.4 Allegheny National Forest. (Photo by Steve Rossman)

Fig. 1.5 Point State Park in Pittsburgh, at the confluence of the Allegheny and Monongahela Rivers, forming the Ohio River. (Photo by Shauna Weyrauch)

Fig. 1.6 Delaware Water Gap, where the Delaware River flows through Kittatinny Ridge, forming the border between Pennsylvania and New Jersey. (Photo by Ben Gelber)

Fig. 1.7 Tobyhanna Lake, where ice was harvested in blocks and shipped by railcars to New York and Philadelphia before the days of electric refrigerators. (Photo by Ben Gelber)

Fig. 1.8 Raymondskill Falls, in southeastern Pike County, tumbling off the Pocono Plateau, is the highest waterfall (150 feet) in the state. There are more than 170 waterfalls in the Keystone State. (Photo by Ben Gelber)

Fig. 1.9 Morning fog trapped in a narrow valley along Interstate 80 east of Lock Haven. (Photo by Ben Gelber)

Fig. 1.10 Average annual precipitation.

Fig. 1.11 Average annual snowfall.

Fig. 2.1 Daily weather observations of Israel Pemberton at Philadelphia in June of 1776. (Courtesy American Philosophical Society)

Fig. 3.1 *Autumn*. Blue Mountain Lake, nestled in a clearing on Brushy Mountain, on a gray fall afternoon. (Photo by Ben Gelber)

Fig. 3.2 *Winter*. McMichael Creek, winding through the southern Poconos, after a snowfall. (Photo by Justin Gelber)

Fig. 3.3 *Spring*. Lush vegetation on Brodhead Creek, a tributary of the Delaware River in the southern Poconos. (Photo by Ben Gelber)

Fig. 3.4 *Summer*. Thunderstorm over Delaware Water Gap, looking north from Portland. (Photo by Ben Gelber)

Fig. 4.1 November 29, 1950, following the great snowstorm. Cars buried on Reifert Street. (Photo by Morris Berman, *Pittsburgh Post-Gazette* Photo Archives, copyright 1950, all rights reserved)

Fig. 4.2 February 17, 1958. A blizzard deposited 33 inches of snow on Honesdale, raising the total depth to 43 inches on Tryon Street. (Courtesy Elizabeth Korb Schuman)

Fig. 4.3 January 8, 1996. Fayette County Head Start program bus snowbound after a blizzard dumped two feet of snow on southwestern Pennsylvania. (Photo by Robin Rombach, *Pittsburgh Post-Gazette* Photo Archives, copyright 1996, all rights reserved)

Fig. 4.4 January 9, 1996. The author's father, Norman Gelber, in East Stroudsburg, digging out from the blizzard that deposited nearly two feet of snow, leaving a record January snow depth of 30 inches. (Photo by Judith Gelber)

Fig. 4.5 January 8, 2005. Residents described the destructive ice storm as a "war zone" on the Pocono Plateau between Effort, Blakeslee and Tobyhanna. (Photo by Michael Pontrelli)

Fig. 4.6 January 23, 2016. Deep snowfall in the southern Poconos leaves 18 to 26 inches in southern Monroe County. (Photo by Ben Gelber)

Fig. 4.7 January 23, 2016. Cars buried in the wake of an 18-inch snowstorm in East Stroudsburg. (Photo by Ben Gelber)

Fig. 4.8 March 15, 2017. Late winter snowstorm deposits two feet of snow on the higher elevations of northeastern Pennsylvania at Resica Falls. (Photo by Ben Gelber)

Fig. 4.9 April 23 1986. Historic late April snowstorm near Saylorsburg on Godfrey Ridge. (Photo by Michael Pontrelli)

Fig. 5.1 January 27, 2014. Snow rollers formed in western Pennsylvania at Clarion, following a surge of arctic air in the wake of an Alberta clipper system and brief thaw. (Photo by Ben Gelber)

Fig. 6.1 October 4-5, 1869. The oldest known photograph of flood damage in Pennsylvania at Stroudsburg, following the flood of October 4, 1869, which appeared in *Picturesque Monroe County* (1897)

Fig. 6.2 June 1, 1889. Stereoscopic views following the Great Johnstown Flood where the Woodvale prison had floated to from a mile away (top), and Main and Franklin Streets, near the junction of Conemaugh and Stony Creeks (bottom). (Courtesy Pennsylvania State Archives)

Fig. 6.3 June 1, 1889. Stereoscopic views after the Great Johnstown Flood, from the Pennsylvania Railroad Depot (top), near the Club House and iron railroad bridge, and at Main and Clinton Streets (bottom). (Courtesy Pennsylvania State Archives)

Fig. 6.4 June 1, 1889. Great Johnstown Flood aftermath near the center of the flood-ravaged community. (Courtesy Pennsylvania State Archives)

Fig. 6.5 June 1, 1889. Homes in Johnstown reduced to rubble following the Great Johnstown Flood. (Courtesy Pennsylvania State Archives)

Fig. 6.6 June 5, 1892. A view of South Franklin Street in Titusville, after the flood and devastating fires of June 4-5, 1892. (Courtesy *Titusville Herald* [Oct. 7, 1897] and Crawford County Historical Society)

Fig. 6.7 March 19, 1936. St. Patrick's Day Flood along the swollen Susquehanna River. (Courtesy Pennsylvania Historical Society)

Fig. 6.8 March 19, 1936. St. Patrick's Day Flood inundated the center of Harrisburg. (Courtesy Pennsylvania Historical Society)

Fig. 6.9 March 19, 1936. St. Patrick's Day Flood damage at the Golden Triangle, at the confluence of the Allegheny and Monongahela Rivers near downtown Pittsburgh. (*Pittsburgh Post-Gazette* Photo Archives, copyright 1999, all rights reserved)

FIGURES   xi

Fig. 6.10 March 19, 1936. St. Patrick's Day Flood covering Liberty Avenue in Pittsburgh. (*Pittsburgh Post-Gazette* Photo Archives, copyright 1986, all rights reserved)

Fig. 6.11 May 23, 1942. Heavy downpours sent the Lackawaxen River out of its banks, triggering a deadly flash flood in Wayne County. Looking north from Willow Avenue, Honesdale. (Courtesy Elizabeth Korb Schuman)

Fig. 6.12 August 19, 1955. The swollen Delaware River, after a floating house split the Northampton Free Bridge connecting Easton, Pennsylvania, and Phillipsburg, New Jersey. At one point water covered the bridge deck. (Courtesy Easton *Express-Times*)

Fig. 6.13 August 19, 1955. Widespread flooding in downtown Easton, with the "World's Fastest Car Wash" living up to its advertisement. The Delaware River rose to a record height of 43.7 feet. Flood stage at Easton is 22 feet. (Courtesy Easton *Express-Times*)

Fig. 6.14 August 19, 1955. Flooding in Easton at Northampton and Front Streets, where the water ignored the one-way sign and surrounded the city. (Courtesy Easton *Express-Times*)

Fig. 6.15 August 19, 1955. The rampaging Brodhead Creek floodwaters tore through the twin boroughs of Stroudsburg and East Stroudsburg, taking scores of lives. (Courtesy Monroe County Historical Association)

Fig 6.16 Bridge at Canadensis into Brodhead Creek in the Poconos. (Courtesy Monroe County Historical Association)

Fig. 6.17 June 25, 1972. The Susquehanna River roared through York, after the passage of Tropical Storm Agnes. (Courtesy Pennsylvania State Archives, from a series of photographs in the *York Daily Record*)

Fig. 6.18 June 25, 1972. Downtown York swamped by the overflowing Susquehanna River, following several days of torrential rain totaling more than 12 inches. (Courtesy Pennsylvania State Archives)

Fig. 6.19 June 25, 1972. The Susquehanna River fanned out more than a mile wide near York. (Courtesy Pennsylvania State Archives)

Fig. 6.20 Widespread flooding swamps Shawnee Golf Course on September 18, 2004, in the wake of the remnants of Hurricane Ivan. (Photo by David Kidwell, *Pocono Record*)

Fig 6.21 Water lapping up at the entrance of Shawnee Inn in following several days of intense rainfall on June 29, 2006. (Photo by Keith Stevenson, *Pocono Record*)

Fig. 6.22 Streets turned into rivers and homes flooded in Portland along the Delaware River on June 29, 2006. (Courtesy *Pocono Record*)

Fig. 6.23 The Delaware River overflowed in Delaware Water Gap, swamping homes on Broad Street after nearly a week of torrential rain, on June 29, 2006. (Photo by David Kidwell, *Pocono Record*)

Fig. 6.24 Soccer fields and clubhouse in Minisink Hills, Smithfield Township, submerged after Hurricane Irene blew past northeastern Pennsylvania on August 28, 2011. (Photo by David Kidwell, *Pocono Record*)

Fig. 7.1 Lightning strike on August 2, 2011, branching out near Lake Wallenpaupack, viewed from Newfoundland, Wayne County. (Photo by Keith Stevenson, *Pocono Record*)

Fig. 7.2 Pittsburgh's National Weather Service radar dome.

Fig. 7.3 May 31, 1985. Two funnels merged into one violent F4 tornado approaching Albion, Erie County. The tornado slammed into Albion, then roared east for three more miles, ending northeast of Cranesville. Twelve persons died and 82 were injured. (Courtesy *Erie's History and Memorabilia*)

Fig. 7.4 May 31, 1985. The large tornado caused total devastation in the community of Albion, where nine persons died. (Courtesy *Erie's History and Memorabilia*)

Fig. 7.5 May 31, 1985. Homes were reduced to rubble along the path of the storm in a one-square mile section of Albion. (Photo by Dick Ropp)

Fig. 7.6 May 31, 1985. A total of 309 homes and businesses were destroyed or severely damaged in the path of the tornado, leaving 1,800 residents homeless. (Courtesy *Erie's History and Memorabilia*)

Fig. 7.7 May 31, 1985. The *Beaver County Times* headline, following the passage of a deadly tornado.

Fig. 7.8 May 31, 1985. Large F3 tornado that destroyed Big Beaver Plaza, taking three lives. The tornado "roared like a freight train at full throttle—bent on a 13-mile trail of destruction." A total of 260 North Sewickley Township homes were damaged or destroyed in the path of the storm. (Courtesy *Beaver County Times*)

Fig. 7.9 May 31, 1985. The aftermath of a 2.2-mile-wide F5 tornado that sliced through Moshannon State Forest. (Photo by Greg Forbes. Courtesy "The Tornado Outbreak of May 31, 1985," Paul Markowski, Pennsylvania State University Department of Meteorology and Atmospheric Science, 2015.)

Fig. 7.10 May 31, 1985. Radar image (WSR-74C) from the top of Walker Building, Pennsylvania State University around 8:00 p.m., revealing a classic hook echo, as a massive tornado tore through Moshannon State Forest. (Courtesy Paul Markowski)

Fig. 7.11 NOAA Storm Data mapping of the tracks of 42 tornadoes recorded on May 31, 1985. (Not shown: the 43rd tornado that touched down after midnight on June 1 near Tobyhanna and a 14th tornado in southeastern Canada, later identified by Environment Canada.)

Fig. 7.12 May 31, 1998. Tornado damage on Cody Street, one block east of Route 219 in Salisbury, Somerset County. (Photo by V. W. H. Campbell, *Pittsburgh Post-Gazette* Photo Archives, copyright 1998, all rights reserved)

Fig. 7.13 June 2, 1998. A rare sight of twin funnels, traveling northeast along the Ohio River approaching Mount Washington, photographed by a KDKA-TV cameraman. (Courtesy *Pittsburgh Post-Gazette* Photo Archives, copyright 1998, all rights reserved)

Fig. 7.14 Aftermath of a tornado that hit Cherry Valley on July 29, 2009. (Photo by David Kidwell, *Pocono Record*)

Fig. 7.15 Home destroyed by a tornado in Cherry Valley on July 29, 2009. (Photo by Adam Richins, *Pocono Record*)

Fig. 7.16 Tornado that formed near New Stanton and moved to near Norvelt and Unity, lifting near the Westmoreland County Fairgrounds around 8:00 p.m. on June 27, 2018. (Photo by John Mark Benson)

Fig. 9.1 Satellite image of Hurricane Sandy approaching the New Jersey coast on October 29, 2012. (NOAA)

Fig. 9.2 National Guard Alpha Troop 2nd 104 assisted the Marshalls Creek Volunteer Fire Department on Mt. Nebo Road. (Courtesy *Pocono Record*)

Fig. 9.3 Pine tree on Gap View Road in Minisink Hills. Thousands of trees whipped down during Sandy's onslaught, toppling power lines. (Photo by David Kidwell, *Pocono Record*)

Tables

4.1 Snowiest winters in eastern Pennsylvania (October-May) 71–72

4.2 Warmest winter months in Pennsylvania 75

4.3 Snowiest months in Pennsylvania (1890–2020) 75

4.4 Record January 1996 snow totals and depths 126

4.5 Philadelphia's top 10 snowstorms (1884-2021) 126

5.1 Low temperatures, January 4-5, 1835 176

5.2 Comparison of mean temperatures at Nazareth, January 1856 and 1857 181

5.3 Comparison of mean winter temperatures at Dyberry 184

5.4 Maximum/minimum temperatures, December 29-31, 1880 187

5.5 Minimum temperatures, January 5, 1904 190

5.6 Minimum temperatures, January 5, 1904 (*Stroudsburg Times*) 190

5.7 Minimum temperatures, January 14, 1912 192

5.8 Maximum/minimum temperatures, December 29, 1917-January 4, 1918 192

5.9 Daylight maximum temperatures, January 17, 1982 201

7.1 Fujita-Pearson Tornado Intensity Scale (F-Scale) 284

7.2 Enhanced Fujita Scale (EF-Scale) 285

8.1 Maximum temperatures, June 1-7, 1925 369

8.2 Maximum temperatures, July 9-11, 1936 371

8.3 Maximum temperatures, August 25-September 4, 1953 374

9.1 The Saffir-Simpson Hurricane Wind Scale 394

9.2 Most powerful Atlantic/Caribbean hurricanes by pressure 399

9.3 Most intense landfalling U.S. mainland hurricanes by pressure 399

Preface and Acknowledgments

Some are weather-wise, some are otherwise.
— BENJAMIN FRANKLIN, *POOR RICHARD'S ALMANACK* (FEBRUARY 1735)

THIS HISTORICAL weather project is designed to provide a chronological compilation of outstanding weather events that have impacted Pennsylvania since the earliest days of settlement through modern times.

I am deeply indebted to the seminal work and encouragement of Dr. David M. Ludlum (1910-1997), founder of *Weatherwise* magazine, for his methodical research and literary contributions to the field of historical meteorology. He enthusiastically supported my initial project, *Pocono Weather* (1992), writing that "we share a love of old snowstorms."

There are so many people I wish to thank, who aided me during the hundreds of research hours I spent in libraries and historical societies around Pennsylvania and neighboring states. Scientists and faculty at several universities provided key guidance in the preparation of the manuscript:

Katherine Glover, Climate Change Institute, University of Maine; Dale Gnidovec, Curator of The Ohio State University's Orton Geological Museum; Jay Hobgood, Department of Geography, The Ohio State University; Peter Jung, Warning Coordination Meteorologist, National Weather Service, State College; Paul Markowski, Department of Meteorology and Atmospheric Science, Pennsylvania State University. I also would like to thank Dr. Gregory Forbes, professor of meteorology at Pennsylvania State University, and later the Severe Weather Expert at The Weather Channel, for reviewing the chapter on severe storms.

I am indebted to historical society and library personnel for their assistance in materials: Ann Andrews, Director, B. F. Jones, Memorial Library, Aliquippa; George Deutsch, Erie County Historical Society, Erie; Michael Sherbon, Pennsylvania State Archives, Harrisburg; Robert Cox, American Philosophical Society, Philadelphia; Jacalyn Mignogna, Carnegie Department of Pittsburgh; Maria Carpico, *Pittsburgh Post-Gazette*; Carol King, Easton *Express-Times*; Sharon Gothard, Easton Area Public Library; Leslie Berger, Elizabeth Scott and Allyson Wind, Kemp Library, East Stroudsburg University; Amy Leiser, Monroe County Historical Association; Sherry Whittaker, Upper Arlington (OH) Public Library; Nancy Amspacher, York County Historical Society. Keith Stevenson, a Pocono Record photographer, generously provided his detailed historical images.

I also wish to acknowledge the generous assistance of Alva Wallis, National Climatic Data Center (NCDC), and Paul Knight and Kyle Imhoff, Pennsylvania State Climate Office state climatologists. Keith Eggleston, climatologist at the Northeast Regional Climate Center, Cornell University, provided considerable data guidance. Thank you to Larry Cottrill, WCMH-TV (NBC4) Vice President and General Manager, in Columbus, Ohio, and my colleagues for their support. The editorial process was aided by kind suggestions from Katherine Glover, Michelle Duda, Katie Millard, and Emily Hitchcock at the Columbus Publishing Lab.

Above all, I wish to honor the memories of my parents, Norman and Judith Gelber, who instilled a love of learning, from my school days through the years of research and a career in television meteorology. Both consummate educators, I am deeply grateful for their assistance in gathering historical information, and will forever cherish our loving conversations.

Introduction

The country itself, its soil, air, water, seasons, and produce, both natural and artificial, is not to be despised.
—William Penn, describing Pennsylvania in a letter to King Charles II
(*Pennsylvania: Birth of a Nation*, Sylvester K. Stevens)

WEATHER HAS BEEN a universal conversation starter probably as far back as the Genesis flood, when someone likely pulled Noah aside and commented, "I thought they said scattered showers ending by the weekend?"

American essayist Charles Dudley Warner, writing in the *Hartford Courant* in 1897, offered: "Everybody talks about the weather, but nobody does anything about it." (Mark Twain is sometimes credited with this line, because it sounds like something he would have said.)

Weather notions posited in Pennsylvania county histories often were quite lyrical. Jackson Lantz, in his *Picturesque Monroe County, Pennsylvania*, published in 1897, wrote: "The climate of this place and region in Monroe County is strong and bracing. The atmosphere in its best condition seems surcharged with electric oxygen, called ozone, which is peculiarly invigorating to jaded and prostrated nerves."

There are ample varieties of weather, and most of us fall into several primary camps: snow lovers, sun worshippers, fair-weather fans and storm chasers. In general, we take a matter-of-fact approach to the elements, while making daily plans and hoping transportation will not be hampered by icy roads or a downpour.

Our choices of home heating and cooling systems, landscaping and vacation plans are dependent upon weather statistics. A cold winter directly impacts the cost of electricity and natural gas just as surely as utility bills jump in a hot, muggy summer. Financial markets are keyed into the effects of weather on energy demands, agriculture, transportation and retail sales.

The inspiration for this book, which catalogues Pennsylvania's weather history and extremes, was the seminal work of the weather historian Dr. David Ludlum (1910–97), founding editor of *Weatherwise* magazine and a prolific author of regional books and articles. After kindly agreeing to read a draft of my initial foray, published in 1992, *Pocono Weather: The Poconos, Eastern Pennsylvania, Northwestern New Jersey*, he encouraged me to keep writing about weather history. And in a previous correspondence, he noted that "we both share a love for old snowstorms."

My next effort came a decade later, *The Pennsylvania Weather Book*, which was more exhaustive in scope by covering the entirety of the Keystone State. Two decades later, after compiling many more major events, the time came to revisit a history of outstanding weather events that have affected Pennsylvania from the early settlement days to the present.

In the early 1980s, as a newly minted television meteorologist, and gripped with more than a touch of anxiety, I was limited to static graphics, magic markers and stick-on letters/numbers that periodically slipped from the porcelain wall map. Fortunately, advancements in graphics and technology, with considerable video sources afford more opportunities to explain the atmospheric antics. We now understand what drives most weather phenomena from the perspective of science, and comprehend the human impacts of weather and climate with a greater appreciation of the vast, awe-inspiring forces of nature.

The panoramic tale of the vicissitudes of Pennsylvania weather history will commence before the arrival of William Penn on October 27, 1682 (Old Style), and span four centuries. The documented weather fluctuations are influenced to a large degree by the diverse landscape, formerly known as Penn's Woods.

CHAPTER ONE

Pennsylvania Landscape and Climate

Pennsylvania is an eastern state on a western scale—not only in sheer size, but also in the variety and grandeur of its landscape.

— JERE MARTIN, PENNSYLVANIA ALMANAC

WALLACE NUTTING, commented in his book, *Pennsylvania Beautiful (Eastern)*, "The beauty of the countryside in Pennsylvania is distinct from that of other regions." The bountiful scenery has inspired various writers and artists to depict the varied landforms in considerable detail.

The Keystone State comprises 46,058 square miles, including a little more than 1,000 square miles of water and 83,000 stream miles. The breadth of Pennsylvania is approximately 300 miles west to east and 160 miles north to south. The geographical center of the Commonwealth is 2.5 miles southwest of Bellefonte in Centre County.

The striking differences in landform, elevation, latitude, and a proximity to the Great Lakes and Atlantic Ocean, account for the array of climate patterns that drive our endless fascination with the weather.

Lay of the Land

The Appalachians are one of the oldest mountain systems in the world, extending more than 1,500 miles from Newfoundland, Canada, to central Alabama. Elongate rows of rough-hewn ridges bisect Pennsylvania trending northeast to southwest, receding into upland plateaus in the northern and western sections. On the eastern side of the mountains, rounded ridges rise above gently rolling hills.

The landscape has been beveled over eons by wind, water and reoccurring ice advances that stripped the topsoil, ground down mountains and reshaped the topography. Incised ridgetops, riparian woodlands and picturesque stream valleys reflect the ceaseless processes of nature's sculpting. Torrents of glacial meltwater later scoured out narrow valleys and passageways flanked by resistant sandstone bluffs.

Weathering and differential erosion of sedimentary rocks created a stunning array of topographic features. The relentless motions of the atmosphere broke up the rock strata into smaller fragments that were washed away or dissolved. Deeply folded layers formed steep-sided ridges that overlook tranquil, scenic waterways. Gently sloping uplands along the western side of the Appalachian ridges are framed by wooded hillsides and variegated foliage.

A diversity of wildlife can be found roaming secluded tracts of forestland, interrupted by

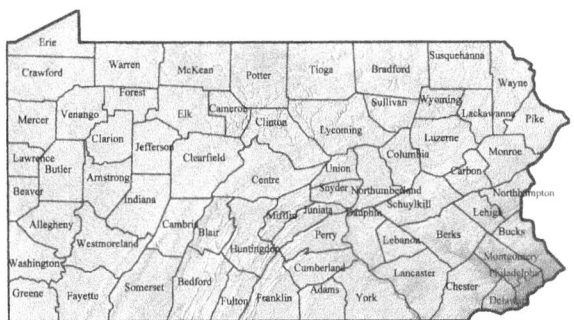

Fig. 1.1 Pennsylvania counties.

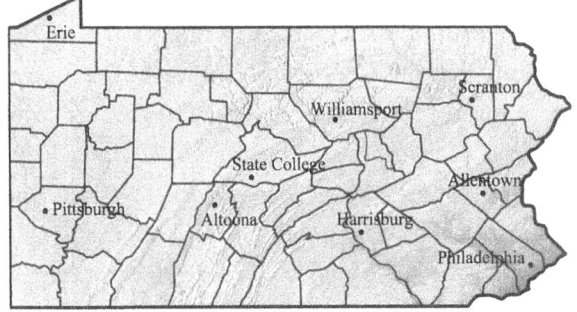

Fig. 1.2 Major Pennsylvania cities.

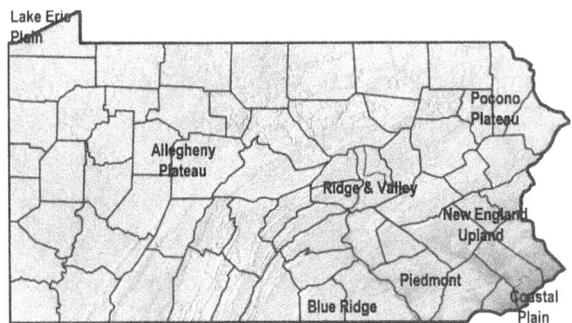

Fig. 1.3 Physiographic regions of Pennsylvania.

narrow stream valleys. The woodlands, brimming with maple, oak, hemlock and hickory, release oxygen into the air and filter pollutants through their deciduous leaves. Hikers and birders flock to meadows and sun-dappled ravines for a view of nature at its finest. Bucolic hills dotted with small farms seemingly connect with the horizon in the southern part of the state. A swath of grassland near the Maryland border boasts tracts of pitched pine and outcroppings that expose ancient bedrock.

Regional Topography of Pennsylvania

Pennsylvania has six primary physiographic provinces that possess a common geologic history and reveal similar topographic features.

The **Atlantic Coastal Plain** province is a small rectangular strip of land along the Delaware River that includes a large portion of the city of Philadelphia. The urban corridor is traversed by streams and tributaries that empty into the Delaware River. The elevation of the coastal plain in Pennsylvania is between 100 and 200 feet, dipping to near sea level at the mouth of Delaware Bay.

The **Piedmont** province in the interior southeast unfolds into rolling hills, hardwood forests and expansive farmland at elevations of 200 to 600 feet. The Piedmont Upland and Gettysburg-Newark Lowland sections off to the north and west reveal higher relief, with isolated ridges rising to 1,200 feet.

A sliver of the New York–New Jersey Highlands represents the western edge of the **New England** province, which extends into east-central Pennsylvania for about 50 miles. Rounded ridges and low hills rising above the landscape form steep slopes that comprise the Reading Prong section, including Mount Penn (elevation 1,120 feet) west of Reading. Elevations range from about 200 to 1,364 feet.

The distinctive **Ridge and Valley** province that bisects Pennsylvania features parallel, even-crested, narrow ridges that rise above 1,000 feet, with peaks topping 2,000 feet. Wide valley floors are traversed by numerous streams and afford pleasant relief from the ridgetops that descend sharply to elevations of 400 to 800 feet.

Southeast of Blue (Kittatinny) Mountain, on the eastern edge of the Appalachian Mountain Range, the verdant Great Valley is about 20 miles wide, known locally as the Cumberland, Lebanon and Lehigh Valleys. In south-central Pennsylvania, South Mountain marks the northernmost portion of the Blue Ridge Mountains lying between the Piedmont and Cumberland Valley.

The rugged Allegheny Front represents the beginning of the **Appalachian Plateaus** province that extends across northern and western Pennsylvania. The striking escarpment once posed a formidable barrier to early settlers. Westward migration was largely restricted to Native American trails along secluded streams shadowed by towering trees, or by way of early Virginia and southeastern Ohio until the railroads arrived.

The scenic Pocono Plateau in the northeastern part of the state is a fairly level and scenic upland that varies in elevation from 1,800 to 2,200 feet. The highest elevation in the Poconos is 2,323 feet in the Moosic Mountains northeast of Scranton. The formerly glaciated region is a lovely mix of natural wetlands and pastoral meadows adorned by wildflowers. Dozens of stunning waterfalls arise from rivulets and springs formed after the glaciers melted in wide depressions, tumbling onto the lower hills.

The Endless Mountains ranging northwest of the Pocono Plateau reach a peak elevation of 2,693 feet at the North Knob of Elk Mountain in Susquehanna County. Forests shade the undulating landscape and provide a sense or pristine beauty. The Deep Valleys section in the north-central part of the state comprises narrow upland plateaus dissected by sharp-edged valleys at elevations mostly between 1,000 and 2,000 feet. This is the most rugged and least populated area, featuring spectacular gorges carved by glacial meltwater that left behind a dendritic drainage pattern.

The Pennsylvania Wilds represents one of the most remote and beautiful areas of the Commonwealth. Bounded by Interstate 80 in the south and Route 6 across the north, the region is home to 27 state parks sprawled across two million acres of public land. Thousands of stream miles and hundreds of miles of trails wend through the Allegheny National Forest that provide a boon for nature explorers.

Fig. 1.4 Allegheny National Forest. (Photo by Steve Rossman)

Pine Creek Gorge in Tioga County is well-known as the Grand Canyon of Pennsylvania. Hundreds of species of birds and waterfowl frequent the woodlots, including ospreys and bald eagles in Sinnemahoning State Park. Eastern elk once roamed the meadows before being driven to near extinction. Today, they are slowly being restored to their ancestral home. Old-growth white pine and American hemlock tower in Cook Forest State Park, a relic of the primeval forests cloaked in oaks and hickories.

The Allegheny Mountain Range merges with the unglaciated plateaus of southwestern Pennsylvania. The Alleghenies comprise the Laurel Highlands and Chestnut Ridge, the westernmost portion of the Allegheny Mountains, and feature the highest elevations in the Keystone State. Mount Davis, at 3,213 feet, nestled on a soft crest in Forbes State Forest near Markleton in southern Somerset County, is the highest point in Pennsylvania.

Northeast of Pittsburgh, the Pittsburgh Low Plateau and Waynesburg Hills sections are hewn by streams and sprinkled with farms. Wide valleys and densely forested hills offer lovely scenery. To the south, narrowing ridges reach heights of 1,500 to 2,500 feet above valley floors dipping to elevations of 400 to 800 feet.

Fig. 1.5 Point State Park in Pittsburgh, at the confluence of the Allegheny and Monongahela Rivers, forming the Ohio River. (Photo by Shauna Woyrauch)

The northwestern corner of the state adjacent to Lake Erie marks the easternmost portion of the **Central Lowlands** province. The Eastern Lake section is a narrow lake plain between two and five miles wide. The elevation at Lake Erie is 570 feet, rising in the south to 1,000 feet. Stepped terraces are a relic of glacial times when the lake was several hundred feet higher than today. After meltwater receded, exposed beach ridges were fortified by an accumulation of sand and sediment.

Crude oil was first distilled into kerosene for illumination in a Pittsburgh refinery by the inventor Samuel Kier in 1853. In the eastern corner of Crawford County, 45 miles southeast of Erie, Edwin Drake drilled the first commercial oil well on the east bank of Oil Creek. Western Pennsylvania quickly became the worldwide leader in the petroleum industry. By the early 1860s, several million barrels of oil were produced from wells scattered across oil country.

Heavy logging dramatically altered the natural landscape that European settlers had encountered in early Pennsylvania, diminishing the number of forested acres from 95 percent to 32 percent by 1900, and leaving only a few hundred acres of old-growth stands in primarily remote areas. Trees were floated down the West Branch Susquehanna River to Williamsport, then known as the lumber capital of the world. Once replete with eastern hemlock, white pine and American beech trees, timbering had largely reduced much of the state's woodlands to patches of rubble.

The timber industry and intensive extraction of oil, gas, coal and minerals caused soil degradation and erosion, which took a heavy toll on the ecology of the region. Forest management helped the forests recover from clearcutting and railroad logging. Second-growth northern hardwood forests are filled with white pine, beech, birch, sugar and red maple, black cherry, northern red oak, white oak and eastern hemlock, and pockets of spruce and fir on the high ridges. The restoration of woodlands comprising hickory, poplar, sycamore and pine woodlands across southern parts of the state offer splendid scenery.

For more than a century, sulfuric acid from coal mines leached into watersheds. Iron residue made streams toxic and foamy with a sterile yellow-orange coloration, the legacy of unregulated mining. Coal and shale exposed to oxygen and running water in abandoned mines created pockets of acidic water containing dissolved heavy metals.

The state legislature of Pennsylvania enacted the Land and Water Conservation and Reclamation Act in January 1968, which focused on mitigating acid mine drainage and commenced environmental reclamation efforts.

Mountain Gaps

The fascinating story of the Delaware Water Gap, a spectacular gorge that splits the forested Kittatinny Ridge, begins at a time when the ancestral headwaters of the Delaware River gathered near Trenton, New Jersey. The magnificent gap is nearly a mile wide at the top and a quarter-mile wide at the river level, where the elevation dips to 280 feet above mean sea level.

The Delaware River, which formed during the Cenozoic Era about 30 to 50 million years ago, was hundreds of feet higher before the land was lowered by erosional forces. The absorption of an ancient creek to the north as lowland rocks gave ground permitted the process of "stream capture" to create the striking geologic feature. Downcut-

Fig. 1.6 Delaware Water Gap, where the Delaware River flows through Kittatinny Ridge, forming the border between Pennsylvania and New Jersey. (Photo by Ben Gelber)

ting forces forged a pathway through a weakened flexure, or fracture, near the headwaters and incised a downward notch in the ridge. The bulk of the work was accomplished by transported pebbles and sand over millions of years.

Farther southwest, where the Lehigh River slices through Blue Mountain, there is another scenic break—Lehigh Gap—between Carbon County to the north and Northampton and Lehigh counties to the south. Not all the gaps currently have rivers flowing through the opening. Wind Gap is one of several dry gaps where a stream was captured and diverted into a neighboring streambed.

Movers and Shakers of the Landscape

The geological story of Pennsylvania is a tale of plate tectonics. Earth's outermost shell (lithosphere) is divided into eight major plates and a half-dozen smaller pieces. Slabs of earth drift like barges on a semi-molten layer of rock (asthenosphere) beneath the outer upper mantle at the rate of one to two inches per year.

The middle of the Atlantic Ocean is separated longitudinally by the Mid-Atlantic Ridge. Along the rift valley, new crust forms as the seafloor separates, widening the ocean basin in the process referred to as seafloor spreading. Trapped gases under pressure melted the buried crust and released magma from fissures on the ocean floor near divergent boundaries.

Tremendous heat and pressure drive convection currents in the mantle that turns rock into a viscous liquid. Continental plates float on the semi-molten crust propelled by Earth's rotation and tidal energy. Lighter rock rides over denser and thinner oceanic crust, powered by gravity and magmatism. Slow-motion collisions cause plates to bend, as pieces are forced beneath the continent in subduction zones, forming deep ocean trenches at the edges.

Meteorologist Alfred Wegener (1880–1930), a polar explorer, postulated in a 1912 paper that the continents had drifted, after noting that the coast of eastern South America would seem to fit with the west coast of Africa. Scorned by geologists at the time, the notion of continental motion would be validated in the 1960s with the discovery that sea-floor spreading was driven by plate tectonics. The theory settled how related species of plants and animals on land were separated by an ocean.

Tectonic forces uplifted the ocean floor long before there was a North American Craton. Folded and thrust-faulted marine layers and ancient volcanic residue deposited a large amount of sediment. The relentless force of flowing water gradually wore down the thickly folded layers of limestones, dolomites and softer shales. In some places, streams met at nearly right angles before joining the main stem river, or trunk. Between the upraised ridges, differential weathering and erosion created a trellis-like drainage pattern.

Earthquakes in the eastern United States reflect accumulated strain slowly released along cracks left over from the ancient crustal jostling. The quakes are deeper and felt more widely in the East in contrast to temblors near the West Coast because seismic waves travel faster through the older, brittle (cooler) rocks. In the Pacific Northwest, continental blocks collide at a relatively low angle, forcing the Juan de Fuca Plate to slide beneath the North American Plate. The Pacific Plate continues to glide past the North American Plate, triggering periodic large earthquakes.

Shaping the Land

A journey 500 million years back in time would reveal that the eastern coastline of proto–North America faced south, turned 45 degrees clockwise from the present orientation.

Early Pennsylvania was situated about 20 degrees south of the equator in a shallow marine basin that bordered the Iapetus (proto-Atlantic) Ocean, which lay to the east of present-day Pittsburgh.

The history of the Appalachian Mountains in eastern North America began 480 million years ago, when an oceanic plate was forced beneath the continental margin. Intense pressure caused crustal melt that spurred volcanic eruptions and expelled magma, which cooled and formed the igneous rock underlying the region. The collision of a volcanic island arc and the ancient continent initiated extensive mountain building.

The upheaval drove materials over the continental shelf as the sea was forced below the craton, melting the basaltic bed and triggering intense volcanic activity. At the end of the

Taconic orogeny around 440 million years ago, a vast mountain chain reached from eastern Canada to the coast to Alabama.

Another island arc from a small section of proto-Africa (Avalonia) departed around 430 million years ago and drifted across the Iapetus Ocean before crashing into proto-North America. The scissors-like closing of the Iapetus Ocean around 410 million years ago resulted from collisions between ancestral Western Europe (Baltica) and North America (Laurentia), which uplifted the Acadian Mountains east of present-day Pennsylvania. Farther west, a great inland sea extended from western New England to the Gulf of Mexico.

Erosion whittled down the lofty peaks that were once higher than the Himalayas. Rivers that formed in the mountain chain transported finer materials into the Appalachian Basin that underlies the Piedmont. The Catskill Delta, which covered an area from New York and Pennsylvania south to West Virginia and Virginia, filled with sediment in the Devonian Period (419 to 359 million years ago).

The accumulation of sand, mud, carbonates and gravel hardened into gray sandstone and conglomerate beneath the rivers and along the coastal margins. Marine sediments composed of mud and clay compressed into shale. Calcium ions transformed into calcium carbonate, and limestone formed in the relatively clear water. Sediment in the river deltas hardened into sandstone at the bottom of the inland sea.

Extreme heat and pressure changed sandstone, limestone, and granite into crystalline metamorphic rocks—quartzite, marble, gneiss—that undergird portions of eastern Pennsylvania. Smaller particulates—silt, clay, pebbles and organic debris—compressed into sedimentary layers of siltstone and shale in slow-moving coastal waters and lake beds.

Chemical weathering of silicate rocks through the interaction with rainwater formed carbonic acid. The rocks slowly dissolved in a decomposition process triggered by reactions with plant and rock acid. Minerals including magnesium and calcium were carried by streams and rivers to the ocean, where they helped remove and store carbon dioxide and temporarily cooled the atmosphere.

At the beginning of the Carboniferous Period (359 to 299 million years ago), North America had drifted north to about 10 degrees below the equator. Crustal fragments of a former supercontinent (Rodinia) crashed into each other. An abundant geological record from the shedding mountain range was deposited on the land and in coastal areas.

The tropical Tippecanoe Sea invaded from the west, sloshing across Pennsylvania to near present-day Pittsburgh and Williamsport during the Mississippian Period (359 to 323 million years ago). Lush vegetation and tropical rainforests covered the elevated shoreline in northwestern Pennsylvania (Barnes and Seavon 2014).

During the Pennsylvanian Period (323 to 299 million years ago), vast swamps filled with plant materials were compressed into coal deposits in the Appalachian Basin. The decaying vegetation accumulated in stagnant, oxygen-starved water, where the woody, organic material (peat) transformed into great coal seams.

Endless cycles of deposition and subsidence of the sediments that piled up in shallow coastal waters caused sinking from the weight of the horizontal strata, driving the shoreline of the Appalachian Sea west to near the modern Ohio River (Barnes and Seavon 1996). Fossil exoskeletons of shelled marine organisms deposited in the lake bottoms and calcium carbonate beneath the inland sea combined with mud and sand to form layers of sandstone, limestone and dolomite.

The grand finale of the great geological upheaval culminated in a series of collisions between proto-North America, proto-Europe, proto-Africa and a small piece of South America early in the Permian Period (299 to 252 million years ago), leading to the formation of a supercontinent (Pangaea).

Continental collisions initiated the Alleghenian orogeny that lifted a soaring mountain chain. The rock layers were compressed and fractured in the thrust-fold belt along the crustal margin. Layers of limestone, sandstone and mud in the Appalachian Basin farther west were deformed. The crumpled sedimentary layers resembled a bunched comforter with upwarps and downwarps shaping the land.

A rapid climate change overwhelmed most of the species near the end of the Permian, referred to as the Permo-Triassic (P-T) extinction, or Great Dying, about 252 million years ago, which claimed approximately 96 percent of the existing marine animals. Massive volcanic eruptions in Siberia released heat-trapping greenhouse gases (carbon dioxide, methane) that likely triggered the greatest extinction in Earth's history.

A rapid global warming would have also heated the ocean water. One theory posited that a substantial increase in methane spawned a dramatic warming that caused a drop in oxygen levels (Gnidovec 2015). The climate during the Triassic Period (252 to 201 million years ago) was hot and dry, with an expansive desert landscape.

Around 230 million years ago, seafloor spreading opened up the young Atlantic Ocean, uplifting and splitting Pangaea into the large continents of Laurasia (north) and Gondwana (south). A small portion of the ancient mountain range became part of Africa. The Triassic-Jurassic extinction a little more than 200 million years ago killed off numerous species, paving the way for the 150-million-year reign of the dinosaurs during the Mesozoic Era.

North America still straddled the equator, surrounded by water that supported a humid tropical climate, modified by a cooler maritime influence in northern coastal areas. The center of North America was mostly underwater and part of the Western Interior Seaway. The very warm Cretaceous Period, the longest portion of the Mesozoic Era, lasted from 145 to 66 million years ago and marked the peak of the "Age of the Dinosaurs."

The vast ancient supercontinents of Laurasia (North America, Eurasia) and Gondwana (South America, a portion of India, Australia, Antarctica) drifted apart and shifted toward their present-day locations. The end of the Cretaceous Period was noteworthy for another mass extinction 66 million years ago, when a large asteroid slammed into an area off the coast of the Yucatan Peninsula, leaving a crater about 112 miles across.

The blast impact threw up an enormous amount of sunlight-filtering dust and unleashed a toxic stew of gases and particles, which combined with water droplets to produce a thick layer of sulfuric haze that fell as acid rain. Sunlight was reduced and the ensuing cooling trend virtually wiped out all non-avian dinosaurs and extinguished three-quarters of all animal and plant species. The demise of the dinosaurs opened the door for the evolution of mammals.

Glacial Stages

Very little is known about two massive glaciations around 650 and 300 million years ago. The seismic movement of the continents from lower (warmer) to higher (colder) latitudes and the warming of the subtropical Pacific likely altered the flow of ocean currents and northward transport of heat in conjunction with ice feedbacks linked to variations in the seasonal distribution of solar radiation at high latitudes.

In the early Eocene, 55.5 million years ago, the global temperature rose 9°F to 14°F in a relatively short period of several thousand years, linked to a large spike in methane and carbon dioxide emitted by volcanic activity. The greenhouse gas content of the atmosphere reached several times greater than today's levels.

During the Paleocene-Eocene Thermal Maximum (PETM), polar ice virtually disappeared from Earth's surface. Southern Canada was a subtropical region and sea levels were 50 to 80 feet higher than today. Crocodiles roamed north of the Arctic Circle and palm trees flourished on the shores of Antarctica. The warmer climate accelerated chemical weathering, which took up carbon dioxide and led to a global cooling in the middle Eocene Epoch that continued in the Oligocene Epoch (33.9 to 23 million years ago). Polar ice sheets formed around Antarctica.

The late Oligocene brought warmer conditions, initially from lower levels of chemi-cal weathering. Global circulation features shifted in the Miocene Epoch (23.5 to 5.3 million years ago), with subtle tectonic movements of the continents. Glaciers in Antarctica expanded when warmer currents that surrounded the polar region were displaced with colder waters, which allowed ice to accumulate, cooling the global climate.

A combination of erosion, weathering and mass wasting lowered the ancient Allegheny Mountains to a broad plain. A period of substantial uplift linked to Earth's thermal properties (convection currents) occurred about 60 million years ago in eastern North America, causing a crustal rebound. The raised topography led to extensive downcutting by rejuvenated streams that sliced through softer shales, siltstones, and limestones.

In the Miocene Climatic Optimum, a period of global warming 18 to 14 million years ago, there was an increase in weathering and erosion caused by changes in the wind and precipitation patterns. A subsequent cooling caused the Antarctic glaciers to grow, which lowered global sea level 150 to 250 feet. Rivers systems in eastern Pennsylvania flowing to the retreating shoreline cut more deeply through sedimentary rock, leaving craggy sandstone bluffs.

Carbon dioxide levels were akin to the present (420 parts per million) during the Pliocene Epoch (5.3 to 2.6 million years ago), and the climate was several degrees warmer than preindustrial times. A distinct cooling trend at the start of the Quaternary Period led to expansive glaciation in the Arctic, as layers of snow compressed into ice sheets over northern Canada and Hudson Bay. The weight of the ice caused Earth's crust to subside. The Atlantic coastline was 50 to 150 miles farther east than today, as great ice sheets advanced southward from Canada in a cyclical fashion.

The Pleistocene Epoch (2.6 million to 11,700 years ago) featured an estimated 30 cycles of ice advancing and retreating in regular 100,000-year periods. Plenty of geological evidence remains for scientists to sift through in the formerly glaciated areas of the northern United States.

Pre-Illinoian glaciation prior to 800,000 years ago reached farther south than the later stages. A Late Illinoian glaciation 196,000 to 128,000 years ago ended a relatively short distance north of the previous advance (Sevon et al. 1999), burying the northern corners of Pennsylvania under up to 3,000 feet of ice.

The most recent glacial stage, or Last Glacial Maximum, commenced after the Eemian interglacial, an interval of warmth 130,000 to 115,000 years ago when the global temperature was several degrees warmer and sea level was 20 to 30 feet higher than today. A reduction in solar insolation at high latitudes in the Northern Hemisphere allowed snow and ice to build up and linger from one season to the next and gradually expand southward.

The Laurentide Ice Sheet moved slowly downhill from the northern mountains of Canada and reached a thickness of nearly two miles around Hudson Bay in eastern Canada. The Late Wisconsinan glaciation entered portions of northern Pennsylvania around 27,000 years ago.

The Last Glacial Maximum peaked about 20,000 years ago, when ice covered one-third of North America, including virtually all of Canada and most of the northern United States, and northern Europe. The average global temperature was about 9°F colder than today, and several times colder on the southern margin of the ice sheet. Areas on the fringe of the frozen landscape had experienced a tundra-like climate. The polar jet stream looped farther south and sustained a favorable storm track for frequent snowfalls.

The weather gradually warmed, causing the ice to slowly retreat by 17,000 years ago. Torrents of glacial meltwater carved new pathways through softer rock and filled depressions. The Great Lakes were carved by glaciers through previously eroded valleys. Eastern Pennsylvania rivers engorged with glacial meltwater rose tens of feet. As the waters receded, fertile farmland reflects ancient drainage patterns created by "rivers of ice" that transported topsoil beyond the southern glacial extent.

The climate in northern Pennsylvania transitioned from treeless tundra to a boreal forest around 13,000 years ago, composed predominantly of spruce and fir, which would be replaced by hemlock and mixed oaks after the Pleistocene Epoch concluded. The climate became warmer and more stable around 8,000 years ago, with swings between colder, drier regimes and warmer, wetter cycles that continue to this day.

Ice Age Relics

Expanding glacial lobes 1,000 to 3,000 feet high advanced like a massive snowplow that scoured the surface and blocked ancient river valleys with ice, forcing waterways to find new outlets. The ice sheet scraped and gouged bedrock, polished rock surfaces and scratched telltale grooves, or striations, boring through softer bedrock. Tons of rocky debris from the glacial peaks formed prominent knobs.

When the ice retreated, meltwater deposited large amounts of sediment—clay, silt, sand, gravel—in U-shaped valleys near the terminus that compacted into low ridges, or moraines. The outwash left behind unsorted till and stony, infertile soils. Out-of-place boulders (erratics) plucked by the glaciers and carried hundreds of miles dot the northern landscape, most notably at Hickory Run State Park in Carbon County.

The abraded surface features and rock debris compressed and hardened into bluffs and hillocks that overlook narrow valleys. The path of meltwater that tunneled through the ice can be traced by sinuous ridges (eskers) comprising stratified sand and gravel. Glacial streams also created distinct channels that formed irregular hills (kames). Rounded depressions (kettle holes) mark concavities where buried chunks of ice melted, creating a string of lovely lakes.

Impounded post-glacial waterways turned into wetlands and swamps. Peat bogs filled with partially decayed plant matter, including tree roots, sank in stagnant swamps. Heating and compression beneath accumulating sediments changed the buried material into soft peat that formed coal seams.

Earth's Orbital Cycles and Ice Ages

Five or six major ice ages have occurred in Earth's history during the past 3 billion years. The global climate drivers for glaciation are complex. Factors include levels of atmospheric carbon dioxide and methane, the movement of tectonic plates, which altered the position of the continents and affected wind patterns and ocean currents, volcanic activity, and cycles in the amount and distribution of solar energy influenced by Earth's orbit.

Rocks high in silicate mineral content took up carbon dioxide, a heat-absorbing gas, through the process of chemical weathering, which gradually lowered Earth's temperature. Mountain-building episodes funneled cold continental air southward that allowed more snow and ice to accumulate at high latitudes shielded from the effects of mild maritime air. As snow congealed into ice under pressure and expanded, the surface became more reflective, reducing the absorption of solar energy in a positive, or reinforcing, climate feedback.

James Croll (1820–1890), a Scottish mathematician and physicist, was the first to proffer a theory of how Earth's orbital characteristics could have correlated with the advance and retreat of the ice sheets and mountain glaciers. In 1875, he published his proposition that Earth's orbital geometry periodically lined up to cause a minimum of solar insolation at high latitudes of the Northern Hemisphere in the cooler months.

Serbian mathematician and scientist Milutin Milankovitch (1879–1958) refined Croll's seminal work, but focused his theory on solar radiation reaching the top of Earth's atmosphere at 65° N, near the Arctic Circle, where the seasonal variations of solar radiation are greatest. Milankovitch worked out an immense number of calculations by hand in the 1920s and 1930s for various latitudes and published his results in 1941.

Milankovitch posited that a reduction in solar radiation during the boreal summer in the sub-Arctic triggered a southward expansion of the ice sheets over tens of thousands of years. Subsequent research decades later, based on the examination of tiny marine fossils extracted in deep-sea cores, provided solid evidence of fluctuations in sea surface temperatures, and the expansion and contraction of ice sheets consistent with variations in high-latitude solar insolation in the Northern Hemisphere.

The Milankovitch cycles formulate how the distribution of sunlight in relation to Earth's orbital parameters varies during periods of approximately 23,000, 41,000 and 100,000 years, based on (1) the direction of Earth's axis in relation to fixed stars (precession), (2) changes in the tilt of Earth on its rotational axis relative to the perpendicular of its orbital plane (obliquity), and (3) changes from a roughly circular to a more elliptical orbit caused by the gravitational pull of Jupiter and Saturn.

Precession reflects the wobble of Earth caused by the pull of gravity, exerted by the sun and moon on the equatorial bulge, and a shift of the seasons. Earth's rotational axis completes a 360-degree circle, likened to a top slowly spinning down. The North Pole presently points to Polaris, but in

11,000 years the rotational axis will line up with Vega as the North Star.

Precession controls the timing of the seasons, which begin progressively earlier. Currently, Earth is closest to the sun in early July. In 13,000 years, perihelion will occur in the Northern Hemisphere summer. Because Earth's orbit drifts relative to the ecliptic plane during a period lasting about 112,000 years, the combined effects result in a cycle of 23,000 years.

The long-term cyclical variation of Earth's axial tilt with respect to its orbital plane affects the seasonal distribution of insolation. The maximum tilt angle of 24.5 degrees, which last occurred 10,700 years ago, accentuated Earth's climate extremes. As more solar heat reached the Northern Hemisphere when it was tipped toward the sun, the intensified energy contributed to a glacial retreat.

The present tilt (23.44 degrees) will reach a minimum (22.1 degrees) in 9,800 years during the course of a well-defined 41,000-year period. The smaller tilt will further reduce the seasonal differences in solar energy, resulting in milder winters and cooler summers.

A longer orbital cycle of approximately 100,000 years affects the variation from winter to summer in the Earth-sun distance of 3.4 percent. Incoming solar radiation varies by 6.8 percent from January 3 to July 6 in the current circular phase. In a highly elliptical orbit, that difference exceeds 23 percent, when Earth makes its closest approach to the sun during a period of about 100,000 years.

In the past 800,000 years, the eccentricity cycle coincides with the interval of periodic ice advances. At perihelion in an elliptical orbit in the Northern Hemisphere summer, deglaciation occurs near the Arctic Circle in response to greater insolation. Yet, precession has a greater effect on glaciation at the peak of eccentricity about 413,000 years in a cycle that influences the timing of the seasons, and the Earth-sun distance is at minimum in the summer in the Northern Hemisphere.

Presently, the sun is farthest from Earth in the Northern Hemisphere in early summer, which is the case for another 4,000 years while the planet follows a more circular orbital cycle, although the axial tilt controls the seasonal heat distribution.

From the standpoint of changing climate zones, research suggests that when the shortest Earth-sun distance coincides with the Northern Hemisphere summer, significantly greater heating in the subtropics generates heavier thunderstorms (rainfall), due to stronger trade winds converging along the Intertropical Convergence Zone (ITCZ) that shift north with the higher sun angle.

A Brown University study added an intriguing wrinkle, focused on the growth of sea ice in the Southern Hemisphere acting as the primary force behind the ice ages. Jung-Eun Lee theorized that when the Earth-sun distance during the summer was farthest, the precession cycle linked to cooler summers contributed to the expansion of sea ice, increasing the reflectivity of Earth's surface. Since the Southern Hemisphere has 20 percent greater ocean coverage than the Northern Hemisphere, a significant climate feedback culminated in global cooling that likely triggered glacial advances.

Post-Glacial Climate of Early Pennsylvania

Archaeological evidence indicates that the first hunter-gatherers arrived in the upper Ohio Valley perhaps as early as 15,000 years ago. Radiocarbon dating of the remains of plants at the Meadowcroft

Rockshelter archaeological site, 27 miles southwest of Pittsburgh, suggested that nomadic hunters had made the arduous trek from eastern Siberia across the Bering Land Bridge to North America earlier than previously thought.

Mercyhurst College researcher Jim Adovasio's team excavated 700 pieces of stone and 50 tools at the campsite seven miles upstream from the Ohio River, where Paleo-Indians procured and later processed food (Scofield 2013). The shelter was situated on the edge of a slowly retreating glacier near Lake Erie.

There were cold interludes during the time even when glaciers were in retreat, notably during the Younger Dryas about 12,900 to 11,700 years ago. A periglacial climate prevailed when the first tiny settlements appeared in Pennsylvania, which was brutal, featuring harsh winters and short, cool summers. The margin of the glaciers regulated the atmospheric circulation, with bitterly cold winds sweeping off the ice sheet.

Although Earth's climate warmed enough for the glaciers to melt 15,000 years ago, cool, damp conditions would have yielded a short growing season for thousands of years. Windblown sediment (loess) ground up by the ice was deposited as topsoil, which washed into streams. Pine and fir trees that comprised the boreal forest gradually yielded to evergreens and leafy deciduous trees beginning around 13,000 years ago.

Holocene Interglacial and Early Civilizations

Climate cycles—alternately cool and wet, then warm and dry—impacted human societal activities throughout the Holocene Epoch.

The climate eventually warmed and stabilized by 10,000 years ago so that agrarian societies would blossom in semi-permanent settlements in the Middle and Near East. A cooler period about 8,200 years ago was followed by warmer conditions constituting the Climatic Optimum 8,000 to 5,000 years ago.

The warming trend caused cultural changes as agricultural practices evolved in favored zones. A downturn in temperature and a distinct trend toward drier conditions around 4,200 years ago in the Late Bronze Age, in the area between Turkey and Egypt (Mesopotamia), has been implicated in the collapse of the Akkadian civilization near the Mediterranean Sea (Kershner 2013).

The swings of the climatic pendulum presented civilizations with fruitful and lean periods. The demise of the Maya and pre-Inca societies has been linked to the stress of climate change, which caused food shortages and social unrest. Proxy evidence obtained from tree rings analyzed in wood in ancient Egypt revealed long periods of dry weather a few thousand years ago.

In North America, archaeologist Bradley Lepper (2014) wrote that "change would have been a big problem for the ancient hunter-gatherers of southern Ohio, who relied on nuts for much of their diet" because fewer trees equated to fewer nuts.

Charcoal dated a millennium later suggested that Indigenous Americans transitioned from hunter-gatherers to farmers, clearing the land by setting fires. Sediment analysis indicated an evolution from predominantly tree to grass pollen in Patton Bog around 1,000 BC, a natural wetland in southeastern Ohio investigated by Ohio University archaeologist Elliot Abrams.

Around AD 1000, a prolonged warmer, wetter regime boosted the production of maize in larger farming communities of Eastern Woodland Indians in western Pennsylvania, who resided primarily along river systems. However, an abrupt climate change brought colder, drier conditions in the latter half of the 13th century, triggering instability as resources diminished, the direct result of truncated growing seasons that likely caused tribal conflict during protracted drought periods.

In northern Europe, the Medieval Warm Period (AD 970 to 1245), linked to a slight increase in the solar output, favored lush vineyards. The Vikings settled an open passageway to Greenland and explored areas as far afield as the shores of North America around Newfoundland by sailing in relatively ice-free conditions.

A sequence of volcanic eruptions interrupted the warmth in the 13th century. Radiocarbon data from trees and geochemical analysis of ash and sulfur trapped in Arctic and Antarctic ice cores pointed to a massive volcanic eruption in Indonesia (Samalas) in the summer of 1257. The following year, London experienced a very cool summer caused by volcanic aerosols that filtered sunlight.

By the end of the 13th century, sea ice had increased in the North Atlantic, which isolated the Greenland outposts. In northern Europe, crops withered in the fields during a series of chilly, damp summers that markedly reduced yields. The Black Death decimated a quarter of the population in Europe between 1347 and 1352, weakened by nutritional shortages and a susceptibility to disease.

The weather took an even deeper plunge into cold and stormy conditions in the Little Ice Age that commenced around 1430 and lasted into the 19th century. The average temperature in northern Europe was 2-3°F colder than present times. Alpine glaciers expanded into high valleys, and crops failed to thrive in cooler, wetter summers and rotted in the fields. Famine and disease accompanied the periodically cooler and damp conditions that prevailed more frequently in some growing seasons.

A rise in the average global temperature commenced in the late 1800s and continued until the 1940s. A significant global cooling trend developed in the middle of the century, lasting through the 1970s. An upswing in warmth in recent decades raised the average world temperature 2°F since pre-industrial times, the warmest conditions in at least 125,000 years.

Evidence of Past Climates

The fossil record contains evidence of the types of plant life in earlier times. Proxy data offers a glimpse into the nature of paleoclimates and an invaluable context for analyzing climate change.

The width of tree rings of ancient trees reveals precipitation and drought conditions over hundreds and thousands of years. Coral sediments provide data regarding sea surface temperature patterns and the amount of oxygen and carbon dioxide in the water. Airborne pollen from trees, plants and crops was deposited in the polar and mountain ice.

The chemical analysis of ice cores drilled hundreds of feet into glaciers reveals soot, fungal spores and pollen sediments carried by the wind and deposited in crystalized snowflakes. The nature of accumulated lake-bed sediments offers evidence of past growing seasons, inferred by temperature and precipitation patterns. Black carbon is an indication of wildfire patterns.

Trapped air bubbles in cores contain oxygen isotopes and trace gases that tell the tale of climatic swings as far back as 800,000 years ago. The ratio of oxygen isotopes offers an indirect measure of Earth's temperature. Water molecules in the ocean contain both oxygen isotopes—O-16 and O-18 (the heavier form of oxygen has two more neutrons).

Danish geophysicist Willi Dansgaard (1922–2011) discovered in the 1960s that heavier oxygen reacts with cold conditions, measured by evaporation, condensation and precipitation. During the ice ages, more of the heavier oxygen molecules were deposited in the ocean at lower latitudes, while lighter oxygen was transported poleward.

Trace elements such as charcoal and volcanic ash in core samples provide indirect evidence of the atmospheric carbon dioxide levels. Large amounts of mineral dust and dirt lodged in the ice are linked to drought. High-resolution analysis of fossilized pollen grains beneath the desert and at the bottom of bogs and ponds reveals the types of vegetation that once flourished, confirmed by radiocarbon dating.

Prolonged drought brought on famine and civil unrest that had a deleterious effect on great civilizations throughout history, such as the Maya in the Yucatan Peninsula during the seventh and eighth centuries. A deteriorating climate in the 14th century in Northern Europe resulted in failed harvests and fallow fields, contributing to pandemics as the population became susceptible to disease from malnutrition.

Climate of the Past 200 Years: A Philadelphia Perspective

The long-term mean temperature in Pennsylvania has risen more than 2°F in the past century. Since 1950, national temperature patterns tracked by the National Oceanic and Atmospheric Administration (NOAA) show a rise of several degrees. Winters at Philadelphia are 3.4°F milder since 1891–1920. The average annual precipitation has increased by several inches at most locations in the state, coinciding with the warming trend.

A look back through the past two centuries reveals warming and cooling cycles. The most reliable set of continuous temperature records commenced in 1825 at Pennsylvania Hospital in Philadelphia. The early data were taken several times a day, including temperature and rainfall, long before the nascent U.S. Weather Bureau city office opened in 1872.

An important caveat when surveying the historical data is to know when a site has changed location and how averages were maintained. Prior to the 1880s, most calculated monthly average temperatures at government sites were based on four (later three) weighted daily observations. The difference amounts to about one degree warmer when compared with the modern method of adding up the daily maximum and minimum readings and dividing by the number of days in the month.

The tail end of the Little Ice Age in eastern North America is generally regarded as the mid-1800s, though the era was not uniformly cold. There were mild intervals, including the 1820s.

An anomalously cold and relatively dry pattern is evident in the 1830s in the Philadelphia records, culminating in three historically cold years in succession—1836, 1837, and 1838. The spring and autumn seasons in the middle of the 19th century were replete with early and late snowfalls, an indication of a persistent northwesterly circulation.

The average annual temperature climbed in the 1860s, which resulted in some heavier snowfalls. The next decade brought a noteworthy cluster of cold years—1872, 1873, and 1875—in a distinctly cooler pattern. Temperatures warmed significantly in the late 1870s and early 1880s. However, following the massive 1883 eruption of Krakatoa (Krakatau) in the Sunda Strait, between Java and Sumatra in Indonesia, a global cooling trend also shows up in the Philadelphia dataset.

The regional pattern turned rather mild in the late 1880s and early 1890s. The pattern flipped shortly thereafter. Generally cold and snowy winters prevailed between 1892–93 and 1898–99. The annual average temperature at Philadelphia dipped noticeably after 1901, with three consecutive chilly years from 1903-1905 that were reminiscent of the lengthy cold periods in the 1830s and 1870s.

The mostly cooler and wetter cycle prevailed through the early 1920s. A substantial warming and drying trend commenced later in the decade and continued through the 1930s, coinciding with the Dust Bowl in the American and Canadian prairies. The 1930s represented the warmest decade since records began. Summers were often hot and desiccating.

The weather globally and regionally cooled in the 1940s, with periods of ample moisture relieving the drought. Later in the decade, and especially in the early 1950s, ridging near the Atlantic Coast contributed to a string of abnormally hot and dry summers between 1952 and 1955. There was a dramatic increase in landfalling hurricanes along the East Coast that were steered by a dominant subtropical ridge. Winters were mild with a dearth of snow-laden nor'easters.

The mean atmospheric flow in North America began to shift in the late 1950s, as the polar jet stream expanded southward in response to high pressure over western North America, likely due to persistent cooler-than-average sea surface temperatures in the northeastern Pacific.

A pattern of predominantly northwesterly winds east of the Rockies in the early 1960s exacerbated a drought in the eastern United States from 1962 to 1966. A colder-than-usual winter pattern contributed to a series of famously snowy winters in the Northeast between 1960-61 and 1970-71.

A decidedly wetter pattern developed in the Northeast that was coupled with milder winters until 1976-77. A dramatic turn of events brought three historically cold and stormy winters in the region and cool summers. A remarkable run of arctic outbreaks defined the winters from 1976–1977 through 1984–1985, with drier-than-average years.

The closing part of the 1980s brought a resumption of warmer summers and mild winters. A wetter cycle developed in 1990 and 1991, which tied 1931 for the warmest calendar year on record at Philadelphia (58.2°F) up to that time.

The eruption of Mount Pinatubo in the Philippines on June 15, 1991, triggered a few years of global cooling. Arctic surges and paralyzing snowstorms were prominent in the winters of 1992–1993, 1993–94 and 1995–96 in the Northeast.

Warmer conditions prevailed during the latter portion of the 1990s, notably 1998 (58.1°F). The decade was the warmest at Philadelphia by a margin of 0.2°F over the 1930s.

Alternately warm and cool years marked the first two decades of the 21st century, with abundant moisture after a widespread drought in 2002. Winters were highly variable in temperature, but moisture was sharply increased in many years. Historically snowy nor'easters were linked to higher sea surface temperature and associated North Atlantic blocking that directed stronger coastal disturbances northward along the Eastern Seaboard.

A new benchmark for annual warmth at Philadelphia was registered in 2012 (58.9°F). The only comparably warm year in the Pennsylvania Hospital records occurred in 1829 (58.9°F). Very warm years occurred in 2024 (58.8°F), 2016 (58.6°F), 2020 (58.0°F), 2011 (57.8°F), 2010 (57.7°F), and 2015 (57.6°F).

Historically wet years occurred in 2011 and 2018 that correlated with a persistent southwesterly flow of moist Gulf air and the passage of several post-tropical cyclones.

Climate Zones of Pennsylvania

The climate of Pennsylvania is classified as humid continental, characterized by reliable to occasionally excessive precipitation. Winters are generally cold, with prolonged bouts of subfreezing weather and snow cover across the north and west. Summers are warm and moderately humid, accompanied by periodic showers and thunderstorms.

The average annual temperature of Pennsylvania is 49°F. However, given the large variation in topography, the Keystone State exhibits a wide mean yearly temperature range—44°F at Pleasant Mount in the northeastern mountains to 56°F at Marcus Hook near sea level.

The average annual precipitation in Pennsylvania is a little more than 42 inches, which is fairly evenly distributed, with a cold season minimum. The driest area is the north-central portion of the state in Tioga County, with a mean of a little more than 36 inches. The wettest areas are the Poconos and Laurel Highlands, with more than 50 inches.

The Keystone State has impressive temperature extremes—111°F at Phoenixville on July 10, 1936, and -42°F at Lawrenceville on February 10, 1899. The mercury has plunged to zero or lower as early as November 16, 1908, at Clearfield (-2°F) and November 16, 1933, at Coudersport (-11°F), and as late as April 8, 1982, (0°F) at Kane.

Maximum readings have soared to 100°F as early as May 27, 1941, at Marcus Hook, and as late as October 5, 1941, at Phoenixville—oddly interesting because both historically hot values occurred in the same year.

Northeastern Highlands

Northeastern Pennsylvania lies squarely in the Ridge and Valley province and exhibits significant variability in average temperature and precipitation, depending on elevation.

Precipitation is ample along the eastern slopes of the Appalachians, with peak values exceeding 50 inches in northern Schuylkill County to the upper Delaware Valley, a result of storms cruising up the Atlantic Coast. Northwest of the Poconos, the average precipitation falls to between 38 and 46 inches.

Cloudiness is more prevalent with the arrival of chilly air masses, coupled with an upslope component. Areas east of the ridge experience downslope flows and compressional warming, which reduces cloud coverage.

Fig. 1.7 Tobyhanna Lake, where ice was harvested in blocks and shipped by railcars to New York and Philadelphia before the days of electric refrigerators. (Photo by Ben Gelber)

The average winter maximum temperatures between 30°F and 35°F sustain snow cover for longer periods of time, particularly in the mountainous areas. Mean winter minimums dip below 15°F on the Pocono Plateau, and subzero readings are not uncommon during periods of persistent snow cover, averaging five to 10 days annually. Seasonal snowfall typically falls between 45 inches in the valleys to more than 70 inches in the northern Poconos.

Summers in the northeastern highlands are moderately warm and humid, though the higher elevations afford relief from the intense heat. Mean summer maximums range from a pleasant 77°F in the highest elevations to 83°F in valleys. Average minimum temperatures are relatively comfortable, between 55°F and 60°F.

Fig. 1.8 Raymondskill Falls, in southeastern Pike County, tumbling off the Pocono Plateau, is the highest waterfall (150 feet) in the state. There are more than 170 waterfalls in the Keystone State. (Photo by Ben Gelber)

Southeastern Lowlands

The influence of the Atlantic Ocean assures adequate year-round precipitation in most years in the Piedmont and Coastal Plain of southeastern Pennsylvania. Annual precipitation averages between 40 and 48 inches, rising to nearly 52 inches at Reading. In the lower Susquehanna Valley, Harrisburg receives an average of 41 inches.

Average winter maximum temperatures range from 35°F to 40°F, and low readings typically fall between 20°F and 25°F. The Piedmont often experiences several years in a row without subzero temperatures, and snow cover is usually short-lived.

Summers in the southeastern lowlands are uncomfortably humid near the Atlantic Coast and around Chesapeake Bay. Average highs range from 84°F to 88°F, with the highest readings near the Maryland border and around Philadelphia. Mean minimums fall between 60°F and 65°F, though barely 70°F in urban Philadelphia.

Central Mountains

The central section of Pennsylvania, which comprises the largest portion of the Ridge and Valley province, is the driest portion of the state, situated between the primary storm tracks along the Eastern Seaboard and across the Great Lakes.

Average annual precipitation amounts range from 36.5 inches in northern Tioga County near the New York border to 44 inches in the central mountains. A northwesterly flow often brings a period of cloud cover accompanied by showers and fog, as moist air ascends a chain of ridges, cools and condenses.

Winter temperatures in the central highlands are comparable to the northeastern counties. The average January maximum is near freezing (30°F to 35°F), depending on elevation, with cold mean minimums (13°F to 18°F), which are lower in a month with persistent snow cover.

Fig. 1.9 Morning fog trapped in a narrow valley along Interstate 80 east of Lock Haven. (Photo by Ben Gelber)

Summers in the central mountains are generally pleasant, without short bouts of high humidity typical of the Piedmont and Coastal Plain. Average maximum readings top out in the low 80s and nighttime lows often fall into the 50s.

Northwestern Highlands

Cloudiness is more prevalent over the Allegheny Plateau, particularly in the cooler seasons, when the winds are blowing orthogonal to the elevated terrain. Precipitation days are frequent in the late fall and winter, though daily amounts are usually lighter compared to eastern areas in proximity to significant coastal storms.

Annual precipitation in northwestern Pennsylvania averages between 42 and 48 inches, with the highest total in eastern Erie County at Union City Filtration Plant (49 inches). Elevated topography, lake breeze boundaries and lake-enhanced snowfalls play a role in the higher mean annual precipitation values. Areas closer to the Ohio border and adjacent to Lake Erie receive an average of slightly less than 42 inches.

Mean maximum temperatures in January are close to or below freezing (30°F to 35°F). Average minimums are lower, compared to most areas of the state (13°F to 18°F), except at Erie (21°F).

The northwestern highlands experience some of the lowest temperatures in the state in the wintertime. The state record low of -42°F was recorded at Smethport on January 5, 1904. Snowfall downwind of Lake Erie averages between 90 and 120 inches, while the remainder of the Allegheny Plateau in northern and western Pennsylvania receives between 50 and 80 inches in a typical winter season, depending on elevation and the prevailing storm track. The highest season snowfalls are observed in eastern Erie County. The most recent 30-year data period awards Union City Filtration Plant (114 inches) with the top figure.

Summers are tolerable in the higher elevations that enjoy comparatively pleasant average maximums (78°F to 83°F) and minimums (51°F to 56°F). Summer storms are common and occasionally severe.

Southwestern Hills

The mean yearly precipitation in the southwestern part of the Appalachian Plateaus province shows a variation from 38 to 44 inches, and 48 to 56 inches in the Laurel Highlands. Snowfall on the Allegheny Plateau averages 30 to 45 inches in the lower hills.

The average season snowfall in the highest elevations of Somerset and southern Cambria counties exceeds 100 inches, with a maximum value of 142 inches at Laurel Summit, where the mean annual precipitation of 56.61 inches qualifies as the wettest spot in the Keystone State.

Cloud cover tends to linger in mountainous areas due to the prevalence of northwesterly winds carrying lake moisture uphill. Proximity to Ohio Valley storms also accounts for slightly heavier precipitation compared to the central sections of the state. Frequent thunderstorms in the spring and summer provide adequate moisture for agricultural activity.

Average maximum temperatures in January in the southwest lie between 33°F and 38°F, depending on elevation, and minimums range from 17°F to 22°F. Allegheny Mountain locations have the lowest values, with maximums near 30°F and minimums below 15°F.

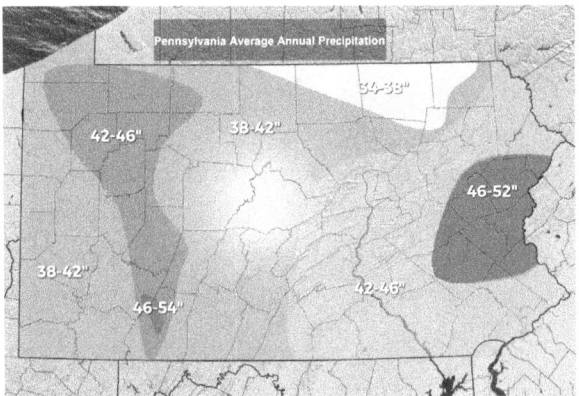

Fig. 1.10 Average annual precipitation.

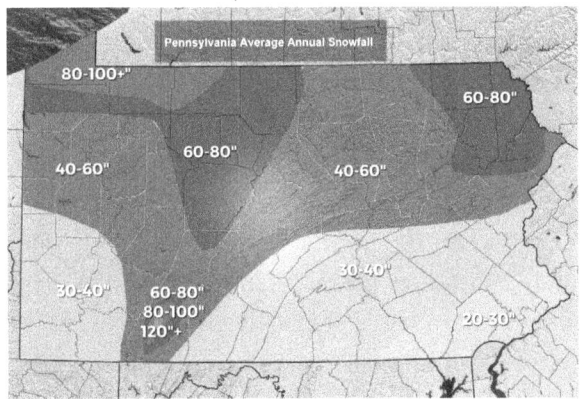

Fig. 1.11 Average annual snowfall.

Summers are warm and sticky in southwestern Pennsylvania, though relatively pleasant over the high Alleghenies. Typical summer highs range from 80°F to 85°F, and minimum values are generally 55°F to 60°F. In the Laurel Highlands east of Pittsburgh, mean maximums dip into the upper 70s.

Growing Season in Pennsylvania

The growing season is defined by the number of days separating the last occurrence of a freeze (32°F) in spring and the first freeze in autumn. The temperature is measured five feet above a grassy surface, so frost is possible when readings are in the mid-30s. This is because on a clear, calm night the ground loses heat faster than air above the surface.

The longest frost-free period in Pennsylvania is observed in the heavily urbanized Delaware Valley, ranging from 210 to 230 days. In the interior southeast, most communities experience an average of 170 to 190 frost-free days.

A lengthy growing season is also common near the Lake Erie shoreline, which is sheltered by relatively warm water in autumn compared to the air temperature a short distance inland. Farmers take advantage of a growing season that averages upwards of 180 days. At Erie, the average date for the first autumn freeze is October 29.

In the central mountains, most locations have a growing period that lasts between 150 and 170 days. Some sheltered Appalachian valley locations average 130 to 150 days between the last and first freezes, because valley floors are susceptible to cold air drainage from the neighboring hillsides.

The Allegheny Highlands in the northwest and Pocono Plateau in the northeast are prone to killing freezes in May and late September. Kane has an average growing season of a little more than 100 days. The high Pocono Plateau averages fewer than 140 frost-free days at Pleasant Mount, Gouldsboro, and Tobyhanna.

The longest growing season at a Pennsylvania weather site came in 2015 at Philadelphia, when temperatures remained above the freezing mark for 251 days from March 30 through December 5.

Pennsylvania Weathers Many Climates

The geological story of Pennsylvania is woven into the diversity of climate patterns, modified by the world's oldest mountain chain that was shaped by both fire and ice. Sifting through the relatively short span of recorded weather history, when compared to eons of seismic upheaval, the imprints of ceaseless motions in the atmosphere are teased out in the dutiful and highly descriptive records of scientists and volunteer weather observers.

Chapter Two

Chroniclers of Pennsylvania Weather

He hath found the climate of the said country very temperate, judging it to be as temperate as that of this country, Holland.

— Cornelius Hendrickson's 1615–1616 exploration of Delaware Bay

EARLY DESCRIPTIONS of the climate of Pennsylvania were compiled by Cleveland Abbe Jr. (1872–1934), son of the venerable founding editor of the *Monthly Weather Review*.

Abbe uncovered some fascinating observations, gathered in *Narratives of Early Pennsylvania, West New Jersey and Delaware, 1610–1707* (Myers 1912). Dutch explorer David Pieterszoon de Vries (c. 1593–1665), born in France, ventured up the Delaware River in the winter of 1632–33 as far north as Camden, New Jersey, which was part of New Netherland.

De Vries offered the first detailed description of the Delaware Valley climate:

[The crew] Did not imagine that we had been frozen up in the river, as no pilot or astrologer could conceive, that in a latitude from the thirty-eighth and a half to the thirty-ninth, such rapid running rivers could freeze. Some maintain that it is because it lies so far west; others adduce other reasons; but I will tell how it can be, from experience and what I have seen inland, stretching towards the north, there are high mountains, covered with snow, and the north and northwest winds blow over the land from these cold mountains, with a pure, clear air, which causes extreme cold and frost, such as is felt … as soon as the wind is southwest, it is so warm that one may stand naked in the woods, and put on a shirt.

On the fourth of six voyages to the New World, de Vries likened the weather he encountered in February 1633 to that of Genoa, Italy, which he believed to be as cold as Holland, expressing surprise that the waters of the Delaware River could freeze over.

At the time the first European settlers arrived in what is now Pennsylvania, there were about 20,000 Native Americans residing in the region. Dutch explorers established trading posts in the Delaware Valley in the 1630s. Swedish immigrants soon followed, constructing the first permanent settlement in the region in 1643 on Tinicum Island southwest of Philadelphia.

Vast tracts of hardwood forests cloaked a large area from near the Atlantic Coastal Plain to the Mississippi River, occasionally interrupted by narrow strips of grasses, wetlands, wild prairies, and patches of land managed by Native Americans. The wide woodlands were described by explorer and surveyor Lewis Evans as "an Ocean of Woods" in his 1755 map.

William Penn

William Penn (1644–1718), when living in England, received a royal charter from King Charles II in 1681 that granted him Penn's Woods, which included Pennsylvania and a portion of Delaware.

The Pennsylvania Colony agreement was initiated by the Duke of York as a repayment for debt owed to the Admiral Sir William Penn. The land was known as "New Wales" and "Sylvania" (Latin for "forest land"), so the King changed the name to "Pennsylvania" to honor Penn's late father.

William Penn first set foot in the New World at New Castle, Delaware, on October 27, 1682 (Old Style). As Proprietor of the Province of Pennsylvania, Penn traveled to Upland (Philadelphia) in early November, a city he personally designed to serve as the seat of the new colonial government. A talented promoter, he gave parcels of land to friends and family, known as manors.

Penn fervently believed in treating the Native American peoples equitably, consistent with his Holy Experiment, while hoping to attract diverse groups, which included English Quakers. Penn became acquainted with the vicissitudes of the elements in the Delaware Valley, which he aptly described in a detailed letter to Lord North: "The weather often changeth without notice, and is constant almost in its inconstancy" (Myers 1912). The original manuscript rests in the Historical Society of Pennsylvania archives.

In the spring of 1683, Penn surveyed the countryside beyond Philadelphia, visiting wigwams and learning how to converse in the language of the early inhabitants (Myers and Penn 1937). He wrote extensively about the fauna, culture, language and laws of the Native Americans, who had coexisted peacefully with the Europeans.

Penn drafted a missive to the Committee of the Free Society of Traders in London (August 1683) that described his perception of the climate:

> I found it from the 14th of October (Old Style), to the beginning of December … rather like an English mild Spring. From December to the beginning of … March, we had sharp, Frosty Weather; not foul, thick, black Weather, as our North-East Winds bring with them in England; but a Skie as clear as in Summer, and the Air dry, cold, piercing and hungry; yet I remember not, that I wore more Clothes than in England. The reason of this Cold is given from the great Lakes that are fed by the Fountains of Canada.

As the days lengthened, Penn observed, "we enjoy'd a sweet Spring, no Gusts but gentle Showers, and a fine skie." Yet not surprising to those familiar with the steamy summers in the Delaware Valley, he remarked that "we have had extraordinary Heats yet mitigated sometimes by Cool Breezese."

Penn also conveyed the notion of variable winds associated with the passage of weather systems and changes in the character of air masses:

> The Wind that ruleth Summer-season, is the South-West; but Spring, Fall and Winter, 'tis

rare to want the wholesome North Wester seven dayes together: And what-ever Mists, Fogs, or Vapours foul the Heavens by Easterly or Southerly winds, in two Hours time are blown away; the one is always followed by the other.

Penn commenced building an estate in Bucks County along the Delaware River, which became Pennsbury Manor. However, his stay would be short, and he returned to England in 1684, after spending only two years in his new colony, to deal with mounting debt. On Penn's voyage to America in 1699, he was accompanied by his wife and daughter and his secretary James Logan (1674–1751). Penn remained in his colony only two years and would never return.

One of the earliest literary descriptions of the weather in Pennsylvania was composed by Thomas Makin, a Latin teacher and clerk to the Assembly of the Province. Makin composed a poem in Latin, dedicated to James Logan in 1728.

"A Discription of Pennsylvania" was translated and forwarded to his friend Israel Pemberton (Proud 1797):

> Twixt hear and cold ye air is temperate;
> Warm Southern winds ye cold does mitigate
> The Northwest wind ye rains & clouds does clear
> Bringing fair weather & a wholesome air

Another of Makin's poems appeared in Watson's *Annals of Pennsylvania and Philadelphia, in the Olden Time*:

> Nay, oft so quick the change,
> —so great its pow'r—
> As summer's heat and winter *in an hour*! …
> But yet so temp'rate are some winters here
> That in the streams no bars of ice appear!

Washington Crosses the Alleghenies

George Washington (1732–1799) experienced the rigors of the Allegheny Plateau during the winter of 1753–54 while serving as a young British officer. Virginia Lieutenant Governor Robert Dinwiddie enlisted 21-year-old Major George Washington for a risky mission designed to deliver a stern warning to the French not to interfere with British rule in the Ohio Country.

The French had established forts near Erie, Waterford and Franklin, showing no intention of giving up any territory to Great Britain. The dispute foreshadowed the French and Indian War (1754–1763).

Washington proceeded from Williamsburg in October 1753, accompanied by Jacob Braam, who spoke French, but with only one experienced tracker—the renowned surveyor, fur trader and French interpreter Christopher Gist (c. 1706-1759), who would later rescue Washington from danger.

Gist was about to undertake his third journey, having kept ample diary notes on the conditions as early as 1750–51, which is the first written record of the weather in the region. In the fall of 1753, Washington's team traversed the rugged forests of the Alleghenies, encountering very difficult travel as conditions deteriorated. His contingent reconnoitered "the Forks of the Ohio" around what is now Pittsburgh, where the Monongahela River merged with the Allegheny to form the Ohio River.

The harsh weather eased up, but conditions would again take a turn for the worse. Along the way, Washington befriended the chiefs of several

Native American tribes, holding a key strategic meeting at Logstown, 17 miles downriver from Pittsburgh, that detailed French military activity along the Allegheny.

Washington's band of trackers and Native American companions encountered what he called in his diary "excessive" rain and snow while heading north from present-day Franklin, Pennsylvania, along French Creek. His assignment was to verbally engage the French commandant at Fort LeBoeuf, on a fork of French Creek near present-day Waterford, 15 miles from Lake Erie, where Washington presented an ultimatum regarding British control of the wilderness between Lake Erie and the Ohio River.

Washington's crew arrived on December 11, 1753, in the midst of a snowstorm. The French commander Captain Jacques Legardeur de Saint-Pierre politely rebuffed Washington after receiving the letter from Dinwiddie. Fortunately for our nation's history, Washington was sent packing without harm.

The weather proved challenging in western Pennsylvania in mid-December 1753 on Washington's return trip home to Virginia. Gist apparently saved Washington's life twice during the arduous return trip, something Gist hardly mentioned in his own diary. However, these events did not escape a grateful Washington. A description of the perilous journey was contained in "The Journal of Major George Washington," which appeared in the *Maryland Gazette* in two installments on March 21 and 28, 1754.

The first episode involved a French Indian guide who raced ahead into a clearing, turned and fired his gun "not 15 Steps" from Washington. Washington escaped harm, charging and subduing the guide with Gist's help, but ultimately let him go. The second harrowing experience occurred several days later along the ice-choked Allegheny River. Washington accidently fell from a jerry-built raft into the frigid waters about 10 feet deep, but managed to hoist himself back onto the boat with help from Gist.

The intrepid pair gave up their raft and spent a frigid night on an island in wet clothes, as Gist battled frostbite. Washington wrote of the escapade, "The Cold was so extremely severe, that Mr. Gist had his Fingers, and some of his Toes frozen." He added that it "rained and snowed incessantly" during the first two weeks of the month in "one continued series of cold wet Weather," and that "the Cold increased very fast, and the Roads were becoming much worse" from the deep snow and subfreezing conditions.

Fig. 2.1 Daily weather observations of Israel Pemberton at Philadelphia in June of 1776. (Courtesy American Philosophical Society)

Early Pennsylvania Weather Observers

The earliest known instrumental weather observations in Pennsylvania were taken by Dr. Cadwallader Colden (1688–1776) of Philadelphia in the winter of 1717–18, using a combination barometer-thermometer (Ludlum 1966). The only evidence of his observations appeared in a letter written by William Douglass of Boston, who thanked Colden for the receipt of the data.

Glenn Schwartz and Jon Nese, co-authors of *The Philadelphia Area Weather Book*, described the earliest surviving set of Pennsylvania weather records by an observer identified as Mr. De S. in Germantown, taken during the winter of 1732–33. Morning and afternoon temperature readings were noted from November 18, 1731, to October 28, 1732 (Old Style), compiled by (Havens 1958).

The renowned American botanist John Bartram (1699–1777), who resided in Philadelphia in the Pennsylvania Colony, kept detailed notes on the weather while collecting plant specimens. Bartram's 1751 publication *Observations on the Inhabitants, Climate, Soil, Rivers, Productions, Animals, and other Matters Worthy of Notice, made by Mr. John Bartram in his Travels from Pennsylvania to Onondaga, Oswego, and the Lake Ontario, in Canada* contained a detailed record of natural observations along the circuitous route.

William Bartram (1739–1823) followed in his father's footsteps, becoming America's leading botanist and taking an interest in recording the temperature, wind direction and nature in his diary, which commenced in the 1790s, after he moved back to the family home outside of Philadelphia.

Israel Pemberton (1715–1779) kept a close eye on the weather from his home two miles west of Philadelphia from 1748–1778. During the Revolutionary War, Pemberton took daily observations of the barometric pressure, wind and temperature at 8:00 a.m., 3:00 p.m. and 9:00 p.m., both inside and outside, at various intervals and provided notes on the sky condition. Portions of Pemberton's records appeared in the *American Philosophical Society Transactions* in 1839, along with the records of Charles Gotthold Reichle in Nazareth (1787–1790).

Englishman Thomas Smith's tract was a site of weather observations along the Schuylkill River in the western part of Philadelphia for a time until 1775. Smith's Labyrinth Garden meteorological records warranted a notice in the *Register of Pennsylvania* (February 2, 1828).

Other prominent weather diarists in southeastern Pennsylvania included farmer and surveyor David Schultze in Goschenhoppen, 40 miles northwest of Philadelphia in the Upper Perkiomen Valley, and Philadelphia resident Jacob Hiltzheimer.

The Reverend Thomas Coombe (1747–1822) of Philadelphia recorded the barometric pressure and indoor and outdoor temperatures at 8:00 a.m. and 3:00 p.m., starting in 1767.

The Reverend Henry M. Muhlenberg (1711–1787) observed the weather at several locations, including at Trappe, about 35 miles west of Trenton. His journals provided invaluable information on the harsh conditions endured by the Continental Army at Valley Forge in the winter of 1777–78.

David Rittenhouse (1732-1796), an enlightened, self-educated mathematician, inventor and astronomer, later served as the first director of the U.S. Mint from 1792–1795. He kept a daily record of the weather observations at his Philadelphia home. Rittenhouse recorded a reading of 94.5°F at the "observatory" on July 24, 1783, noting that the temperature was two degrees higher than any day in the city since at least 1769.

The observation elevated Rittenhouse to the position of a "leading authority" on Philadelphia weather, according to the *Pennsylvania Gazette and Freeman's Journal* (July 30, 1783). Rittenhouse family members continued the tradition of taking weather records after his death in 1796.

On March 4, 1797, John Adams took the oath of office in Federal Hall in Philadelphia in the chamber of the House of Representatives, on a rather pleasant late winter day: "Cloudy, some rain a.m., fair p.m." The early afternoon temperature on Inauguration Day was 53°F, with a mild southwest wind.

Peter Legaux kept lengthy weather records at Spring Mills, near Norristown, for the Pennsylvania Vine Company (1786–1828). Robert Cathcart maintained observations at York from 1801–1808 (Ludlum 1966).

Joseph Price (1752-1828) kept a diary, with many detailed weather references, in Montgomery County in present-day Wynnewood, from the age of 35 until shortly before his death. According to the Lower Merion Historical Society, Price purchased 156 acres from William Penn and bought another 49 acres later for his home.

Charles Peirce (1770–1851) compiled *A Meteorological Account of the Weather in Philadelphia, From January 1, 1790, to January 1, 1847, Including Fifty-Seven Years*, the first detailed weather summary for the Delaware Valley. Peirce, a New Hampshire native and founder of *The Portsmouth Oracle*, became a bookseller and moved to Germantown.

Peirce's description of significant weather events appeared in other journals, "collected in a good form for reference and preference," which was noted in his obituary in the *Portsmouth Journal* and reprinted in the *New York Daily Times* on October 3, 1851. He first gathered weather data in 1790 at Morrisville, across the river from Trenton, New Jersey, before settling in Germantown in 1813.

Peirce was celebrated as "an unwearied chronicler of the weather," who faithfully prepared "monthly reports of meteorological matters, interspersed frequently with remarks on passing events."

As of January 1800, there were 12 weather stations in six states that included three in Pennsylvania: Bethlehem, Morrisville and Philadelphia (Brown 1940). An excellent summary of observations and historic weather events appeared in *The Pennsylvania Magazine of History and Biography* (1891), entitled "Pennsylvania Weather Records, 1644–1835," compiled from various press sources and journals.

Benjamin Franklin

Benjamin Franklin (1706–90) was born in Boston, but spent much of his life in Philadelphia, where he contributed enormously to the early science of meteorology. A renowned printer, publisher, inventor and philosopher, Franklin presciently laid the groundwork for the fundamentals of a science that would be more fully developed in the next century.

Franklin and others formed a scholarly group in Philadelphia which was named the American Philosophical Society in 1743 with John Bartram and other luminaries. Rosenberg (2009) related that Franklin sought to create an institution "that collected specimens, accumulated scientific libraies, discussed topics, and published secular knowledge."

As the publisher of the *Pennsylvania Gazette*, Franklin's investigations into astronomy, storms, air pressure, tornadoes and the Gulf Stream were even more remarkable, given the limited observational data. In 1769, his Gulf Stream chart was printed from sketches prepared by Captain Timothy Folger, one of many meteorological contributions described in "A Chronological Outline of the History of Meteorology in the United States and of North America," a *Monthly Weather Review* series in March and April 1909.

On the evening of November 2, 1743 (New Style), Franklin intended to observe a lunar eclipse in Philadelphia, but his view was obscured by clouds that rolled in before the start of the eclipse (Ludlum 1963). He quickly learned through correspondence with his brother in Boston and an article in the *Boston Evening Post* that the eclipse was visible 300 miles northeast of Philadelphia, before "a storm of wind and rain" arrived "with great violence."

Franklin, who was appointed postmaster of Philadelphia, comprehensively gleaned information from travelers and newspaper accounts about the nature of the tropical storm that proved costly to southeastern New England, whipping up the highest tide in 20 years and causing extensive damage to wharves and shipping interests.

He composed a letter to the Reverend Jared Eliot in Connecticut, dated July 16, 1747, positing that a coastal storm traveled from southwest to northeast, accompanied by a northeasterly surface wind (Abbe 1906). Franklin reasoned correctly that the fuel for the storm came from a compensating inflow of warm, moist air from the Gulf of Mexico. He surmised that dense, cool air generated a return flow from the north to northeast, influenced by regional topographical features.

Franklin put down his thoughts in greater detail on the nature of nor'easters in a letter to Alexander Small, dated May 12, 1760, which appeared in *Experiments and Observations on Electricity* (1769, 381–383):

> Thus to produce our North-East storms, I suppose some great heat and rarefaction of the air in or about the Gulph of Mexico; the air thence rising has its place supplied by the next more northern, cooler, and therefore denser and heavier, air; that, being in motion, is followed by the next more northern air, &c. &c. in a successive current, to which current our coast and inland ridge of mountains give the direction of North-East, as they lie N. E. and S. W.

Franklin had access to a thermometer at least as early as 1749, writing down a temperature of 100°F on June 29, 1749, remembered as "Hot Sunday" in the colonies. He had already begun to explore the nature of electricity as early as 1745–1746 (Cohen 1990).

Franklin ultimately described a painful shock in 1747, which led to his theorizing two years later on the efficacy of iron lightning rods that could protect buildings from strikes. His famous—and risky—kite-flying experiment was conducted in a thunderstorm on June 15, 1752, which could have proved fatal.

After tying a silk string to a kite that had an iron key on the end, Franklin successfully drew a spark into a Leyden jar to contain the passing electrical charge. He held onto the experimental apparatus from inside a shed, his arm extended through a window. The dangerous experiment proved his hypothesis that lightning was indeed a form of electricity.

Later that year, Franklin built a lightning rod in Philadelphia. This conception challenged the ingrained belief, hammered home in church sermons, that lethal lightning bolts manifested the wrath of God when steeples were struck and buildings burned to the ground. Public sentiment quickly would swing over to Franklin's scientific view that lightning-related fires were potentially avoidable.

Franklin was a timely witness to a "whirlwind" in the Maryland countryside in August 1755, while traveling with his son and a group of gentlemen (Van Doren 1938). He followed the curious spinning column for some distance, and even tried to crack a whip through the swirling air, until he could no longer keep up with the whirl. Franklin subsequently composed letters on the subject of vortices, which were sent to members of the Royal Society of London and garnered the attention of natural philosophers.

Franklin was the first to propose a link between volcanoes and climate change. He had pondered for months whether a persistent dry fog, which appeared in northern Europe in the latter half of 1783, was linked to severe winter weather in 1783–84. He surmised that the mammoth eruption of the Laki volcano in Iceland in June 1783 was responsible for the persistent haze that filtered sunlight, triggering global cooling and widespread crop failures.

Independence Day Weather

Thomas Jefferson (1743–1826) was a devoted weather observer. In 1797, he wrote letters to friends encouraging a network of observations that would entail outfitting instruments in each county of Virginia. His *Garden Book*, covering a period from 1766-1824, contained horticultural and bird migration records, along with weather observations. He convinced James Madison at nearby Montpelier to keep weather records in 1784, continued by James Madison Sr. during 1794 to 1801, when he was away (Betts 1944).

William Hogeland, author of *Philadelphia in Declaration: The Nine Tumultuous Weeks When America Became Independent, May 1–July 4, 1776*, described the weather in the days leading up to the adoption of the United States Declaration of Independence. The beginning of July 1776 was very warm and sticky in Philadelphia, as the Second Continental Congress debated the wording of the historical document. Diarist Christopher Marshall noted on July 1, 1776, a sunny day "grew very warm," with a southerly wind that contributed to a typical muggy day in the Delaware Valley that culminated in an afternoon thunderstorm.

Hogeland (2010) discovered subtle differences in accounts of the timing of the storm. Marshall wrote in his diary, "At 4 came a thunder gust with rain ... cleared up by six." Observers on the scene commented that "the sound of rain just starting to hit the windows, another with roaring thunder. Others say heat and humidity mounted, the storm held off till late afternoon."

For a glimpse of the weather on the fateful days leading up to the approval of the Declaration of Independence on July 4, 1776, the December 1895 *Monthly Weather Review* drew upon the observations of Thomas Jefferson (1743–1826) published in his "Weather Memorandum Book." Jefferson wrote that he had purchased a thermometer from local merchant John Sparhawk.

Although Jefferson had remarked on the weather as early as 1772 in private notes, the purchase of a thermometer on July 1, 1776, commenced his nearly unbroken span of weather observations that concluded by the end of 1816 at Monticello in Charlottesville, Virginia.

Weatherwise magazine contributor Sean Potter (2011) reviewed the setting on July 2, 1776, when a heavy rain fell from before 10:00 a.m. until about 2:00 p.m. Inside the Pennsylvania State House,

Congress voted to declare America's independence from Great Britain as delegates debated the wording drafted by Jefferson.

Potter assessed the available accounts on the weather in Philadelphia: "The debates took up the entire day of July 3—which Marshall described as having had a 'fine clear & very cool morning,' with a morning low of 71.5°F recorded by Jefferson's thermometer, with mild weather spilling well into the following day."

On July 4, 1776, Jefferson recorded a temperature of 68°F at 6:00 a.m., hours before the formal adoption of the Declaration of Independence. Prophetically, Jefferson noted a temperature of 76°F at 1:00 p.m. "about an hour before a final agreement on the Declaration was reached," as the sunshine gave way to clouds.

Early Climate Change Debate

A perceived climate change in late-18th century America was linked to the clearing and cultivation of forests. The notion of a warmer, more extreme shift in the weather caught the attention of some distinguished scientists, sparking a spirited academic debate (Fleming 1990).

Modern land-use studies confirmed how agricultural practices impact the exchange of water vapor and the circulation of air. Shrinking forestland promotes a greater daily temperature range. The widespread irrigation of cropland coupled with transpiration release water vapor into the air, which foments thunderstorms in the Corn Belt that are responsible for about half of the summer precipitation in the Midwest.

Learned people in colonial times who engaged in the climate debate largely depended upon the recollections of the "oldest inhabitants" across the region and anecdotal remarks culled from sketchy memories, aided by a collection of private journals (Gelber 2009).

Swedish botanist Pehr (Peter) Kalm (1716–1779) journeyed to Philadelphia during the period of 1748–1751. He consulted "aged Swedes" to ascertain the nature of the purported climate change (Watson 1868). In his *Travels into North America*, published in the early 1770s, Kalm wrote that longtime settlers believed winters were generally becoming milder, with less rain and snow and frequent thaws. Spring seasons, on the other hand, seemed to have shifted backward, with chilly conditions lingering into May.

On August 17, 1770, Philadelphia physician Hugh Williamson (1735–1819), an educator and legislator, weighed in on the topic, reading from his essay "An Attempt to Account for the Change of Climate Which Has Been Observed in the Middle Colonies in North America," which was published in issues of the *Transactions of the American Philosophical Society* between 1769 and 1771.

Williamson acknowledged the tropical easterly trade winds and influence of convection (heating) when little was known about such topics. He later added his scholarly heft to the growing sentiment that the climate had become less extreme. He reasoned, "as clearing the country, will mitigate the cold of our winters, it will also increase the heat of our summers … the same cause in those seasons will appear to produce different effects, and that instead of more heat, we will presently have less in summer than usual."

Williamson lauded the wholesome effects of draining the swampy soils when woodlands were removed near Philadelphia because "the air was constantly charged with a gross putrescent fluid" and charging physicians "to be careful to trace the history of every disease."

Williamson concluded that development had improved the health and climate of Philadelphia:

> The salutatory effects which have resulted from clearing and paving Philadelphia, are obvious to every inhabitant … a series of irregularities, nervousness, bilious, remitting and intermitting fevers … are now evidently on the decline … as our winters grow more temperate.

Several American physicians took an interest in the role of heat and humidity in the outbreaks of yellow fever, malaria and smallpox in early America. Benjamin Rush (1746–1813), a distinguished Philadelphia physician, put down his thoughts in an "Account of the climate of Pennsylvania, and its influence upon the human body," gleaned from "medical enquiries and observations," which he penned in 1789 (revised in 1805).

In an original essay that appeared in *American Museum* (1789, 252), Rush made note of the generally dry autumn weather, which he interspersed with Native American beliefs:

> These rains are the harbingers of the winter, and the Indians have long ago taught the inhabitants of Pennsylvania that the degrees of cold during the winter, are in proportion to the quantity of rain which falls during the autumn.

Rush's thoughts were reprinted in *The Register of Pennsylvania* in September 1828, which later appeared in *A Gazetteer of the State of Pennsylvania* (1833):

> There is another circumstance … which contributes very much to mitigate the heat of summer, and that is, it seldom continues more than two or three days, without being succeeded by showers of rain, accompanied sometimes by thunder and lightning, and afterwards by a north-west wind, which produces coolness in the air, that is highly invigorating and agreeable.

Rush further observed: "We have no two successive years alike … There is but one steady trait, and that is, it is uniformly variable." He also commented on the perceived climate change:

> From the accounts which have been handed down to us by our ancestors, there is reason to believe, that the climate of Pennsylvania has undergone a material change. Thunder and lightning are less frequent: the cold of our winters, and the heat of our summers, are less uniform, than they were forty or fifty years ago. Nor is this all: the springs are much colder, and the autumns more temperate, than formerly, insomuch that cattle are not housed so soon, by one month, as they were in former years.

Dr. John Redman Coxe (1773–1864), who studied under Rush, treated dozens of patients during the yellow fever epidemic of 1793 in Philadelphia. Coxe received his medical degree from the University of Pennsylvania in 1794, commencing a detailed weather journal starting in 1798.

Historian David Laskin, in his delightful *Braving the Elements*, covered the acrimonious exchanges defending the climatic character of America, a sore spot because some European scholars used every opportunity to malign the climate. The notion of an unhealthy climate was authoritatively countered by Benjamin Franklin, who in 1760 extolled the "salubrity of the air, the healthiness of the climate, the plenty of good provisions" in his "Information to Those Who Would Remove to America."

The great educator and lexicographer Noah Webster (1758–1843) weighed in on the relationship between land use and climate change. Webster was primarily interested in the relationship between diseases such as yellow fever and climate (Fleming 1990).

In 1799, he gave a presentation at the Connecticut Academy of Arts and Sciences, remarking that "the clearing and cultivation of our country, has moderated the rigour of our cold weather, that the cold of our winters, though less steady, has been most sensibly increased."

He continued: "It appears that all the alterations in a country, in consequence of clearing and cultivation, result only in making a different distribution of heat and cold, moisture and dry weather, among the several seasons." His argument was that opening of the lands by clearing the forests removed moisture, resulting in warmer summers and colder winters.

Some European observers clung to the unsupported belief that the climate of early America was deleterious for human activity—cold and damp, with underdeveloped fauna. This provable calumny was challenged vigorously and passionately by Thomas Jefferson in his *Notes on the State of Virginia*, first put forth in 1781 and expanded in a small Paris printing (1785). His book garnered a wider London publication in 1787.

French naturalist, cosmologist and author Georges-Louis Leclerc, Comte de Buffon (1707–1788), ignited the climate controversy in his *Histoire Naturelle (1749–88)*, with eight additional volumes (44 total), published after his death between 1789 and 1804. Buffon posited that the coldness of the New World climate was responsible for the "degeneration" of animal and plant species.

The French philosopher Constantin-François Chassebœuf, Comte de Volney (1757–1820), published *A View of the Soil and Climate of the United States of America* (1804). Pennsylvanian Charles B. Brown (1771–1810) annotated his study while using the opportunity in his "remarks" to set the record straight. Brown challenged Volney's notion that a northeast wind in Philadelphia "oppresses the brain, and produces torpor and head-ache."

Brown concurred that a northeast wind can be "rather uncheering," but maintained it was not "directly hurtful to the health." He acknowledged a quaint cultural custom that favored "the almost general disuse of the bath" while "clothed in cloth flannel, and black fur hats, and lying on a feather bed at night, drinking but wine and porter, and eating strong meats three times a day, and never allowing water to touch any part of them but their extremities." In this manner, Americans adapted to "the great heats."

Naturalist and physician Benjamin Smith Barton (1766-1815), a professor at the University of Pennsylvania and nephew of David Rittenhouse, considered the role of climate change and the timing of the emergence of spring flora. He enjoyed surveying the western fringe of Pennsylvania with Rittenhouse as a teenager. During a lecture given in 1807 at the short-lived Linnean Society in Philadelphia, Barton sought to establish a link between botany and climatology through research.

William Darlington (1782–1863), a respected physician, botanist and weather observer in the 1820s at West Chester, 25 miles west of Philadelphia, opined in *Silliman's Journal* (April 1828) that "the quantity of snow, in this region, is much diminished within half a country; … probably owing in a great measure to the clearing of the forests, and the extended cultivation of the country."

James Pollard Espy

Weather historian James R. Fleming, in his fascinatingly detailed *Meteorology in America, 1800–1870*, credits Pennsylvanian James Pollard Espy (1785–1860) with establishing Philadelphia as the central location for weather research in the United States.

James Espy requested the exchange of daily records from local observers for publication in the *Journal of the Franklin Institute*, forming a meteorological committee that commenced publishing monthly data in January 1831, which represented the first attempt at formalizing a cooperative weather data network in the United States, fulfilling the dream of Thomas Jefferson.

Espy was born in Westmoreland County in the western frontier. After studying at Transylvania Academy in Kentucky, then teaching and practicing law in several states, he settled in Philadelphia in 1817, joining the Franklin Institute of the State of Pennsylvania for the Promotion of the Mechanic Arts in 1824, where he taught classical languages and mathematics.

Espy wrote his first scientific paper on atmospheric science for the American Philosophical Society in Philadelphia in 1821. His next foray into the fledgling field of meteorology would not come until nearly a decade later, when he took up the concept of convection—a rising column of water vapor and the subsequent release of latent heat—during the life cycle of a thunderstorm. Espy correctly described the nature of latent heat in an 1829 treatise, explaining how a column of air heated near the surface becomes less dense, causing it to rise and expand as the environmental pressure decreases, eventually condensing into clouds capable of producing a storm.

In September 1834, the Joint Committee on Meteorology, chaired by Espy, convened in Philadelphia under the auspices of the American Philosophical Society and the Franklin Institute to "confer together on the best means of promoting the advancement of meteorology." One of the primary goals was to advance the study of storms, which would entail the formation of a nationwide weather network.

In his *History of American Weather* series, historian David Ludlum called this event "the first effort of the meteorologically minded men in the United States to advance the science of cooperation." Espy, a dynamic lecturer and self-promoter, passionately joined in the study of tornadoes, after visiting the scene of the deadly New Brunswick, New Jersey, tornado in 1835. He reviewed similar damaging wind events between 1834 and 1839.

Espy commenced on the lecture circuit in 1837 in Philadelphia, and would eventually travel the country and abroad to promote and debate his theories on storm generation. He published a slew of academic papers on astronomy and meteorology, including pressure and dew point, and refined his concept that an upward flow of warm, moist air powered a thunderstorm. In his publications, he asserted that thunderstorms were produced by winds flowing toward the center of a low pressure sustained by convection.

His hubris made it impossible to consider a competing scientific concept, proffered by the

Connecticut engineer William Redfield, which incorporated the role of centrifugal force in the circulation of larger storm systems, rather than being solely driven by an inward flow of winds. There would be rancorous discourse, mostly conducted by Espy himself, since Redfield opted to stay out of the limelight.

Understandably, a limiting factor was the lack of systematic data. Espy understood this, especially after creating the first American weather map by plotting a handful of surface wind observations months after a December 1836 storm, published in the *Journal of the Franklin Institute* the following year (Nese and Schwartz 2005).

Encouraged by Espy, the Pennsylvania State Legislature approved a sum of $4,000 on April 1, 1837, allowing the Franklin Institute to manage a statewide weather network. The lofty goal was to establish one observer in every county, who would receive a barometer, rain gauge and several thermometers to log daily observations.

The task of finding dependable observers was another matter, but the records appeared in monthly installments in the *Journal of the Franklin Institute* beginning in 1839 and continued for several years. Espy put together a detailed wind analysis of a storm that crossed the Midwest and East on March 16–18, 1838, based on data collected by 40 field correspondents (Monmonier 1999).

He published his conceptual framework of natural laws governing storms in *The Philosophy of Storms* (1841), a book that further enhanced his stature in the scientific community. It also helped to secure his appointment as the first government meteorologist employed by the U.S. War Department in 1842, and later the Navy Department in 1848 in cooperation with the Secretary of the Smithsonian Institution, Professor Joseph Henry.

Nonetheless, Espy's detractors took exception to his megalomaniac demeanor and grandiose plans for weather modification. The press anointed him the title of the "Storm King."

Espy proffered three proposals to Congress that included funding a large network of observers. He requested a more-than-modest salary to manage this national weather system. More controversially, his theory of latent heat release, building up clouds into storms, would lead to an experiment in artificial rainmaking. The weekly experiment he proposed in 1838 entailed sparking forest fires across a 600-mile corridor of the Rockies from north to south, divided by 20-mile intervals, over drought-stricken Western land. His propositions would be funded only if the plan proved to be successful.

Espy was totally invested in the notion that he could initiate condensation by heating the air and filling the atmosphere with smoke to induce rain. His grand scheme sparked widespread ridicule in the Senate and "greatly damaged his standing in the scientific community" (Ludlum 1970).

In 1842, John Quincy Adams made mention of his skeptical exchange with Espy. Adams referred to "Mr. Espy, the storm breeder" in remarks before the Congress. Adams would say, "I told him with all possible civility that it would be of no use to to memorialize the House of Representatives in behalf of his three wishes."

Robert Hare

Robert Hare (1781–1858), a chemistry professor at the University of Pennsylvania's medical school from 1818 through 1847, was a lifelong Philadelphia resident, who focused his early inventions and research on the nature of electricity.

Dr. Hare had a passion for storms, joining other scientists of the day at the site of the damage caused by the great Norfolk and Long Island Hurricane in early September 1821. The following year, Hare published a paper describing in some detail the northeasterly flow that accompanied a coastal disturbance, taking a large-scale approach that included a return flow carrying energy back to the tropics. Hare favored the concept of latent heat released in a column of rising tropical air.

However, in the late 1830s, the irascible Hare shifted his energy to the role of huge currents of opposing electrical activity acting as a conduit in a tornado. He firmly believed that electrical discharges sustained a funnel, a concept that was controversial and unpopular, and was eventually discredited.

William Ferrel

The American Storm Controversy would ultimately be settled by another scientist born in Pennsylvania. From humble beginnings in Bedford County, William Ferrel (1817–1891) derived the mathematics necessary to explicate the physical laws which incorporated the rightward deflection created by the Coriolis force (Northern Hemisphere) of air in motion—the wind—and the effect of friction.

Ferrel applied the principle of the conservation of angular momentum on the general circulation, and centrifugal force on a mass when it is rotated in outward flight, in an 1856 article. He conceived the concept of air flowing from high pressure and spiraling inward toward lower pressure.

The genius of Ferrel closed the gap between Espy's straight-line inflow of air and Redfield's circular wind pattern surrounding low pressure. By then the science field had largely discarded storm theories based solely on electricity and magnetism.

James Henry Coffin

James H. Coffin (1806–1873), a prestigious Lafayette College mathematics and natural philosophy professor in Easton, published the first detailed wind studies that would provide a framework for weather forecasting based on the natural laws of motion.

The historian Frank R. Millikan (1997), who worked on the Joseph Henry Papers Project at the Smithsonian Institution, characterized the role of Coffin as a contractor in analyzing and interpreting the vast dataset, "as many as half a million separate observations in a year" and confusing symbols. Coffin employed up to 15 assistants to manage the data.

In 1861, Coffin published the first of a two-volume set of climatic and storm data for the period of 1854–1859. His other meteorological publications included *Winds of the Northern Hemisphere* (1853) and *Winds of the Globe* (1875), the latter completed by his son, Selden J. Coffin, who managed Lafayette College weather observations in the 1880s.

Lorin Blodget

Lorin Blodget (1823–1901) was born in Jamestown, New York. He initially served as a volunteer weather observer in his hometown, before moving to Philadelphia (Ludlum 1970). In 1857, the "youthful statistician" wrote his masterly 536-page *Climatology of the United States and of the Temperate Latitudes of the Northern American Continent*.

This seminal work used meteorological data from a national network of observers established by the Smithsonian Institution. Blodget later served as editor of the Philadelphia *North American* and secretary of the Philadelphia Board of Trade (1858–1864). His detailed description of the early American climate, based on sparse data compiled for the U.S. War Department, became a framework for the fledgling field of American climatology.

Notable 19th Century Pennsylvania Observers

Academic institutions periodically served as a local center for meteorological observations in the 1800s. Professor Michael Jacobs was a mathematics professor at Pennsylvania College (now Gettysburg College) from 1832–1866, where he provided records to the Franklin Institute, and later the Smithsonian Institution national weather collection agency (Ludlum 1984).

Professor Jacobs made the following simple observation from his vantage point of his beloved natural preserve on July 3, 1864: "Great battles of Gettysburg fought." Warm and humid weather prevailed during three bloody days from July 1–3, 1863. Rain began falling late on July 3, and he measured 1.39 inches at 4:00 p.m. on the Fourth of July.

A British correspondent added: "The night [of July 4–5] was very bad—thunder and lightning, torrents of rain, the road knee-deep in mud and water." Soggy conditions slowed the progress of General Robert E. Lee, commander of the Confederate Army, as he headed south across Maryland. Heavy rain returned on July 8–9, 1863, raising the Potomac River further, which delayed the troop withdrawal (Ludlum 1960).

An interesting story involves a local amateur weather observer named Stodgell Stokes, born into a Quaker family in Bucks County in 1788. He moved to the Pocono foothills around 1815 and opened a general store in Stroudsburg. An early opponent of slavery, he joined the Society of Friends and assisted in the Underground Railroad that helped secure the safe passage of former slaves to Canada.

Stokes later moved to Moorestown, New Jersey, near his Bucks County roots. At the age of 94, he built a new type of rain gauge, described in a family paper that was read at the Monroe County Historical Society in Stroudsburg in 1928 (Walters 1965):

> He then conceived that the ordinary rain gauge was wrong in principle, and invented a new one he thought was better. No storm was too severe for him to go out in and experiment with this gauge, and he soon took a severe bronchitis from which he died. I fully believe had it not been for this rain gauge he would have lived to well over a century.

Stokes's son, John N. Stokes (1826–1875), also took an interest in the weather. In a biographical index created by the Monroe County curator, Mrs. Horace G. Walters wrote that John did not possess a keen business acumen, though for a short time he partnered with his father in the family store. The younger Stokes was regarded as "a good horseman, a splendid swimmer, and an inveterate practical joker. His kindly disposition and his inability to say 'no' led to his listening of many a hard-luck tale and to his endorsement of many a note which he afterward had to pay."

Until his accidental drowning at Lamberts Rock on McMichael Creek in July 1875, Stokes supplied useful weather information to the local

Jeffersonian. Another member of the Stokes family, Anna Maria (1819–1878), daughter of Dr. Samuel Stokes, became the first formal weather observer in Stroudsburg in 1839, using instruments supplied by the Franklin Institute in Philadelphia.

Theodore Day, born in 1837 in Elizabethtown, New Jersey, was a schoolteacher and naturalist in southern Wayne County. His grandfather Stephen settled north of Honesdale in October 1816. Day's father lived in New Jersey, before returning to the area of Dyberry Township in 1843. After a stint in the Civil War, Theodore Day switched to farming, sending his data to the Department of Agriculture on "grafting and pruning trees" (Haines 1902).

Day's work led to a lifelong appreciation of the elements and a long record as a weather observer. His data commenced in April 1854, and he joined the volunteer network in 1860, under the auspices of the Smithsonian Institution. Day's records from 1865 to 1903 are preserved. He filled readers with his phenological knowledge until his death (1916).

Organizing a National Weather Service

A unified climate record would not be possible without the voluminous data from first-order weather stations and second-order airways locations, coupled with data taken by thousands of volunteer observers.

The first systematic national weather network dates back to 1814, when Dr. James Lovell, Surgeon General of the U.S. Army, issued an order that instructed surgeons at forts around the country to start keeping methodical weather data (Lawson 1840). In 1817, the General Land Office issued a memorandum that its 20 offices commence routine weather observations (temperature and precipitation) at prescribed times. A similar decree from the Surgeon General was implemented in 1819 at U.S. Army posts, which grew to a network of 13 forts across the Midwest by 1838.

The Army records actually date back to the late 18th century, when a few far-flung outposts in the Northwest Territory, such as Fort Defiance in the Ohio Country, sporadically made note of wind and weather conditions. The first published military weather records covered the period from 1822–1825 and reside in the National Archives in College Park, Maryland.

During the decades that followed, records were haphazard. In 1847, Joseph Henry (1797–1878), the first Smithsonian Secretary proposed "a system of extended meteorological observations for solving the problem of American storms." The next year, Henry budgeted $1,000 for a Smithsonian project that would establish the early national volunteer weather network.

The invention of the telegraph in 1848 offered the technological breakthrough for transmitting information rapidly over long distances. A small network of telegraph operators across the country began reporting basic weather observations normally taken at sunrise, 9:00 a.m., 3:00 p.m. and 9:00 p.m.

Weather observers were issued standardized forms, with guidelines regarding the necessity of maintaining routine daily observations of temperature, barometric pressure, humidity, winds, cloud cover and precipitation. There was a section termed "casual phenomena" for storms and natural events (meteors, auroras, earthquakes).

One of the strangest reported phenomena occurred in Chester County on June 6, 1869. Snail shells fell to the ground mingled with raindrops in a light shower, an observation that was forwarded to the Academy of Natural Science in Philadelphia (Watson 1993). Y. S. Walter, editor of the *Delaware County Republican*, was a witness to the event, calling it "a storm within a storm." The shells that swirled in the rain appeared to be living organisms, as described in *Scientific American* (1870). Some of the specimens were broken but others were intact.

The citizen-observer network that started with about 150 volunteers grew to about 500 in the 1850s, and managing the data became unwieldy for the Smithsonian Institution. Some of the data collection and the distribution of registers were taken up by the Patent Office's agricultural division in the late 1850s, and later the newly established Department of Agriculture in 1862.

The Civil War disrupted the program, and only a handful of stations were reporting regular weather readings by 1865. However, the need for a national storm-warning service had become more pressing, especially for shipping interests around the Great Lakes, after too many marine disasters in surprise storms.

On February 9, 1870, Congress passed a resolution, signed by President Ulysses S. Grant, squarely placing the role of data-gathering, and the issuance and dissemination of weather forecasts, under the direction of the Secretary of War and U.S. Army Signal Service Corps.

Commencing on November 1, 1870, meteorological reports were taken at 24 cities across the United States three times daily and sent by telegraph to Washington, where meteorologists hand-plotted the data and issued daily forecasts. The edited information was used by various outlets, including newspapers and railroads.

The U.S. Signal Office commenced the publication of weather maps on January 1, 1871. In January 1874, the Army Signal Service fully absorbed the original Smithsonian volunteer weather network created by Secretary Joseph Henry, though not all observers continued to furnish observations to the military and simply ceased taking observations.

After years of criticism from members of Congress, who had discovered ample evidence of fraud and cronyism, the national weather service was transferred to the U.S. Department of Agriculture. The Signal Corps data-gathering responsibilities were transferred to the new Weather Bureau in 1891, which took over the production of the *Daily Weather Map*, later part of the *Weekly* series beginning in 1969. The early maps plotted observations from the continental United States observations of each day's "air temperature, barometric pressure, wind velocity and direction" (NOAA).

By the turn of the century, more than 160 stations operated by the Weather Bureau supplied daily measurements of meteorological variables—temperature, pressure, wind direction and speed, dew point and precipitation intensity—twice a day based on daily measurements taken around 8:00 a.m. and 8:00 p.m. (Kocin 1988).

Data streamed in from around the country by telegraph, telephone and radio to the Weather Bureau, military and aviation interests with information on the surface and upper-air conditions, which improved local forecasts and the selection of the safest air routes.

Pennsylvania State Weather Service

The Pennsylvania State Weather Service was organized on December 15, 1886, under the direction of the Franklin Institute in Philadelphia.

On May 13, 1887, the State Legislature appropriated $3,000 for the establishment of a statewide network, with additional aid coming from the Franklin Institute Advisory, supplementing the U.S. Army Signal Service Corps contribution. Volunteer observers supplied with instruments were required to take daily maximum/minimum temperatures, precipitation, snowfall and additional observational notes.

The first state weather summary appeared in September 1887, augmented by the weekly agricultural bulletin, first published on July 11, 1888. Weather data sent in by observers appeared as an appendix to the *Journal of the Franklin Institute* from 1887–1895.

In July 1895, a collection of 89 weather stations comprised the Pennsylvania weather network, part of a larger field that consisted of about 3,000 volunteer observers scattered across the United States and its territories, according to the *Monthly Weather Review*. In March 1896, all observations fell under the auspices of the U.S. Department of Agriculture, with state data published under *Climate and Crops: Pennsylvania Section*. A later iteration was labeled *Climatological Data: Pennsylvania Section* and is currently called *Climatological Data—Pennsylvania*. The existing records are on file at the National Centers for Environmental Information (NCEI) in Asheville, North Carolina.

The U.S. Weather Bureau became part of the Department of Commerce in 1940. The Environmental Science Service Administration (ESSA) absorbed the Weather Bureau and a few other agencies in 1965, which was reorganized in 1970 within the National Oceanic and Atmospheric Administration (NOAA). The following year, the modern National Weather Service (NWS) was established.

Cooperative Weather Observer Program

The task of assembling and editing a large volume of historical data for the purpose of presenting a chronological account of our weather history is an ongoing project, because as we all know, tomorrow could bring another noteworthy event.

In March 1891, the *Monthly Weather Review* proudly announced that 2,406 volunteer weather observers were involved in the United States Weather Bureau data-collection program. The observer network had expanded to 90 communities throughout Pennsylvania in 1925, according to the *Altoona Mirror* (October 9, 1925), for the purpose of gathering climatological data.

At the national level, the number of volunteer observers in the NWS Cooperative Observer Program (COOP) peaked at about 13,500 participants. Currently, more than 11,000 volunteer observers maintain daily records. The lengthy climate database is essential to provide an understanding of the local risks severe weather poses to life and property, and to provide a benchmark to analyze regional climate change.

Weather stations are situated in urban, suburban and rural settings, national parks and academic institutions. Instruments are housed in well-ventilated weather shelters. Daily readings consist of maximum and minimum temperatures, 24-hour precipitation and snowfall totals, and the mean total snow depth. Some agricultural extension sites provide hydrological and soil temperature data.

Long-time rural stations offer a dataset not directly impacted by urbanization that extends beyond a half-century. A compilation of records has been produced and authenticated by the Office of the State Climatologist of Pennsylvania at the Pennsylvania State University in State College. Monthly and seasonal state averages are also available through the Northeast Regional Climate Center in Ithaca, New York.

Some stations are listed with an additional notation in reference to the distance in miles and compass direction from the nearest post office; for example, Clermont 4 NW is four miles northwest of the Clermont Post Office.

The highest network site in Pennsylvania is located at Hidden Valley in Somerset County at an elevation of 2,876 feet. The lowest reporting station is at the Philadelphia International Airport, situated at an elevation of 7 feet above sea level in Delaware County.

Pennsylvania currently has about 200 volunteer observers who forward their daily observations to the nearest NWS office. Data are published monthly by the National Centers for Environmental Information (NCEI). Although all sections of the state are well-represented by weather observations, the greatest concentration of stations is in the more populous southeastern and southwestern parts of the state.

COOP observers come from various walks of life and are usually longtime residents of a community who simply enjoy taking weather readings in their backyard. In addition, thousands of private observers stream weather data online through sites such as Weather Underground and the NWS link to Community Collaborative Rain, Hail & Snow Network (CoCoRaHS).

In 1971, Pennsylvania state climatologist Paul W. Dailey Jr. determined that the 65-year span of weather records (1855–1920) maintained by Dr. Jesse C. Green at West Chester comprised the longest period of record for a single observer. Green took daily readings until he passed away at the age of 102 in 1920.

The longest serving volunteer weather observer was Richard G. Hendrickson, a resident of Bridgehampton, Long Island, who died on January 9, 2016, at the age of 103. Hendrickson took observations at least twice daily, totaling more than 150,000 individual temperature, precipitation and other readings, starting on July 1, 1930, and continuing until he turned 103 (Barron 2014). Earl Stuart of Cottage Grove, Oregon, maintained records for 77 years from 1917 to 1994.

John Clyde LeBar, of Stroudsburg, provided weather information for the local press prior to becoming the government observer in December 1910. At the end of a decade-long tenure, the *Daily Record* took the opportunity to recognize LeBar in an editorial on the state of volunteerism dated March 4, 1920:

For ten years, he has been very faithful in making the reports of the weather to the bureau during which time his only compensation had been the thanks of the department ... It is unfortunate that the government cannot see its way clear to offer a reasonable compensation for a service which requires considerable time of the observer as well as requiring the service of a person who has patience to be very exacting in manner of keeping records, for the weather records are worthless if they are not letter perfect.

Dixon R. Miller (1911-2003) was raised west near Harrisburg in Summerdale. He ascended the 172 steps of the Pohopoco Fire Tower every day at sunrise until shortly before his death at 91. He was assigned to keep rainfall data by the state Department of Forests and Water in 1934, becoming an official U.S. Weather Bureau observer in 1935 (private letter, May 17, 2000).

In the spring of 1938, Miller added the title of fire tower warden to his duties. After passing the civil service examination in 1940, Miller maintained airways observations at Harrisburg, and later moved to other sites as a towerman. He became the resident hydrographer in 1948 for the Division of Forest and Water at Blakeslee, under the auspices of the Swiftwater office (now the Pennsylvania Department of Conservation and Natural Resources).

Miller (1958) published an article that detailed the historic Pocono snowfalls in the long winter of 1957–58, when the snow depth reached 52 inches at his western Monroe County home. In 1963, he assumed fire tower duties at the Pohopoco Fire Tower (elevation 2,215 ft.) atop Pimple Hill, two miles southwest of Long Pond.

He took an abiding interest in the Dilldown Frost Pocket, a mile below the fire tower (1,905 ft.), submitting detailed observations to the "Forest Fire Specialist Supervisor" on September 5, 1996. Dixon witnessed frost in every summer month: 23° on June 11, 1988; 28°F on July 2, 1982; 30°F on July 4, 1986, when at sunrise "the vegetation was white with frost." He also noted a low of 23°F on August 22, 1982. The lowest temperature observed in the frost hollow was -40°F.

A better-known frost pocket is situated four miles northwest of State College. The Barrens area is cloaked by white pine and scrub oak growing on sandy soils in a formerly mined and deforested narrow valley. A favorite haunt of Pennsylvania State University meteorology students and faculty, minimum temperatures are as much as 30 degrees cooler than at the university weather station on a clear, calm night.

SKYWARN Program

The NWS SKYWARN Storm Spotter Program has upwards of 400,000 trained volunteer severe-weather spotters who provide eyes on the ground to keep communities safe during severe weather. Annual spotter training sessions are conducted by county emergency managers and NWS personnel in the early spring before the start of the storm season.

The program commenced in the 1970s with the aim to provide critical "ground truth" in severe weather. Trained spotters include ham radio operators, law enforcement members, public utility workers, academics and private citizens who receive detailed instructions on how to report field observations during potentially dangerous weather.

Chapter Three

Seasons on Parade

Like a reconnoitering force of an invading army sent to feel out positions and get the lay of the land, the advance cohorts of winter swept down on the east on Saturday (October 10, 1925), bringing snow and bitter cold.

—Morning Press (East Stroudsburg), October 12, 1925

THE TIME-HONORED aphorism frequently attributed to Mark Twain, "If you don't like the weather in New England, just wait a few minutes," cited by Bennett Cerf in his book *Try and Stop Me*, is a fitting description of the Keystone State.

"Everybody talks about the weather, but nobody does anything about it," the famous aphorism that appeared in a *Hartford Courant* editorial on August 24, 1897, was more than likely penned by Twain's fellow Hartford resident Charles Dudley Warner. Given that they were fast friends, there could have been a philosophical collaboration.

Weather refers to the state of the atmosphere—sunshine, cloud cover, rain, snow and wind—in all its glorious varieties. Extreme weather takes a substantial economic toll on agriculture, infrastructure and the transportation and tourism sectors. Prolonged heat waves raise the cost of keeping cool in summer while cold outbreaks focus the attention of energy traders in the commodities markets.

Climate is the long view of weather—a statistical treatment of meteorological variables that quantify and classify common meteorological characteristics. Average temperature and precipitation data are primarily based on the most recent 30-year period (1991–2020), published by NOAA's National Centers for Environmental Information (NCEI), with data from more than 15,000 reporting stations in the United States.

The period is considered a sufficient baseline for comparative analysis of climate trends. Notably, the average annual temperature in the contiguous United States has risen 1.7°F since 1901–1930.

Solar Energy Drives the Weather and Climate

The sun is the center of our solar system—one of approximately 100 billion stars clustered in the Milky Way. Solar energy drives Earth's weather and climate patterns, providing warmth, light and sustenance through photosynthesis.

The volatile cauldron of luminous gases generates an enormous amount of heat in the solar core, where hydrogen atoms are compressed and consumed in the production of hydrogen nuclei. Tremendous pressure generated by gravitational attraction spurs nuclear reactions that fuse hydrogen into helium with a force equivalent to 10 trillion one-megaton hydrogen bombs. Coupled with other nuclear reactions, a vast amount of energy heats the solar core to about 27 million °F.

The sun is a middle-aged star and the closest to Earth, at an average distance of 93 million miles. Earth is presently nearest to the sun on January 3 (91.4 million miles) and farthest away about July 6 (94.5 million miles).

Earth orbits the sun in a period of 365 days and 6 hours. The tilt of Earth's rotational axis (23.5 degrees) relative to its orbital plane determines the seasonal distribution of insolation.

The solar zenith angle, which is the difference between the altitude of the sun and the vertical, varies 47 degrees between late June and late December at 40° N. Two and a half times more solar radiation reaches the surface in early summer in the Northern Hemisphere compared to early winter at the latitude of Pennsylvania.

Earth's axis is tipped toward the sun during the summer months. The higher solar elevation angle is more effective at heating a wide surface area. In the wintertime, the Northern Hemisphere is tilted away from the sun and the peak solar rays are more oblique.

Solar Energy—Straw that Stirs the Drink

Radiant energy emitted by the sun reaches the planet's surface in eight minutes 20 seconds, mostly in the ultraviolet (UV), visible and near-infrared portion of the electromagnetic spectrum as shortwave energy.

Earth's energy budget is modulated by incoming solar radiation balanced by outgoing thermal radiation, maintaining a hospitable average global land and ocean surface temperature of 57°F (NOAA).

A little less than half of the total solar energy striking an imaginary plane perpendicular to the top of Earth's atmosphere is absorbed by the land and ocean. About 23 percent of incoming sunlight is absorbed by atmospheric gases, clouds and particles (aerosols). Another 30 percent is reflected to space by bright clouds, aerosols and Earth's surface features.

Unequal heating between the tropics (surplus) and high latitudes (deficit) is responsible for the large-scale circulations. The resultant pressure gradient sets the air in motion to equalize the difference. Large mid-latitude weather systems, tropical cyclones and ocean currents redistribute excess planetary heat from the tropics to higher latitudes.

The general circulation is defined by large cells of upward and downward motions. Equatorial low pressure that promotes rising bubbles of hot air is directed poleward. The trajectory of the air is deflected to the right in the Northern Hemisphere by the Coriolis force, which is imparted by the turning of Earth on its rotational axis.

Poleward-moving air descends in the subtropics, forming sprawling belts of high pressure and most of the world's great deserts. The clockwise circulation around subtropical high pressure drives the easterly trade winds that flow equatorward and converge along the Intertropical Convergence Zone (ITCZ). The northward migration of the ITCZ with the more direct rays of the sun is an integral part of the wet monsoon in the tropics.

Subtropical and polar airstreams meet in the middle latitudes along the meandering polar front on the southern margin of the upper-level westerlies. The frontal boundary is an active weather zone, where low pressure areas spin up in response to jet stream dynamics and the lifting of unstable air. Mid-latitude waves exchange heat energy and moisture by drawing warm air northward and cold air south. The subsequent clash forms large areas of precipitation (rain, snow), and in zones of great contrast the potential for severe or violent storms.

Earth's water, or hydrologic cycle is modulated by heating and the evaporation of surface moisture that encourages parcels of air to rise. Buoyancy is enhanced by moisture transport and convergence. The growth of water droplets and ice crystals returns beneficial moisture to the surface. Precipitation runs off into lakes and rivers, percolating in the soils and replenishing essential groundwater supplies.

Microscopic particles provide a surface for water vapor to precipitate. Tiny particles of sea salt, dust and bioaerosols released by plants and transported by the wind play a critical role in precipitation processes. If there is sufficient lift, cooling and ambient moisture, static energy is added to the atmosphere.

Emissions from industrial sources, vehicles and nature affect the global energy balance by absorbing outgoing terrestrial radiation, some of which is radiated back to Earth. Particles released from aircraft exhaust scatter and absorb solar rays and act as condensation nuclei, which increases high cloud cover. Atmospheric turbidity from pollution and the ejection of volcanic sulfate particles filter sunlight, which has a cooling effect. Fine particulates contribute to atmospheric haze on a humid day.

Climate Controls

The climate of Pennsylvania is influenced by the diverse topography and variations in elevation, land-use patterns, latitude, and proximity to the large bodies of water, primarily the Great Lakes and Atlantic Ocean.

Latitude plays a significant role in the climate of a region. The southern part of Pennsylvania has freer access to a southerly flow of warm, humid air. The eastern portion of the state is susceptible to an easterly flow of maritime air that enhances precipitation totals. Shallow cold air masses moistened while crossing the Great Lakes face a physical impediment crossing the Appalachians, limiting the intrusion of cold air masses east of the mountains.

State College National Weather Service (NWS) meteorologist Peter Jung has attributed an apparent mini tornado alley in south-central Pennsylvania to the proximity of humid air streaming northward from Chesapeake Bay, interacting with favorable wind shear. The flatter terrain east of the mountains and lee troughs are conducive to sustained rotating updrafts.

The northwestern portion of the state is more directly influenced by the Great Lakes. In late autumn and early winter, the water is substantially warmer than the air, inducing instability that gives rise to lake-enhanced rain and snow showers. In the summer, cooling lake breezes develop as the adjacent land warms faster than the water and the pressure lowers slightly inland, establishing a pressure gradient. The circulation reverses in the evening, when the land surface cools more quickly than the water. A sea breeze from Delaware Bay locally cools the southeastern corner of the state.

Terrain plays a significant role in the prevalence of cloud cover and precipitation over the Allegheny Plateau. The environment cools with height at an average rate of 3.5°F for every 1,000 feet. An upslope wind often creates a persistent deck of low clouds and fog cloaking the ridgetops in a relatively cold, moist air mass. The northwesterly airflow moves downslope east of the Alleghenies, warming by compression and drying, providing a greater number of sunny to partly cloudy days in southeastern Pennsylvania.

The average annual percentage of possible sunshine in the Keystone State, based on historical NOAA data, ranges from 45 percent at Pittsburgh to 56 percent at Philadelphia and 58 percent at Harrisburg. In terms of cloudy days, Pittsburgh (203 days) and Erie (205 days) experience mostly cloudy skies an average of 56 percent of the year, about 10 percent more than Philadelphia (160 days), Allentown (161 days), and Harrisburg (165 days).

The cloudiest month of the year is December, when the mean possible sunshine dips to 28 percent at Pittsburgh, 41 percent at Scranton, and 49 percent at Harrisburg and Philadelphia. In the summer, the average possible sunshine in July peaks at 57 percent at Pittsburgh, 61 percent at Philadelphia, 62 percent at Scranton, and 68 percent at Harrisburg.

Mountain-valley breezes modify the weather on bright sunny days. A south-facing hillside and valley absorb more sunlight and warm the air, which rises and moves through the valley floor, creating an upslope circulation. The airflow changes direction in the evening with nocturnal cooling; higher pressure on the cooler hilltops forces dense air to flow downhill, cooling the valley floor on a mainly clear, calm night faster than the ridgetops.

Subtle circulations are observed between forestland, agricultural tracts and urban areas, where surface roughness imparts modest low-level convergence. Vegetated rural areas add moisture to the air through evaporation and transpiration that contributes to moist thermals and building cumulus clouds that grow into afternoon thunderstorms.

Polar Jet Stream and Circulation Features

The polar jet stream is a river of air that encircles the globe high above Earth between 25,000 and 35,000 feet. The upper-level steering currents move air masses and weather systems around the world, with cold air north of the high-altitude flow and warmer air residing equatorward.

The prevailing westerlies are found above a zone of greatest temperature contrast. The core winds are usually between 75 and 125 mph, though jet segments can exceed 200 mph. Diverging air east of a trough, or dip in the jet stream, creates compensating rising motions conducive to storm development and areas of precipitation.

Spinning cloud filaments on a satellite loop reflect a shortwave perturbation or trough that serves as a catalyst for storms that break out beneath the whirlpool of cold air. The embedded waves of upper energy interact with a low-level return flow of warm and humid air streaming north from the Gulf of Mexico to initiate showers and thunderstorms fueled by increasing instability.

Dynamic upper-level lows supply energy for cross-country storm systems that draw cold, dry air in contact with tropical moisture. Less commonly, low pressure aloft becomes separated from the polar jet stream and forms a slow-moving cut-off low that brings an extended period of unsettled weather until the meandering system is ingested by another trough.

Primary Circulation Features

The prevailing mid-latitude westerlies develop pronounced undulations spurred by Earth's rotation and the Coriolis effect. Semi-permanent weather features have a considerable impact on the orientation of the jet stream and are responsible for large-scale patterns of rising and sinking motions.

The Aleutian Low, located in the Gulf of Alaska near the Aleutian Islands, breeds storms. Relatively warm ocean water astride frigid continental air is a natural birthplace for storm systems that traverse the North Pacific and impact western North America. Disturbances cause low-pressure areas to reorganize east of the Rocky Mountains and bring widespread precipitation.

Tropical forcing plays an important role in the formation of storms impacting the mid-latitudes. Atmospheric waves develop over warm waters in the Indian Ocean and western Pacific Basin. The injection of moisture and pulses of energy into the polar westerlies fuels storm complexes. Groups of thunderstorms pull moisture upwards, releasing a large amount of heat, as water vapor condenses. Disturbances rotating around an upper-air trough provide instability, lift and spin to support a series of mid-latitude cyclones.

A landfalling "atmospheric river" periodically sweeps a corridor of moisture into western North America. The plume acts like a fire hose at an altitude of one to two miles. Lift over the mountains contributes to instability, heavier rainfall, and deep, high-elevation snowfall in the Olympic Mountains and Cascades in Washington and Oregon, and the Sierra Nevada and Coast Ranges of California.

An atmospheric river can transport about 25 times the volume of water than that of the Mississippi River. Flash floods and debris flows cause catastrophic damage, which are worsened in burn scar areas devoid of vegetation, after months of wildfires that degrade the soil and leave little absorptive plant growth.

The "Pineapple Express" emanates from the tropical Pacific near Hawaii, bringing excessive rainfall totals of four to locally 12 inches, when sopping air ascends the west-facing mountains. Prolific snowfalls of three to six feet blanket the Sierra Nevada, and occasionally the mountains of Southern California. In the spring, snowmelt accounts for 30 percent of the yearly supply of water in California. A "Maya Express" from the Central American Gyre can reach northward to the Lower Mississippi Valley.

The semi-permanent Bermuda-Azores High in the subtropics dominates the circulation near the Atlantic Coast. The stout high-pressure area pumps warm and humid air in a clockwise flow into the eastern portion of the country and steers hurricanes westward. When the angle of the sun lowers, the center of the North Atlantic High will slowly shift southward during the fall and winter.

Ludlum (1983) characterized the primary patterns across North America and weather associated with different jet stream configurations.

A *zonal* flow spreads mild Pacific air eastward and keeps cold air masses from penetrating beyond the northern portion of the country. The broadly west-to-east jet stream orientation moves disturbances along at a good clip, too fast to connect with moisture from the Gulf of Mexico, which limits precipitation opportunities east of the Rocky Mountains.

A *western trough-eastern ridge* pattern is associated with a strong low-pressure trough in the northeastern Pacific. East of the Rocky Mountains, the upper-air flow buckles and sustains a subtropical ridge, confining cold air masses to the western United States and northern Plains. A clockwise southwesterly flow of air originating over the Sonoran Desert is responsible for summer heat waves and dry conditions in the East.

A *western ridge-eastern trough* configuration promotes high pressure over the Intermountain West and mild, dry weather. A building ridge discharges cold continental air into the northern Plains and Midwest. In the colder months, the clash of frigid arctic air with a southerly flow loaded with moisture sets up a pattern of widespread precipitation and thunderstorms east of the Rocky Mountains, especially when energy in the northern and southern branches of the jet stream merges or phases.

The *western ridge-central trough-eastern ridge* alignment is a relatively stable regime. Upper-level disturbances slide across the southern Rockies and Upper Midwest, drawing warm, humid air into the Central states that bring inclement weather, while the East generally experiences fair and mild conditions beneath an upper-level ridge.

An anomalous *western trough-central ridge-eastern trough* is often referred to as an "omega block" because it is shaped like the Greek letter. This stubborn atmospheric layout is more common in the springtime. A blocking pattern anchored by high pressure in the middle of the nation is flanked by two upper-level lows. Active weather occurs from the southern Rockies to the High Plains. Quiet weather prevails beneath high pressure in the midsection of the country, and wet weather lingers in the East.

In a split flow regime, the northern and southern branches of the jet stream diverge around a stout ridge in the West, which inhibits storm formation due to a lack of phasing and moisture transport and strong dynamics.

Autumn (September–November)

The start of autumn represents a transition from hot, sticky weather to crisp days and myriad pleasures: corn mazes, apple-picking, pumpkin patches, hayrides, decorative gourds and those glorious fall festivals.

The poetically termed autumnal equinox (Latin for "equal night") happens on September 22–23, when the most direct rays of the sun fall on the celestial equator following a southward trajectory. The sun angle declines to 50 degrees at solar noon at the autumnal equinox over Pennsylvania, a time when day and night are approximately 12 hours long.

The primary storm track enters North America near Vancouver Island and Puget Sound. Low pressure crosses the northern Rockies, reorganizing in Wyoming and heading east across the northern Plains. A Pacific front accompanied by showers sweeps across the northern tier of states, carrying mild air across the nation with limited moisture east of the Rockies.

The weather is generally delightful in early autumn. Chilly Canadian air masses tend to skirt the Northern states, deflected by the western arm of the North Atlantic High. Subtropical high pressure over the southern Appalachians pulls balmy southwesterly winds in a clockwise flow, meeting up with a bubble of fair weather building south from eastern Canada.

In some years, the remnants of a tropical system impact the Eastern Seaboard in September and early October, or take an arcing path from the Gulf of Mexico across the central or northern Appalachians, accompanied by heavy rain, damaging winds and the threat of flash flooding.

The autumn season favors azure skies streaked with feathery clouds. Soft southerly breezes waft through the trees and rustle the leaves. The average high temperature in Pennsylvania slips back into the 60s after a few days of summerlike warmth. Morning lows more frequently dip into the crisp 30s and 40s in the northern Appalachians, bringing a touch of frost in the highland valleys.

Fig. 3.1 Autumn. Blue Mountain Lake, nestled in a clearing on Brushy Mountain, on a gray fall afternoon. (Photo by Ben Gelber)

The seasonal inflection point carries the scent of fallen leaves and decomposing vegetation. The faint fragrance of Canadian pines seems to come from the boreal forests of northern Canada. The air is sweetened by terpenes, among other gas molecules classified as organic hydrocarbons. Chemical reactions involving nitrous oxide and sunlight produce ozone that has a distinct odor.

A shallow temperature inversion is common in the morning as nighttime lengthens with the decreased solar angle (declination), allowing the ground more time to chill, especially under clear skies and light winds. Nighttime radiative cooling results in an inversion, a stable layer of milder air just above the colder surface.

A pattern of high pressure with a light airflow contributes to a larger buildup of fine particulate matter from industrial emissions and automobile exhaust with little wind to disperse soot that can enter the lungs and bloodstream. Smoke particulates from western wildfires that circulate eastward in the mid-level flow all the way to the Atlantic Coast contribute to the haze.

Shorter days and diminishing sunlight trigger a natural cessation of food production in the leaves in the early fall, and resources are directed to the tree trunk. The reduction of green chlorophyll reveals hidden pigments in the leaves, creating splashes of color in the northern woodlands that will soon be ablaze in flaming foliage.

The panoply of color arrives in mid-October beneath a crystal-blue autumn sky in a seemingly synchronized pattern. Bright sunny days and long crisp nights cause sugar in the sap to become concentrated and react to form anthocyanins that are responsible for the dazzling red and purple shades. The vivid fall color palette is diminished by an abnormally warm, wet fall pattern that delays the shutdown of chlorophyll.

Leaves absorb and reflect different wavelengths of sunlight, which helps determine the color, along with the tree type and available soil moisture. The northern hardwoods turn early, starting with the colorful red maple and followed by the sugar maple, unveiling rich yellow and orange pigments.

Xanthophyll pigments are responsible for the bright yellow, and carotenoids add dashes of orange to the fiery foliage. Gold and amber appear in the hickories and some oaks. Riparian stands lining the bottomlands include silver maple, elm and sycamore that offer splendid gold, orange and red hues.

The southern portion of Pennsylvania, where peak foliage season arrives in late October with the southward progression of color in the mixed oaks, features mustard yellow, russet and reddish-purple. The panorama is enhanced by dogwood (red), black gum (maroon), sassafras (orange), black birch (blazing yellow) and beech (gold). Conifers, such as pine and hemlock, add green contrast.

Along the coastal plain, sweetgum, willow oak and southern red oak offer a complementary menu of glorious fall landscapes. The stunning color sequence is completed by Halloween. Another familiar visitor is early morning ground fog, especially in the highland valleys. Ghostly vapors drift across the fields and adjacent country roads on a clear, windless night. Daytime heating evaporates the fog that sometimes lifts to form a stubborn deck of low stratus clouds until the moisture mixes with daytime heating.

The main storm track in early autumn brings low-elevation rain to the Columbia River Valley and Willamette Valley, and the first high-country snow. Snow whitens the peaks of Mt. Hood, Mt. Rainier and Mt. Shasta in a winter garb.

The atmosphere is less active east of the Mississippi Valley in October. Expansive high pressure centered in the southeastern United States provides stretches of tranquil weather, only briefly interrupted by a storm system tracking across the Upper Midwest that pulls moisture north along a trailing cold front.

A warm spell that follows the initial freeze of autumn is often referred to as a Second Summer, a notion dating back to colonial America and the Native American practice of preparing for winter during a late period of balmy weather, gathering winter stores, perhaps associated with campfires.

Seasonally strong low pressure is established in the Gulf of Alaska during the autumn, sending disturbances ashore in the Pacific Northwest. A series of rain and high-country snowfalls reflects the early wet season in the West. Systems head east to the Mississippi Valley. An easterly upslope flow north of the storm track produces the first snows of the season from the central Rockies to the Black Hills of South Dakota.

A sharper plunge in the jet stream along the West Coast triggers thunderstorms and isolated tornadoes in Central and Southern California. The first in a series of rainstorms after six months of dry weather brings relief from the intense heat and wildfires. The dynamic upper-level low translates across the Great Basin, accompanied by rain and pockets of heavy snow in the Mountain West.

A Colorado Low develops downstream of the upper system to the east of the Front Range and tracks across the central Plains to the Upper Midwest. Rain and embedded strong storms march across the midsection of the country, administered by a southwesterly flow that transports abundant moisture northward.

The pace of weather systems in North America quickens in November. The primary storm track enters North America near the coast of Oregon and Northern California. Low pressure organizes in the Four Corners region and brings snow and wind across the southern Rockies and Upper Midwest. In the East, widespread rain and storms interrupt an otherwise tranquil autumn pattern, ending with a brief shot of chilly air.

The storied "gales of November" are notorious cyclones that intensify rapidly over the Upper Midwest, producing windswept snow. Hurricane-force wind gusts have caused historic shipwrecks on the Great Lakes, and travelers caught in the icy grip of an early-season snowstorm face powerful blasts of wind that draws in January-like cold.

The western end of a sinking arctic boundary across the Tennessee Valley and Ozark Plateau occasionally becomes the focus for a secondary wave that taps into cold air settling in over the Midwest and Great Lakes. A southern track through the Mid-Atlantic sets the stage for an early snowstorm in the Northeast before or during the Thanksgiving holiday, primarily affecting the colder interior sections.

Winter (December–February)

There is a magical quality to the winter season. Youngsters and winter sports enthusiasts relish the opportunity to dig out the sleds or skis buried in the garage as the excitement builds ahead of the first predicted snowfall.

In the minds of commuters, a forecast for snow and ice instills a feeling of dread. Even a light coating will test the mettle of experienced motorists and often results in a series of spinouts and fender benders. In the wake of a full-blown snowstorm, frigid temperatures drive up energy demands and make snow removal more challenging.

Meteorological winter begins on December 1. The conventional choice provides the basis for a convenient statistical temperature period closest to the average coldest three-month period from December 5 to March 5.

The sun's path continues to shift southward in the autumn months due to the tilt of Earth's axis. Between December 20 and 23, the sun's vertical rays strike the latitude of the Tropic of Capricorn (23° S), at the most southerly point, which marks the shortest day of the year across the Northern Hemisphere.

At the precise time of the winter solstice, the sun appears to stand still (from the Latin *sistere*). The noon sun climbs to an elevation of 26.5 degrees above the southern horizon at 40° N, marking the start of astronomical winter in the Northern Hemisphere. In late December, the region experiences about 9.5 hours of daylight.

The average maximum temperature in Pennsylvania in early winter dips into the 30s across most of the state, and low 40s in the southeast, although there is considerable variability depending on the weather situation. Nighttime lows frequently fall into the 20s and 10s, especially if the ground is snow-covered. The "January thaw" that generally occurs around the third week of the month is not a certainty. Yet the real statistical signal is probably a reflection of a temporarily diminished supply of arctic air.

The North Pacific High retreats southward by early winter, paving the way for a parade of disturbances emanating from low pressure in the Gulf of Alaska. A surface storm that comes ashore in the Pacific Northwest travels across the northern Rockies and central Plains, blanketing portions of Montana and Idaho with snow that extends across the northern Plains. A swath of rain falls from the southern Plains to the Upper Midwest.

A sharp dip in the jet stream brings valley rain and high-elevation snow to California. Beneath the swirling pool of cold air aloft moving ashore, thunderstorms with small hail pummel Southern California, and snow blankets the mountains east of Los Angeles and the high desert as snow levels lower.

The parent storm becomes disjointed over the Four Corners region, and a lee-side low develops east of the Rockies that tracks through the Great Plains to the Lower Great Lakes. A northerly flow produces a band of snow from the central Rockies to the northern Plains. An impressive gradient kicks up strong winds and near-blizzard conditions. In the open warm sector east of the low-pressure area, gusty thunderstorms are forced by an advancing cold front.

A southern disturbance ejected from an upper-air trough in the Southwest organizes over the southern Plains, before traveling northeast through the Mid-Mississippi Valley. Snow breaks out across the Intermountain West to the northern Plains and Upper Midwest. Bands of heavy rain and strong thunderstorms develop in the warm sector when southerly winds converge on a front. Occasionally, cold air interacts with low pressure crossing the Ohio Valley and northern Appalachians, bringing snow to the interior Northeast, as a secondary low forms near the Delmarva Peninsula.

A typical cross-country winter storm sweeps inland from the coast of Washington and Oregon and dives southeast into the Great Basin. A subtropical connection induces low pressure over the southern Plains, turning northeast across the Ozark Plateau and up the Ohio Valley to western New York. Heavy snow falls from Kansas City to Des Moines, Chicago and Detroit. Freezing rain creates hazardous conditions west of the shallow arctic boundary underneath a southwesterly flow of mild air aloft.

A blockbuster Atlantic coastal storm develops when a cold upper-level system in the Midwest phases with a southern storm, which becomes absorbed in the northern branch of the flow. The transformation occurs where jet stream winds converge over the warm Gulf Stream waters adjacent to cold continental air, as moisture streams northward.

Warm air is forced to rise over the cold dome, resulting in a layer of mid-level frontogenesis. Jet stream dynamics promote enhanced atmospheric lift, visible as an expansive, comma-shaped precipitation shield.

Energy is transferred over to the Mid-Atlantic Coast, setting the stage for a major winter storm to wallop the Northeast. A moist easterly conveyor belt undergoes orographic lift in the Appalachians, producing bands of heavy snow.

A storm trajectory near the "40/70 benchmark" position, defined as a point 100 miles southeast of Montauk Point on Long Island (40° N/70° W), places much of eastern Pennsylvania in the sweet spot for a heavy snowfall. Cold high pressure over eastern Canada slows the forward progress of a nor'easter, prolonging the storm's impacts, including strengthening northeasterly winds in response to the pressure gradient.

Fig. 3.2 Winter. McMichael Creek, winding through the southern Poconos, after a snowfall. (Photo by Justin Gelber)

A southern system sliding east along a frontal zone sometimes heads out to sea instead of turning north, when northern and southern stream support merges farther east. A narrow band of wintry precipitation in the interior Mid-Atlantic region falls as a cold rain along the coastal plain, possibly ending as a wintry mix once the system moves offshore.

Another prominent winter feature is the familiar Alberta clipper storm, named after the

19th-century ships, reflecting a compact system crossing northwestern North America. A surface reflection in the Northwest Territories and Prairie Provinces drops southeast around high pressure in western Canada.

A clipper system lives up to its name, sailing across the northern United States. The compact storm is moisture-starved, producing a narrow band of light-to-moderate snow along and just north of the track, with blustery conditions. Rain showers fall farther south, where southwesterly winds introduce milder air.

Sometimes a lobe of the polar vortex over Hudson Bay in eastern Canada directs a spoke of energy southward that brings snow squalls and poor visibility in the Northeast. An Alberta clipper swinging farther south pushes a band of snow across the Midwest and northern Appalachians, often followed by a reinforcing blast of arctic air.

In late winter, a higher sun angle mitigates the impact of most winter storms, because more insolation penetrates even thick clouds and facilitates melting on pavement. Snowfall becomes more elevation dependent in March, where a drop in temperature of several degrees can make the difference between a slushy few inches in the valleys to double-digit totals in mountainous areas.

Mild days with a hint of spring pop up in late winter. Birds begin to sing in the morning and a few shoots emerge from a winter slumber. Of course, highs in the 50s and 60s are merely a false spring.

Spring (March–May)

Charles Dickens aptly described the fickle nature of March weather in *Great Expectations*: "It was one of those March days when the sun shines hot and the wind blows cold: when it is summer in the light, and winter in the shade."

Spring arrives in fits and starts because lingering cold air over eastern Canada bumps into the burgeoning warmth in the Southern states in the mid-latitudes. A back-and-forth drama features balmy spring weather with a touch of early summer, only to backslide into unpleasantly chilly weather, leaden skies and a wintry mix of precipitation.

The vernal equinox is determined by the time the sun crosses above Earth's equator, where the most direct solar rays shine overhead at noon on March 20–21. On the equinox, the planetary axis is not tilted toward or away from the sun. At solar noon, the sun rises 50 degrees above the horizon in Pennsylvania.

The average high temperature on the first day of astronomical spring ranges from the low 40s in the Allegheny Mountains in the northwest to the low 50s in the southeast—a jump of more than 10 degrees in a month's time driven by increasing daylight. Grudgingly, patterns of persistent low clouds give way to more hours of sunshine, as nature rejuvenates after the long winter dormancy.

Fig. 3.3 Spring. Lush vegetation on Brodhead Creek, a tributary of the Delaware River in the southern Poconos. (Photo by Ben Gelber)

Crumpled leaf litter from the previous season contributes to the rebirth after gathering and storing energy. Life first rouses in the rural wetlands. Songbirds and spring creepers chirp out a message that signals spring is near.

Shoots poke though the mud and buds emerge on flowering trees. Colorful wildflowers spring up in the woodlands and along trails as the soil warms. In hollows and ravines, red and silver maples begin to leaf out amidst blooming woody plants.

The polar jet stream brings a series of disturbances ashore in the Pacific Northwest, carving out a dynamic upper-level storm. A large, slow-moving storm arrives along the coast of Oregon and Northern California, trekking through the Intermountain West, with rain and high-elevation snow.

Low pressure reorganizes in a favorable zone of cyclogenesis east of the Front Range. A storm that forms over the Texas Panhandle is referred to as a Panhandle hook, pivoting northeast from the southern Plains to the Ohio Valley. Northern systems track northeast across the central Plains.

The sprawling storm system is a reflection of the annual clash of the seasons—a battle between early summer warmth and winter's lingering chill. Rounds of rain and strong thunderstorms lash the heartland, and heavy snow falls from the Rockies to the northern Plains. Low pressure on the frontal boundary triggers widespread showers and storms in the Ohio Valley and Southern states.

Thunderstorms that tap into pockets of strong winds aloft potentially turn severe, bringing down pulses of wind and spawning tornadoes where the winds aloft are turning with height. Steady rain reaches the Eastern Seaboard, but the stronger dynamics usually remain farther northwest, causing a line of storms to weaken crossing the Appalachians. A plume of tropical air east of the mountains sometimes reinvigorates the frontal system, resulting in an outbreak of strong storms from the Carolinas to southern New England.

A late winter blast occasionally plasters the Northeast with heavy wet snow when stalled upper-level low pressure over the Canadian Maritimes feeds cold southward, which collides with a frontal zone draped over the Mid-Atlantic region.

A quaintly described spring snowfall, referred to as an "onion snow" in the olden days, cloaks lawns and gardens and clings to trees like paste. The jet stream position determines whether heavy snow is centered over the Alleghenies or closer to the coast, falling a few hundred miles west of the track of low pressure.

Lengthening daylight allows plants to begin to manufacture food on sunny days. Chlorophyll is the green pigment that takes in light energy from the sun in a process dictated by an internal clock. Plants absorb carbon dioxide through their leaves and draw soil moisture from the roots to produce sugar (glucose), which releases essential oxygen into the atmosphere. The converted starch moves upwards through the branches, nourishing stems, leaves, flowers, root systems and woody trunks.

A panoply of color and bursting blooms excites the senses. The springtime air is infused with the sweet smell of grass wafting over the verdant landscape. Tulips are among the early spring comers, as growing degree days (GDD) accumulate. Annuals usher in a striking array of yellow, cream and pink daffodil flowers, joining colorful tulips, hyacinths and crocuses. Blossoms are frequented by pollinators flitting about in search of nectar and pollen to deposit.

Rhododendrons and azaleas appear alongside blue Quaker ladies and an assortment of radiant spring wildflowers adorning the trails and woodlands. Orchard owners pay close attention for the inevitable late freeze, though growers have limited options when the temperature is forecast to fall below 25°F. Planting tomatoes, peppers and tender plants is not recommended until after Mother's Day or later in Pennsylvania.

Early spring is the windiest time of the year because the difference in temperature between the surface and a few miles overhead is greatest. Insolation heats the surface and warms the air, as winter chill lingers at a higher altitude. Eddies reflect rising currents or thermals that are replaced by sinking cool air, generating gusty winds with daytime heating.

In a "backward" spring, a lobe of the polar vortex settles near Hudson Bay in eastern Canada, delivering an unwelcome blast of cold air and wet snowflakes as late as May. At times, a chilly northeasterly circulation around high pressure in New England sponsors a "backdoor cold front" that sends temperatures tumbling from the balmy 60s into the 40s in a matter of hours, with a deck of low clouds interrupting a burst of spring warmth.

More frequent visitations of balmy southwesterly breezes issue strong hints of summer without the stifling humidity. Greenery abounds, and the fragrant scent of blooming flowers and trees permeates the air, wafting along in the southerly breeze.

Daylight increases to 13–14 hours, prompting the inexorable retreat of the polar jet stream and putting an end to parting shots of chilly air. Summer beckons by Memorial Day in most years, powered by a flexing subtropical ridge in the Southeast.

Summer (June–August)

The summer solstice occurs on June 20–21, when the direct rays of the sun fall over the Tropic of Cancer (23.5° N). At solar noon, the sun reaches its highest position in the sky (73.5 degrees at 40° N), providing a little more than 15 hours of daylight.

Summer-flowering bulbs planted in late spring dot the landscape. An array of colorful wildflowers is visible from the roadside in rural sections. The weather is generally ideal for swimming, boating, baseball and cookouts, except for generally short-lived pop-up storms that provide essential plant and soil moisture.

The jet stream enters North America near the Canadian border in early summer, curving northeast around the Sonoran ridge in the Southwest. Energetic mid-level impulses buckle the flow in the northern Plains and Great Lakes, escalating the threat of thunderstorm complexes forming on the periphery of a heat dome in the Southeast.

The wettest three-month period in Pennsylvania straddles late spring and early summer (May–July). The mean monthly precipitation totals are in the four- to five-inch range. Occasionally, a frontal system intercepts a southerly flow of moisture from the Gulf of Mexico, resulting in frequent showers and thunderstorms.

Fig. 3.4 Summer. Thunderstorm over Delaware Water Gap, looking north from Portland. (Photo by Ben Gelber)

The summer upper-air flow is sluggish and the main storm track shifts north of the Great Lakes and across northern New England as the north-south temperature gradient diminishes. The primary summer weather traffic controller is a resilient Central Atlantic high-pressure area. The weather is relatively tranquil, and storms are sporadic. This does not rule out an occasional upper-level disturbance that brings a few rounds of thunderstorms and tropical downpours.

The summer months in Pennsylvania are reliably warm and humid, but with generally pleasant conditions in the mountainous areas. The average July maximum temperatures range from the upper 70s in the highest elevations of the Northern Tier to the mid- and upper 80s in the southeast, where the nights tend to be stuffy and late nighttime readings often remain above 70°F.

The westward extension of the Bermuda High attains its northernmost position in midsummer. The traditional sweltering dog days of summer—July 3 to August 11—coincide with the appearance of Sirius, the Dog Star, in the constellation Canis Major. Orion the Hunter, with three stars in a line comprising Orion's Belt, appears before dawn. Sirius is nearby and lower in the southeast. Fireflies flitting across backyard lawns and the amplified chorus of the annual dog-day cicadas are joined by the metronomic chirping and clicking of crickets.

A few tropical systems in the western Atlantic make landfall along the Southeast Coast and northern Gulf of Mexico shoreline during the summer months. Tropical cyclones drawn northward on the western edge of a Bermuda High bring wind and rain, and often flash flooding along the path up the Eastern Seaboard or west of the Appalachians and across the Great Lakes region.

Atmospheric Pressure Records in Pennsylvania

A typical surface weather map displayed on television and other sources is dotted with blue "highs" and red "lows" that reflect maximum and minimum pressure centers. High pressure is almost always associated with fair weather, and low pressure is usually associated with inclement conditions when sufficient moisture is available.

Atmospheric pressure is the collective weight of gases in a column of air. Standard sea-level pressure (SLP) is defined as 29.92 inches of mercury (1013.25 millibars); expressed another way, it exerts a pressure at sea level near 14.7 pounds per square inch (psi).

Temperature variations in the atmosphere give rise to differences in pressure from lower to higher latitudes that sets air in motion on a broad scale to equalize the pressure gradient, moving from areas of higher to lower pressure. Transient weather systems and diurnal heating create a pressure gradient. The circulation is clockwise around highs and counterclockwise around lows in the Northern Hemisphere.

Low pressure close to the surface develops in response to warming that causes the less dense air to rise, facilitating moisture in a convergent airflow that enhances the upward motion of air parcels. If the ambient moisture and lift are sufficient, trillions of tiny droplets aggregate into clouds bearing rain and snow.

Sinking air increases the pressure, drying of the column of air. High-pressure areas reflect a broad zone of air forced downward by convergence aloft, promoting generally fair conditions and lighter winds at the surface. Persistent subtropical highs over the ocean and desert regions are especially stable, with seemingly endless stretches of clear skies. Arctic high pressure reflects an expanse of dense, frigid air building over snow-covered terrain, where sunlight is minimal at best in the wintertime.

On January 26–27, 1927, under a big dome of high pressure, all-time high barometric pressure readings were noted at Scranton (31.08 inches), Harrisburg (31.04 inches) and Erie (31.04 inches, tying February 9, 1934). Pittsburgh recorded its highest mark (30.97 inches) on February 9, 1934. On February 13, 1981, Philadelphia eclipsed the city's 1927 earlier record (31.02 inches), setting a new pressure maximum (31.10 inches).

Record low barometric pressure readings were observed during the height of the March 13, 1993, blizzard. A new standard was set at Philadelphia (28.43 inches), surpassing the earlier low mark of 28.54 inches on March 7, 1932. New record lows occurred at East Stroudsburg (28.50 inches), and at Scranton (28.58 inches), where the older value stood since February 25, 1965 (28.72 inches).

In western Pennsylvania, Pittsburgh's all-time lowest barometer reading (28.59 inches) occurred during another blizzard on January 12, 1964. At Erie, the barometer on January 26, 1978, plunged to 28.34 inches in the Blizzard of '78, surpassing the "Great Lakes Hurricane" blizzard nadir (28.61 inches on November 9, 1913).

Record low-pressure readings occurred during the passage of Superstorm Sandy shortly after landfall on the evening of October 29, 2012, surpassing the March 1993 marks in several instances: 28.12 inches at Philadelphia; 28.46 inches at Harrisburg (28.62 inches on January 3, 1913); 28.49 inches at Allentown. The second-lowest barometric pressure levels were observed at Scranton (28.69 inches), Stroudsburg (28.56 inches), and Mount Pocono (28.64 inches).

Weather Data-Gathering Technology

The United States National Centers for Environmental Prediction (NCEP) in College Park, Maryland, provides national and global weather forecasts, drawing on 20 terabytes of data to provide a wide array of short- and long-term forecasts to meet the needs of the public and private sectors.

NCEP Centers are one of several large global data-gathering agencies. Nine NWS centers offer a wide range of guidance that includes aviation outlooks, severe weather analyses, space and oceanic weather forecasts, and longer-range climate projections. The NWS staffs 122 forecast offices in the United States, Puerto Rico, and Guam, issuing weather forecasts and data products for public dissemination.

A fleet of weather satellites observes changes in polar ice, vegetation, surface water, terrain and global sea level. Surface observations taken manually by automated stations over land and sea (buoys) monitor current conditions.

During the formative days of modern meteorology, weather forecasters did not have the luxury of a steady and instantaneous flow of information. Predictions were primarily focused on surface features, extrapolated upper-air data and experiential knowledge.

In the 1930s, weather balloons unraveled some of the mysteries of the upper atmosphere by transmitting temperature, pressure, moisture and wind data to surface stations. Primitive computers in the 1950s opened the door for generating weather forecasts based on mathematical equations that simulated the workings of the atmosphere over longer periods of time.

The next huge step forward in observational data came in April 1960, when the first weather satellite (TIROS-1) commenced a 78-day orbit as a remote-sensing instrument. In December 1966, the ATS-1 launch marked the first meteorological satellite to be put into a geostationary orbit.

Geostationary weather satellites follow Earth from west to east at an altitude of about 22,236 miles above the equator, moving at the same speed and direction. The temperature of the cloud tops is inferred by measuring thermal (infrared) radiation emitted from Earth and reflected sunlight from cloud tops. Satellite instruments provide more than 16,000 global measurements transmitted daily to Earth that are assimilated in the analysis, or initial computer forecast model data, critically filling in gaps over the ocean and remote areas.

Polar-orbiting satellites follow a north-south route twice daily, passing very close to the poles while circling Earth in a sun-synchronous orbit at an altitude of 512 miles. The Suomi NPP, a NASA polar-orbiting weather satellite launched in 2011, collects vital data, including temperature, pressure, clouds and sea surface temperature and salinity. A polar-orbiting satellite acquires data 14 times daily from above Earth's surface.

In May 2014, NOAA assigned Suomi NPP as its primary operational polar-orbiting satellite observing the dynamic Earth. NASA sent the Joint Polar Satellite System-1 (JPSS-1) rocketing into space in November 2017 to improve forecast accuracy during the three- to seven-day period to better cover hurricane tracks, storm systems and wildfires. Supplemental data from the European Space Agency's (ESA) Sentinel-5P satellite provides a wider perspective of global weather systems.

The November 2016 launch of the Geostationary Operational Environmental Satellite (GOES-R) Series marked the first of the next generation of satellites. GOES-R is part of an international constellation of satellites that provide free weather data shared by 191 governments, under the auspices of the World Meteorological Organization (WMO). The fleet of United States GOES spacecraft, operated jointly by NOAA, NASA and the Department of Defense, offers striking high-resolution images of clouds and storm systems, snow cover, atmospheric rivers, dust storms and plumes of wildfire smoke.

The advanced GOES-R Series provides 60 times more satellite imagery than earlier programs to provide continuous coverage of weather and the environment.

Instruments onboard monitor ozone, ultraviolet radiation, temperature, water vapor and radiation in the lower and middle atmosphere. The geostationary cameras are equipped to detect lightning, which provide useful data for severe weather forecasters.

Satellite data arrives every five minutes, five times faster than earlier versions, with up to four times the spatial resolution. Rapid-refresh models provide images in intervals of less than a minute, which helps forecasters expedite alerts that reach the public and emergency management personnel.

The number of imaging bands increased from five to 16, or three times the spectral information as before. The finer, clearer cloud formations and moisture details are plainly visible, depicting subtle features such as vortices, gravity waves and moisture transport. Global weather observations are augmented by more than 7,000 ships, 700 ocean buoys and about 3,000 aircraft.

Surface observational data pours in from about 1,000 Automated Surface Observing Systems (ASOS) that include air temperature, relative humidity, dew-point temperature, cloud heights, precipitation type and character, wind direction and speed, and lightning detection.

National Weather Service (NWS) and Federal Aviation Administration (FAA) sites are supplemented by military and private stations totaling more than 10,000 sites, feeding data hourly or several times a day that are incorporated into NOAA forecast model systems. NWS offices release weather balloons at the same time from 92 stations in the U.S. and its territories, plus 10 Caribbean sites.

A suspended instrument package (radiosonde) contains sensors that register temperature, wind, air density, pressure and the amount of moisture, as the balloon filled with hydrogen or helium gas ascends through the atmosphere. Weather data are transmitted by radio signals to the launch site.

The vital information is incorporated in computer forecast models that provide a steady stream of numerical weather guidance.

Hydrogen or helium balloons are sent skyward from nearly 900 locations worldwide twice daily, though some countries stage just one launch due to the cost. The weather balloon usually attains an altitude of up to 115,000 feet before disintegrating, with component parts falling back to Earth.

Numerical Weather Prediction

In the fast-paced, high-tech world, we demand to know what the weather is going to be with a high degree of accuracy almost down to the hour.

Weather forecasts are hardly foolproof, as everyone knows. Even the most sophisticated computer models miss small-scale events and the intensification of larger weather systems that emerge from slow-moving upper-level features. Small disturbances that migrate between sampling sites are sometimes missed entirely unless captured by one of several short-term high-resolution models.

In 1963, mathematician Edward Lorenz studied model sensitivity caused by data initialization discrepancies. He determined that minuscule data errors magnify over time and distance, resulting in inaccurate forecast outcomes. Chaos theory was dubbed the "butterfly effect" to reflect how the connectivity between the random flapping of wings in one location influences the weather thousands of miles away. The work of Lorenz provided an early understanding of the inherent randomness in deterministic guidance used by weather forecasters.

Running a series of computer model forecasts requires upwards of four billion bits of data daily. The model simulations are initialized in three-dimensional global grid boxes stacked vertically and arrayed in a way to capture the interactive workings of the atmosphere that affect adjacent grids.

Daily weather forecasts are prepared by meteorologists who review more than a half-dozen global and regional forecast models run up to four times every 24 hours that extend in timeframes of several days to as long as two weeks. Derived statistical model outputs complement the model simulations to localize forecast outcomes.

Some computer weather models are statistical while others are based on an amalgamation of dynamical processes. Analog models incorporate persistence or climatology to derive scenarios based on past events. In addition to a single deterministic run with an outcome based on the input, an ensemble suite provides a probabilistic forecast by perturbing the initial conditions to manage uncertainty in forecast simulations. The ensemble packages are especially helpful in the three- to seven-day time frame.

The Weather Research and Forecast Innovation Act approved by Congress in 2017 allowed NOAA to join forces with the National Center for Atmospheric Research (NCAR) to raise accuracy levels in computer modeling.

Numerical weather prediction (NWP) has improved substantially with vastly increased computing capacity and the better assimilation of data crunched through dynamical equations. Smaller spacing between data sites offers a higher resolution of modeled grid points that better initialize vertical motions, mid-sized convective systems and topography.

The High-Resolution Rapid Refresh model (HRRR) is primarily used to evaluate the potential for thunderstorm development in the short term. The convection-allowing model (CAM) assimilates radar data every 15 minutes and resolves cloud cover, which impacts differential heating and instability. The NOAA Rapid Refresh has a smaller grid spacing (3 kilometers) to initialize real-time data hourly in an evolving severe weather situation.

Tropical models focus on one of the three wind levels to analyze wind shear, while more complex models combine statistical data that incorporate past storms and upper-level patterns. Dynamical models couple surface observations with atmospheric physics. General Circulation Models (GCM) assimilate considerable atmospheric and oceanic data to evaluate conditions that provide crucial information for aviation, oceanography, tropical cyclones, severe storms, space programs and climate change.

Global models are run every six or 12 hours. Vertically aligned cubes represent a small section of the atmosphere several miles wide. Incoming weather data obtained from Earth-observing systems are recalculated every two to 20 minutes, depending on the model. Equations are then solved over time for cloud growth, radiation exchanges, precipitation, and vertical or turbulent motions. Data over the oceans are sparse, with limited sampling that delays model forecast consistency until the system is over land.

The American model (GFS), and NOAA's Global Ensemble Forecast System (GEFS), with 30 model runs, or members, are among the greater than 30 various modeling systems analyzed by forecasters. The European Centre for Medium-Range Forecasts (ECMWF) has 50 ensemble members. The UKMET and CMC (Canadian), which has 21 ensemble runs (GEPS), are among the best-known global model solutions that are run several times daily.

The 2021 upgrade to the American Global Forecast System (GFS) model added computing power and higher resolution, with 127 vertical slices, compared to 64 previously. The improvements from the near-surface boundary layer to the stratosphere incorporated a range of atmospheric drivers, including sudden stratospheric warming events, tropical cyclone intensification and tracks, excessive rainfall projections and the impact of wind-driven ocean waves on coastal areas.

The ECMWF modeling system earned widespread acclaim in late October 2012 after accurately predicting a nearly 90-degree left turn of Hurricane Sandy seven days in advance of landfall along the New Jersey coast, and four days ahead of the GFS. Superstorm Sandy swamped low-lying coastal areas and roadways, inundating the New York City subway system.

The massive hurricane turned into a post-tropical rain and windstorm and continued west across southern New Jersey and Pennsylvania, felling trees and power lines, leaving 8.5 million households without electricity for days to weeks. The storm caused $77 billion in damages (2020 dollars) and claimed more than 200 lives.

The ECMWF (Euro) model outperformed its American counterpart in projecting the out-to-sea track of Hurricane Joaquin in October 2015. The Euro also excelled in projecting the westward track of Hurricane Irma, days before the storm turned northward and pounded the Florida Keys on September 6, 2017, while the GFS was late in correcting for the change in trajectory.

The regional North American Mesoscale Forecast System (NAM) has outdistanced other modeling systems, notably capturing the northward progression of a severe nor'easter that blanketed the Mid-Atlantic region under 10 to 40 inches of snow on January 22–24, 2016.

Following Sandy's tragic impacts, and in light of the inferior performance of the GFS, U.S. Congress increased funding in 2013 to upgrade the system. The $44.5 million investment came to fruition in 2015 with the addition of the Cray supercomputer housed in Reston, Virginia. The Cray upgrade processes three quadrillion calculations per second, a 10-fold increase in computing power compared to the earlier GFS (up from 213 trillion operations per second to 2,600 teraflops).

Despite major success in predicting the precise track of a massive winter storm that pummeled southeastern New England in late January 2015, and in other coastal storms, the American model sometimes lags the ECMWF numerical model regarding data assimilation used to generate a weather forecast.

The Euro model usually excels in the medium-range simulations. Greater computing power and higher-resolution data provide a better handle on the physical processes and dynamics in the ocean-atmosphere interplay than NOAA's more than two dozen global and regional models. However, recent updates to the GFS modeling system have produced good results that have proved more accurate than other global models in significant winter storm simulations across North America.

Chapter Four

Winter Storms and Out-of-Season Snowfalls

Onion-skins very thin / Mild winter coming in;
Onion-skins thick and tough, / Coming winter cold and rough.
—Lt. H.H.C. Dunwoody, Signal Service Notes, Issue 9 (1883)

ON OCTOBER 22, 1842, the Pottsville *Miners' Journal* opined on the upcoming winter: "The weather-wise decide that we shall have a rousing winter of it; we put great faith in this prediction, and advise our friends abroad to put in a good stick of anthracite …"

In the olden days, rugged winters spelled serious coal and potential food shortages. Historical newspapers were replete with tales of hardship and scarce fuel for heating. Those stories were balanced by joyful accounts of horse-drawn sleigh rides and winter fairs. Great snowstorms that buried parts of the region drew reminiscences of bygone winters, often in consultation with the "oldest inhabitant" in search of a precedent.

Our preoccupation with winter predictions and the prospect of a white Christmas is likely a throwback to the anticipation of a snow day filled with sledding and hot chocolate. As we grow older, the serene beauty of a heavy snowfall clinging to branches and plastering trees and decks is balanced by an expected dicey commute on treacherous roads.

Schools and businesses close in major snow and ice storms. Powerful winds increase the number of power outages. Smothering snowstorms disrupt travel and snarl transportation for days. The situation worsens when paired with an arctic outbreak that drives up heating bills and clean-up costs. After a winter storm slams the region, blowing snow conceals roads and mounts large drifts in open fields.

Blinding snow squalls can reduce the visibility to near zero and make travel hazardous, as roads become slick in a thick swirl of wet snowflakes. On March 28, 2022, six people died and 24 were injured in a chain-reaction pileup on Interstate 81 in Schuylkill County that involved 41 passenger vehicles and 39 commercial vehicles around 10:30 a.m. More than 40 vehicles crashed in icy conditions along the same freeway in Schuylkill County on February 19.

Groundhog Day in America: "Six More Winters There Will Be!"

Despite the limited reliability of long-range forecasts, an endeavor that is far from an exact science, generalized winter outlooks garner attention every preceding autumn. We eagerly wish to glean any hint of what Mother Nature might be planning for the months ahead as soon as the leaves start to turn.

Meteorologists search for clues in faraway places that link large-scale weather patterns separated by thousands of miles. The 90-day seasonal temperature and precipitation outlooks published by NOAA's Climate Prediction Center (CPC) are based on a statistical analysis of variables that include sea surface temperatures paired with historical outcomes or climate analogs.

Popular winter folklore comes from scouring early autumn signs in nature: the number of acorns sprinkled beneath trees, a thicker coat of hair on animals and the width of the black band on a woolly bear caterpillar, among many other lesser-known nuggets. Naturalists point out that the folklore is not predictive regarding the upcoming winter, but instead reflects pre-existing conditions.

Once winter gets going, many of us look for indications of an early spring. Every year, in the dead of winter, the nation turns its lonely eyes to the behavior of a glorified member of the rodent family—an astute marmot—to gauge how much longer we will have to put up with the cold and snow.

Pennsylvania German farmers offered reliable weather expressions at the ready: "It is raining to to keep in the cats and bring out the ducks." This reflects a mistranslation of "It is snowing for cats and ducks," according to Eric Sloane's *Folklore of American Weather*. The true meaning of the saying implied a sufficient snow for tracking.

European Celts originated the idea of animals offering a glimpse into the future at the halfway point of winter as far back as the fifth century. Early German immigrants carried the folklore with them to the United States, laying the groundwork for the most famous celebrated seasonal seer: a groundhog residing in the small western Pennsylvania community of Punxsutawney.

Since the late 1800s, anxious spectators have gathered on Gobbler's Knob, a small hill just outside of town, in the chilly predawn hours every February 2, to await the peerless proclamation of the famous prognosticator himself—the prophetic Punxsutawney Phil.

Tradition dating back to medieval France and Germany stipulated bears and marmots strolling out of hibernation too early were frightened by the sight of their shadow and promptly scurried back into their winter dens for another four to six weeks.

Christianity adopted Candlemas as the midway date between the winter solstice and spring equinox, when clergy passed out blessed candles to light in the windows in medieval Europe. The midpoint of winter would have been based on the Julian calendar around 1,000 years ago, when the spring equinox occurred in the middle of March.

Folklore dictated that if the sun shined on Candlemas Day, six more weeks of winter lay ahead, a quaint notion that German (Pennsylvania Dutch) settlers brought to early America in the 18th century. The first known reference to Groundhog Day in American history exists in the archives of the Historical Society of Berks County in Reading.

The diary of shopkeeper James Morris, dated February 4, 1841, stated:

> Last Tuesday, the 2nd, was Candlemas day, the day on which, according to the Germans, the Groundhog peeps out of his winter quarters and if he sees his shadow he pops back for another six weeks nap, but if the day be cloudy he remains out, as the weather is to be moderate.

Clymer Freas, city editor of *The Punxsutawney Spirit*, is credited with the first formal conception of Groundhog Day. Taking a cue from the groundhog hunters, dubbed the Punxsutawney Groundhog Club, he declared on February 2, 1886: "Today is groundhog day, and up to the time of going to press the beast has not seen his shadow."

The first official trip up to Gobbler's Knob, a community 64 miles northeast of Pittsburgh, occurred the following year on February 2, 1887. The annual pilgrimage to consult the "Seer of Seers, Sage of Sages, Prognosticator of Prognosticators, and Weather Prophet Extraordinary" in the "Weather Capital of the World" soon became a national storyline.

A look at Phil's track record, maintained by the Punxsutawney Groundhog Club, proves that winter is invariably alive and kicking well into March. During the period from 1887–2025, not counting 10 events with no known record, the groundhog predicted an early spring on 21 occasions due to overcast skies, compared to 107 years when Phil saw his shadow, which meant six more weeks of winter in the region, according to lore.

The groundhog-in-residence, with his family, are ensconced in warm digs 364 days out of the year in the Punxsutawney Memorial Library. His humble abode is made of fiberglass and carved in wood. Every February 2, the renowned furry forecaster emerges from the movie-prop stump, attended by thousands of eager visitors hanging on his pronouncement on the state of the winter. The prediction is declared by a Fair Weatherman of the Groundhog Club's Inner Circle, all dressed in formal attire and black hats at Gobbler's Knob.

The weather has rarely been a major issue on Groundhog Day in Punxsutawney, except for one high wind event, which happened on February 2, 1983, which ripped across western Pennsylvania and nearly blew away Phil's shadow. A wind gust at Erie was clocked at 85 mph.

Snowflakes: A Wintry Mix

Most snowflakes form near or below 10,000 feet, acquiring myriad forms that are determined by the amount of moisture in the air, lift and ambient temperatures as ice crystals fall through different layers. Snowflakes comprise a conglomeration of hundreds of snow crystals that group together through aggregation and fall at 1–6 feet per second, unless picked up by an updraft.

The formation of ice crystals begins when water vapor transforms into a solid without a liquid phase (deposition). Liquid water can coexist with ice at temperatures down to -40°F. A layer of drier air entrained in the cloud reduces the number of snowflakes because ice nucleation stops with the loss of seed crystals, turning leftover precipitation to freezing drizzle.

Snow crystals require tiny nuclei that provide a surface to bond with water droplets. Aerosols, such as bacteria, dust, pollutants, sea spray and particles, act as the freezing nuclei that support seed crystals. Water molecules evaporate faster because saturation vapor pressure over water is greater than that over ice, causing the ice crystals to grow at the expense of cloud droplets. Over time, snow crystals become heavy enough to fall to Earth.

Snowflakes change shape in response to varying atmospheric conditions. An individual snowflake can take the shape of a plate, column or needle, depending on the temperature and moisture content of the environment. As snowflakes float, water freezes around the edges and creates unique shapes and sizes that are altered by mid-air collisions and wind currents.

Every snowflake develops from a six-armed embryonic ice crystal that slowly falls. The most common snowflake is a branch-like dendrite that forms at cloud temperatures mostly between 14°F and -4°F. The distribution of water molecules composed of oxygen and hydrogen atoms is responsible for the intricate six-sided crystal lattice structure. The presence of air between the branches creates the fluffiness.

Snowfall rates depend on a combination of lift, moisture transport and saturation in the dendritic growth zone (DGZ). Snow crystals adhere to other crystals and grow to .02–0.2 inch in diameter. Light winds support the formation of even larger flakes at near-freezing temperatures (0.5–1 inch) that tend to stick together.

Snowflakes are transformed during the long descent, which can take up to 45 minutes. If the environmental temperature warms above 32°F, a thin layer of water forms that acts like glue, clumping finer flakes into large ones.

The average snow-to-liquid equivalent ratio (SLR) is usually given as 10:1 (.10 inch of liquid equates to about an inch of snow). A wet snow can have a SLR of 5:1, while an Alberta clipper gliding by in a frigid air mass can yield a ratio as high as 30:1.

Snowstorms

Researcher James E. Miller outlined two common winter storm tracks in 1946, "from a study of 208 cyclones over a period of ten years." A Miller Type A storm develops over the Gulf of Mexico or along the Atlantic Coast. A Type B system reflects a western low-pressure center that gives way to a secondary coastal disturbance, when mid-level energy is shunted to the coast and interacts with relatively mild maritime influences.

Computer model guidance typically offers a range of snowfall scenarios 24 hours of an impending winter storm. The devil is invariably in the details, such as a vacillating rain-snow line that has large ramifications on precipitation types and snowfall totals.

A snowstorm can dynamically manufacture cold air in a marginal event through evaporative cooling, making the snowfall projections more challenging. In heavier bursts of precipitation, a wintry mix can change to snow for a time, and then back to rain when the precipitation rate eases up.

Strong winter storms moving north along the Eastern Seaboard are referred to as nor'easters, which reflects the direction of the surface wind on the colder western side of the storm track. Powerful coastal systems are responsible for wind-driven snow, often mixed with rain in the coastal plain, and battering ocean waves along the shoreline.

Rapidly strengthening nor'easters can deliver prolific snowfalls. On February 8–9, 2013, an offshore coastal storm deepened explosively (28.60 inches/968 millibars), dropping 30.6 inches of snow at Upton, New York, Long Island, and 40 inches at Camden, Connecticut. Boston received 24.9 inches, and Portland, Maine, had a record 34.9 inches, as winds reached 84 mph in eastern Massachusetts and 102 mph in Nova Scotia.

An exceptional pressure fall of 24 millibars (.71 inch) within 24 hours is called bombogenesis. On January 4, 2018, a sprawling storm system deepened dramatically east of Long Island (28.02 inches/949 millibars), when the pressure plunged 1.74 inches (59 millibars) in 24 hours. Warmer sea surface temperatures have contributed to a greater number of "bomb cyclones" in recent years.

Official Philadelphia snowfall city records that commence in the winter of 1884–85 show 20 snowfalls of 12 inches or more. Half of those events occurred between 1996 and 2016. Heavy snowfalls have become more common in the Middle Atlantic region since 2000 for two primary reasons: sea surface temperatures trending 1°F to 2°F warmer than a half-century ago, and a recent tendency toward more persistent blocking high pressure in the North Atlantic associated with a high-amplitude jet stream shape.

A rise in water temperature provides more energy through evaporation for storms. A strong ridge of high pressure near Greenland causes the polar jet stream to buckle near the Eastern Seaboard, setting up a favorable trajectory for a snow-laden nor'easter with a stream of cold air supplied by high pressure over eastern Canada and a cold conveyor belt of easterly winds interacting with the warm Gulf Stream.

The seeds of a winter storm are evident on satellite imagery as a swirl of clouds or embedded wave of energy. Moisture convergence is focused near the "warm nose" of a low-level jet that impinges on dense cold air, setting up an "overrunning" zone up and over the cold dome, causing condensation and precipitation.

Dynamical support is essential for a winter storm. A mid-level jet accompanying a disturbance causes upward motions that lowers the surface pressure and increases moisture transport. Instability increases as warm, moist air drawn into the system rises and cold air moves southward behind the area of low pressure sinks.

A moist easterly low-level jet enhances the mid-level frontogenesis, with the spoke of energy that strengthens the temperature gradient. If the upper-level circulation develops a northwest-to-southeast orientation (negative tilt), a period of intense banded snow accumulates at a rate of one to three inches per hour. A dual, or coupled jet structure with overlapping jet streaks—zones of fast-moving winds—increases the lift in the right entrance and left exit regions, doubling down on the storm totals.

Wraparound bands of snow rotating around the northwestern side of a low-pressure system form in a deformation or confluence zone where northerly and easterly winds collide, as warm air is shunted aloft. Mesoscale bands typically align in a narrow corridor about 25 to 50 miles wide and 50 to 200 miles in length that sometimes persists over the same area for several hours.

Mesoscale banding developed on the morning of February 2, 2009, from a storm located 500 miles east of Philadelphia. A 10-mile-wide swath of intense snowfall rates developed at the intersection of a moisture plume and arctic air, totaling nearly 12 inches at Manheim in Lancaster County, including 7.7 inches in a couple of hours.

Lake-Effect Snow

The main Pennsylvania snowbelt is downwind of Lake Erie in the northwest. Arctic air passing over the warmer waters creates lake plumes of fluffy snowflakes, steered by the wind direction. Snow bands, energized by the warm, moist air ascending inland terrain, dump heavy snow on areas from Corry and Bear Lake to Sugar Grove. Lake-effect snow bursts sweep eastward across the ridges, causing hazardous travel conditions.

The highly dynamic lake-effect process ramps up when a block of frigid air pours over the water, establishing a large vertical temperature gradient. The lake-induced instability increases by drawing warmer, moist air upwards beneath much colder air, until reaching a temperature inversion or cap. The ideal temperature profile for snow squalls is air 23°F or more colder a mile above the lake.

The cooling Great Lakes in the autumn causes chilled surface water to sink, until the upper layer of the lake attains maximum density at 39°F and mixing diminishes. The surface layer slowly cools to freezing and ice forms.

The wind direction plays a key role in snow bands, as frictional convergence and orographic lift wring out a significant amount of moisture in a cold air mass. The greater the distance that the wind travels over open water (fetch), the heavier the snow bands, which depends on the vertical temperature difference and rising motions enhanced by a mid-level disturbance.

The ideal near-surface wind speed for the formation of lake-effect snow bands is 15–25 mph. Snow streamers sometimes put down snow at the rate of up to two to four inches an hour, accompanied by flashes of lightning and rumbles of thunder with sufficient low-level instability, creating white-out conditions.

Lake-enhanced snow bands can travel a considerable distance, reaching the Pocono highlands, northern Piedmont and Hudson Valley in southeastern New York before dissipating. Downslope motions feather out snow squalls into scattered flurries east of the Appalachian ridges.

Historic Lake-Effect Snow Events

A massive lake-effect snow event occurred on November 9–14, 1996. In Erie County, 55 inches fell at Edinboro. The NWS office at Erie International Airport recorded 26.8 inches, with a maximum daily fall of 11.7 inches on the 10th. However, a spotter on the outskirts of Erie reported 41 inches. In Crawford County, 21 inches fell at Conneautville, the same total at Lottsville in Warren County. Locations in Warren and McKean counties received upwards of two feet of snow.

The snow bands extended as far east as Lycoming County and the north-central mountains. In neighboring northeastern Ohio, the snowfall from November 9–14, 1996, reached 68.9 inches at Chardon and 56.2 inches at Kirtland. The monthly accumulation at Chardon (76.5 inches) set an Ohio snowfall record, surpassing December 1962 (69.5 inches).

One of the most prolific lake-effect snowfalls commenced near the Lake Erie shoreline on Christmas Eve 2017 at 7:00 p.m. The snowfall amounted to only 1.1 inch by midnight at Erie International Airport, but became intense on Christmas Day, falling at the rate of four inches an hour. A record 24-hour total of 32.6 inches fell from 7:00 a.m. on Christmas morning until the daybreak observation on December 26. (The previous daily record at Erie International Airport was 20 inches on November 22, 1956.)

Daily totals included 20.9 inches on Christmas Day 2017 and 21.8 inches on December 26, capped off by 5.1 inches on the 27th. The two-day snowfall of 42.7 inches on December 25–26, 2017, occurred in about 30 hours. The only comparable 24-hour snowfall in the Erie city records was 26.5 inches on December 11–12, 1944, with an event total of 30.2 inches from December 11–14. The Albion observer measured 30.8 inches.

Erie International Airport closed on Christmas night 2017 at 10:00 p.m. and did not reopen until noon on the 27th. Two dozen Pennsylvania National Guard members brought nine vehicles to assist in the dig-out, after a declaration of disaster emergency was signed by an Erie County official. Communities at the eastern end of Lake Ontario in New York were buried under upwards of five feet of snow.

Another 18.2 inches of snow fell at Erie in a surge of brutally cold air December 28–30, 2017, raising the monthly accumulation to 93.8 inches, which broke the previous December record of 66.9 inches in 1989. The snowfall set a state December record by eclipsing Blue Knob in 1890 (86 inches). The full winter of 2017–18 brought 166.3 inches of snow at Erie, surpassing 2000–01 (149.1 inches).

Wet snow began piling up in the northwestern part of Pennsylvania on November 29, 2024, as arctic air flowed over the record warm Lake Erie waters. Erie International Airport picked up 22.6 inches. Girard had 63.8 inches by December 4.

Ice Storms

Pennsylvania is prone to ice storms because of the topography. The higher elevations are susceptible to icing due to colder temperatures. The interior valleys have prolonged bouts of ice accretion, when cold air gets trapped beneath a layer of mild air streaming north and is difficult to dislodge. Until the shallow layer of dense cold air is scoured out by southerly winds, icy conditions create hazardous conditions for drivers and pedestrians.

The making of an ice storm begins with a layer of mild air about a mile above the surface. Snowflakes falling from the cloud melt on the way down. If the liquid droplets encounter subfreezing temperatures a few thousand feet above the ground, moisture refreezes into ice pellets or sleet (.16 inch in diameter). A deeper "warm nose" of above-freezing temperatures extending nearly all the way down to the cold surface, where ground temperatures are below freezing, creates a treacherous glaze.

A prolonged ice storm develops when mild air steadily glides over a shallow cold air mass north of the storm center. Ice accretions of .25-.50 inch bend tree limbs and power lines, and untreated roads become skating rinks. A thicker buildup of ice on branches and transmission lines increases the weight, toppling trees and utility poles, which results in widespread power outages for days.

Raindrops that rupture upon impact release trapped air bubbles. In this instance, ice does not glisten and is especially deceptive for drivers—the dreaded transparent "black ice." Elevated surfaces such as bridges, ramps and overpasses are especially susceptible to icing because the surface loses heat more readily in the open air compared to the ground.

A rarer hazard is freezing fog, which is a cloud on the ground that coats a surface with a thin layer of rime ice, when tiny droplets condense in a subfreezing environment. Another common winter problem is snow melting during the day and the slush refreezing after nightfall as the temperature drops, making travel slippery.

Heavy icing caused widespread power outages and transportation problems across parts of Pennsylvania on the following dates: January 5, 1873; February 21–22, 1902; January 1, 1948; December 16–17, 1973; January 5, 1983; January 7–8, 1994.

A prolonged storm on December 29–30, 1942, in the northeastern counties encased tree limbs and power lines in an icy sheath up to three inches thick. Another great ice storm on January 8–11, 1953, followed the Great Southern Glaze Storm of January 1951, which laid down a 100-mile-wide swath of ice from Louisiana to West Virginia, resulting in $100 million in damages in the Southeast, described in a February 1952 *Climatological Data* report by Ben Harlin of the Nashville Weather Bureau.

Rain turned to freezing rain on January 8–9, 1953, in eastern Pennsylvania, interior New Jersey and parts of Connecticut and persisted for several days, knocking out 90 percent of the region's telephone and electricity service. Cold air streaming south met up with a southwesterly flow of mild, very moist air transported by a deep upper-air trough.

At elevations above 1,000 feet in northeastern Pennsylvania, glaze reached a thickness of 4 to 4.4 inches in Carbon and Monroe counties. In the Alleghenies, 1 to 3.5 inches of ice accumulated on frozen surfaces and up to 2.5 inches coated the central mountains. Some places had more than 40 continuous hours of freezing rain. Tens of thousands of trees were bent to the ground by the weight of the ice, and breakage increased as the winds accelerated.

The Pennsylvania Power and Light Company reported at least 50,000 customers were without power. As many as 125 line crews battled the miserable elements to restore electricity, according to the *Climatological Data* summary. Two miles of power poles collapsed near Kunkletown, and downed trees and utility lines blocked roads in many areas.

A disastrous ice storm caused havoc in the interior northeastern counties of Pennsylvania on January 5–6, 2005. Precipitation began as a cold rain early on the 5th and turned to snow for a time before switching back to freezing rain on the high Pocono Plateau and upper elevations of the Lehigh Valley.

A second ice event on January 8, 2005, made things worse. Monroe and Carbon Counties were placed under a state of emergency due to so many fallen trees caused by ice accumulating an inch or more. Observer Michael Pontrelli reported one to two inches of ice encasing downed limbs on Godfrey Ridge 10 miles southwest of Stroudsburg. Red Cross shelters were set up, providing more than 6,000 meals. Power was not fully restored until January 16, 2005.

According to the monthly NWS *Storm Data* report:

> At one point, nearly three-quarters of Carbon County was without power and nearly 46,000 homes and businesses in Monroe County lost power. System-wide 238,000 Pennsylvania Power and Light customers lost power and the total repair cost was estimated at $25,000,000. Metropolitan Edison reported about 40,000 of its customers lost power. Thousands upon thousands of trees were knocked down or damaged. The number of outages cascaded … that increased to 27,000 the afternoon of the 6th and 91,000 the evening of the 6th.

Ice wreaked havoc from the Lehigh Valley southward to Philadelphia on February 5, 2014, as snow turned to freezing rain. Broken tree limbs weighted down by wet snow and ice crashed onto power lines in southeastern Pennsylvania, resulting in a record power loss to 715,000 Pennsylvania Electric Company (PECO) customers in the Philadelphia area (45 percent), and a total of 849,000 in eastern Pennsylvania (*Storm Data*).

Pennsylvania Governor Tom Corbett quickly declared a state of emergency, joined by President

Barack Obama's federal declaration. Power was out in some areas for a week, and most schools and colleges were closed for the entire time. Water treatment plants were without power, and residents were urged to boil water in an advisory. The Red Cross established shelters to help residents cope with the loss of power and water.

Ice storms are relatively rare before mid-December because the ground retains the autumn warmth. The earliest significant ice storm in modern Pennsylvania history occurred on November 14, 1997. Nearly an inch of mixed precipitation fell over interior sections in subfreezing nighttime conditions that resulted in an unprecedented mid-November ice storm that extended south to the Lehigh Valley.

Ice storms are unlikely in the state after February because the higher sun angle allows greater diffuse solar radiation to filter through the leaden skies. However, an unusual spring ice event coated the northern highlands on April 15–16, 1929.

Winter Cycles

Winters in the 1960s and 1970s in Pennsylvania averaged several degrees colder when compared to recent decades. Statewide, the mean winter temperature from 1961–1970 (24.8°F) was sharply lower than for the period of 2001–2010 (28.3°F).

Nine of the 10 warmest winters (since 1895) occurred between 1997–98 and 2022–23. There is an interesting clustering of unseasonably warm winters with little snowfall: 1929–30 to 1932–33; 1936–37 to 1938–39; 1948–49 to 1953–54; 1971–72 to 1974–75; 1988–89 to 1991–92; 1996–97 to 1999–2000; 2004–05 to 2007–08. The explanation may lie in the large heat capacity of the ocean, which favors a northward displacement of the jet stream.

Interannual variability is sometimes dramatic. The exceptionally harsh "War Winter" of 1917–18 was followed by a very mild, nearly snowless winter in 1918–19. Similarly, the severely cold winter in 1993–94, with 17 snowstorms and historic chill, was followed by the polar opposite, literally, in 1994–95, when Scranton had a January record 10 consecutive days with readings above freezing, day and night.

The winters of 2013–14 and 2014–15 were exceptionally cold and snowy. The term "polar vortex" gained currency in January 2014, though the pattern was described in literature as early as the 19th century. A frigid February 2015 was the second coldest February in records back to 1895. The script flipped dramatically in the winters of 2015–16 and 2016–17, which rank among the warmest on record.

Two unusually snowy 15-year periods are evident since 1850 (Table 4.1): 1859–60 to 1874–75 and 1957–58 to 1970–71 (Gelber 1998).

Table 4.1
Season Snowfall in Eastern Pennsylvania (October–May)

1862–63 to 1880–81

North Whitehall (350 ft.)
1862–1863	77.3
1867–1868	78.2

Egypt (450 ft.)
1872–73	86.4
1874–75	79.5
1880–81	55.5

Blooming Grove (1000 ft.)

1867–1868	92.0
1872–1873	93.0
1873–1874	86.4
1874–1875	100.0

Dyberry (1,100 ft.)

1867–1868	115.4
1872–1873	113.2
1873–1874	112.2
1874–1875	98.0

Stroudsburg (480 ft.)

1957–1958	109.5
1960–1961	98.6
1962–1963	67.8
1963–1964	75.9
1966–1967	100.6
1969–1970	63.9
1970–1971	56.1

Hawley (890 ft.)

1957–1958	104.6
1960–1961	86.3
1962–1963	70.2
1963–1964	68.6
1966–1967	96.5
1969–1970	76.1
1970–1971	68.8

Gouldsboro (1,875 ft.)

1957–1958	147.1
1960–1961	126.0
1962–1963	100.0
1963–1964	101.0
1966–1967	104.0
1969–1970	113.0
1970–1971	94.6

A letter to the *Stroudsburg Times* published on January 21, 1910, admonished the "reckless individual who predicted early in the season that we were going to have an open winter" because "the birds were not flying South as early ... fur on the squirrels and rabbits was not near as thick."

Exclaiming, "Open Winter indeed! ... the snow is piled up three feet high in front of our door steps" and that "gold medals and University degrees are being handed out for rewards for mendacity."

Historic Warm Winters

Mild winters often bring on early hints of spring. On February 11, 1828, a *Pennsylvania Magazine of History and Biography* (1891) article remembered "peach trees in bloom at Bellefonte, Centre County."

The remarkably "open" or relatively warm winter of 1889-90 (Table 4.2) ranks as the warmest in state history since official records commenced in 1888. December 1890 (39.0°F) averaged 10.1°F above the long-term mean (1901–2000).

On Christmas Day 1890, the mercury soared into the upper 60s, with maximums of 73°F at Centre Valley and 70°F at Uniontown. Pittsburgh had its mildest December (45.6°F), with nary a snowflake. The closest approach to this outstanding anomaly in the Steel City occurred in December 2015 (44.6°F).

January 1890 (36.9°F) averaged 11.6°F above the period mean in the Keystone State and brought only a trace of snow at Pittsburgh. The state maximum temperature in January 1890 was a decidedly spring-like 77°F at Coatesville. Harrisburg did not record its first measurable snow of the season until January 23, 1890, one day later than the modern mark of January 22, 2007.

February 1890 (36.4°F) was temperate—9.9°F above the long-term state average—and second in warmth only to February 2017 (36.6°F). The mean temperature at Pittsburgh (42.0°F) bests the anomalous warmth of February 2017 (40.6°F).

The warmest winter in the 20th century in Pennsylvania happened in 1931–32 (35.2°F), well above the long-term mean (26.9°F). January 1932 (39.2°F) attained an anomaly of historic proportions—13.9°F above the historic average.

Table 4.2
Warmest Winter Months in Pennsylvania (°F)

	Winter of 1889–90	Warmest (1895–2020)	
December	39.0	38.4	1923
January	36.9	39.1	1932
February	36.4	35.5	1998
Average	37.4	36.7	1931–32

Snowfall Season

The average seasonal snowfall (October–May) in Pennsylvania varies with elevation. The southeast typically receives less than 30 inches. Areas east of Lake Erie in the snowbelt and the Laurel Highlands east of the Pittsburgh experience winter totals often topping 100 inches. Most of Pennsylvania averages between 30 and 50 inches of snow in a season.

Historical records reveal a wide range of winter snow totals. The Philadelphia International Airport (elevation 5 ft.) had a trace in 1972–73, and 0.3 inch in the winters of 2019-20 and 2022-23. The highest snowfall at a government station was 225 inches at the Blue Knob site (elevation 2,440 ft.) in the winter of 1890–91. The Blue Knob ski resort, southwest of Altoona in Bedford County, has a summit elevation of 3,146 feet, the state's highest skiable mountain.

A NWS snow spotter at Edinboro in Erie County reported 245 inches in the winter of 2000–01. The Corry observer tallied 228 inches in 2006–07, and 223.4 inches piled up in 1959–60.

In the Alleghenies, the Seven Springs Mountain Resort in northwestern Somerset County reported 222.5 inches in 1992–93 (elevation 2,800 ft.). The winter of 2009–10 brought 218.3 inches at Laurel Summit (elevation 2,730 ft.).

Heavy lake-enhanced snowfalls occur in warm winters, when the lakes are largely unfrozen. The winter of 1998–99 brought a total of 172.3 inches, reported to the Cleveland NWS by a northwestern Pennsylvania spotter. The NWS Snow Observation Program entails a network of volunteer observers.

Table 4.3
Snowiest Months in Pennsylvania (1890–2020)

Monthly Total		Location	Year
September	1.0	Edinboro	1895
		Strongstown	1993
October	19.0	Corry	1962
November	60.5	Corry	1933
December	93.8	Erie	2017
January	84.4	Ebensburg	1978
February	114.2	Laurel Summit	2010
March	88.0	Seven Springs	1993
April	38.5	Somerset	1928
May	16.5	Monroeton	1945
June	Trace	Multiple sites	1902, 1903, 1907, 1918, 1945

Early Pennsylvania Winters

One of the earliest compilations of severe winters in Pennsylvania appeared in John Watson's *Annals of Philadelphia and Pennsylvania, in the Olden Time.*

Interviews were compiled by Pehr (Peter) Kalm (1716–1779), a visiting Swedish botanist and professor during the years of 1748–51. Kalm interviewed a 91-year-old resident, who recounted crossing the frozen Delaware River during the frigid winter of 1697–98 on wagons loaded with hay.

An account of the harsh winter of 1704–05, attributed to statesman and merchant Isaac Norris (1701–1766), described the rigorous conditions that beset Philadelphia:

> We have had the deepest snow this winter that has been known by the longest English liver here. No traveling; all avenues shut; the post has not gone these six weeks; the river fast; and the people bring loads over it as they did seven years ago—[as in 1697–98, aforementioned]. Many creatures are like to perish.

The distinction for the harshest winter weather in colonial America likely belongs to 1740–41. The depth of the snow in eastern Pennsylvania was "in general more than three feet deep," forcing settlers to subsist on the remains of animals found frozen in the woods, as the severe cold dragged on through February 1741 (Watson 1868).

Roger Brickner's *New York City Weather* listed daily highlights from scant journal records. Cold and light snow arrived on November 15, 1740, though the first major snowfall would not come until December 28 that winter. One storm after another accompanied by periods of unrelenting cold continued through the first three weeks of January 1741 and lingering at times in February.

Joshua Hempstead of New London, Connecticut, mentioned snow in his diary on January 19, 22 and 27, 1741. Snow fell on five days in early February, followed by a series of heavy snowfalls in the middle of March 1741.

A long-time Delaware Valley inhabitant, Mrs. Shoemaker, recalled that "all the tops of the fences were so covered, that sleighs and sleds passed over them in every direction" (Watson 1868). A notice in the *United States Gazette* on April 5, 1843, listed a snowfall of three feet on April 18–19, 1741, without specifying the region.

James Logan, William Penn's former secretary, wrote a letter in 1748 about "the hard winter of 1741," long-remembered as "one of remarkable severity" that lasted from the third week of December 1740 through the middle of March 1741.

A string of forts to protect farmers from Native American attacks was constructed in 1755–1756, commissioned by the Pennsylvania Assembly and directed by Benjamin Franklin. The "Franklin Forts" amounted to a cluster of log buildings and a stockade fence that provided supplies and manpower, according to Monroe County Historical Association executive director Amy Leiser.

The region was rife with hostility, dating back to the Walking Purchase of 1737, regarded by Native Americans as a blatant theft, which led to entanglements with encroaching settlements. Hundreds of settlers were killed in raids after some tribes aligned with French troops, who were battling the British and American forces in western Pennsylvania.

Captain James Van Etten was tasked by Benjamin Franklin with building the northern-most in a series of forts during the French and Indian War, near present-day Bushkill in Middle Smithfield Township, during the winter of 1756–57. He assumed command of Fort Hyndshaw, and then was sent to Fort Hamilton in what is now downtown Stroudsburg in the spring of 1757.

Van Etten kept a diary of his experiences, commenting on the difficult conditions his troops faced following snowfalls on December 17 and 20, 1756, related in Rupp (1845):

> The corporal, with men assisting, hauled firewood to the fort, and I went on scout with three Men; found the snow about knee deep; but went four miles, made no discovery; returned to the fort after dark.

1764–65 Severe Winter

The winter of 1764–65 was widely judged to be "the snowiest since 1748 and the coldest since 1741" by preeminent weather historian David Ludlum, in *Early American Winters: 1604–1820*.

A major snowstorm lashed Philadelphia on December 25–26, 1764, followed by several days of bitter chill. Another heavy snowfall came on January 5–6, 1765. A particularly frigid period occurred January 26–28, 1765, with the thermometer hovering near or below zero at Kings College (now Columbia University).

In early January 1765, the snow cover was a respectable two feet at Germantown, according to the diary of the Reverend Henry M. Muhlenberg. The weather during the last week of January 1765 was intensely cold. Samuel Hazard, the venerable editor of the *Pennsylvania Register*, collected notices on winters from 1681–1800, which were reprinted in Watson's *Annals of Philadelphia*.

Hazard wrote that the Delaware River was closed to navigation due to ice from December 24, 1764, until February 28, 1765. On February 7, "an ox was roasted whole on the river Delaware, which, from the novelty of the thing drew together a great number of people." John Watts, the secretary of the New York Historical Society, wrote on February 11, "For this is weeks we have been in Greenland, the ground covered near three feet of snow" (Brickner 2006).

After the advent of astronomical spring, a furious snowstorm buried Philadelphia and the vicinity on March 24, 1765. The *Pennsylvania Gazette* (March 28, 1765) wrote: "A very severe snowstorm, the wind blowing very high … two feet, or two and a half, on the level, and in some places deeper," regarded it as "the greatest quantity of snow that had been known (considering the advanced season) for many years past."

Trees were uprooted by the weight of the snow, roads were impassable for days, and battering winds caused numerous shipwrecks. Surveyors Charles Mason and Jeremiah Dixon kept records near the Pennsylvania–Maryland border during the winter of 1764–65. On March 24, 1765, the "snow was near 3 feet deep." The *United States Gazette* (April 5, 1843) called this the heaviest March snowfall since two feet fell in 1725. Such massive depths have been witnessed in interior southeastern Pennsylvania rarely—notably in January 1831, February 1899, March 1958, January 1996 and January 2017.

A noteworthy storm struck south of Pennsylvania on January 27–28, 1772, known as the "Washington and Jefferson Snowstorm" because it was recalled by both of our Founding Fathers in their diaries. Thomas Jefferson returned home to Monticello with his new bride in the great snowstorm after their New Year's Day wedding.

The *United States Gazette* on June 27, 1818, carried an item from the *Winchester Gazette* in the Shenandoah Valley that recalled a snowfall of "2 feet 9 inches deep" (33 inches), also citing the *Maryland Gazette* (January 30, 1772) of "upwards of three feet" around Annapolis.

Revolutionary War Winters

The periodic severe winter weather conditions during the American Revolutionary War between 1775 and 1783 were described in vivid detail by the historian and meteorologist David Ludlum (1966).

The Continental Army barely escaped the port city of New York in a heavy fog on August 30, 1776. However, further losses forced Washington's army to retreat to Pennsylvania on December 7–8, 1776, with British forces in pursuit. Cold air settled in on December 18–19, causing the river to freeze, which set the stage for an inspiring victory (Ludlum 1983).

The enduring image of the Revolutionary War, immortalized in Emanuel Leutze's iconic painting *Washington Crossing the Delaware*, portrayed the harrowing journey on Christmas night of 1776 led by General George Washington. The Continental Army departed at McConkey's Ferry in conditions even worse than the oil painting depicted, as the troops faced a combination snow, sleet and wind at nightfall.

On Christmas night, about 2,400 troops would begin crossing the ice-covered Delaware River at 6:00 p.m., working under the assumption that two other divisions had successfully reached the New Jersey side of the river below Trenton as darkness fell. Unbeknownst to Washington, the deteriorating conditions prevented the southern flank from completing the journey.

The celebrated Christmas night 1776 crossing is reenacted every year at Washington Crossing Historic Park, formerly a small ferry operation in the early 19th century at Taylorsville. The next morning, Washington's contingent reached Trenton, achieving the element of surprise, and defeated the unsuspecting Hessian army.

For the weather conditions during the stormy early winter, we turn to the journal of Reverend Henry M. Muhlenberg (1945) at Trappe, 35 miles west of Trenton, who noted that the snow was "a foot deep and it's bitter cold." Thomas Jefferson monitored the snowfall at his Monticello home in central Virginia, where he recorded "22 or 24 inches," after a true nor'easter passed southeast of Pennsylvania.

Washington entered New Jersey to join his troops on January 3, 1777, who engaged a smaller British force in the Battle of Princeton in advance of General Lord Cornwallis. Later in the day, Washington correctly surmised that a clear sky and light northwesterly breeze was favorable for a hard freeze. He created a diversion by ordering the setting of bonfires, effectively outmaneuvering the British and returning safely to winter quarters.

A larger snowfall hit southeastern Pennsylvania on February 24, 1777, which Muhlenberg called "an extraordinary snow storm. ... Perhaps as much as two feet fell." The observations of Israel (Phineas) Pemberton, taken near Philadelphia and preserved in the American Philosophical Society archives, mentioned an 18-inch snowfall.

Philadelphia was later captured by the British in September 1777. Washington's troops were forced to vacate Germantown following a second crushing defeat to the British army directed by Sir William Howe. A ragtag, ill-equipped Continental Army eventually settled in an encampment at Valley Forge, 20 miles northwest of Philadelphia, on December 19, 1777.

Cold and stormy conditions were a harbinger of the terrible ordeal awaiting the beaten-down American soldiers in the winter of 1777–78, who would suffer terribly with few rations and weather-worn clothing. Disease, exposure and malnutrition would take a heavy toll on the 12,000 troops, and nearly 2,000 perished. The Continental Army departed Valley Forge on June 19, 1778.

The following winter of 1778–79 was noteworthy for the Hessian Storm of December 25–26, 1778, named for the battering sustained by British mercenaries harbored at New York City and Newport, Rhode Island. Snow, high winds, and low temperatures on Christmas Day hampered the movements of Hessian soldiers, and many ultimately died from exposure to subzero temperatures (Ludlum 1966).

1779–80
Hard Winter

The legendary Hard Winter of 1779–80 was the worst in the region since 1740–41.

The Delaware River at Philadelphia froze around December 1, 1779, and retained a layer of ice two or three feet thick until March 14, 1780, described in Edward Garriott's *Cold Waves and Frosts in the United States*. Citing Thomas Jefferson's *Notes on Virginia*, Garriott (1906) stated that "the Chesapeake Bay was frozen solid from its head to the mouth of the Potomac."

Ludlum (1966) compiled several accounts of the winter of 1779–80. Conditions failed to thaw until February 11, 1780. After an interminable cold period, the Delaware River did not reopen for navigation until March 4, 1780.

The onset of the Hard Winter could be traced back to late November 1779, when a cold pattern settled in and two heavy snowfalls left a substantial cover. Henry M. Muhlenberg entered in his journal, on November 26, that there was "a dreadful wind and snow-storm." Another storm on December 5 followed several days of rain, after which the snow "lay a foot and a half deep and there was no going out of doors."

Another heavy rainstorm on December 12–13, 1779, was followed by an arctic blast on the 14th, communicated as "such a terrible and bitterly cold windstorm as we have not had in a long time. Men and beasts tremble." Several days later, heavy snow changed to rain at Trappe on December 17–18, washing away much of the earlier snowfall, although Muhlenberg noted that a deep blanket of snow lingered in the north.

The first in a series of snowstorms during a 10-day period arrived on December 28, 1779, beginning as snow at Trappe and ending as a chilly rain. A second powerful winter storm struck on January 2–3, 1780, accompanied by gusty winds. Here, Muhlenberg commented on January 3 that "the house and yard are so circumvallated that one can scarcely get out or in, and the snow is still falling."

A third snowstorm struck on the night of January 4–5, 1780. Wind-blown snow fell for several days, prompting Muhlenberg to write on January 6: "Violent storm from the northwest the whole day with intermittent snow." A few weeks later, on January 23, Muhlenberg mentioned the snow was "over three feet deep" in Lancaster County.

During the Hard Winter of 1779–80, American troops were encamped in Morristown, New Jersey, about 25 miles west of New York City, after December 31, 1779, while keeping a watchful eye on the British, who enjoyed considerably more comfortable quarters in Manhattan. American troops would cross from New Jersey to Staten Island on foot over the frozen bay to do battle with the British as the cold deepened.

General George Washington, in addition to his duties as commander in chief of the Revolutionary Army, still found the time to keep a record of the weather, noting an 18-inch snowfall in northern New Jersey on January 5–6, 1780, in his diary.

The best source of historical weather information in Pennsylvania and New Jersey during the Hard Winter of 1779–80 appears in Oscar Harvey's *A History of Wilkes-Barre*:

> By the middle of December [1779] the snow was about two feet deep in central and northern New Jersey and eastern Pennsylvania. … In the woods and other sheltered places it lay for many weeks at least four feet deep upon the level. The weather was intensely cold during a greater part of that period, and harbors, rivers and creeks and brooks were all frozen over.

A diary entry by Revolutionary War officer John Jenkins Jr. on January 11, 1780, stated: "A party of men set out to go through the swamp—across the Pocono range—on snow-shoes; the snow about three feet deep."

Brutally cold weather froze inland bays and harbors as far south as the Virginia–North Carolina border. In a passage from David Ludlum's *American Weather Book*, the Somerset militia in central New Jersey tried to create a road between Hackettstown and Princeton but encountered impossibly deep snow.

Andrew Mellick Jr. vividly captured the forbidding landscape in his reportage: "The whole face of the country lay buried from three to five feet deep; roads, fences, and frozen streams were obliterated … in places the drifts were piled ten to twelve feet high."

A great tragedy occurred near Lancaster on February 5, 1780, which was covered in the diary of Thomas Hughes: "Forty people crossing the Susquehanna in sleighs—being on their return from a wedding—the ice broke, and six and thirty were drowned—amongst the unfortunates the new married couple." More snow fell on February 8, 1780, according to Washington's diary, after the temperature had moderated. This was the only significant recorded snowfall during the month.

Colonel Daniel Brodhead, stationed at Fort Pitt, at the confluence of the Monongahela and Allegheny Rivers, declared in a letter to General George Washington, dated February 11, 1780: "Such a deep snow and such ice has not been known at this place in the memory of the eldest natives; Deer & Turkies die by hundreds for want of food, the snow on the Alleghany & Laurel hills is four feet deep."

Washington noted that two inches of snow and sleet fell on March 12, 1780, citing the "snow about 9 inches deep" at Morristown, New Jersey, on March 16. On March 17 there was "cold penetrating." On March 19: "Another heavy snowstorm in the evening." A further remark mentioned "Bitter cold" the next day. A final snowstorm on March 31, 1780, lasted till the first day of April: "9 or 10 inches deep upon a level … Pretty good sleighing in the forenoon."

Flurries were noticed in northern New England on May 1, 1780, and frost was observed in some low interior places on June 5–6 and June 9. A rare summer frost was witnessed in the cold hollows of the Northeast on July 12, 1780.

1783–84
Winter of Deep Snows

The infamous Long Winter of 1783–84 came only four years after the Hard Winter. A persistent bluish haze was observed in the summer of 1783 over North America, Europe, western Asia and North Africa that filtered sunlight and led to a global cooling trend.

Benjamin Franklin correctly surmised that the massive explosion of the Laki eruption in Iceland contributed to an extended period of unusually cold conditions in northeastern North America and northern Europe. A veil of fine particulates and sulfate aerosols circling the globe would filter sunlight for a few years.

The most comprehensive snowfall record for the winter of 1783–84 comes from the diary of Joseph Lewis of Morristown, New Jersey. An early snowfall blanketed a wide area from New Jersey to Maine on November 12–13, 1783, depositing 2.5 inches at Morristown, noted in Lewis's diary, published in *Proceedings of the New Jersey Historical Society* (1943, 1945). A second winter storm that struck on November 27–28, 1783, dumped a hefty 11 inches.

The weather in December 1783 at Morristown was marked by "periods of alternate cold and warmth," with frequent snows. Moderate snowfalls occurred on December 2–3 (2 inches); December 4 (3 inches); December 18 (5 inches); Christmas Eve December 24 (2 inches); December 25–26 (1 inch); and December 29 (1 inch).

But the heaviest snowfall of the entire winter arrived on December 30–31, 1783, accompanied by much drifting. Snow began falling at Morristown around 3:00 a.m. and continued for 25 hours, accumulating "about 20 inches."

Navigation was halted on the Delaware River on December 26, 1783, which froze a few days later, and did not open up until March 12, 1784, based on notices compiled by editor Samuel Hazard, in his weekly *Register of Pennsylvania* (July and December 1828).

The snow reached a considerable depth in northern New Jersey and eastern Pennsylvania by early January 1784, and the storms kept on coming. Significant storms blanketed Morristown on January 11 (5 inches), January 18–19 (10 inches) and January 20 (3 inches). On January 26, 1784, snow commenced at 9:00 p.m. and accumulated to a depth of 18 inches on the 27th.

January 1784 brought 36 inches of snow at Morristown, coupled with a very snowy November (13.5 inches) and December 1783 (34 inches), totaling 83.5 inches for the three-month period. No winter since in the region has produced so much snow so early in the season. There was a sudden thaw on January 13, 1784, "so that within a few hours, we have experienced a transition from heat to cold, of at least 53 degrees," according to Watson's *Annals of Philadelphia*. Yet true to form, in a protracted severe winter, bitter cold returned in February.

A deep freeze from February 10–17, 1784, noted in Garriott's *Cold Waves and Frosts*, brought a minimum of -12°F near Philadelphia, probably on the 14th. At Hartford, Connecticut, the mercury dipped below -10°F on eight consecutive mornings during February 10–17, with minimum readings of -20°F on February 14 and -19°F on February 19, 1784.

Another late frigid snap brought three consecutive bitter subzero mornings at Philadelphia, commencing with February 29, 1784, an especially wintry Leap Day. Two moderate snowfalls in early March 1784 were followed by a thaw "during the second half of the month." The Great Flood of 1784 followed a torrential rainstorm on March 13–14, 1784, that caused the snow to melt rapidly. Ice jams formed as the ice broke up and lodged against the riverbanks. As those impediments gave way, torrents of water rushed over the banks of the Susquehanna River and caused a devastating flood in the Wyoming Valley plains.

Snow would again cover the ground in April 1784 in portions of Pennsylvania. Ludlum's *Early American Winters: 1604-1820* noted the records of surveyor John Fitch in the "western country." He observed the final breakup of ice along the Monongahela River on March 10, 1784, and the Youghiogheny River the following day. Fitch's records (1784–1791) were preserved and can be found in the Library of Congress.

The account in *Washington Weather* (Ambrose et al. 2005) leaves little doubt that the winter of 1783-84 was one of the most severe seasons. On March 5, 1784, George Washington noted in his diary that he "arrived at this Cottage on Christmas Eve, where I have been locked up ever since in frost and snow." James Madison wrote to Thomas Jefferson that the winter of 1783–84 had exceeded all previous winters for severity.

December 1786
Triple Storms

A barrage of three major snowstorms in early December 1786 brought unprecedented early-season snow depths to northern New Jersey, though little is known about the snowfall totals in eastern Pennsylvania.

Again, the Morristown, New Jersey, records of Joseph Lewis, spanning the years of 1783–1795, published in the *Proceedings of the New Jersey Historical Society* (1943, 1945), provide the best perspective on the meteorological triple-play in December 1786.

The first of three major snowstorms deposited 18 inches at Morristown on December 4–5, 1786. The diary of a local farmer in Goschenhoppen, 40 miles northwest of Philadelphia, confirmed "a very deep snow," in *The Journals and Papers of David Schultze*. Three days later, another eight inches of snow would fall at Morristown. A third storm added 15 inches on December 9–10, 1786, that totaled 41 inches in one week.

We now turn to surveyor Winthrop Sargent, who happened to be in Philadelphia in December 1786, while taking an active role in the Ohio Company before settling in Marietta, Ohio.

Sargent left us with a good description of the "Triple Storms" encountered in his transit from Philadelphia to New York City (Ludlum 1966):

> December 4–5: At 10 oClock [December 4] some light snow, & by 12 we had a severe storm which continued through the day. ... This morng [December 5] at 10 oClock it abated & the wind backened in to N W from which quarter it blew in force all Day—it ceased to snow after noon but here are drifts

of more than four feet in Height which very much impedes our travelling.

December 7–8: In the last night there fell five or six inches of snow at very least, which may make tolerable sleigh'g.

December 9–10: A N East wind last night, which was this morng [December 9] increased to a violent gale attended with a great fall of snow—A continuation of the storm through the Day. ... Full two feet of snow on a level over the whole country.

By December 14, 1786, Sargent noted, "snow dissolving fast." The lowest temperature in his diary was 12°F on December 6. The weather moderated considerably by mid-month, dissolving much of the deep snow cover, which was nearly all gone by early January 1787.

1798–99
Long Winter

The Long Winter of 1798–99 began with an early snowstorm that started as rain at New York City on November 19, 1798, transitioning to heavy snow shortly before midnight.

Longtime local observer Hugh Gaine, who maintained a daily weather journal, measured 18 inches at New York City, easily exceeding all modern November snowfalls. Joseph Price, in Lower Merion Township northwest of Philadelphia, wrote in his diary: "Snow near a foot deep in the woods." An uncommonly low reading of 22°F occurred after the storm (Ludlum 1966). Up to three feet of snow fell over interior New England.

Four significant snowfalls followed during the first 19 days of December 1798. At Philadelphia, the accumulation of snow was thought to be the greatest ever known up to the first of January, according to Samuel Hazard, in his summary of past winters, which appeared in the *Register of Pennsylvania* (July 1828).

Bouts of bitterly cold weather accompanied by light to moderate snowfalls contributed to an extended period of rough, wintry weather across the Mid-Atlantic through March 1799. Subzero temperatures were recorded on March 5 along the Connecticut coast, and a substantial snow cover lingered as late as March 12 at Philadelphia.

The *United States Gazette* (April 5, 1843) carried a list of great March snowstorms that included a fall of 20 to 24 inches on March 14, 1799. Two histories verified the exceptional late cumulative snow cover in sparsely settled western Pennsylvania. Alfred Creigh, in his *History of Washington County, Pennsylvania*, wrote that the depth of the snow was three feet. In *Pioneer Outline History of Northwestern Pennsylvania*, W. J. McKnight stated that the snow depth reached five feet on the Allegheny Plateau.

In the tradition of protracted winters, snowflakes were flying over the highlands in April and early May 1799, justifiably earning the title of the Long Winter.

March 30–April 1, 1807
Great Spring Snowtorm

A massive late-season snowstorm traveled from the Tennessee Valley to southeastern Pennsylvania on March 30–April 1, 1807. The disturbance drew considerable moisture from the Atlantic Ocean, resulting in a spectacular snowfall (Ludlum 1966).

Heavy wet snow began falling over interior Pennsylvania on the evening of March 30, 1807, piling up to historic depths by the first day of April. The Philadelphia *Democratic Press* reported 36 inches at Huntingdon in the south-central hills, and 36 to 42 inches in the Nittany Valley, near the present site of Pennsylvania State University. Up to five feet of snow piled up in Upstate New York and interior New England. The top total was an astounding 60 inches at Danville, Vermont.

In his *History and Geography of Bradford County, Pennsylvania*, Clement F. Heverly wrote that "snow fell continuously three days and was between four and five feet deep." Despite the "cold, blustery weather," April sunshine caused the massive snow accumulation to melt rapidly, causing "a great flood, one of the most notable in the Susquehanna river."

Twenty-nine years later, the editor of the Montrose *Independent Volunteer* compared a huge January 1836 snowstorm to the spectacular April 1, 1807, snowfall that left a total of 54 inches, noted the *United States Gazette* (January 20, 1836).

1816
Year Without a Summer

Farmers in the northeastern United States were stunned by the turn of events in the late spring of 1816. The anomalous cold summer of 1816 would forever be remembered as the "year without a summer" and "eighteen hundred and froze to death."

Weather records were sparse in 1816, but Charles Peirce (1847) assembled an invaluable account of the conditions in Pennsylvania in each of month of 1816:

April 1816: Jack Frost came along mounted upon a cold boisterous northwester, and made everything tremble and shiver before him. The blustering snow squalls would have been more suitable for January than April.

Unprecedented late freezes ruined crops in an exceptionally short and unproductive growing season throughout the Northeast, resulting in serious food shortages and great economic hardships. The mean monthly temperature at Morrisville, Pennsylvania, during June, July and August 1816 (64°F, 68°F, 66°F), averaged about seven degrees below the contemporary means at nearby Trenton, New Jersey.

Peirce described August 1816 as "such a cheerless, desponding, melancholy summer month, the oldest inhabitant never, perhaps, experienced." On August 10, 1816, the editor of *Niles' Weekly Register* took notice of the Cold Years of 1812–1816, stating that "it has been observed by the most careless observer, that since 1812 the seasons have been very unlike what they formerly had been."

In "Climate of the United States," the *Register* reported, "This summer has been hitherto extremely cold," noting the exceptionally dry conditions "in the central parts of Pennsylvania, and in the great range of Appalachian mountains."

The cause of the historically unseasonable late spring and summer in 1816 was the mighty eruption of Tambora in Indonesia in April 1815. The massive explosion, perhaps the greatest in human history and four times larger than Krakatoa in 1883, emitted 12 cubic miles of volcanic matter that reached an altitude of 25 miles or more. The fallout covered an area more than six times the size of the ash cloud created by the June 1991 eruption of Mount Pinatubo (Wood 2014).

Tambora was the last in a series of volcanic eruptions from 1809–1815, and the most violent in 74,000 years. The explosion sent about 50 million tons of sulfur dioxide into the stratosphere, which converted in sulfate aerosols. Chemical reactions involving atmospheric moisture turned sulfur dioxide into sulfuric acid clouds that filtered sunlight. The volcanic haze was swept around the world by the high-altitude winds.

Pleasantly warm weather in early June 1816 had given farmers renewed hope for crops severely damaged in the late freeze on May 29–30, 1816. Instead, a lingering pool of polar air near Hudson Bay settled southward and caused a devastating freeze, accompanied by the rare occurrence of snow in June in the interior Northeast.

Although there are no crude weather maps to view in the early 19th century, the polar vortex must have plunged historically far south. A compilation of reports in the *Weekly Register* mentioned snow "in the neighborhood of Pittsburg," but "very little in Pennsylvania east of the mountains."

An account from Homer, New York, in the *United States Gazette* (May 17, 1834), reprinted from an 1816 edition, provided a summary of the unseasonably cold weather:

> The 30th of last March [1816] was colder, in this part of the country, than at any other time during the preceding winter. The 16th of April, the ground was firmly frozen, and the snow over our shoes. The 18th of May, the ground was covered with snow, which lay nearly through the day. The morning of the 30th of May, the ground was white with frost, and ice on water in vessels, thicker than window glass. 6th of June—winter-like weather, and snow-squalls—apple trees in bloom. June 7th—This morning the ground frozen, and ice about one-fourth of an inch thick.

Light snow fell in parts of New York and northern Pennsylvania on the morning of June 6, with some accumulation, and again on June 8, 1816, accompanied by intense chill that required fires to keep warm. The diary of Elisha Risdon, at Hopkinsville, New York, was provided to the *Nelsonville* Tribune by a local resident, Florence Spencer of Lancaster, Ohio, which appeared in an undated newspaper citation in 1971.

The June 7, 1816, entry stated, "Very cold. Froze ice one-half inch thick." On July 11, she continued: "We are alarmed, not only to the present want but future stores. Crops are very backward and promise but little. Our present necessities are great. Many of our neighbors are without bread."

An Erie observer was quoted in the *Weekly Register* on June 14, 1816: "The season has been dry and frosty for weeks together. It appears that we should have no crops in these parts—the corn has been all killed by the frost of the 9th [of June], and until very lately Lake Erie was not navigable for the ice."

The *Bethlehem Daily Times*, in a blizzard story on March 14, 1888, dug into the archives for an account of the arrival of the polar outbreak on June 5–6, 1816: "From 'a record of unseasonable snows' in this part of Pennsylvania, being extracts from old diaries and note books, dating as far back as 1774, we extract the following: 'In 1816, the memorable 'year without a summer, heavy snows fell on several days in June.'"

W. J. McKnight's *A Pioneer Outline History of Northwestern Pennsylvania* had this to say about the melancholy conditions:

> In 1816, the year, frost occurred in every month. Ice formed half an inch thick in May. Snow fell to the depth of three inches in June. Ice was formed to the thickness of a common window-glass on July 5. Indian corn was so frozen that the greater part was cut in August and dried for fodder, and the pioneers supplied from the corn of 1815 for the seeding of the spring of 1817.

In an entry on June 30, 1924, the *Philadelphia Inquirer* provided some additional history of the calamitous growing season of 1816. A press clipping lodged in an old Bible belonging to Mrs. Joseph Mitchell of Lansdale recalled "heavy frosts from the middle of May to June 12, resulting in several crop losses."

At Downingtown, there were "severe frosts" on June 10–11, 1816, that killed the beans and "destroyed whole fields of corn" (Watson 1868). The corn and hay crops had been badly stunted by the late summer of 1816, although vegetables replanted in late June fared reasonably well farther south.

But the summer of 1816 would include one more indignity—a freeze on August 24—that would be the last straw for farmers. A widespread killing freeze on September 27, 1816, put an exclamation point on the disastrous growing season. Serious food deficits persisted through the winter of 1816–17, leading to a westward migration from New England and across the northern Appalachians called "Ohio Fever" that motivated farming families to head west in search of a more productive growing region.

January 14–16, 1831
Great Snowstorm of 1831

The winter of 1830–31 in northern Pennsylvania was deemed by Heverly (1926) "very severe, hardest experienced since that of 1779–80."

Harsh chill arrived on December 21–22, 1830, which was followed by a thaw and very heavy rain on December 30–31. *Hazard's Register of Pennsylvania* reported that an "unprecedented freshet" inundated the low-lying sections of Wilkes-Barre and and Kingston.

In late 1830s, James Espy formed a committee of meteorology at the Franklin Institute that was designed to learn more about storm tracks "by simultaneous observations over a wide extent of territory." Espy wanted observers to record the temperature, pressure and wind twice a day, along with the daily precipitation, and general weather remarks.

Espy's timing was ideal. A period of seasonably cold weather gave way to a thaw and heavy rain on January 4–5, 1831. The temperature rose into the 50s at Philadelphia on January 4–5, 1831, with 1.90 inches of rain, according to the "Meteorological Register" in the *Journal of the Franklin Institute*.

The weather turned cold again after the rainstorm, followed by a heavy snow and sleet storm on January 8–9, 1831. The *Gettysburg Sentinel* reported snowfalls ranging from 10 inches at Philadelphia, upwards to about a foot in the southeastern part of the state (*Register of Pennsylvania*).

The daybreak temperature at Philadelphia on January 13, 1831, dipped to 8°F and only recovered to 16°F at 2:00 p.m. The next day, a system storm took shape in the Gulf of Mexico which would become one of the greatest Mid-Atlantic snowstorms.

The storm churned northeast across Georgia early on Friday, January 14, 1831, on a day that dawned fair and frigid in Pennsylvania, with temperatures straddling 0°F in the interior. Heavy snow blanketed sections of Georgia and the Carolinas.

Snow arrived in Philadelphia around 8:00 p.m. and continued for more than 40 hours, increasing in severity on Saturday, January 15, 1831, and not letting up until midday on Sunday. Totals ranged from 13 inches at Washington to 18 inches at Baltimore and Erasmus Hall in Brooklyn. The slow movement contributed to the deep interior snowfalls. At Gettysburg, the accumulation reached 30 inches (Ludlum 1968). The snowfall at Harrisburg was "upwards of two feet," with high drifts that blocked mail deliveries and stagecoaches from Philadelphia and Baltimore, noted by the state librarian William Musgrove in the *Register*.

The *Lancaster Journal* stated that the average depth of the snow was "variously estimated at from two to two-and-a-half feet," including what remained of the "nine to twelve inches" that fell on January 9–10, 1831. Drifts were as high as five to six feet, "rendering the main roads impassable for several days." The librarian at the State Library in Harrisburg judged the snow to be the deepest since the winter of 1783–84.

Editor Samuel Hazard took it upon himself to provide a thorough account of the "Great Snowstorm of January 14–16, 1831" in his *Register of Pennsylvania* (April 9, 1831). The hardships caused by a "continued gale of wind" that whipped the enormous snowpack to depths as high as fences buried cattle and snapped tree limbs. Some buildings were crushed under the weight of the snow. The *Bedford Gazette* called it the "King of Snow," and mentioned depths of three to four feet.

In the next issue of the *Register* on April 16, 1831, even more accounts had trickled in from correspondents. The *West Chester Republican* reported that the "average depth of three feet" was the greatest since the winter of 1804–05 in southeastern Pennsylvania. In the Susquehanna Valley, the snow depth at Lewistown was between 36 and 42 inches in the country, which "exceeds that of any period," according to the oldest inhabitants.

The *Columbia Spy* reported "about three feet, and the drifts to be twelve feet high." The Pottsville *Miners' Journal* said the snow depth reached three feet, including a portion of a 10-inch snowfall that occurred on January 8–9, 1831. After a one-inch coating on January 18, the editor declared, "The oldest inhabitants declare that the quantity of snow on the ground exceeds that for any period within their recollection."

The *Easton Argus* listed a fall of 30 inches and a total depth of "at least three feet on the level." Farther north, a notice in the *Register* reported 20 inches fell at Milford. In the west, the *Pittsburgh Gazette* reported a very substantial snowfall of 22 inches. The Lewistown correspondent noted that three significant snowfalls came after the Great Snowstorm of January 14–16, 1831, ranging from six to 12 inches in the region.

The monthly *Journal of the Franklin Institute* also listed two inches at Philadelphia on January 21, a "slight snow" on the 29th, and a moderate storm of "snow, hail, and rain in night" beginning on January 31, 1831. The impressive snowfalls did not extend to the northwestern corner of the state.

The *Erie Observer* remarked that there was "scarcely enough to make good sleighing."

On April 30, 1831, Hazard's *Register* carried a dispatch from the *Crawford Messenger* that mentioned -24°F at Meadville on February 7, 1831.

December 28–30, 1834
Southern Pennsylvania Snowstorm

A massive snowfall on December 28–30, 1834, caught the attention of the press. The Pottsville *Miners' Journal* reported 22 inches. The *Easton Sentinel and Argus* recorded 15 inches, which laid the groundwork, literally, for the historic Cold Week of January 1835

The measurement by Alfred Creigh in Carlisle, published in *Hazard's Register of Pennsylvania* on April 11, 1835, noted a hefty snowfall of 21 inches. The storm at Washington brought 18 to 24 inches (*Richmond Inquirer*, January 13, 1835).

January 8–10, 1836
Big Snow of 1836

The winter of 1835–36 would be long remembered for the duration of cold weather and the single largest snowstorm in the history of eastern Pennsylvania. This winter was probably the coldest on record in Philadelphia. The mean December–February temperature recorded at Pennsylvania Hospital (27.7°F) was the coldest on record for the period commencing in 1825.

On January 18, 1918, while in the grip of another memorably severe winter, the *Stroudsburg Times* consulted nonagenarian Henry Biesecker for his memories of the winter of 1835–36: "There was a fall of four feet of snow that lay for days and people would find deer floundering in it, unable to escape."

Pocono historian Luther Hoffman was consulted by the East Stroudsburg *Morning Sun* on April 28, 1928, from his unpublished historical notes on the winter of 1835–36:

> There had been a very mild fall until during November, when a storm brought several inches of snow while freezing set in. Piles of corn in the ear were on the ground, unhusked, and were frozen fast … remaining there until spring. In the early part of December, snow started to fall ... traffic was tied up for many weeks. … Rural roads were blocked practically all winter. Farmers afoot and on horseback went to the boroughs to have grist around or to secure food. On many farms the trees were cut down nearest the house to supply wood …

An analysis of the storm by David Ludlum in *Early American Winters II, 1821–1870* described a stalled coastal low-pressure trough on January 7–8, 1836. The impending pileup of snow resulted from the collision between a steady stream of arctic air and a maritime flow, setting the stage for the "Big Snow" (Ludlum 1960).

Rain or mixed precipitation changed to wet snow on January 8, 1836, in the interior, but not until the following morning on the coastal plain. The *Sussex Register* in Newton, New Jersey, reported 18 inches of wet snow falling in the first 24 hours of the snowstorm.

As temperatures dipped to freezing on January 9, 1836, rain turned to snow, growing to a depth of 12 to 15 inches in Philadelphia and New York City. The *United States Gazette* reported a slushy 15-inch accumulation in Philadelphia, but estimated that the fall would have been closer to 36 inches had temperatures been a few degrees colder. The total melted precipitation at Pennsylvania Hospital in downtown Philadelphia during the four-day period beginning on Thursday, January 7, 1836, through the 10th measured an amazing 4.50 inches.

The *New York Herald* wrote: "The quantity of snow lying on our streets is beyond anything that ever appeared in our time. About four or five years ago [January 1831], we had a tolerable snowstorm, which afforded fine sleighing for six weeks in succession. But the quantity then was only half what it is at present."

The slow-moving snowstorm nearly stalled over southeastern New England, prolonging the intense snowfall rates. Spectacular accumulations ranging from three to five feet occurred in the highlands of northern Pennsylvania and south-central New York. The greatest falls of nearly 60 inches occurred at Ithaca and Rome, New York (Ludlum 1968).

The editor of the *Independent Volunteer* in Montrose declared: "We are literally surrounded on all sides by mountains of snow, which peer through their snow-capped peaks above one another in a style quite magnificent." Dubbed the Big Snow by the press, the snowfall of 42 inches was compared to another great snowstorm on March 30–April 1, 1807, which buried the community under 54 inches. On January 20, 1836, the *United States Gazette* affirmed that the 1807 snowfall had left "a trifling greater depth" than in January 1836.

In the southern Poconos, Luther Hoffman's unpublished historical notes described the East Stroudsburg area snowfall: "In January [1836], snow fell continuously for three days to a depth of about four feet in what is known as the Big Snow." Twenty-five miles to the south, the *Easton Argus* reported 34 inches. Both totals easily exceed modern records at Stroudsburg (35.4 inches in March 1958) and in the Lehigh Valley (31.9 inches in January 2016).

Relatively mild conditions prevailed in the Northeast before much colder weather arrived January 15–16, 1836. Snow returned on January 17, totaling 6.5 inches, according to the records of William Swift at the U.S. Naval Hospital in New York City (Brickner 2006). Frigid weather persisted on January 20–21, 1836, in the city (-6°F). Another heavy snowfall on January 24–25 deposited six more inches (Brickner 2006), raising the monthly total above 30 inches. The *New Jersey Journal* reported that six inches fell in central New Jersey.

Subzero conditions prevailed again on January 29–30, 1836, preceding an eight-inch snowstorm on January 31, 1836, giving New York City 38.5 inches—greater than any single month in the city's modern records. More subzero cold occurred on the mornings of February 2–3, with an absolute minimum of -8°F on the 5th. Five snowfalls deposited 18.5 inches in February at Fort Columbus (Governors Island), accompanied by an icy low (-2°F) on February 18.

The records at Pennsylvania Hospital ranked February 1836 (24.0°F) as the third coldest February in Philadelphia weather history, after 1838 and 1934. Notices in the *United States Gazette* described snow depths of four to eight feet, including drifts. Heavy snow fell again on February 28, 1836, marking the 16th snowstorm of the winter, comparable to the winter of 1993–94 that produced 17 storms between December 1993 and March 1994.

Substantial negative-temperature departures from the long-term average helped preserve the snow cover well into spring. Snow fell at Philadelphia and New York City on April 13, 1836.

December 1839
Triple Storms II

Three major storms struck the Northeast in December 1839, accompanied by powerful winds that caused great destruction in coastal sections, especially in New England.

A major storm crossed eastern Long Island on a northeasterly course on December 14–16, 1839, bringing a cold rain to eastern Pennsylvania that ended as a little snow on December 16–17. North of New York City, 20 inches of snow fell at New Haven, Connecticut, and 24 inches blanketed Westchester County, New York (Ludlum 1968).

A period of sunny and cold weather gave way to a disturbance that took a southerly route across North Carolina, passing west of Norfolk, Virginia, on December 22, 1839. The coastal track came close to Cape May, New Jersey, placing the heaviest swath of snow from northeastern Maryland to interior southern Pennsylvania.

The snow reached the city of Washington at 3:00 a.m. on December 22, 1839, accumulating to a depth of 10 inches, and 16 inches at Baltimore, mixed with sleet. Hazard's *Register* put the Philadelphia total at 10 inches. Peirce (1847) reported 18 to 20 inches of snow fell northwest of Washington and Philadelphia.

The *Gettysburg Sentinel* reported "at least 24 or 25 inches on the level," with much drifting (Ludlum 1968). At Chambersburg, a fall of two feet completely blocked roads in all directions. Farther north, the *Easton Sentinel* reported the "average depth of the snow to be two feet." Various press accounts mentioned accumulations of 18 inches at Lancaster and in Chester County, but only seven inches fell at New York City.

A third snowstorm struck the Northeast on December 27–28, 1839, but forged an interior path over Virginia, causing the snow to quickly change to rain by late on the 27th. However, New England was hit hard again with deep snow and high winds. Two feet of snow fell from Hartford, Connecticut, to Worcester, Massachusetts, and powerful gales were responsible for numerous shipwrecks along the New England coastline.

The largest reported fall occurred at Montgomery, New York, where 32 inches fell 55 miles north of Manhattan. Three feet of snow piled up in northwestern Connecticut in Litchfield County. The barometer dipped to 28.77 inches in Boston, indicative of the severity of the storm.

Winter of 1842–43
Another Long Winter

The *Stroudsburg Times* duly reached out to Henry Biesecker on January 18, 1918, for recollections of a few details about memorably harsh winters. He recalled that the 1842–43 season featured "sixteen weeks of uninterrupted sleighing," adding, "lumberers, engaged in logging, used their sleds for that entire period of time."

The winter of 1842–43 commenced early, with a heavy snowfall over the Upper Midwest and Ohio Valley on November 8–9, 1842, that remained on the ground until early April 1843 in places. A particularly violent snowstorm hit western New York on November 18–19, 1842, dropping two feet of snow. A few days later, the Pottsville *Miners' Journal* reported "quite a snowstorm" on November 23, 1842, and ice in the canals by the 24th.

A widespread snowstorm moved up the coast on November 29–30, 1842. The snow turned to rain at Washington, Philadelphia and New York City after an accumulation of several inches. Pottsville received nine inches. The mean temperature at Philadelphia in November 1842 was 38.6°F, nearly nine degrees below the modern normal and qualifying as the second coldest November in city history.

The weather stayed moderately cold in December 1842, preserving the snow cover in the interior. A heavy snowfall on December 28–29 seemed to lock in winter. However, a remarkable turnaround commenced during the second week of January 1843 that lasted several weeks and provided a genuine thaw.

Cold weather and frequent snow returned in February and March. A severe winter storm developed over the western Gulf of Mexico on March 15, 1843, leaving an extensive path of snow from the Louisiana coast to Canada (Ludlum 1968). Six to 12 inches blanketed the Southern states from Little Rock to Memphis and Nashville.

The snowstorm swept northward along the Eastern Seaboard on March 16-17, 1843, depositing a foot of snow on Washington and New York City. Peirce (1847) described the March nor'easter as "the most severe and violent snowstorm which had occurred for many long years, in the month of March."

At Philadelphia, the snow commenced around 2:00 p.m. on March 16 and continued through the following day, dropping 12 inches on the city. The Pottsville *Miner's Journal* reported 18 inches, followed by frigid readings of 5°F (17th) and 10°F (22nd). The Stroudsburg *Jeffersonian Republican* commented on March 22, 1843: "The snow, in many places, is still three or four feet deep" and "air as cold as December."

The *Miners' Journal* reported a two-inch snowfall on March 23, 1843, accompanied by conditions "resembling a complete hurricane." Another even larger snowstorm blanketed the Deep South on March 24–25, 1843. Low pressure charged up the East Coast on March 27–28, 1843. At Pottsville, the snow changed over to rain shortly before midnight on the 27th, though with "little melting." A few days later, the *Miners' Journal* lamented, "Four feet of snow in Schuylkill County on the first of April." Philadelphia recorded its coldest March in 1843 (30.0°F). Several inches of snow fell on April 4–5, 1843, and a little more fell on the 9th (Peirce 1847). Even more unusual, a rare mid-May snowfall in 1843 cloaked the Cumberland Mountains in eastern Kentucky and Tennessee.

Snow fell again in northern Ohio and Pennsylvania on June 1, 1843, which was followed by a widespread late freeze on June 2 (Ludlum 1968).

April 14-17, 1854
Easter Weekend Snowstorm

Two distinct low-pressure systems brought heavy snowfalls down to the coastal plain in the middle of April 1854, spanning Easter weekend.

The *Belvidere Intelligencer* reported falls of 18 to 24 inches in northern New Jersey. In the first wave, four inches fell at New York City on April 14–15, 1854, according to W. I. Clark, in the *New York Times* (May 12, 1904). Near Norristown, the "ground was white." From a Philadelphia dispatch: "A furious snowstorm commenced here this morning, and was still raging at sundown."

A second low-pressure area on April 16–17, 1854, caused the snow to pick up on Easter Sunday, April 16, adding another four inches at Norristown and eight inches at Pocopson in the southeastern part of the state. A little west of New York City, "some two feet of snow fell," noted Clark.

In the central mountains of Pennsylvania, the *Clearfield Republican* reported "a bed of snow some twelve or fourteen inches deep." The Pottsville *Miners' Journal* reported 12 inches, and 18 inches in the higher elevations of Carbon County. The Harrisburg *Daily Patriot* remembered a fall of 20 inches (April 20, 1875).

In the southern Pocono region, historian Luther Hoffman listed a snowfall of 18 inches in his records near East Stroudsburg. The local *Jeffersonian* reported 36 inches on the high Pocono Plateau. The *Easton Argus* spoke of drifts five feet deep "making the roads almost impassable."

January 1856
Two Deep Snowfalls

A massive snowstorm blanketed southeastern Pennsylvania and northern New Jersey on January 5–6, 1856. The storm began at Philadelphia on Saturday afternoon, January 5, and continued until 4:00 a.m. on the 6th. Nearly 15 inches of snow piled up, according to the *Public Ledger*, and "thousands of laborers" armed with "shovels and broom" were called to clear the sidewalks. Accumulations reached 20 inches at Newark, New Jersey, and 18 inches at Boston, but with relatively light falls farther inland.

Another heavy snowfall followed a frigid blast of air. The *Tioga County Agitator* in Wellsborough (now Wellsboro) reported a snowfall of 20 inches, which raised the overall depth to nearly 30 inches. Another significant snowfall occurred on January 26–27, 1856.

January 18–19, 1857
Great Cold Storm

The Cold Storm of January 1857 stands out for a combination of windswept snow, driving winds and near-zero temperatures at unusually southern latitudes.

A frigid mass of arctic air had settled over the Northeast and Middle Atlantic states on January 17–18, 1857, setting the stage for a great clash of arctic air advancing southward along the spine of the Appalachians. Low pressure developed on the arctic boundary in the Gulf of Mexico and proceeded quickly northward along the Atlantic Coast, spreading heavy snow from the mountains of northern Georgia to the interior Carolinas late on January 17, 1857. Eight inches fell at Athens, Georgia.

The snow arrived in southern Virginia shortly after midnight on January 18, 1857. Accumulations over the interior exceeded 12 inches. A Raleigh, North Carolina, dispatch in the *Washington Post* (February 14, 1899), recounted a 24-inch snowfall in the January 1857 storm, which exceeds the current record of 20.3 inches on January 24–25, 2000.

Heavy snow broke out over Pennsylvania by early afternoon on January 18, 1857. The snow accumulated rapidly. The Philadelphia *Evening Bulletin* reported 18 inches at Washington, D.C., and up to 24 inches near Philadelphia. At Baltimore, the depth was reportedly "on the level about two feet," piled into drifts six to 10 feet high.

Professor Michael Jacobs at Gettysburg had a reading of 0°F at 9:00 a.m. as snow commenced to fall. The temperature continued to drift downward, reaching -3°F at nightfall. Frigid northeast winds piled the estimated nine-inch snowfall into large drifts.

The Smithsonian site observer at Norristown, northwest of Philadelphia, estimated a little more than 12 inches of snow had accumulated, probably a conservative figure based on 1.20 inches of melted precipitation, adding, "The snow lies in drifts from 6 to 12 feet."

Up at Lafayette College in Easton, Professor James Coffin measured 17 inches in 23 hours, commencing at 3:00 p.m. on January 18, 1857, and ending around 2:00 p.m. on the 19th as the thermometer hovered slightly above 0°F. The Smithsonian observer at Berwick measured 17 inches in the Susquehanna Valley.

On December 29, 1887, the keen editor of the Stroudsburg *Jeffersonian* combed the archives to give a more precise fall—then deemed to be 18 inches. The *Warren Journal* in Belvidere, New Jersey, reported drifts up to 10 feet. A poetic postscript to the storm appeared in the *Belvidere Intelligencer*: "The very sunbeams froze, or whatever of them found their way through the clouds. The wind blew furiously, driving the snow through every nook and cranny."

Winter of 1867–68
Frequent Snows, Cold Waves

The prolonged snowy winter of 1867–68 brought frequent snows that intensified the cold. A heavy snowfall on January 20–21 in the eastern half of the state dropped 12 inches at Dyberry. Northern Tier amounts in Bradford County were as high as 18 inches (Heverly 1926), though incorrectly listed as January 17. The persistent snow only reinforced entrenched arctic air.

The snowfall in January 1868 at Dyberry of 31.6 inches hardly melted. In the Lehigh Valley, North Whitehall accumulated 17.6 inches, 10 miles northwest of Allentown. On January 30, the editor of the Stroudsburg *Jeffersonian* commented, "The ground has been covered with snow, and the sleighing in this neighborhood has been excellent since the 12th of December."

An early March storm deposited up to 15 inches of snow at Blooming Grove on March 1–2, 1868. The Stroudsburg *Jeffersonian* said, "The wind blew like a regular hurricane" during the "severest snowstorm of the season."

A thaw during the second week of March ended with a second massive nor'easter on March 20–21, 1868. In Chester County, the Pocopson observer measured 20 inches. Notices in the *New York Times* reported that south of Philadelphia the snowfall increased to 24 inches. The greatest catch was 32 inches in northern Delaware, the bulk of which fell in 16 hours, accompanied by drifts 15 feet high, noted in the *Philadelphia Evening Traveler*, which exceeds the modern state mark of 29.3 inches at Greenwood 2.9 SE (February 5–6, 2010).

The winter of 1867–68 finished strong and "lingered into the lap of spring." Heavy snow fell on April 4–5, 7–8, 9–10 and 12, 1868. A surprise late-season winter storm deposited three to six inches on April 25 in the eastern highlands, north of Blue Mountain. The high elevations added a dusting on May 8, 1868.

How much snow fell in 1867–68 in eastern Pennsylvania? The *Doylestown Intelligencer*, north of Philadelphia, gave a total snowfall of 94 inches, exceeding all modern winters. The North Whitehall weather site received 78.2 inches in the winter of 1867–68, padded by a record April snowfall contribution (17.3 inches). The only comparable snowy winters at the North Whitehall Township location produced 73 inches in 1859–60 and 77.3 inches in 1862–63.

Dyberry observer Theodore Day measured 115.4 inches in the long winter of 1867–68, including 25.8 inches in April 1868—the greatest April total in the 38-year data span (1865–1903). The snowy season at Central Park in New York City (88.3 inches) in 1867–68 was not even approached until 1995–96 (75.6 inches). Trenton (75 inches) and Newark (75.3 inches) also had historic snow totals.

December 25–27, 1872
Post-Christmas Blizzard

A severe arctic outbreak gripped the Northeast several days before Christmas 1872. Temperatures tumbled far below zero on Christmas morning. Meanwhile, a mighty winter storm was gathering strength along the southeastern Atlantic coast.

At Columbia, South Carolina, a dispatch in the *New York Times* called it "the heaviest storm of snow and sleet ever known in this section." The canopy of snow spread rapidly northward, reaching Washington, D.C., around noon on Christmas Day.

The snow began at Philadelphia a little after sunset and came down furiously within hours, described in the Philadelphia *Public Ledger*:

> The most terrific snowstorm prevail here today that we have had for many years. The snow is two feet deep and is still falling rapidly. The wind blows a fearful gale, drifting so as to render travel almost impossible. The mercury is 6 degrees above zero.

The snow depth at Philadelphia reached 10 inches early on December 27, 1872. The western regions received more, where a Chester County Smithsonian observer measured 15 inches. The observer at Egypt, northwest of Allentown, recorded a hefty 22 inches. Amounts climbed to the north. The *Easton Express* reported a snow depth of 30 inches on December 27, 1872, possibly including what remained of a heavy snowfall of a little more than six inches on December 18–19, 1872. The 30-inch depth exceeds modern Lehigh Valley records.

The Stroudsburg *Jeffersonian* reported "a little over two feet" of snow on the ground at the end of the storm on December 27, 1872, then later clarified in an article on December 29, 1887, that the actual snowfall was 18 inches. There apparently was a sharp cut-off to the north, because only 10 inches fell at Blooming Grove in eastern Pike County. Egypt had 32.2 inches of snow in December 1872, the heaviest on record in the Lehigh Valley in December.

Across New Jersey, observer P. V. Spader in New Brunswick measured 20 inches between 2:00 a.m. on December 26 and 7:00 p.m. on the 27th. The New York City official total was 18 inches at Central Park. However, the King's College in Brooklyn recorded 27 inches.

Bitter cold waves followed in late December 1872 and recurred in January 1873. The nadir occurred on January 29, 1873, with some Pennsylvania stations reporting lows below -30°F.

The long cold, snowy winter of 1872–73 lingered well into the lap of spring. Light to moderate snows continued during the second week of March, followed by a heavy fall at Dyberry on March 21–22 (10 inches). Snow came again on March 24–25, prompting the mercury to plunge to a late-season frigid mark of -5°F on the 28th at Dyberry in the Poconos.

Longtime Dyberry observer Theodore Day remarked on March 31, 1873: "Snow from 1 to 3 feet." Heverly (1926) wrote the snow was "so deep in the woods that sugar-makers were unable to gather sap."

The snowy regime continued through much of April 1873. Heverly (1926) wrote that snow "fell to the depth of two feet" in Bradford County on April 21–22. The northeastern highlands were coated on the 25th, with valley flakes. The backward spring pattern brought a final one-inch snowfall at Dyberry on May 3, 1873.

At the conclusion of the long winter of 1872–73, the observer at Egypt had tallied 86.4 inches, which exceeds all modern season snowfall records in the Lehigh Valley.

April 28–29, 1874
Latest Major Snowstorm

The first in many blasts of winter in the most backward April in state history arrived in the Northeast on April 3, 1874, kicking off the coldest April on record.

In the northern Poconos, the mercury at Blooming Grove tumbled to 31°F at 9:00 p.m. on April 3, 1874, and plunged to 22°F the next morning, holding steady in the mid-20s all day on the 4th. A light snowfall (two inches) was followed by a reinforcing surge of arctic air, as the mercury slid to 17°F at 9:00 p.m.

On April 5, 1874, a wintry blast came on Easter that featured subfreezing conditions throughout the Northeast and Mid-Atlantic. The readings at Blooming Grove ranged from 18°F at 7:00 a.m. to only 32°F at 2:00 p.m. Baltimore recorded a maximum of 38°F—the coldest Easter on record.

A few days later, a rare April snowstorm struck Memphis, Tennessee, that commenced at 11:00 a.m. on April 9, 1874, and continued for 12 hours, "astounding the oldest inhabitants of this latitude," according to the *New York Times*. The swath of heavy snow extended northward to Louisville and Cleveland.

In northern Pennsylvania, rain changed to snow on April 9–10, 1874, with falls of three inches at Blooming Grove, six inches at Scranton, more than six inches at Erie, and totals in excess of 12 inches in the mountains. The *Jeffersonian* mentioned a fall of 14 inches at Binghamton, New York. Snow squalls continued on the 11th. On April 12, the daybreak temperature at Egypt, northwest of Allentown, plunged to 19°F and only rose to 36°F at 2:00 p.m.

Another wintry blast on April 16–17, 1874, left two inches of snow at Dyberry. The thermometer at Blooming Grove registered just freezing (32°F) on the afternoon of the 17th.

The next in a remarkable series of late snows came on April 25, 1874. A mixture of rain and snow changed to all snow north of the Lehigh Valley, with a mixture at New York City. The snowfall at East Stroudsburg totaled 4 inches (Luther Hoffman) and 6 inches at Blooming Grove. At Harrisburg, the *Patriot* reported that it "commenced snowing vigorously in the morning" and changed to rain, with several inches coating the ridges.

The big event came on April 28–29, 1874. Rain overspread southern Pennsylvania during the morning of April 28, turning to sleet and snow north of Philadelphia. Accumulations ranged from three inches at Baltimore and Harrisburg to six inches at Reading and 15 inches at Hazleton, according to the Harrisburg *Patriot*.

The *Williamsport Sun* reported 14 inches at Buttonwood in Lycoming County.

The unpublished records of Luther Hoffman north of East Stroudsburg recounted a fall of 12 inches, which was also the total at Blooming Grove. The accumulation lessened to six inches north of Honesdale at Dyberry.

A notice in the *Stroudsburg Daily Times* on May 13, 1907, recalled the historic snowfall: "In the latter part of April [1874], sleighing was very good, the snow being several feet deep." The snow stayed on the ground until May 8, 1874, in sheltered areas. Press citations in the *Philadelphia Inquirer*, *New York Times* and *Daily Tribune* mentioned falls of 15 inches at Wilkes-Barre and Mauch Chunk (now Jim Thorpe) and 18 inches in the higher elevations.

Farther south the rain did not change to snow until the early morning of April 29, 1874. From the nation's capital to Philadelphia and New York City, a burst of heavy snow was accompanied by thunder and lightning, indicative of a dynamic system. The *Philadelphia Inquirer* stated: "Snow fell here this morning [April 29] for three or four hours, covering the ground." The April 29, 1874, coating at Philadelphia surpasses the city's latest modern measurable snow (0.1 inch on April 27, 1967).

New York City weather historian Stephen Fybish uncovered the official Central Park observations taken during the April 1874 storm (Barron 2010): "Snowing hard and one inch of slush on the ground," adding, "Bleakest of mornings, with great patches of snow on roofs and sidewalks." The event is the latest measurable snowfall in city data. The New York *Daily Tribune* noted "a light fall of snow at Richmond, Virginia," and historic late accumulations at sites around Baltimore (four inches) and Washington (two inches).

A widespread hard freeze followed on April 30, 1874, which ensured that April 1874 would be the coldest of all time at Philadelphia (42.4°F) since 1825.

March and April 1875 Snowy Onslaught

The cold and snowy winter of 1874–75 took a severe turn in the first week of March 1875, when 33 inches of snow fell at Blooming Grove in Pike County. Snow fell generally across the state on March 1, 3 and 5, leading up to a huge snowstorm.

The biggest blow of the winter arrived on March 7–8, 1875, when a severe coastal storm blanketed much of Pennsylvania under 10 to 20 inches of snow, raising snow depths to three feet in the northeastern highlands.

The *New York Times* and *Philadelphia Inquirer* reported deep falls in the interior at Altoona (25 inches), Sunbury (22 inches), Wilkes-Barre (20 inches) and Williamsport (15 inches), tapering off to less than 10 inches in the southeast. The diary of Luther Hoffman noted 18 inches. The *Jeffersonian* stated that 15 inches fell in Stroudsburg, and 18 to 20 inches fell in outlying areas to the northwest across the Poconos. The press reported a "superabundance" of snow—six feet deep on Main Street. The snowfall tapered off to seven inches at New Castle in the northwest.

At the end of March 1875, Wayne County observer Theodore Day noted in his records: "Sleighing from January 8 to March 31, 1875. … Snow 2 ½ to 3 feet deep in the woods." Ice jams developed as the heavy snowpack began to melt, resulting in serious flooding. However, a blast of arctic air halted the runoff, accompanied by a week of subfreezing weather in the north from March 17–25, 1875.

A late-season coastal storm swept the northern Mid-Atlantic on April 12–13, 1875, depositing six to 10 inches of snow across much of the state. The largest observer totals (12 inches) were reported at East Stroudsburg (Luther Hoffman), and at Dyberry and Blooming Grove. The Harrisburg *Daily Patriot* called it a "furious storm" and noted falls up to 12 inches between Harrisburg and Milton in the Susquehanna Valley.

A frigid blast of air on April 16, 1875, sent temperatures plummeting to historic April lows, accompanied by some accumulating snow over the interior of Pennsylvania. Another improbable snowstorm on April 24–25, 1875, brought 0.8 inch in Cameron County and 3 inches at Dyberry. New York City had its latest measurable snowfall (3 inches). Even that snow was not the end of the long winter. A coating fell in the northern Poconos on May 4, with 2 inches at Blooming Grove.

The season snowfall totaled 100 inches at Blooming Grove and 98 inches at Dyberry. The Smithsonian observer at Egypt measured 79.5 inches, comparable to totals observed in 1867–68 in the Lehigh Valley, exceeding all modern winters. A Minersville resident in Schuylkill County tallied 105 inches (*Jeffersonian*, April 29, 1875).

January 8–9, 1884
Big Pittsburgh Snowstorm

On January 8, 1884, snow began falling in Pittsburgh around 1:00 a.m. at the rate of one to two inches per hour, as measured by the Signal Service office. By the time the storm ended around 12:30 a.m. on the 9th, the official accumulation reached 18 inches. Outlying districts reported as much as two feet of heavy wet snow.

The *Pittsburgh Commercial Gazette* called the January 1884 storm the "greatest on record in this section." At Titusville, the storm brought the "heaviest fall of snow in a single day ever known here." Streetcars were halted in their tracks, and wagons and sleds did not fare any better.

In eastern Pennsylvania, the snow changed to rain early on January 9, 1884. The central and northern mountains received a heavy snowfall, mixed with sleet and freezing rain. The central pressure deepened below 29 inches near the coast, drawing mild Atlantic air inland that resulted in a soaking rain at Philadelphia.

High winds lashed eastern Pennsylvania through the day and night of January 8–9, 1884. A notice in the *Philadelphia Inquirer* from Shenandoah relayed: "Telegraph and telephone wires were blown down, colliery stacks and breaker roofs destroyed, and plate-glass windows in this place and Shamokin were blown in."

The January 1884 blizzard extended far and wide, with reports of heavy snow from Atlanta, Georgia, to Harrisonburg, Virginia. Dispatches in the *New York Times* from Wheeling, West Virginia (20 inches), Richmond, Kentucky (14 inches), and Lockport, New York (36 inches), reported prolific snow.

January 8–9, 1886
Another Big January Snowfall

Two years later to the date on January 8–9, 1886, another disruptive snowstorm set in at Pittsburgh about 5:00 p.m. By noon the next day, 13.5 inches of snow had fallen. The Brookville measurement in Jefferson County was 18 inches. The barometer reading plunged to 28.72 inches at New York City.

In eastern Pennsylvania, a Tamaqua dispatch in the *Jeffersonian* declared the storm "one of the strongest blizzards that ever visited the region." A notice from Reading reported 20 inches of snow, commencing at 9:00 p.m. on January 8, until the snow tapered off at noon on the 9th, as the low center crossed central Long Island.

Press reports from around the state listed the large snowfalls: 24 inches at Mount Carmel and Columbia; 20 inches southwest of Pittsburgh, the heaviest fall since January 14–16, 1831; 14 inches at Chambersburg. More than 1,000 workers were called in to clear the Pennsylvania Railroad tracks.

March 11–14, 1888
Blizzard of 1888

A sudden intense snowstorm swept the Great Plains on January 12, 1888, a blizzard accompanied by rapidly falling temperatures. The "Schoolchildren's Blizzard of '88" claimed more than 200 lives, including students and teachers in the Dakota Territory who were unprepared for the fury of wind and extreme cold as the day wore on.

Two months later to the day, few could have anticipated that a Western-style blizzard would strike the Northeast on the eve of spring. The "White Hurricane" that engulfed the northern Mid-Atlantic and Northeast had no precedent at the time.

Weather historian David Ludlum, in his *New Jersey Weather Book*, declared: "Nothing in the folklore traditions of colonial America or the formally recorded meteorological history of the northeastern United States equals the storm that enveloped the area from Virginia to Maine on March 11–14, 1888."

Maximum wind gusts are reduced to account for the employment of the more accurate three-cup anemometers used by the U.S. Weather Bureau beginning January 1, 1928. In a *History of Weather Bureau Wind Measurements* (1963), an appendix presented a table with reduced maximum velocities at government weather stations by a factor of 22.5 percent (50 mph) to 27.7 percent (75 mph). Regardless, wind gusts reached hurricane force near the coast on March 12, 1888, in the midst of the blizzard.

The weather map on Saturday, March 10, 1888, revealed an innocuous-looking low-pressure trough drifting southeast across the Great Lakes. In the next 24 hours, a secondary wave developed on the southern end of the trough over eastern Georgia, which was evident on the national daily weather maps published for March 11.

Now there were two distinct low-pressure areas of equal intensity on the weather map—one near Wilmington, North Carolina (29.68 inches), and the other north of Lake Ontario (29.70 inches). As energy translated eastward, the coastal disturbance became the main center, drawing bitterly cold arctic air into the circulation. The clash of air masses caused the surface pressure to deepen rapidly, accounting for the violent winds that blew the rain sideways in the coastal plain.

At Washington, D.C., rain changed to heavy wet snow around 3:00 p.m. on March 11, 1888, totaling six inches by midnight. The changeover to snow at Philadelphia occurred around 11:15 p.m. and on March 12 at 12:10 a.m. at New York City.

Rapidly falling temperatures and increasing wind brought down electric and telegraphic wires, putting cities in the storm's path in the dark, save for the relatively few gas streetlights. Although the heavy snow began to taper off around daybreak on

March 12, 1888, swirling gale-force winds continued piling the snow into ever-deepening drifts.

Urban centers from the nation's capital to Philadelphia and New York City were virtually cut off from all communications. The U.S. Army Signal Service observer in Philadelphia called the storm "the most disastrous that has ever visited this locality." Twenty-seven of 40 vessels in the harbor along the Delaware River, near the mouth of the bay, were damaged, and total marine losses were put at $500,000, according to the *Monthly Weather Review*.

The snowfalls at Washington (10 inches) and Philadelphia (10.5 inches) were substantial, but comparatively light relative to areas to the northeast. A northerly wind blew furiously for several days, with powerful gusts making accurate snowfall measurements nearly impossible.

The Stroudsburg *Jeffersonian* offered a helpful storm timeline:

Between 9 and 10 o'clock [March 11] it began a very gentle drizzle, and this kept on until 2 o'clock in the afternoon when the rain had turned into sleet, and continued until about 5 o'clock when it began to snow. On Monday [March 12] we had a full twelve inches of snow. ... The wind from Monday morning until Tuesday morning blowing great guns.

The blizzard reached its peak intensity on March 12, 1888, off the tip of southern New Jersey in the morning while churning slowly north. On March 13, the center of low pressure dipped to 28.92 inches (979 millibars) on the U.S. Signal Service Daily Weather Map at 3:00 p.m. near Block Island, Rhode Island, after performing a looping motion east of New York City. A blocking ridge of high pressure caused the system to stall for many hours off the southeastern tip of New England, whipping up fierce winds.

Nearly all telephone, telegraph and electric wires were toppled in the great storm. The large pressure difference brought relentless hurricane-force gusts, burying the region in mountainous drifts as high as the second stories of buildings in New York City and the countryside. The highest observed drift was 40 feet at Bangall, Dutchess County, New York. The New Jersey Historical Society described a derailment of a Sussex River Railroad passenger train that required two days for rescue crews to reach the survivors.

Snowfall totals across eastern Pennsylvania ranged from 11 to 31 inches at the official government weather sites. The greatest total was 31 inches at Blooming Grove in Pike County. In Monroe County, Luther Hoffman recorded in his notes a fall of 18 inches at East Stroudsburg. Theodore Day also measured 18 inches at Dyberry in Wayne County.

The temperature plummeted through the teens and single digits overnight on March 12–13, 1888, dipping to 3°F at Stroudsburg and 6°F at New York City, where the high only reached 12°F on March 13. Over the interior, daytime temperatures hovered in the single digits, the coldest weather so late in the season.

The NOAA *Climate Narrative* published for the Wilkes-Barre/Scranton Airport listed a snowfall of 15 inches, based on early city records, coupled with huge drifts as winds gusted to 65 mph (53 mph adjusted).

The *Bethlehem Daily Times* detailed the blizzard's impacts on the Lehigh Valley: "The blizzard was terrific for nearly 48 hours, not abating until about 9 o'clock last evening [March 13] and the temperature still hovering about freezing point this morning, and snow falling lightly." Snow drifts were reportedly up to 20 feet deep. The peak wind speed at Philadelphia reached 66 mph (54 mph adjusted).

The *Pennsylvania State Weather Service Bulletin* summarized the great blizzard:

> Telegraph wires and poles were soon prostrated, railway trains were blockaded by huge snow banks, and all communications by wire or rail was completely cut off for a few days, during which time the high winds kept the snow a moving mass.

Western Pennsylvania was largely spared from the storm's fury because the storm tracked well offshore and was fairly compact. The snowfall at Pittsburgh in March 1888 amounted to only two inches.

At the conclusion of the Blizzard of '88, spectacular snow totals were measured over southwestern New England, eastern New York, and northern New Jersey. The greatest totals: 25 inches at Rahway, New Jersey; 20.9 inches at New York City; an estimated 36 inches in parts of Brooklyn, Queens and at Hartford, Connecticut; 44.7 inches at New Haven; 46.7 inches at Albany; 55 inches at Troy, and 58 inches at Saratoga, New York.

Communications were lost from Virginia to Maine, as snowbound communities struggled to cope with huge drifts and downed electrical wires and telegraph poles. Dozens of trains were snowbound between New York City and Albany and in Connecticut.

The death toll from the Blizzard of '88 was estimated at around 400 persons, including 200 in the New York City area. Many lives were lost attempting to travel to or from work during the height of the storm, and victims were found in the deep snowbanks, where they had perished after wandering exhausted in the bitter and numbing cold.

December 1890
Deep Interior Snowfalls

Several big snowfalls made weather history in an otherwise mild winter in 1890-91 at the beginning at end of the season, like bookends.

A huge snowfall buried extreme western and northern Pennsylvania on December 16–18, depositing 25.9 inches at Pittsburgh, with the lion's share of snow falling on the 17th (22 inches). The *Pittsburgh Commercial Gazette* called the storm "the most destructive of the kind that has ever been experienced in this city," taking a heavy toll on "operations of railroad, telegraph, and street-car companies."

The *Salem Daily News* in Ohio carried dispatches from Bellefonte and Lock Haven in the central mountains experiencing "two feet" of snow in a 36-hour snowstorm. The heaviest falls reported to the State Weather Service included 32 inches at Troy, 25.5 inches at Blue Knob, and 21.8 inches at Lock Haven. Press dispatches indicated a general canopy of heavy snow ranging from 18 inches at Wilkes-Barre to more than two feet at Bradford. The Dyberry observer noted 12 inches.

A second big blow came on Christmas night, leaving another 10 to 15 inches on the ground from the nation's capital northward through Pennsylvania on December 25–27, 1890. Blue Knob (elevation 2,500 feet), 20 miles southwest of Altoona, picked up an additional 26.5 inches, which raised the December snowfall to 86 inches.

The Pittsburgh Weather Bureau measured 13.2 inches in the post-Christmas storm, plumping up the record December snowfall to 41.3 inches. In the northeast, Dyberry received 15 inches. New York City had a substantial fall of 14 inches.

The winter of 1890-91 ended with another extensive major snowfall on March 27–29, 1891. Large falls occurred from Baltimore (9.5 inches) north to York (14 inches), Harrisburg (18.1 inches) and Brookville (17 inches), and east through the northern Poconos at Dyberry (14 inches).

The *Washington Post* reported tremendous falls to the west on March 27–28, including "two and a half feet" at Winchester, Virginia, which fell in 41 hours (*Philadelphia Inquirer*) and 30 inches at Martinsburg, West Virginia. The *Inquirer* also reported impressive totals in southern Pennsylvania at Huntingdon (28 inches), Lancaster (19 inches), and Norristown and Collegeville (12 inches). The *New York Times* listed substantial snowfalls of 15 inches at Reading and 20 inches at Charleston, West Virginia.

April 10–12, 1894
Great April Snowstorm

A moderate snowfall on April 6–7, 1894, left a coating of three to six inches in northern parts of Pennsylvania. The Centre Hall *Centre Reporter* noted six inches. Less than a week later, one of the state's greatest spring snowstorms struck on April 10–12, bringing rain and sleet that changed to snow.

Snowfall totals in southeastern Pennsylvania were generally four to seven inches, except for 1.2 inches at Philadelphia, where mostly mixed precipitation occurred. Farther inland, historic April snowfalls mounted from east to west: 17 inches at Lancaster; 18 inches at Harrisburg and State College; 18.5 inches at Carlisle; 25.9 inches at Lock Haven; 29.2 inches at Coatesville.

In the northeast, the snowfall at Salem Corners of 31.5 inches in Wayne County counted toward a remarkable April total of 38.2 inches—the second greatest April figure in state history (Somerset recorded 38.5 inches in April 1928). The western Poconos picked up "about 19 inches," noted the Stroudsburg *Jeffersonian*. On April 7, 1898, the newspaper recalled that 12 inches fell at Stroudsburg, in the upper Delaware Valley, in April 1894.

Press reports confirmed widespread snow totals in the central mountains of two to three feet that caused significant damage. The *Philadelphia Inquirer* carried notices of stunning falls as high as 36 inches near Bellefonte, 27 inches at Williamsport, and 24 inches at Danville.

February 11–14, 1899
Great Eastern Blizzard

The month of February 1899 would be remembered for heavy snowfalls and a frigid arctic outbreak of historic proportions.

Three distinct snowfalls were associated with waves of low pressure tracking along a frontal boundary that extended from eastern North Carolina to the western Gulf Coast. The first in a series of low-pressure waves developed over South Carolina on February 5, 1899. Up to six inches of snow blanketed eastern Pennsylvania (Kocin 1988). A second storm formed over Alabama on February 6, but remained farther offshore, with a lighter snowfall. A third system off the Georgia coast tracked northward on February 7–8, depositing six to 10 inches of snow from the Mid-Atlantic to eastern New England.

In the wake of the trio of storms, an extraordinary cold air mass descended on the Plains and the Midwest, where readings plunged as low as -30°F to -40°F.

The brutally cold air advanced into Pennsylvania on February 8–9, 1899, sending the temperature tumbling to historic lows. The meteorological setting featured a vast high-pressure system (31.15 inches) over the Great Plains and a new area of low pressure forming over the Gulf of Mexico along the arctic boundary on February 10. Absolute minimum readings in Pennsylvania ranged from -6°F at Philadelphia to -39°F at Lawrenceburg.

A lead surface disturbance near the coast of northern Florida spread a mixture of snow, sleet and rain north across the Carolinas. The main system cranked up in the northern Gulf of Mexico on February 12, 1899, and rumbled up the coast, reaching the Virginia Capes early on February 13. Snowfall rates reached maximum intensity in the afternoon and evening, accompanied by sustained winds of 35 mph and gusts near 50 mph, creating blizzard conditions coupled with brutally low single-digit temperatures.

Raleigh, North Carolina, received 18 inches of snow—the greatest fall since 24 inches fell on January 18–19, 1857, according to the *Washington Post*. Massive totals of 30 inches were reported at Keedysville, Charlotte Hall and Easton on the Eastern Shore of Maryland.

The blizzard was crowned the "Snow King" by the *Washington Post*. Washington (20.5 inches) and Baltimore (21.4 inches) scored big falls. The nation's capital established an all-time snow depth of 34.2 inches, including what remained of the 14 inches that fell from February 4–8, 1899. Baltimore had a depth of 30 inches at the conclusion of the storm, which was not exceeded until February 2010.

The Great Eastern Blizzard of February 1899 was exceptional for having the greatest longitudinal snowfall along the Eastern Seaboard. A trace of snow was observed at Fort Myers, Florida, the most southerly occurrence in the United States until January 1977 when snowflakes fell near Miami Beach. In the grip of a vast arctic air mass, Tampa had 0.1 inch, and 1.9 inches fell at Jacksonville. Charleston, South Carolina, recorded 3.9 inches of snow, the second heaviest of record.

Milford, Delaware, reported 24 inches, and in southern New Jersey, the Cape May fall was 34 inches, with a depth of 40 inches. The greatest accumulation in New Jersey was 36 inches at Little Egg Harbor, setting a state record. On the northern fringe of the massive winter storm, the snowfall at New York City totaled 14.6 inches (total snow depth of 23 inches) and Boston received 16 inches in the storm.

Eastern Pennsylvania felt the brunt of the blizzard. Philadelphia tallied 18.9 inches and a cumulative depth of 26 inches (tying February 13, 1885), a city record that stood until January 1996 (27.6 inches). The Kennett Square observer, 15 miles west of Philadelphia, reported 32 inches and York had 29.5 inches; both locations reported a snow depth of 36 inches. The heaviest snowfall monthly totals would be turned in at York (47.1 inches) and Hamlinton (46.2 inches).

The published Coatesville measurement of 35 inches on February 13, 1899, and the storm total surpassing 50 inches appear implausible, based on nearby sites. The melted liquid equivalent storm figure topping five inches should be disregarded.

The *Philadelphia Inquirer* chronicled the misery brought on by the Great Eastern Blizzard that left a total of 30.3 inches of snow between February 5 and 14, 1899:

There is no way of adequately describing the storm which raged in and through and over and around Philadelphia yesterday, and drove to the woods that class of 'oldest inhabitants' who are fond of indulging in the reminiscences of winters we used to have. Not since the city was a baby has such an avalanche of congealed moisture descended upon it.

The snowfall at Lafayette College in Easton was estimated near 16.5 inches. Reports farther north ranged from 10 inches at Wilkes-Barre to 18 inches at Dyberry. A little to the southeast, 24 inches fell at Hawley. The *Stroudsburg Daily Times* reported 25 inches, with drifts as high as 18 feet whipped up by winds above 35 mph. The Stroudsburg thermometer reading on February 12, 1899, during the height of the storm, registered only slightly above 0°F at noon.

April 1901
Two Big April Snowstorms

Two record-worthy snowfalls occurred in April 1901 in the western part of the state. The first storm on April 2–3, 1901, established an April single snowfall record at Pittsburgh, with 13.5 inches of snow, with higher totals in the Laurel Highlands. State College had 11.3 inches and Brookville had 10.9 inches. In the northeastern highlands, the *Stroudsburg Times* reported 5 inches at Canadensis.

A massive storm struck western Pennsylvania and eastern Ohio on April 19–21, 1901, bringing historic snowfalls as a slow-moving storm crossed the southern Appalachians, reaching the North Carolina coast on April 21. The combination of rain, sleet and snow caused significant flooding along the tributaries of the Ohio River. More than three inches of precipitation fell in some areas.

The U.S. Department of Agriculture's *Climate and Crop Service* report for Pennsylvania in April 1901 noted, "Damages to buildings, telephone and telegraph lines, and railroads were immense," and total losses were estimated at more than $2 million. At Pittsburgh, "miles of wires were down, and business by telephone and telegraph was greatly hampered."

Huge snowfalls occurred in northwestern Pennsylvania. The weather station at Saegertown in Crawford County had a massive 26-inch accumulation. The *Philadelphia Inquirer* carried a dispatch from Erie stating that "a blizzard has been raging in this section," leaving "two feet on the level," with up to three feet in the outlying areas. The snowfall at Meadville was estimated at 18 inches, and 10 inches fell at Waterford. Buildings collapsed at Corry under the weight of the heavy wet snow.

In eastern Ohio, a state record of 42 inches at Gratiot and Canton exceeded all storms outside the snowbelt. Warren and Niles noted 35 inches. Hundreds of buildings collapsed, and travel and communications were disrupted for three days. Farmers suffered a major setback from the cold and damp soil, which delayed seed germination and forced many to replant during a "backward season for crops" (Schmidlin 1996).

February 1902
Back-to-Back Snowstorms

Two heavy snowfalls hit Pennsylvania in less than a week in February 1902. The first storm lashed the eastern counties on February 16–17, dumping

11.4 inches at Philadelphia and 18 inches at Atlantic City, New Jersey. Amounts tapered off to eight inches at Dyberry. Little snow fell in the west. High winds piled the snow into drifts up to 15 feet at Pen Argyl. Winds gusted to nearly 50 mph (adjusted) at New York City.

A second major storm on February 21–23, 1902, started as heavy snow but soon turned into freezing rain. Philadelphia had three inches of snow and sleet, followed by many hours of rain. The storm was described in *Climate and Crops: Pennsylvania Section* as "the most destructive to telegraph and telephone lines ever experienced."

Broken tree limbs encased in ice littered city streets and brought down electrical lines throughout the city and neighboring communities in southeastern Pennsylvania. Communications were severely disrupted and transportation was paralyzed for several days after the storm.

Dyberry received 11 inches, and the interior was buried under up to two feet of snow. Williamsport received 23 inches in 24 hours, and a dispatch in the *Philadelphia Inquirer* noted a total depth of up to 29 inches. (The modern record for a February storm is 22.8 inches on February 19-20, 1972. The greatest 24-hour snowfall for any month is 24.1 inches on January 12-13, 1964.)

The *Stroudsburg Daily Times* reported, "The snow lays between two and three feet deep on the Pocono." The snow depth at Tamaqua was estimated to be "about two feet" on February 23, 1902, with light snow still falling. A press notice from Altoona stated, "All Friday night [February 21–22] the white flakes dropped in great profusion. The railroad and telegraph companies are the most seriously crippled."

A thaw several days after the snowy onslaught resulted in damaging floods in eastern Pennsylvania, which were made much worse by the breakup of ice and snow, releasing a large volume of pent-up water all at one time.

Yet another major snowstorm blasted Pennsylvania on March 5–6, 1902, bringing an additional 12 to 16 inches to the northern and western sections. The top figure was 17.7 inches at State College, and Williamsport collected 14.8 inches.

January 24–25, 1905
Massive January Snow in East

The winter of 1904–05 was exceptionally cold and snowy in Pennsylvania. Scranton received 88.6 inches of snow, establishing a historic season record.

A heavy early snow arrived November 13–14, 1904, totaling six to 12 inches in eastern Pennsylvania, which gave way to a frigid December with frequent snows.

The first of two heavy January snowfalls struck the Northeast on January 3–4, 1905, depositing another six to 12 inches, accompanied by strong winds. A few weeks later, a full-fledged blizzard on January 24–25, 1905, dumped one to two feet of snow.

At Milford, in eastern Pike County, the longtime observer Alla K. Doughty wrote down 26.6 inches, adding in her detailed report: "The woods were so blocked with snow drifts ... that no mail was received for two days in the county." A *New York Times* entry mentioned that 20 inches fell at Stroudsburg. The Dover, New Jersey, observer recorded 26 inches.

The January 1905 total at Milford of 43 inches was remarkable for a valley site.

December 25–27, 1909
Post-Christmas Blizzard II

Another great post-Christmas blizzard swept the Middle Atlantic region in December 1909 that was similar to the 1872 storm, but the deepest snow fell farther south.

The *Monthly Weather Review* described a "weak cyclonic disturbance" in southern Arizona that was located in the Texas Panhandle early on December 24, 1909, and "extensive precipitation." In the next 24 hours, the low traveled some "1,200 miles, at the rate of 40 miles per hour," as a new storm formed along the Mid-Atlantic Coast.

An event summary in *Climatological Service of the Weather Bureau: District Number 1* (December 1909) reported a low barometer reading of 28.57 inches at Cape May, New Jersey, on Christmas morning. The historically low pressure indicated rapid deepening from an injection of warm, moist Atlantic air.

When light snow began falling shortly after 9:00 a.m., there was palpable excitement. As the day wore on, Philadelphia residents knew they were in for a big storm, though the evening snowfall had only amounted to about five inches.

The *Philadelphia Inquirer* described the intensifying storm as "hampering railroads and trolley traffic, breaking down wires, blocking navigation on the river … and sending scores to the hospitals, victims of falls on the ice." Railroad and streetcar tracks were sidelined for four days by the storm. After the storm ceased, the 24-hour Philadelphia snow accumulation reached 21 inches, a record that stood until January 1996.

The heaviest snowfall blanketed Delaware, where 25 inches fell at Milford and 23 inches at Wilmington. The snowfall amounts north of Philadelphia totaled 22 inches at Reading and 20 inches at Allentown, according to the *Morning Call*. The *Stroudsburg Times* reported 15 inches in the upper Delaware Valley and a frigid -4°F on the 28th. The *Monroe Record* noted 14 inches fell at Brodheadsville.

Less than three weeks after the huge post-Christmas storm, another major snowstorm buried a large portion of Pennsylvania on January 13–14, 1910. Press reports listed accumulations of 18 inches at Stroudsburg (*Stroudsburg Times*), 16 inches at Bethlehem (*Philadelphia Inquirer*), and 10 inches at Milford.

The month ended with a storm on January 29–30, 1910, that brought howling winds. Twelve inches of snow fell at Pocono Lake, pushing the monthly fall to 38 inches, a January figure not exceeded in the Pocono Plateau section until 1987, when 48 inches piled up at Tobyhanna.

November 9–11, 1913
Great Lakes Hurricane

The Great Lakes Hurricane, dubbed the "Freshwater Fury" by the press, brought exceptionally strong winds and large snowfalls, commencing on November 7, 1913, from Ontario, Canada, southeast to the upper Ohio Valley on November 9–10, 1913.

The storm originated in Colorado and was joined by an Alberta clipper, evident on the daily maps published in the *Monthly Weather Review*. A secondary low developed over the interior Mid-Atlantic region and tracked northwest to Erie, with a central pressure dipping to 29.32 inches.

The storm consolidation directed intense counterclockwise winds, which howled over the unfrozen lakes and increased the moisture content and energy of the system that turned into a monster passing over Lake Erie.

The storm devastated shipping around the Great Lakes. Ten large ships were lost, and 10 others were heavily damaged, coupled with even more shipwrecks on reefs. A dozen ships lost their entire crew. In all, an estimated 235 to 300 sailors perished in the storm (Schmidlin 1996).

Transportation and communications in western Pennsylvania were halted for days. Wind gusts reached 80 mph at Buffalo and 90 mph along parts of the Lake Erie shore, felling telephone, telegraph and electric wires encased in ice and snow. Six to 12 inches of snow blanketed much of Ohio and western Pennsylvania, and heavier bands deposited up to 18 inches, driven into huge drifts by wind gusts clocked at 60 mph.

March 1-2, 1914
Billy Sunday Blizzard

The winter of 1913–14 brought a historic number of deep snowstorms over the interior of Pennsylvania.

A holiday storm deposited 10 inches at Montrose and 11 inches at State College on December 25–26, 1913. A major snowfall followed on January 3–5, 1914, burying the western half of the state under 24 inches at State College and 21 inches at Altoona.

A Valentine's Day snowstorm deposited substantial totals in the north on February 13–15, 1914: Binghamton, New York, 21 inches; Towanda 20 inches; Scranton 17 inches; Williamsport 16 inches. The *Altoona Mirror* reported 18 inches fell at Johnstown, capped off by another six inches February 16 in a separate system that deposited several inches in many areas around the state.

But the most severe storm of the winter—judged to be the worst since the Blizzard of 1888—churned up the East Coast on March 1, 1914, accompanied by a blinding wind-driven snow that accumulated more than a foot deep, with huge drifts.

William Ashley "Billy" Sunday, a former baseball National League outfielder, who became a hugely popular evangelist in the early 20th century, had a scheduled revival in Scranton on March 1, 1914, at the corner of Washington Street and Walnut Avenue.

The ensuing storm was dubbed the "Billy Sunday Storm" because the inclement weather trapped thousands in a temporary tabernacle. The blizzard buried the Northern Electric Trolley tracks in 18-foot drifts in Lithia Valley. The *Scranton Times* headline the next day declared: "BLIZZARD HELD NO TERROR FOR A CROWD THAT FLOCKED TO HEAR SUNDAY PREACH," adding, "Thousands Made Tabernacle A Haven In All Night Storm."

Railroad lines were blocked, and telephone and telegraph communications were cut off by powerful winds gusting to at least 45 mph (37 mph adjusted). Four persons died in the Scranton region in the storm. The *Stroudsburg Times* reported drifts of up to five feet in town, and 15 feet in the higher elevations to the northwest at Wooddale.

Snow totals were highest in the east. Twelve inches fell at George School, Gordon and Stroudsburg; 15 inches at Scranton; and up to 17 inches in the Poconos. Intense banding brought totals of one to two feet in New Jersey. Long Branch reported 24 inches, where the observer measured a wind gust of 72 mph (58 mph adjusted).

The barometric pressure dipped to a record low of 28.38 inches at New York City, with a gust of 84 mph (67 mph adjusted). Mixed precipitation in southeastern Pennsylvania held accumulations to six inches.

January 2–3, 1925
Big Lehigh Valley Snowfall

A deep snowfall kicked off the snowiest January on record in east-central Pennsylvania. Heavy snow began falling before dawn over eastern Pennsylvania on January 2, 1925, and would reach historic depths: Pottsville (25.8 inches); Freeland (21 inches); Bethlehem (21 inches); Allentown (20.2 inches).

The *Pike County Press* stated that "20 inches [fell] at Stroudsburg, though only 10 inches fell at Milford thirty miles to the northeast." New York City caught 11.1 inches before the precipitation mixed with sleet and rain.

Heavy snow fell frequently during the closing days of the month that averaged four to eight inches on January 20, 27, and 28–29, 1925. All-time monthly snowfall records were established at Freeland (56.5 inches); Phoenixville (48 inches); Altoona (44 inches), Allentown (43.2 inches), and Carlisle (41.5 inches). In northwestern New Jersey, Belvidere tallied 50 inches, nearly overtaking the monthly state snowfall record of 51.2 inches at Freehold (December 1880).

Surprisingly, Allentown did not have measurable snow over the remainder of the winter of 1924–25 after the record-breaking January snowfall. February 1925 was exceptionally warm at Philadelphia (42.2°F), a substantial 6.3°F above the contemporary normal.

March 28–30, 1942
Deep Central Snowfall

A heavy snowfall on March 2–3, 1942, commenced a record snowy March in parts of central Pennsylvania. Up to 19 inches blanketed the southwestern highlands, including 16.2 inches at Pittsburgh.

An even bigger storm of historic proportions for late March started as rain on the 28th before changing to snow inland. Polar air wedged in beneath a mild southerly airstream set the stage for a prolonged heavy wet snowfall over the interior. Eastern Pennsylvania had a rain/snow mix, with only minor accumulations, and the western portion of the state got off lightly as well. However, the central portion of the state was not so lucky.

In the early evening of March 28, 1942, a cold rain transitioned to heavy wet snow in the central mountains at the Pennsylvania State University in State College. The next morning, 13 inches had fallen, and during the succeeding 24 hours another 17.5 inches piled up—a record single-storm total of 31 inches. The monthly accumulation established a record for any month (47.5 inches). East Waterford also had a monthly snowfall of 47.5 inches.

The heaviest snowfall on March 28–30, 1942, occurred at an official station six miles south of Renovo (34 inches), in the north-central mountains. Huntingdon and Shippensburg recorded 26 inches of snow in 24 hours on the 29th (Shippensburg had a total fall of 28 inches).

In western Maryland, a 24-hour total of 31 inches at Clear Spring and storm total of 36 inches at Edgemont together established new state snowfall records up to that time. At Washington, D.C., the urban snowfall (11.5 inches) on March 28–29,

1942, nearly matched the city's March record (12 inches) on March 16–18, 1892. The northwestern suburbs of the nation's capital received up to 18 inches of sloppy, wet snow. Baltimore's fall of 22 inches set an all-time March record.

November 24–27, 1950
Great Appalachian Storm

The Great Appalachian Storm of November 24–27, 1950, brought record-breaking snowfalls to western Pennsylvania and torrential rains accompanied by hurricane-force wind gusts east of the mountains.

The seedling for the massive storm was low pressure forming over western North Carolina as arctic air spilled across western Pennsylvania. A persistent Thanksgiving Day rain on November 23, 1950, turned to snow west of the mountains, but without a hint of the extremely dynamic weather scenario unfolding southeast of Pennsylvania.

On November 24, the tremendous thermal contrast became a boundary for low pressure sliding north toward central Pennsylvania. The key to the explosive storm development was an upper-level trough over Wisconsin that merged with the southern system and closed off over eastern Ohio. High pressure over eastern Canada hemmed in the storm and halted the eastward progression of frigid air.

By the next morning, the powerful low had only reached southern Virginia, passing Washington, D.C., before curving northwest into central Pennsylvania. The temperature at State College plummeted from 51°F at midnight to 10°F in 24 hours on November 25, and 5 inches of snow piled up in blizzard conditions.

On the opposite side of an extraordinary frontal boundary lying across the middle of Pennsylvania, torrential rain pelted areas to the east. Rainfalls of three to five inches caused extensive flooding in eastern Pennsylvania. Destructive hurricane-force winds battered the region throughout the day, snapping trees and toppling power lines, causing widespread power outages that lasted for several days.

Winds gusted to 76 mph at Stroudsburg, 88 mph at Allentown, 94 mph at New York City (LaGuardia Airport), 100 mph at Hartford, Connecticut, and 108 mph at Newark. Gusts reached 160 mph atop Mount Washington, New Hampshire. Even sheltered Central Park in New York City was buffeted by a gust clocked at 70 mph. Homes, businesses and resorts suffered extensive losses.

In western Pennsylvania, a broad mass of arctic air oozed eastward, as readings plummeted into the teens and single digits on November 25, 1950. Roads were completely blocked by mounting drifts, and rail lines and public transportation came to a halt—buses, trolleys and mail deliveries. Businesses would be shut down for days, leaving only small corner grocery stores to handle basic needs.

A look at the daily snow totals in western Pennsylvania showed that on November 24, 1950, 14 inches of snow fell at Pittsburgh's Allegheny County Airport, seven miles southeast of Pittsburgh, followed by another 10.1 inches on the 25th and 2.3 inches the next day, for a record total of 27.4 inches in three days. The monthly catch of 32.3 inches also established a record, with all but one inch falling in the last eight days of the month.

Eventually, rain changed to snow on the night of November 25–26, 1950, over the northeastern highlands, as the monster storm turned and was absorbed by a vast upper-level cold trough. The surface low later stalled over Lake Erie on November 27, prolonging the snow for several more days, aided by lake-enhanced snow showers.

Fig. 4.1 November 29, 1950, following the great snowstorm. Cars buried on Reifert Street. (Photo by Morris Berman; Pittsburgh Post-Gazette Photo Archives, copyright 1950, all rights reserved)

Light snow continued to fall across northern and western Pennsylvania as the storm filled. The greatest falls during November 23–28, 1950, totaled 37 inches at New Castle (33-inch depth), and 35.3 inches at Meadville. Erie received 19.5 inches on the 24th, with a storm total of 28.4 inches on November 24–27, 1950—second only to December 11–14, 1944 (30.2 inches). In all, 46.9 inches of snow fell at Erie in November 1950.

Spectacular snowfall totals were also reported in the mountains of northern West Virginia westward across eastern Ohio. Pickens, West Virginia, received 57 inches, and a state-record 62 inches fell at Coburn Creek. In eastern Ohio, Steubenville logged 36.3 inches November 24–26, 1950, and a storm total of 44 inches through the 28th, establishing a state snowfall mark at the time.

Severe drifting in eastern Ohio and southwestern Pennsylvania reached historic depths of 15 to 25 feet. The Great Appalachian Wind and Snowstorm claimed at least 50 lives in Pennsylvania and resulted in more than $1 million in damages. The death toll in the eastern United States reached 353, and damages totaled $66.7 million nationally in 1950 dollars.

February 15–17, 1958
Great Mid-February Blizzard of 1958

The months of February and March 1958 are the snowiest consecutive months in northeast Pennsylvania weather history. Four major snowstorms deposited an incredible 88.5 inches of snow at Stroudsburg (elevation 480 ft.) in eastern Monroe County, and 92.1 inches at Gouldsboro (elevation 1,890 ft.) in southern Wayne County.

The winter of 1957–58 had already produced a respectable snowfall on December 3–4, 1957, followed by a record late-season rainstorm on December 19–20 that totaled 5.97 inches at Stroudsburg.

A significant storm on February 7–8, 1958, dropped upwards of a foot of snow on northeastern Pennsylvania. This would merely be the first volley in a series of four heavy snowfalls in a five-week period between mid-February and late March 1958.

Weather historian David Ludlum put the Great Mid-February Blizzard of 1958 in perspective in his *Vermont Weather Book*:

No other snow event of the twentieth century can match the huge dimensions displayed by the Great Mid-February Storm of 1958, extending as it did along the entire Atlantic seaboard and inland to the Appalachians. In the nineteenth century, only three snow storms rate as equals: The Great Snowstorm of January 1831, The Cold Storm of January 1857, and the Great Eastern Blizzard of February 1899. Even the Blizzard of '88 did not meet the above requirements since its geographical extent was limited to the Northeast.

The incipient blizzard appeared as a weak wave on February 14, 1958, along the Gulf Coast. The storm followed a classic path across the Deep South on February 15, picking up abundant moisture before heading up the Atlantic Coast.

A deep mantle of snow blanketed the interior Southeast from Alabama to Virginia on February 14–15, 1958. The burgeoning winter storm assumed blizzard proportions as it churned past the New Jersey coast on February 16, piling the fluffy snow into mountainous drifts. The largest falls were reported over the higher elevations in northeastern Pennsylvania, leaving most routes impassable, stranding hundreds of travelers. Western Pennsylvania escaped by comparison, with two to eight inches of snow in most areas.

The Stroudsburg *Daily Record* reported huge falls of 36 inches at Buck Hill Falls and 30 inches at Mount Pocono. Thirty inches fell at the official site in Gouldsboro, with a cumulative depth of 48 inches. Farther north at Hawley, the snowfall was 35 inches, with a record 48-inch depth. Rescue teams from the Tobyhanna Army Depot dug out trapped motorists.

The greatest snowfall—42.5 inches—was measured four miles southwest of Hawley at Lakeville 1 SSE. A 24-hour fall of 41 inches on February 15–16 at the site along Lake Wallenpaupack established a state 24-hour record. The *Wayne Independent* in Honesdale reported that wind gusts of 40 to 60 mph caused massive drifting, effectively shutting down transportation in many areas for days.

Fig. 4.2 February 17, 1958. Blizzard deposited 33 inches of snow on Honesdale, raising the total depth to 43 inches on Tryon Street. (Courtesy Elizabeth Korb Schuman)

Snow accumulations in southeastern Pennsylvania varied from 13 inches at Philadelphia to 17 inches at West Chester and 20 inches at Lancaster. In the Lehigh Valley, 15.8 inches fell at Allentown, tapering off to 11.2 inches at Scranton. The Stroudsburg observer measured 24.3 inches and a record February depth of 33 inches.

The urban centers were hit hard by the storm, averaging 10 to 15 inches from Washington (14.4 inches) to Baltimore (15.5 inches) and New York City's LaGuardia Airport (10.4 inches). Heavier falls occurred over eastern New York and southern New England, totaling 17.2 inches at New Haven, Connecticut, 26.5 inches at Syracuse, and 38.7 inches at Calicoon, New York. Boston was buried under 19.4 inches.

The weather was bitterly cold on February 17, 1958. In the Poconos, Pimple Hill (elevation 2,212 ft.) had a subzero day, with a maximum of -1°F, and Stroudsburg only reached 5°F. Two days later, Governor George M. Leader declared a state of "extreme emergency" because so many Pennsylvania roads were impassable. Helicopters were used to supply aid to persons in need as food and fuel supplies dwindled the next day. The February Blizzard of 1958 claimed 43 lives and caused $50 million in property losses nationally.

March 18–21, 1958
Deep March Snowfalls

The stormy pattern in February and early March 1958 left a huge amount of snow on the ground in interior sections of eastern Pennsylvania. The third heavy snowfall in five weeks dropped 10.7 inches at Stroudsburg and 16 inches at Gouldsboro in the Poconos on March 13–14, 1958. The building snow cover reached 30 inches on the high Pocono Plateau.

Yet, as snow-weary residents of Pennsylvania looked forward to an early spring after being battered by a succession of winter storms, a fourth storm was brewing in the Gulf of Mexico, which would encounter a blocking high-pressure area on March 18, 1958.

Unprecedented water equivalent (melted) precipitation that commenced late on March 18 and lingered into March 21 exceeded three inches. The combination of powerful winds and moisture-laden wet snowfall toppled trees and brought down power lines, leaving millions in the dark. Traffic accidents and heart attacks claimed 27 lives in Pennsylvania during the storm.

The greatest snowfalls in Pennsylvania occurred in Berks and Chester counties. At Morgantown, 40 miles northwest of Philadelphia, a state record fall of 50 inches came fast and furiously—38 inches piled up in 24 hours on March 20, 1958. In Chester County, the Devault observer measured an astounding storm total of 49 inches. A little farther southwest, York had a record snowfall of 33 inches, which raised the total depth to a record-tying 36 inches (tying February 1899). Holtwood also received 33 inches and the West Chester observer measured 32 inches.

The storm total at Allentown of 20.3 inches established a March record. North of the Lehigh Valley in the southern Poconos, amounts were massive. Stroudsburg observer William Hagerty logged 35.4 inches and measured a record March depth of 32 inches on March 20. The storm intensity was evident in his station record 24-hour catch of 29 inches that day—6.8 inches in one hour established a record intensity fall.

On the Pocono Plateau about 20 miles to the northwest of Stroudsburg, Gouldsboro had a nearly identical storm total of 35.3 inches, setting a state snow depth record of 60 inches that had piled up atop a remnant snow cover. The monthly accumulation of 56.3 inches at Gouldsboro was also record-worthy. In western Monroe County, longtime observer Dixon Miller recorded 24 inches at Pimple Hill near Blakeslee, and a depth of 52 inches. In the northern Poconos, Hawley had 28 inches (48-inch depth). The top snowfall in neighboring northwestern New Jersey was 30 inches at Branchville.

A band of exceptionally heavy accumulations buried the suburbs of Washington, D.C., Baltimore, Philadelphia and New York City. The coastal plain was just warm enough for a portion of the soggy wet snowfall to melt as it fell at times, as spectacular totals were recorded just to the west.

The snowfall at Philadelphia was 11.4 inches, which was typical for the metropolitan areas from the nation's capital to New York City (11.8 inches), though a little inland a substantial 22.5 inches fell at Eastchester, New York. March snowfall records were observed in Delaware at Middletown (27 inches) and Wilmington (17.9 inches).

The 1957-58 season snowfall was historic in the southern Pocono region. In the lower elevations, Stroudsburg had record season snowfall of 109.5 inches. On the Pocono Plateau, totals reached 141.7 inches at Long Pond and 147.1 inches at Gouldsboro. In the southeast, the final count at West Chester of 73.5 inches exceeded all previous winters at the time. Philadelphia (41.8 inches) had its snowiest winter since 1904–5 (43.8 inches).

March 3–5, 1960
March Blizzard

A great nor'easter was in its formative stage over the northwestern Gulf of Mexico early on March 2, 1960, while the East was in the grasp of a remarkable late-season cold wave. A new storm developed off the South Carolina coastline during the predawn hours of March 3, spreading a canopy of heavy snow northward.

Snow began falling over southeastern Pennsylvania shortly before daybreak on March 3, 1960, becoming heavy by late morning. The storm moved rapidly northeast to a position about 150 miles off the New Jersey coast by early evening. The system slowed, generating persistent easterly gales and blizzard conditions at times.

The 28-hour snowfall at Stroudsburg totaled 21 inches, and Gouldsboro had 20 inches. The large totals extended farther inland to Freeland (24.5 inches), Scranton (16.9 inches) and Williamsport (13.7 inches). The total snow depth at Freeland on March 6, 1960, reached 33 inches.

Allentown had 14.2 inches, tapering down to 8.4 inches at Philadelphia. Even the western counties were hit hard: Pittsburgh measured 12.8 inches. The coastal plain was also hit hard. Newark, New Jersey, received 17.4 inches, and Boston had 19.7 inches. The greatest totals were 31.3 inches at Nantucket, Massachusetts, and 27 inches at Dobbs Ferry, New York.

Winter of 1960–61
Three Deep Snowfalls

The winter of 1960–61 brought three noteworthy East Coast snowstorms: December 11–12, 1960; January 19–20, 1961; and February 3–4, 1961.

The early December storm hit the coastal sections hardest, with winds gusting to 55 mph. Snowfall mounts ranged from 10 to 15 inches over eastern Pennsylvania. A total of 14.6 inches piled up at Philadelphia, and 20.4 inches fell at Newark, New Jersey.

Bouts of arctic chill mixed with brief thaws prevailed through the middle of January. On the eve of President John F. Kennedy's inauguration on January 19, 1961, heavy snow and strong winds accompanied a coastal disturbance that deposited 7.7 inches of snow on the nation's capital. The snow ended by daybreak on Inauguration Day, January 20, but conditions were bitterly cold for the outdoor festivities.

Farther north the snow totals increased to one to two feet. Winds topping 30 mph whipped the snow into deep drifts. Snowfall totals included 9.9 inches at New York City, increasing to 13.2 inches at Philadelphia, 16 inches at Allentown, and 18.7 inches at Harrisburg. To the north, Stroudsburg picked up 23 inches in a little more than 24 hours.

By the end of January 1961, Harrisburg had established a new monthly record snowfall of 34 inches that would last until 1996 (38.9 inches), surpassing the previous snowfall record for an entire season (1892–93). But there was more to come.

Low pressure traveled east from the southern Plains to the Ohio Valley on February 1–2, 1961, and in familiar fashion yielded to a system along the Georgia coast. Heavy snow broke out across the Mid-Atlantic on February 3, 1961.

Near-blizzard conditions enveloped much of the eastern two-thirds of Pennsylvania on February 3–4, 1961. Amounts increased steadily northward: 10.3 inches at Philadelphia; 17.3 inches at Allentown; 18.5 inches at Stroudsburg; 20 inches at Montrose; 22 inches at Gouldsboro; 22.6 inches at Newark, New Jersey. Although the northwestern corner of the state received less than six inches, a pocket of deep snowfall on the Allegheny Plateau left 28 inches at Galeton and 24.5 inches at Kane.

Wind gusts ranged from 30 to 50 mph, pushing snow into deep drifts close to 10 feet high in places. The NOAA monthly *Climatological Data* summary for Pennsylvania added: "Transportation facilities were nearly paralyzed as airports and all except major highways were closed. Some secondary roads remained closed for nearly a week," shutting down schools in rural areas for days. Fifteen deaths in the Commonwealth were attributed to "over-exertion."

On February 4, 1961, the snow depth at Stroudsburg established an all-time station record (34 inches). The greatest snow cover was observed in Wayne County at Lakeville (49 inches), along the shore of Lake Wallenpaupack. In northwestern New Jersey, historic snow depths were measured at Canistear Reservoir (52 inches) and Layton (50 inches).

The two-week cold wave finally broke at the conclusion of the storm. Many locations enjoyed above-freezing temperatures for the first time in 17 days, signaling a reprieve from the arctic assault. A mild pattern in late February and March 1961 replaced the harsh conditions that had prevailed since early December 1960.

However, true to form in most severe winters, Mother Nature threw in a few wet snowfalls in April. Heavy snow fell over the northern counties on April 9–10 and April 12–13, 1961, with totals topping 10 inches in the mountains in both storms. A historic late cold snap was accompanied by flurries in the northern highlands on May 27, 1961. Traces of snow were reported at Stroudsburg in the upper Delaware Valley and at New York City's LaGuardia Airport—the latest occurrences on record.

The 1960–61 seasonal snowfall at Harrisburg (81.3 inches) and State College (91.8 inches) set new records, though the State College mark would be erased in 1993–94. Stroudsburg (98.6 inches) had its fourth snowiest winter (tied in 1995–96). The greatest winter total on the Pocono Plateau occurred at Long Pond (124.4 inches). Newark, New Jersey (73.5 inches), had its snowiest winter of the century.

In the tall Laurel Highlands of southwestern Pennsylvania, Somerset recorded 178.8 inches of snow in the winter of 1960–61, with the largest amounts coming in December (41.3 inches), January (47 inches) and February (46.5 inches). In the central mountains, 136 inches were measured at Lewis Run, and 131 inches fell at Bloserville.

January 12–13, 1964
Eastern Blizzard

A major Eastern blizzard in January 1964 made headlines for a combination of deep snowfalls and relentless northeasterly gales that drove the powdery snow into mountainous drifts across the Keystone State.

The weather map on the evening of January 11, 1964, showed two areas of low pressure. The primary low was located over southwestern Missouri, while a secondary storm took shape along the western Gulf Coast. At daybreak on January 12, a large storm was centered near the South Carolina coast and the western system was in Tennessee.

The coastal storm intensified rapidly, slowed by a blocking ridge of high pressure over eastern Canada, while the initial low drifted to Kentucky, setting up a conveyor belt of moisture reaching back to the upper Ohio Valley.

Snow began falling over Pennsylvania during the afternoon hours on January 12 and gained intensity rapidly. Strong winds gusting up to 45 mph created blizzard conditions as the storm deepened east of New Jersey early on the 13th. More than 20 inches of snow covered the northern and central mountains by late in the day.

Some of the larger falls included: State College (27.5 inches); Williamsport (24.1 inches, a 24-hour record); Gouldsboro and Pleasant Mount (24 inches); Scranton (21.1 inches); Stroudsburg (20.6 inches), Harrisburg (18.1 inches), Pittsburgh (15.6 inches). Major cities in southern Pennsylvania were also virtually shut down by the wind and blinding snowfall.

Much of Pennsylvania received eight to 16 inches of snow, except in the extreme southeastern and northwestern counties, where the snowfall was closer to six inches. A pocket of 20-plus inches covered the northwestern mountains.

A second large snowstorm dropped more than a foot of snow on February 18–20, 1964, with totals of 20.3 inches at Harrisburg and 18.1 inches at Stroudsburg.

January 29–30, 1966 Blizzard of '66

A powerful midwinter storm paralyzed the Mid-Atlantic region in late January 1966. Two snowfalls left moderate accumulations late in the month over a 10-day period. However, the blizzard on January 29–30, 1966, proved to be the worst by far, accompanied by sustained winds of 40 mph and gusts of 50 to 60 mph.

The morning weather map on January 28, 1966, showed a wave over the Big Bend region of West Texas. The following day, low pressure deepened and was positioned near Montgomery, Alabama. A new storm center formed along the Southeast Coast and tracked to a position east of Atlantic City on the evening of the 29th.

Snow totals mounted to more than one foot from Washington, D.C., northward to eastern Pennsylvania. Harrisburg received 12.2 inches, with snowfalls in some areas totaling 15 to 20 inches. The western counties of Pennsylvania received three to nine inches, except where lake-enhanced snows pushed amounts to over a foot in Erie County.

The state's *Climatological Data* report for January 1966 estimated that the cost of snow removal in Pennsylvania exceeded $10 million. Even though the blizzard commenced on the weekend, many schools in eastern Pennsylvania remained closed for the entire week, with travel clogged by deep drifts in outlying areas as high as 15 feet.

The Blizzard of '66 brought days of wraparound snows to Upstate New York. The primary portion of snowstorm at Oswego (52 inches) on January 27–30 totaled 103 inches at the end of the month. The storm-related death toll in the Northeast and Mid-Atlantic was 103 persons by early February.

December 24, 1966
Christmas Eve Blizzard

Heavy snow fell across parts of Pennsylvania on December 13, 20 and 24, 1966, creating the whitest of all possible Christmases in the Keystone State.

The third in a series of coastal disturbances spread a canopy of snow rapidly up the East Coast on the morning of December 24, 1966, forming in the Gulf of Mexico. The snow intensified in the afternoon, falling at the rate of two to three inches per hour in eastern Pennsylvania, accompanied by peals of thunder in an intense six-hour period. As the storm marched up the coast, moisture-rich Atlantic air rode inland and intensified, bringing near-whiteout conditions for hours. Totals ranged from 12.7 inches at Philadelphia to 21.2 inches at Coatesville and 22 inches at West Chester.

Stroudsburg residents dug out from under a record December snowfall of 18.8 inches and a record December depth of 23 inches on Christmas morning in the wake of a six-inch fall three days earlier. Longtime observer William Hagerty measured 23 inches at Neola in the hills, 10 miles west of Stroudsburg. Record December totals were observed at Stroudsburg (36 inches) and Allentown (28.4 inches).

After a nearly snowless January 1967, February and March returned to form. Major snowfalls occurred in Pennsylvania on February 6–7 and 9–10, 1967. Heavy snow fell again on March 5–7 and 15–17, 1967. Montrose received 25.5 inches on March 5–7 and 15.9 inches on March 15–17, setting an all-time March snowfall record (52.9 inches).

Historic chill followed, with high temperatures confined to the teens in the north on March 18, and minimums as low as -11°F at Stroudsburg and -20°F at Milanville on the 19th. Subzero lows were registered as far south as Allentown and Quakertown (-1°F).

A first-day-of-spring snowstorm on March 21–22 was accompanied by thunder and lightning, with much of the eastern counties receiving another five to 10 inches. East Stroudsburg collected 24.5 inches in February and 33.5 inches in March 1967. Two late-April snowfalls on the 24th (2.6 inches) and 27th (2 inches) pushed the season total to 100.6 inches, second only to 1957–58 (109.5 inches).

The northern mountains of Pennsylvania were whitened in places on May 6–7, 1967, in the coldest May in state weather history (50.7°F).

December 1969
Another Snowy December

The winter of 1969–70 began with an elevation snowstorm November 5–6, 1969, that deposited eight inches at Mount Pocono and 11 inches at Tobyhanna.

A moderate snowfall on December 14, 1969, was followed by a bigger storm on December 21–22, with record December snowfalls at Scranton (12.2 inches) and Harrisburg (12.1 inches). On the Pocono Plateau, 15 inches of snow fell at Gouldsboro, and 10 inches accumulated in the nearby valleys. In the southeast, the snow turned to rain and then back to snow, with moderate accumulations.

A light snow that accompanied a disturbance racing eastward from the central Plains on Christmas Eve 1969 brought a shot of arctic air, accompanied by three inches of snow at Hazleton, on top of the 15-inch cover, according to the Hazleton *Standard-Speaker*.

Christmas Day in 1969 dawned bitterly cold across the Northeast, with temperatures dipping to near -10°F in the rural interior. But as Pennsylvania residents settled in to enjoy a sunny, crisp holiday, a major storm was brewing in the Gulf of Mexico.

On Christmas afternoon, a steady snow reached Philadelphia, quickly spreading north around nightfall. The storm center passed near Philadelphia the next morning. The inland track caused the precipitation to change to sleet and freezing rain in the Delaware Valley, after 5.1 inches fell at Philadelphia.

Farther north and west, the storm brought all snow, accumulating up to 15 inches at Hanover and Gouldsboro and 17.2 inches at Williamsport. The Hazleton *Standard-Speaker* reported 18 inches—a total of three feet in six days.

December 1969 brought record December snow totals in the northeast: Gouldsboro 46.1 inches; Tobyhanna 43.1 inches; Hawley 42 inches; Neola 36.8 inches; Stroudsburg 36 inches; Williamsport 35.5 inches; Harrisburg 28.3 inches. The maximum depth at Gouldsboro of 35 inches set a December record.

The massive winter storm slowed over southeastern New England, dumping historic snows in Vermont at Burlington (29.8 inches) and Montpelier (39 inches) northward to Quebec, Canada, at Montreal (28 inches).

Snow cover lingered in the northern mountains of Pennsylvania into early April. An inch or more of snow covered the ground at Gouldsboro for a record 116 consecutive days from December 14, 1969, to April 8, 1970. The next winter brought 111 consecutive days of snow cover at Gouldsboro (December 12, 1970, to April 12, 1971).

April 3–5, 1975
April Windstorm Brings Snow

A potent storm traveled from the Middle Mississippi Valley to New England on April 2–3, 1975, then intensified dramatically (28.75 inches), with hurricane-force wind gusts reaching Pennsylvania, while pulling in arctic air to generate snow squalls.

Peak wind gusts associated with a vast pressure gradient were historic: Lancaster 91 mph; State College/Pennsylvania State University and Mifflin County Airport 90 mph; York 86 mph; Chambersburg 77 mph; Harrisburg 66 mph; Philadelphia 65 mph. High winds caused substantial damage to roofs, trees and outbuildings. One woman in Wrightsville was killed when a tree crashed onto her mobile home.

Snowfall, in the form of bands and squalls, was heavy downwind of Lake Erie and over the northwestern highlands, ranging from 14.5 inches at Bradford 4 W to 24 inches at Corry through the 6th. In the north-central mountains, 6.9 inches fell at Montrose.

February 5–7, 1978
Blizzard of 1978

The winter of 1977–78 had already established a pattern of substantial snowfalls.

Three heavy snowfalls occurred in the Mid-Atlantic between January 13 and 20, 1978. The third in a series of snowstorms on January 19–21 brought more than a foot of snow in a broad swath of Pennsylvania from Philadelphia to Pittsburgh northward to the central mountains. Record snow depths of 30 inches at State College and 26 inches at Pittsburgh were observed on January 22, 1978.

A few days later, a blizzard of historic proportions swept the eastern Midwest on January 25–26, 1978, but brought heavy rain to Pennsylvania as the center of the low crossed Ohio in the early morning hours of January 26. A record non-tropical low pressure was observed at Cleveland—28.28 inches. (The national record was exceeded on October 26, 2010, when a pressure of 28.21 inches was registered at Bigfork, Minnesota.)

The winter of 1977–78 was far from done. The morning weather map on February 5, 1978, showed a low-pressure trough pivoting through eastern Ohio. The northern low passed over Lake Erie, while a secondary storm developed a few hundred miles east of the Virginia coast. Upper-level energy contributed to rapid intensification as the winds aloft steered the coastal storm northwestward.

Intermittent snow became steadier and heavier during the morning of February 6 and expanded northward to New England by the afternoon. Philadelphia was blanketed with 14.1 inches, and two to three times as much covered the Poconos. East Stroudsburg had 23.7 inches (30-inch depth), with drifts of up to six feet, noted in the East Stroudsburg University station records. On the high plateau, Tobyhanna received 32.5 inches (45-inch depth) and Gouldsboro 38 inches (52-inch depth). Pleasant Mount reported 22.5 inches (49-inch depth).

The coastal plain experienced snowfalls, ranging from 9.1 inches at Baltimore to 14.5 inches at Wilmington, Delaware, and 17.7 inches at New York City/Central Park. Riverhead, Long Island, recorded 25 inches, and Rochester, New York, had 25.8 inches.

The storm stalled over the eastern tip of Long Island early on February 7, 1978, prolonging the heavy snow over eastern New England that reached record proportions at that time. Boston received 27.1 inches, with a peak wind gust of 61 mph. In Rhode Island, Providence (28.6 inches) and Woonsocket (38 inches) had historic snowfalls.

The ground at Pittsburgh remained snow-covered for a record 64 consecutive days through March 12, 1978, breaking the previous mark of 62 days (1976–77). Allentown had an inch or more of snow cover on 59 consecutive days through March 13, 1978, exceeding the previous record of 49 days through February 12, 1977, tied with February 19, 1925. (This value was shattered by 19 days in 1993–94.)

In the southern Pocono region, East Stroudsburg had an inch or more of snow on the ground for 101 days in the winter of 1977–78, and a season snowfall totaling 76.2 inches. On the high plateau, Gouldsboro received 131.2 inches in the winter of 1977–78.

Erie amassed a record accumulation of 142.8 inches of snow at the time, which was not surpassed until 2000–01 (149.1 inches). Philadelphia tallied 54.9 inches, missing the 1898–99 record at the time by only 0.5 inch.

February 11–12, 1983
Blizzard of '83

The setting for the great snowfall of February 11–12, 1983, featured a frigid dome of high pressure over Quebec and a jet stream disturbance streaking toward the Mid-Atlantic.

The phasing of upper-level energy and a low-pressure wave developing along a sharp thermal boundary off the coast of South Carolina provided the necessary ingredients for a snowstorm of historic proportions along the Eastern Seaboard. The first snowflakes reached Washington, D.C., around midnight on February 11, 1983. Snow was falling at Harrisburg before dawn on February 12, reaching southern New England in the afternoon.

As the storm marched up the coast, moisture-rich Atlantic air rode over the top of dense cold air, causing the snow to come down at record rates in eastern Pennsylvania. During the evening hours, the snowfall achieved its maximum intensity. The rate at Allentown peaked at five inches in an hour, and four inches fell at East Stroudsburg in an hour. High winds whipped the snow into drifts six feet deep, as blizzard conditions enveloped the region, punctuated by lightning and thunder.

Storm totals were record-worthy at the time: Harrisburg 25 inches; Allentown 25.2 inches; Byberry 24.5 inches; Philadelphia 21.3 inches (breaking the December 1909 mark by 0.3 inch). In the southern Poconos, the totals were also massive, ranging from 23.5 inches at East Stroudsburg and 24.5 inches at Bartonsville to 25 inches at Kresgeville and 27 inches at Tannersville. The northwestern portion of the state largely missed out on most of the action, with a light snowfall, while the southern counties received a hefty 10 to 20 inches.

December 10–12, 1992
Early Historic Nor'easter

A powerful nor'easter that developed over Georgia on December 9, 1992, intensified dramatically over the eastern Carolinas, churning northward to Chesapeake Bay.

Hurricane-force winds and record snows occurred during a three-day period from December 10–12, 1992. Blocking high pressure over eastern Canada slowed the forward motion of the storm, causing winds to strengthen dramatically.

The strongest blasts were reported on the morning of December 11, 1992, blowing down trees and limbs and damaging roofs and signs. Power was lost to hundreds of thousands of customers in Pennsylvania. Gusts were clocked at 82 mph at New Holland (Lancaster County), 77 mph at New York's LaGuardia Airport and 60 mph at Scranton.

Rain alternated with snow in the eastern part of the state, while huge snow accumulations piled up over the interior. Snowfall totals reached 36 inches at Ogletown in the Laurel Highlands and 33 inches at DuBois. Snowfalls at State College (18.1 inches) and Mount Pocono (19.5 inches) established December records. A high-elevation site four miles northeast of Gouldsboro collected 26 inches.

Spectacular December falls were similarly observed at high spots in New Jersey (25 inches), Maryland (42 inches) and Massachusetts (48 inches).

March 13–14, 1993
Storm of the Century

The Blizzard of '93 caused $11 billion in damages (2020 dollars) and took more than 270 lives in 13 states, including 49 related deaths in Pennsylvania (NOAA).

On March 11, 1993, a Pacific disturbance migrated to the western Gulf of Mexico, sparking thunderstorms off the Texas coast. The forecasting models had already latched onto the potential of a huge winter storm impacting the entire Atlantic coast. A NWS statement on March 12, 1993, warned that "this storm may be unusually severe and perhaps record-breaking," and "could be the worst storm of the century."

Arctic and subtropical airstreams would soon meet up to spawn a monster winter storm over the northeastern Gulf of Mexico, instigating violent thunderstorms and tornadoes in Florida, coupled with damaging winds and heavy rain as far south as western Cuba.

On March 12, the arrival of a strong jet streak caused a disturbance to explode over the Gulf of Mexico after moving off the Texas coast. Violent straight-line storm gusts exceeding 100 mph and 11 tornadoes raked the Florida Peninsula early on March 13, as rapidly deepening low pressure over the Florida Panhandle generated up to 12 feet of storm surge that battered the coastal areas. Peak gusts reached 115 mph in Hernando County.

As the storm traveled northeast across Georgia toward the Carolina coastline, winds gusted to 101 mph on Flattop Mountain, North Carolina. Huge snowfalls piled up across the southern Appalachians, reaching 60 inches in Tennessee at Mt. LeConte (elevation 6,493 ft.), and 50 inches blanketed Mt. Mitchell, North Carolina (elevation 6,240 ft.).

The storm crossed over southern Georgia on Saturday morning, March 13, picking up copious Atlantic moisture. Snow broke out over southeastern Pennsylvania around midnight and reached the northern counties during the wee hours of the morning. Businesses had already closed in the face of the impending storm. Winds increased to 40 mph, reducing the visibility to near zero, slowing transportation to a crawl.

Pennsylvania Governor Robert Casey declared a state of emergency at 3:00 p.m. on March 13, 1993, effectively shutting down all airports and closing Interstate 80 and the Pennsylvania Turnpike. The National Guard rescued stranded motorists and cleared state roadways. Several homes and buildings collapsed under the weight of the snow in Venango and Potter counties.

In the evening of March 13, 1993, the center of the monster blizzard was over the southern tip of the Delmarva Peninsula. Snow fell at the rate of two to four inches per hour over Pennsylvania. Claps of thunder resounded during the height of the blizzard. Record low barometric pressure readings were an indication of the severity of the storm and accounted for the extreme winds. A peak gust of 75 mph was recorded at Philadelphia and Cape May, New Jersey. The barometer plunged to an all-time low reading of 28.43 inches, breaking the previous record of 28.54 inches on March 7, 1932.

In western Pennsylvania, winds gusted to 50 mph during the afternoon and evening of March 13, 1993, dropping visibilities to near zero as snow fell at the rate of two to three inches per hour. Snow totals on the Allegheny Plateau ranged from 10 inches in the northwest to 36 inches at Latrobe and 32 inches at Donegal in the southwest. Drifts were six to 10 feet high.

The greatest snowfall in the state was recorded in the Laurel Highlands at Seven Springs at an elevation of 2,800 feet, where the final total of 47 inches included a 24-hour fall of 40 inches on March 13–14, 1993, only an inch shy of the February 1958 state 24-hour record at Lakeville 1 SSE.

The snowfall was massive at Pittsburgh, totaling 25.3 inches by the morning of March 14. A calendar-day record of 23.6 inches on the 13th broke the previous daily record of 22 inches observed on

December 17, 1890. Some of the highest totals in the western half of the state were: 36 inches at Latrobe; 27.7 inches at State College; 24 inches at DuBois; 18 inches at Bradford; 14 inches at Erie.

In the northeast, 21.4 inches at Scranton (20.6 inches in 24 hours) edged out the previous snowfall mark (20.5 inches) measured on November 24–25, 1971. (This total was exceeded in January 1996 and March 2017.) A peak wind gust of 56 mph was clocked at the Wilkes-Barre/Scranton airport in Avoca during the early morning of March 14, 1993. Swirling winds drove the snow into drifts eight feet high.

In the upper Susquehanna Valley, 20.8 inches fell at Towanda, the greatest fall since February 19–20, 1972 (30.4 inches). Intense banding brought historic totals to western sections of the plateau: 42 inches northeast of Gouldsboro; 37 inches at Wellsboro; 34 inches at Hollisterville; 33.1 inches at Long Pond. In the eastern Poconos totals were uniform: 20.1 inches at East Stroudsburg and 22 inches at Tobyhanna.

A total of 17.6 inches fell at Allentown, but much larger amounts were reported just to the west. A whopping 32 inches fell at Lehighton. In the south-central part of the state, Shippensburg had 25.7 inches, exceeding the great snowstorm of March 6–7, 1962 (22 inches), and Harrisburg received 20.4 inches. Amounts tapered off to six to 12 inches from Washington, D.C., to New York City.

Wind gusts were clocked at 81 mph at Boston. The lowest continental pressure (28.38 inches) occurred at White Plains, and New York City recorded a very low reading (28.43 inches) and peak wind gust of 75 mph clocked at LaGuardia Airport. Historic low-pressure readings were logged at Scranton (28.58 inches) and East Stroudsburg (28.50 inches) during the height of the blizzard.

Dramatic snow totals were recorded up and down the Eastern Seaboard, where an estimated 50 billion tons of snow were deposited in the Storm of the Century. Other noteworthy snowfalls: 44 inches at Snowshoe, West Virginia; 42.9 inches at Syracuse, New York; 35 inches at Lincoln, New Hampshire; 29.3 inches at Beckley, West Virginia; 24 inches at Mountain City, Georgia; 13 inches at Birmingham, Alabama (17 inches fell south of the city).

In Maryland, 40 inches fell at Frostburg (record depth of 54 inches), and 47 inches piled up at Grantsville, where the 24-hour fall of 34 inches broke the state record of 31 inches at Clear Spring that occurred on March 28–29, 1942.

The snow tapered off over Pennsylvania early on March 14, 1993. Daytime highs on the 14th were extremely cold, especially in the northeast, ranging from 13°F at Tobyhanna to 22°F at Stroudsburg. Bitter cold invaded the East in the wake of the storm. Historic late-season cold records occurred at Kane (-22°F) and Pleasant Mount (-20°F).

Temperatures slumped to the single digits as far south as Birmingham, Alabama (2°F) and 33°F at Orlando, Florida.

A parting shot of frigid air on March 18–19, 1993, brought remarkably low morning readings of -14°F at Tobyhanna and -11°F at Hawley. Tobyhanna received 38.8 inches of snow in February and 43.2 inches in March 1993, raising the winter accumulation above 100 inches.

The "Storm of the Century" shut down virtually all interstate highways from Atlanta to Boston and closed the major airports. According to NOAA, "nearly 10 million people and businesses lost electricity" and hundreds of coastal homes from North Carolina to Long Island were damaged.

1993–94
Three Major Snowstorms

A brief thaw greeted Pennsylvanians on New Year's Day 1994, but a blast of winter was waiting just around the corner. A storm that formed east of the Colorado Front Range and central Plains on January 2, 1994, yielded to a Gulf Coast storm the next day.

On January 3, 1994, a secondary low-pressure system along the coast created a wide band of snow across the crest of the Appalachians. Historic snowfall rates of five to six inches per hour were observed at Pittsburgh and Williamsport in the morning (*Storm Data*). Twenty inches fell in downtown Pittsburgh, though the total at Allegheny County Airport was less (13.9 inches). Derry received 22 inches.

In northern Pennsylvania, six to 12 inches blanketed the region. The southeastern counties eastward across New Jersey were dealt a nasty blow in the form of an ice storm. Hundreds of thousands of residents lost power. The snow fell so quickly that traffic came to a stop along the Pennsylvania Turnpike and Interstates 70 and 79.

On January 6, 1994, Governor Robert Casey declared a state of emergency for Fayette, Greene, Washington and Westmoreland counties located in southwestern Pennsylvania, and the Pennsylvania National Guard was called on to assist in the arduous task of snow removal. The sheer weight of the snow caused roofs of barns and outbuildings to collapse. A NOAA storm summary, citing a Pittsburgh newspaper, reported 10 fatalities from heart attacks and 185 injuries, mostly sustained in falls.

The next winter storm arrived just a few days later, bringing moderate snow in the north and an ice storm of historic proportions to the southeastern counties. Low pressure traveled from the northern portion of the Rockies to the Middle Mississippi Valley and Kentucky early on January 7, 1994. A new storm formed along the Virginia coast on the 8th, but ran up against arctic high pressure over Ontario, Canada.

A morning glaze thickened to a half-inch in Philadelphia, and an inch of ice coated the northwestern suburbs, causing severe outages as tree limbs and lines snapped under the weight of the ice. Thousands of traffic accidents and many injuries from falls were reported in the greater Philadelphia area, though no deaths occurred.

The *Storm Data* summary estimated that 590,000 PECO customers lost power—a record 40 percent. The previous worst outage occurred after the great March 20, 1958, wet snowfall (400,000 customers). The heaviest losses were in Chester, Bucks, Montgomery and Philadelphia counties. Full power was not restored until January 11, 1994.

On January 17, 1994, an arctic boundary and traveling wave of low pressure brought another round of wintry weather to the Commonwealth that commenced before daybreak. Snow came down at the rate of four inches an hour at Chambersburg, with a total accumulation of 20 inches. The observer at Greencastle reported 17 inches, though most areas across the state reported six to 12 inches. Several barns and small buildings in Franklin and Mifflin counties collapsed in the storm.

Snow depths on January 18, 1994, in parts of Pennsylvania averaged close to two feet. Some representative depths ranged from 22 inches at East Stroudsburg to a then-record 27 inches at Scranton, surpassing the historic 26-inch depth on February 4, 1961. The snow depth reached 30 inches at Lehighton and 40 inches at Long Pond on the 18th.

Exceptional cold arrived in the wake of the storm. The high Allegheny Plateau region had a subzero day on January 19, 1994. Communities in eastern Pennsylvania established all-time record lows between -15°F and -30°F on January 21, 1994.

More snow fell on January 26–27, 1994, and by the end of the month, records for January were notched at East Stroudsburg (37.2 inches) and Scranton (42.3 inches, establishing an all-time monthly record). Many monthly snowfalls soared past 40 inches throughout the Poconos: 42.8 inches at Mount Pocono; 47 inches at Honesdale and Long Pond; 59.1 inches at Hollisterville.

In the Lehigh Valley, 33.9 inches of snow fell at Allentown in January 1994—the greatest monthly snow total since January 1925 (43.2 inches). In the Susquehanna Valley, a new record snowfall at Harrisburg of 34.2 inches edged out that of January 1961 by 0.4 inch, a record that would fall just two years later. Williamsport recorded 34.5 inches.

Back-to-back disturbances brought heavy snow and ice again on February 8–9 and February 11, 1994, with large snowfalls extending northward through New York City to Boston. A low-pressure trough extended to the Gulf on February 8, as waves of low pressure formed near South Carolina and rippled northward. A second storm over Tennessee transferred energy to the coast on February 10, becoming the main system.

Areas to the south were again paralyzed by several inches of sleet and ice. Delaware suffered its worst ice storm in modern memory in the first storm. On February 10, the Deep South was hit with a glaze storm as the second disturbance rounded the bend and headed northward, depositing four inches of sleet around Washington, D.C.

Coastal cities bore the brunt of the second storm on February 11, 1994, with major falls at New York City (12.8 inches) and Newark, New Jersey (18 inches). Allentown received 9.2 and 9.8 inches, respectively, in the snowstorms.

March 1994 came in like the relentless winter. A severe blizzard hit Pennsylvania on March 2–4, 1994, dropping more than a foot of snow over much of the state, except in the far western and southeastern counties, where accumulations ranged from three to six inches.

On March 2, 1994, low pressure that traveled from West Texas to the South Carolina coast gained intensity over the relatively warm Atlantic, strengthening off the New Jersey coast early on March 3. The polar and subtropical jet streams converged near the Eastern Seaboard, setting up intense bands of snow in central Pennsylvania.

Snow began falling shortly before midnight on the 2nd, continuing without a let-up for about 28 hours. The heaviest falls occurred over the central mountains, leaving 27.7 inches at State College, equaling the March 1993 blizzard and establishing a new 24-hour snow intensity mark (26.6 inches).

Other large falls in the central mountains included: 27 inches at Rector; 24 inches at Lock Haven; 20 inches at Wellsboro; 18 inches at Johnstown; 17.2 inches at Altoona; 14 inches at Punxsutawney; 7.8 inches at Pittsburgh. On the Pocono Plateau, amounts reached 21 inches at Long Pond, tapering to 12.1 inches at East Stroudsburg.

Blowing snow brought rare blizzard conditions in Centre and Clinton counties. Deep drifts and an avalanche trapped 25 motorists between Renovo and Lock Haven (Clinton County) on Route 120 under upwards of 20 feet of snow. PennDot crews and the State Police used helicopters and snowmobiles to help locate and rescue stranded motorists. A number of roofs and outbuildings collapsed under the combined weight of the snow and ice from previous storms. Property damage was assessed at $5 million in central Pennsylvania.

The snowflakes fell so intensely that roads in north-central Pennsylvania were shut down as drifts reached eight to 10 feet along a three-mile portion of Route 120 near Baker Run. Snow depths achieved late season records in some northern mountain locations, as high as 49 inches at Philipsburg (March 4) and 40 inches at Pleasant Mount (March 6).

Snow blanketed the higher elevations again on March 18 and March 29, 1994, wrapping up an exceptional season. There was a great deal of variability in snowfall totals in 1993–94, from 23.4 inches at Philadelphia to 154.9 inches at Bradford in the northwestern mountains. In northern Wayne County, an inch or more of snow remained on the ground at Pleasant Mount for 110 consecutive days from December 19, 1993, until April 10, 1994.

Modern snow-cover records were established at many eastern cities. Scranton experienced 93 days of continuous snow cover in 1993–94. Allentown retained continuous snow cover through March 22 (78 days), and Harrisburg had 76 consecutive days with snow on the ground through March 21. In the upper Delaware Valley, snow remained at East Stroudsburg for 81 straight days until March 25. Mount Pocono had snow cover for 113 straight days (December 19, 1993, to March 31, 1994), comparable to the 116-day area record at Gouldsboro in 1969-70.

Scranton accumulated 90.4 inches of snow by the end of the winter of 1993–94, edging out 1904–5 by 1.8 inches. The season snowfall at Allentown established a new record of 75.2 inches. Farther north, the East Stroudsburg observer tallied 80.4 inches.

In the northeastern highlands, WNEP-TV meteorologist Tom Clark at Mountaintop (elevation 1,360 ft.) measured 106.7 inches for the season. The largest totals were tallied at Montrose (136.5 inches) and Hollisterville (132.1 inches). In the southern Poconos, 115.8 inches accumulated at Long Pond. Mount Pocono observer Michael Sandler measured 105.4 inches.

In the central counties, the Harrisburg snowfall in 1993–94 (75.9 inches) was second at the time to 1960–61 (81.3 inches). State College established a new seasonal mark of 109.3 inches, and Altoona recorded 99.9 inches. The seasonal accumulation at Pittsburgh of 76.8 inches was a little shy of the record set in 1950–51 (82 inches). Erie tallied 131.3 inches, the city's second snowiest winter on record at the time.

January 6-8, 1996
Blizzard of '96

The severe winter of 1995-96 was heralded by an early blast of arctic air on November 11-12, which brought a light coating of snow that blanketed the northern highlands.

A new disturbance formed on the southern boundary of the arctic air along the North Carolina coast on November 13, causing heavy snow to break out over the interior. The storm eased early on the 15th, after large accumulations buried the Allegheny Highlands.

Massive early-season falls of 32.2 inches at Somerset and 24 inches at Kane set November records. Pittsburgh received 10.6 inches on November 13–15, 1995, though with a drop in elevation of 500 feet to the Ohio River near the Ohio border the snowfall was only a trace.

Central Pennsylvania received a substantial snowfall. Amounts ranged from 7.6 inches at Williamsport to 16 inches at Mansfield and Troy, farther north, and 24 inches in parts of Clearfield County. Five hundred trees on the Pennsylvania State University campus at State College were damaged under the weight of a 17.7-inch snowfall—the second greatest November storm after 22.5 inches fell on November 12, 1968. Student volunteers joined maintenance workers in clearing Beaver Stadium for the Saturday football game.

In the northeast, Scranton picked up 8.9 inches, with amounts rising to 12.3 inches at Hollisterville, 14.3 inches at Mount Pocono, and 17 inches at Tobyhanna and Honesdale. In the eastern and south-central valleys, mixed precipitation held accumulations to two to four inches. A widespread snowstorm on November 29, 1995, brought four to seven inches. Monthly snowfall totals reached 9.2 inches at East Stroudsburg, 13.8 inches at Williamsport (November record), and 22.1 inches at Mount Pocono.

A cold December brought moderate snowfalls on December 9 and 14, 1995, followed by a bigger storm on December 18–20, which dropped six to 12 inches of snow over the interior, and up to 16 inches at Lehighton and Tobyhanna State Park.

Another winter storm on January 2–3, 1996, brought three to nine inches of snow, with a top fall of 11.2 inches at Hollisterville. But the biggest blow of the winter was just beginning to show up on the long-range computer models.

The seedling of the Blizzard of '96 came from a wave near the Louisiana coast on the southern end of a trough of low pressure that extended northeastward to West Virginia. A vigorous impulse diving southeast of the Rocky Mountains on January 5 became the catalyst for a massive winter storm. A strong corridor of winds aloft and two additional pieces of upper-level energy initiated the rapid development of a storm system that would soon grow into a blizzard of historic proportions.

A burgeoning low over the Gulf of Mexico on January 6, 1996, reached the South Carolina coast early the next day. Two centers of low pressure, one over eastern Tennessee and another in southeastern Georgia, consolidated along the North Carolina coast.

On January 6–7, interior snowfalls quickly piled up over Virginia: 24.9 inches fell at Roanoke; 24.5 inches at Washington Dulles International Airport, 26 miles west of downtown; and 17.1 inches at the nation's capital. Baltimore received 22.5 inches, and 19 inches piled up at Charleston, West Virginia. A stunning total of 49 inches fell at Big Meadows, Virginia (elevation 3,539 ft.), in Shenandoah National Park—a state record. Cincinnati, Ohio, also had a record 14.4 inches. (This mark was eclipsed on February 4–5, 1998, after a 17.9-inch snowfall.)

The snowy assault was underway in southern Pennsylvania shortly before midnight on January 6, 1996, as low pressure intensified east of the Virginia Capes. The slow-moving storm drifted to a position east of Delaware Bay early on the 8th. Winds gusted between 40 and 50 mph for several hours, reaching blizzard criteria.

The maximum snowfall in the Keystone State was measured at 38 inches in Adams County, according to the NOAA *Storm Data*. A spectacular record snowfall of 30.7 inches was tallied "by estimation" at Philadelphia International Airport, establishing a 24-hour record snowfall. The snow depth of 27.6 inches edged out the February 1885 mark (26 inches).

A retrospective in the *Philadelphia Inquirer* on the 25th anniversary of the Blizzard of '96 by Anthony Ward described the paralyzing effects of the storm in the city, including more than 30 emergency vehicles stranded in the deep drifts. The water equivalent total (1.55 inches) at Philadelphia International Airport led to inquiries regarding the 30.7-inch estimated snowfall, based on a 20-to-1 snow-to-liquid ratio. Yet, the Franklin Institute had a snow water equivalent of 2.00 inches. Following a four-year federal study, which included an analysis of surrounding measurements, the total was finally accepted by the government.

Huge snowfalls were reported in the southeastern counties at Spring Grove (36 inches), Reading 4 NNW (34 inches) and Perkasie (34 inches). The Reading observer noted a wind gust of 43 mph. Other noteworthy snowfalls included: York 31.8 inches; Marcus Hook 30.7 inches; Graterford 30 inches. In the Lehigh Valley, 25.6 inches at Allentown surpassed the February 1983 blizzard by 0.4 inch. (This total was eclipsed by the January 22–23, 2016, storm that deposited 31.9 inches.)

Excessive snowfalls extended from Harrisburg (24.2 inches) to Scranton 21 inches—including a record 20.6 inches in 24 hours. (This mark held until March 13–14, 2017, when 22.4 inches fell in a late-season storm.). In the southern Poconos, snowfalls topped 20 inches at East Stroudsburg (21.4 inches), Mount Pocono (22 inches) and Hawley (22 inches). The greatest accumulation occurred on the Pocono Plateau, with the aid of upslope lift: Gouldsboro 4 NE (36.5 inches); Long Pond (36 inches); Tobyhanna State Park (33 inches).

On the western fringe of the blizzard, unexpectedly high snowfalls occurred at State College (17.5 inches) and Pittsburgh (9.8 inches), which bogged down travel for days. Some roads were closed for up to 32 hours due to blowing and drifting. The northwestern corner of the state was largely spared by the storm, with only a few inches.

Fig. 4.3 January 8, 1996. Fayette County Head Start Program bus snowbound after a blizzard dumped two feet of snow on southwestern Pennsylvania. (Photo by Robin Rombach, Pittsburgh Post-Gazette *Photo Archives, copyright 1996, all rights reserved)*

A snow emergency was declared by Pennsylvania Governor Tom Ridge for 47 of 67 counties, closing roads except for emergency travel. Thirteen deaths were recorded in the state, all but one due to heart attacks from shoveling snow. Schools and airports closed, and flights were cancelled. The death toll in the Blizzard of '96 was 86 (NOAA).

Fig. 4.4 January 9, 1996. The author's father, Norman Gelber, in East Stroudsburg, digging out from the blizzard that deposited nearly two feet of snow, leaving a record January snow depth of 30 inches. (Photo by Judith Gelber)

In New Jersey, 30 inches fell at Hazlet and Moorestown, 27.8 inches at Newark (record depth of 31 inches), 20.6 inches at New York City and 23.7 inches at LaGuardia Airport in Queens. A reported total depth of 27 inches at Manhattan's Central Park exceeded the previous record of 26.4 inches (December 27, 1947). In southern New England, 22.8 inches fell at Providence, Rhode Island, and 18.2 inches buried Boston, where Logan Airport reported an all-time record snow depth of 32 inches.

The third in a series of heavy snowstorms hit on January 12–13, 1996, blanketing Pennsylvania beneath four to eight inches of snow, but with a corridor of very heavy totals in the northeast: 17.2 inches fell at Hollisterville and 25 inches at Montrose.

The snowy stretch raised snow depths to historic levels (Table 4.4) of two to five feet across much of eastern and central Pennsylvania, not widely experienced in the state, except for January 1780, January 1831 and February 1958.

Historic monthly totals reached 59.4 inches at Montrose, 58.5 inches at LaPorte and 57 inches at the Tobyhanna State Park office. Record falls for January were registered at Reading (46.6 inches), State College (41.2 inches) and Harrisburg (38.9 inches). Allentown (35.5 inches) had its snowiest January after 1925.

Table 4.5
Philadelphia Top 10 Snowstorms
(1884–2021)

January 7–8, 1996	30.7
February 5–6, 2010	28.5
December 19–20, 2009	23.2
January 23–24, 2016	22.4
February 11–12, 1983	21.3
December 25–26, 1909	21.0
April 3–4, 1915	19.0
February 12–14, 1899	18.9
February 16–17, 2003	18.7
January 22–24, 1935	16.7

The calamity of a month only got worse when an improbable series of three soaking rainstorms (January 18–19, 24–25 and 27–28, 1996) not only washed away the mountains of snow but inevitably resulted in disastrous floods. Adding insult to injury, heavy rain on January 18–19, 1996, totaling up to three inches, coupled with temperatures soaring into the 50s and 60s, washed away the huge snowpack and triggered disastrous flooding. Thousands of residents were evacuated and forced to flee the rising water. Eighteen people died in the flooding in Pennsylvania.

Moderate to heavy snowfalls on February 2–3, March 7–8 and March 28–29, 1996, affected most of the state and extended the harsh winter. A coastal

Table 4.4
January 1996 Snow Totals
and Record Depths

Monthly Snowfall		Snow Depth		Date
Montrose	59.4	Beavertown	55	Jan. 13
Beavertown	59.0	Laurelton Cntr.	52	Jan. 12
Laporte	58.5	Pleasant Mount	51	Jan. 16
Hollisterville	55.5	Hollisterville	50	Jan. 13
Laurelton Cntr.	51.5	Honesdale	47	Jan. 13
Shippensburg	48.7	Safe Harbor	45	Jan. 13
Reading	46.6	Paupack	42	Jan. 13
State College	41.0	Montrose	40	Jan. 13
Harrisburg	38.9	Wilkes-Barre	40	Jan. 14

disturbance interacted with a northern impulse to bring a rain/snow mixture on Easter, April 7, leaving a coating in the valleys and 7.6 inches at Mount Pocono, ending on the 8th.

Southeastern Pennsylvania picked up two to four inches of snow on April 9–10, 1996, and Atlantic City notched a record April snowfall of 5.7 inches. A final visitation of snowflakes left a dusting in the northern mountains on May 12—Mother's Day—to cap off a memorable season.

The winter of 1995–96 was the snowiest on record at Scranton (98.3 inches), Williamsport (85.9 inches) and New York City (75.6 inches). Snowfall records were also toppled at Philadelphia (65.5 inches) and Baltimore (62.5 inches), though these figures were surpassed in the winter of 2009–10.

Allentown (75.4 inches) had its snowiest winter, edging out 1993–94 (75.2 inches). Harrisburg received 77.6 inches, the second heaviest season total after 1960–61 (81.3 inches). East Stroudsburg observer Judith Gelber (the author's mother) tallied 98.6 inches of snow for the season, tying 1960–61 for the third snowiest winter since 1910.

Erie (129.2 inches) had its third snowiest winter up to that time, and Binghamton, New York, (133.5 inches) eclipsed the 1993–94 season by 3.2 inches. Totals in Somerset County at two high-elevation sites ranged from 162 to 167 inches.

The greatest snow total in the winter of 1995–96 in Pennsylvania was reported by a Montrose spotter (184.7 inches), according to WNEP-TV meteorologist Tom Clark in Scranton. A Gouldsboro weather spotter reported 161.6 inches, exceeding all previous season snowfalls in the official station records (1914–1987), formerly established in 1957–58 (147.1 inches). Lehigh Township observer Dennis Rinaldi recorded a spectacular total of 176.4 inches four miles northeast of Gouldsboro (elevation 2,140 ft.).

December 1996
Three Elevation Snowstorms

The first of three elevation storms struck northeastern and central Pennsylvania on December 5–6, 1996, depositing anywhere from a slushy coating in the valleys to 2.8 inches at State College, 3 inches at Tannersville, 6 inches at Scotrun and 15.5 inches at Pocono Mountains Municipal Airport (elevation 1,916 ft.). Even greater falls occurred above 2,000 feet: 22.5 inches at Camelback Mountain (elevation 2,131 ft.) and 21.5 inches at Pimple Hill (elevation 2,215 ft.).

Another strong impulse buckled the jet stream on December 7, 1996, inducing a second wave of low pressure that brought three to seven inches of snow to eastern Pennsylvania on higher terrain. Interior Massachusetts and New Hampshire were clobbered with as much as 18 to 30 inches of snow as the storm intensified off the New England coast.

A third blow in less than a week came on December 11–12, 1996, as a coastal storm virtually stalled offshore under a deep upper-level trough, bringing another round of heavy snow to northern Pennsylvania. Four inches fell in one hour on Godfrey Ridge, east of Saylorsburg, with a storm total of 8.5 inches. A foot of snow blanketed the Pocono Plateau. Branches snapped under the weight of the wet snow, bringing down power lines along Route 209 southwest of Stroudsburg.

The Mount Pocono observer, Michael Sandler, measured 34.6 inches in December 1996, exceeded only twice in December at the longtime Tobyhanna site (1966, 1969).

Winter of 2002–03
Two Great Holiday Snowstorms

The winter of 2002–03 kicked off early with a big Christmas holiday snowstorm. An upper-air system in the Southwest on December 21 interacted with a strong subtropical jet stream to form a deep storm that lifted from South Texas to the Tennessee Valley, drawing moisture northward. Thunderstorms broke out across the Southern states in the warm sector.

Low pressure over Mississippi on Christmas Eve gave way to a secondary storm near the Virginia coast situated east of Atlantic City on the morning of December 25, 2002. Rain turned to snow in the eastern counties, piling up quickly to depths of five to 10 inches atop a layer of ice.

A faster changeover and the upslope winds buried the higher elevations. Storm totals in the northeast reached 19 inches at Montrose and 20 inches at Hawley. The largest amounts were 26 inches at Tobyhanna State Park (elevation 2,212 ft.) and 24 inches at Hollisterville. Snow showers added another 1.5 inches at Tobyhanna through the 27th. Persistent chill and periodic light to moderate snows brought the feel of a traditional old-fashioned winter in 2002-03, with a perpetually snow-covered landscape.

The combination of a polar high drifting south to the Great Lakes and a prolonged flow of moisture generated a much more widespread deep snowfall on February 15–17, 2003, that stretch from the Midwest to New England. Low pressure formed in the lee of the southern Rockies on February 14 and tracked east to the Tennessee Valley on the 16th. The strengthening storm reached the Virginia coast the next day, coming under the influence of blocking high pressure and a cold northeasterly circulation.

Light snow broke out over western Pennsylvania early on February 15, 2003, with the first batch of upper-level energy. The snow became steadier and heavier overnight on February 16–17, as low pressure organized along the coast. A heavy band extended from Columbus, Ohio (15.5 inches), to the western Appalachians. One to two feet of snow blanketed east-central Ohio, with as much as 30 inches at Woodsfield near the West Virginia border, while an ice storm crippled transportation in the southern part of the state.

Snowfall totals mounted through the long Valentine's Day weekend from the upper Ohio Valley to the Mid-Atlantic Coast, lasting into Presidents' Day. Accumulations exceeded two feet in southern and eastern Pennsylvania. Some of the largest falls were: 31 inches at Enola; 26 inches at Reading 4 NNW; 25 inches at Biglerville; 24.1 inches at East Stroudsburg; 23 inches at York; 22 inches at Reading; 20.7 inches at Philadelphia; 20 inches at Allentown. In New Jersey, Green Pond collected 27 inches in Morris County.

The snowfall was impressive inland: 15.1 inches at Pittsburgh; 18 inches at State College; 20.3 inches at Harrisburg. Enhanced lift over the Allegheny Mountains generated massive accumulations in the Laurel Highlands, where 35 inches piled up at Buckstown (elevation 2,460 ft.), with a total depth of 49 inches and monthly accumulation of 65 inches. The resort community of Seven Springs (elevation 2,550 ft.) in Somerset County recorded 41.5 inches in 37 hours at 10:00 p.m. on February 18, 2003.

Keysers Ridge, Maryland, set a state snowfall record (49.5 inches). The total February snowfall at Eagle Rock (81.5 inches) achieved a new state monthly snowfall mark at the time (surpassed in

2010). Baltimore set a snowfall record with 28.2 inches (topped by an inch in January 2016). Bear 2 E in Delaware reported 27.6 inches.

The snowy winter of 2002–03 brought 99.8 inches at Mount Pocono and 110.6 inches at Montrose. The final figure of 189 inches at Laurel Summit (elevation 2,730 ft.) included impressive contributions of 62.3 inches in January and 59 inches in February.

Fig. 4.5 January 8, 2005. Residents described the destructive ice storm as a "war zone" on the Pocono Plateau between Blakeslee and Tobyhanna. (Photo by Michael Pontrelli)

February 13–15, 2007
Valentine's Day Storm

The winter of 2006–07 had been very mild and relatively snow-free in the Northeast during an El Niño season that promoted a brisk westerly flow of Pacific air—until the pattern broke down in February 2007 with an amplified western ridge.

On February 11–12, 2007, a large upper-level low-pressure system advanced into the southern Plains, spinning up a storm over Texas that traveled northeast across Arkansas and Kentucky. A large polar high over the Canadian Prairies drifted into the northern Great Lakes, feeding a stream of arctic air into the eastern half of the country, accompanied by strengthening winds that reflected a deepening pressure gradient.

Snow broke out early in the day across southern Pennsylvania, with a mixture of sleet and freezing rain later dominating the precipitation pattern, as upper energy fed a secondary slow-moving wave along the Mid-Atlantic Coast. The nor'easter became the dominant system on Valentine's Day 2007, bringing excessive wintry precipitation whipped by strong winds. The warm nose was prominent at 10:00 p.m. on the 13th, when East Stroudsburg had driving sleet at 11°F and Reading reported freezing rain at 17°F.

Interstates 78, 80 and 81 were forced to close during the afternoon and evening hours on February 14, 2007. A perilous traffic jam developed on Interstate 78 in the eastern part of the state because it took more than a day to close all the freeway entrance ramps. Vehicles became trapped for more than 24 hours and motorists were rescued by National Guardsmen in Humvees, who provided fuel, food and essential supplies to stranded motorists.

The holiday storm brought strong winds and blizzard conditions across the northern portion of the state, where a powdery snow piled up to depths of 10 to 20 inches. The largest reported snowfalls were: East Benton (Lackawanna County) 23 inches; Pleasant Mount (Wayne County) 22 inches; Wyalusing (Bradford County) 21.5 inches; Meadville (Erie County) 20 inches. Scranton was on the cutting edge of the snow-ice boundary and received 17.1 inches; Pleasant Mount recorded 22 inches.

A prolonged period of sleet and freezing rain limited the snow totals to four to eight inches in the eastern and southern counties. A wind gust of

52 mph was recorded at the Mount Pocono airport just before 7:00 p.m. on the 14th. The storm caught state road crews off guard, who did not anticipate the intense snowfall rates. Cleanup costs were estimated at $20 million, according to PennDOT officials.

Airports from Chicago to Philadelphia were closed for one to two days, caught in the swirl of snow and ice, costing the airlines $153 million. Burlington, Vermont, set a 24-hour snowfall record (25.3 inches) on February 14, 2007, with a storm total of 25.7 inches. The only heavier snowfall at the time was 29.8 inches on December 25–28, 1969.

Overall, winter storm damages totaled $1 billion, and 41 died in storm-related accidents and from hearts attacks in the Midwest and Northeast.

Winter of 2009–10
Three Historic Snowstorms

The historically snowy season of 2009–10 will be long remembered throughout the Mid-Atlantic region for prodigious snows near and south of the Mason-Dixon Line. At the same time, the ground in Upstate New York and Vermont lay bare of snow because the primary storm track passed south of Pennsylvania and directly off the coast.

North American Blizzard I, dubbed "Snowpocalypse" in the media, swept up the Eastern Seaboard on December 18–20, 2009. Low pressure formed along a cold front in the Gulf of Mexico and combined with a weak low in the Upper Mississippi Valley, before the energy shifted to the coast.

Historic December snowfalls piled up at Baltimore (21.1 inches) and Philadelphia (23.2 inches) on December 19–20, 2009, surpassing Philadelphia's previous record December fall of 21 inches in 1909, though the snow bypassed much of northern and western Pennsylvania.

On February 4, 2010, low pressure in the western Gulf on a cold front tracked across the Mid-South to Kentucky, before yielding to a deeper secondary storm along the North Carolina–Virginia coast on February 5–6. The storm nicknamed "Snowmageddon" exploited entrenched arctic air over the Northeast and a broad fetch of Atlantic moisture.

Snow began falling heavily midday on February 5, 2010. Massive snow totals piled up across southern Pennsylvania, but again evading the northern counties. The Philadelphia area received its biggest snow since the city was settled—a whopping 28.5 inches bested the historic January 1996 snowfall (27.6 inches). Impressive totals were: Pittsburgh 21.9 inches; Lancaster 21 inches; York Haven 20 inches; Altoona 19 inches.

An upslope flow component contributed to an impressive fall of 33 inches at Laurel Summit, with 29.5 inches falling in 24 hours in the southwest. Massive accumulations were observed throughout the Mid-Atlantic: Colesville, Maryland 40 inches; Washington, D.C. (Dulles Airport) 32.4 inches; Wilmington, Delaware 25.8 inches; Baltimore 25 inches; Estell Manor, New Jersey 24 inches; Atlantic City 18.2 inches.

Another batch of Pacific energy combined with a disturbance dropping southeast from Canada a few days later and phased over Ohio to form another slow-moving winter storm on February 9, 2010. Thick snowflakes were flying again across much of Ohio and Pennsylvania, as the energy shifted to a coastal disturbance.

The largest snowfall on February 9–10, 2010, in Pennsylvania was 26 inches at Landisville, leaving a record depth of 38 inches. Hanover recorded

23 inches, the third major snowfall of the winter. Snowfall totals nearly matched the event days earlier: Baltimore 19.5 inches; Allentown 17.8 inches; Philadelphia 15.8 inches; Harrisburg 15.7 inches. The total depth of 36 inches measured at 7:00 p.m. at Baltimore on February 10, 2010, shattered the February 13, 1899, mark (30 inches). The next morning, snow depths ranged up to 46 inches at Savage River, Maryland, and 31.5 inches at Dover, Delaware.

Wind gusts between 30 and 50 mph severely hampered travel and the dig-out from snow drifts as high as five feet. A total of 5,900 airplane flights were cancelled. New York City called out 1,000 snowplows and 4,200 workers in a snow removal effort that cost the city an estimated $1 million per inch.

Meanwhile, another winter storm was already sweeping across the southern Plains on February 11–12, 2010, which deposited an improbable 12.5 inches on Dallas, Texas, and dusted northern Florida. Satellite data indicated that snow covered the ground in every state, including a small patch atop Mauna Kea on the Big Island of Hawaii.

The historic winter of 2009–10 was not finished. A persistent upper-air trough held in check by an abnormally strong blocking high in the North Atlantic kept the cold air locked into place, causing storms to rotate north along the East Coast.

A surface wave originating over the Gulf of Mexico was captured by northern energy near Lake Erie as it moved up the Atlantic Coast. The first wave brought rain on February 23–24, 2010, that turned to snow inland, depositing 8.9 inches at East Stroudsburg and 10 inches in the southern Poconos.

But that was only a prelude to a monster snowstorm that took shape off the Virginia coast on February 25, 2010, eventually slowing to a crawl south of Long Island, as the center deepened to 28.70 inches (972 millibars)—an extraordinarily intense system. Bombogenesis resulted in an extreme pressure gradient that generated high winds. Intense snowfall buried southern New York and northern sections of Pennsylvania and New Jersey northward to southwestern New England.

The Northern Tier of Pennsylvania received upwards of two feet of snow on February 25–26, 2010, with the heaviest fall at Bear Creek (28 inches). The northeast was hard-hit—22.1 inches fell at Pocono Peak Lake and 20.5 inches at Mount Pocono, with a few additional inches falling on February 27–28. The greatest snow depth reported in the Poconos was 52 inches at Gouldsboro 4 NE (elevation 2,140 ft.).

In the Allegheny Mountains, upslope snowfalls were enhanced by a pocket of upper-level energy. A tremendous snowfall of 34.6 inches was measured at Laurel Summit from February 25–27, 2010, with additional light snow raising the total depth to 50 inches on the 28th. The monthly accumulation of 117.8 inches established a single-month Pennsylvania record. Other spectacular monthly catches included 96.5 inches at Chalk Hill 2 ENE and 81 inches at Ebensburg Sewage Plant. In York County near the Maryland border, Hanover logged 53 inches.

Historic snowfall marks were widely established in February 2010: Mansfield, Ohio, 52.3 inches; Philadelphia 51.5 inches; Baltimore 50 inches; Pittsburgh 48.7 inches; Harrisburg 42.1 inches; New York City (Central Park) 36.9 inches; Atlantic City 36.6 inches.

Despite little or no measurable snow in March and April 2010, a seasonal snowfall record was easily toppled at Philadelphia (78.7 inches), surpassing 1995–96 (65.5 inches). The East Nantmeal Township observer in Chester County reported 86.6 inches. In the Alleghenies, Laurel Summit tallied 218.3 inches, and Somerset had 168.9 inches. Oakland, Maryland, reported a season total of 214.3 inches, with a huge February contribution (111.3 inches). Both figures set Maryland snowfall records.

December 26–27, 2010
Powerful Post-Christmas Snowstorm

An intense nor'easter brushed past the Atlantic Coast on December 26–27, 2010, with a low pressure of 28.38 inches (961 millibars). Deep snows and fierce winds pelted the coastal plain, bringing some of the greatest snow totals ever recorded in December in the region.

Philadelphia received 12.4 inches of snow, with amounts soaring to 20.1 inches at Atlantic City, 20 inches at New York City, 24.2 inches at Newark, and 31.8 inches at Elizabeth, New Jersey. Mount Pocono recorded a wind gust of 59 mph at daybreak on the 27th.

October 29, 2011
Snowtober

Snow in October in Pennsylvania is not rare, especially over higher terrain and in the northwestern highlands.

Yet, nothing in the weather history books compares to the Halloween Weekend snowstorm of October 29, 2011, which brought historic early-season falls across much of Pennsylvania on the Saturday before Halloween. The weight of the compacted moisture-laden snow bent and toppled tree limbs and power lines, leaving 500,000 customers in eastern Pennsylvania without power for several days.

A low-pressure wave formed along a cold front over North Carolina on October 28, 2011, and then headed up the Eastern Seaboard. Cold, dry Canadian air lay to the north and proved sufficient in depth to reduce temperatures to near freezing when the snow increased in intensity in the morning hours of October 29.

Measurable snowfall reached all the way to Philadelphia (0.3 inch) and totaled six to 12 inches over the interior of the state. The heaviest snowfall piled up over the higher elevations in the eastern sections: 16 inches fell at Springtown (elevation 860 ft.) in Bucks County and Hazleton (elevation 1,800 ft.). In the central mountains, Dunlo picked up 12.3 inches.

Many places roughly doubled their previous October snowfall marks: Scranton (9.1 inches); Allentown (6.8 inches); Harrisburg (5.5 inches). Totals in the Poconos reached 13.3 inches at Tobyhanna, 11.5 inches at Bushkill, 8 inches at Tannersville and Delaware Water Gap, and 5.2 inches at East Stroudsburg.

February 13–14, 2014
Valentine's Day Deep Snow

As if the polar vortex visit in January 2014 was not enough, winter revved up again in February. An intense low-pressure system developed over the western Gulf of Mexico on February 12, 2014, and took a familiar path across the Florida Panhandle to the South Carolina coast early on the 13th.

The big storm chugged northward, spreading heavy snow from central Virginia and the Washington—Baltimore metropolitan areas to southern Pennsylvania before daybreak. Strong jet stream dynamics would give the storm a boost as it passed over the eastern tip of North Carolina to a position east of Delaware later in the evening.

A first wave brought up to 16 inches of snow to Chester County by early afternoon on the 13th, and an average of six to 12 inches in eastern Pennsylvania. After a break of about six hours, when the temperature edged above freezing and the precipitation changed to light rain and a mountain glaze, the atmosphere geared up for round two.

A dynamic upper low pivoted northeast from Virginia to southeastern Pennsylvania, bringing a second burst of heavy wet snow toward evening. Thunder and lightning accompanied this feature across central New Jersey. Snow fell at the rate of more than three inches an hour. East Stroudsburg observer Justin Gelber (author's son) measured 4 inches between 9:00 and 10:30 p.m., with huge flakes swirling in the wind. This onslaught almost equaled the February 1983 blizzard snowfall rate (four inches per hour).

The snowfall associated with the upper low finally tapered off between 2:00 and 4:00 a.m. on Valentine's Day 2014, after depositing a total of six to 10 inches and more than a foot in northeastern Pike County, all on top of the earlier fall. Storm totals recorded by spotters ranged as high as 25 inches five miles west of Milford, 20.1 inches at Dingmans Ferry and 19.6 inches at East Stroudsburg. The depth at the end of the storm ranged from 24 inches at East Stroudsburg (greatest since February 19, 2003—26 inches) to 30 inches at Bushkill (deepest since January 13, 1996).

Impressive storm totals included: Allentown 19.2 inches; Saylorsburg 17 inches; Blakeslee 16 inches. In the southeast, more than 20 inches fell northwest of Philadelphia. The heaviest snowfall was 21.6 inches at Morgantown, Berks County. However, CoCoRaHS spotters reported more than two feet, with the highest figure at East Nantmeal in Chester County (27 inches). Some additional totals included: Honey Brook 20.2 inches (Chester County); Birdsboro 20 inches (Berks County); Albrightsville 19 inches (Carbon County); Perkasie 17 inches (Bucks County).

Northwestern New Jersey had as much as 24 inches at Washington (Warren County), Rockaway (Morris County) and Holland Township (Hunterdon County).

February 2014 set monthly snowfall records at East Stroudsburg (43.9 inches) and Allentown (36.5 inches). Philadelphia amassed 68 inches of snow by the end of the month—the second most in any winter, after 2009–10 (78.7 inches). A dusting of snow (0.3 inch) on April 15 at Allentown increased the season total to 68.1 inches—the third heaviest accumulation in a single winter. Farther north at East Stroudsburg, the observer tallied 77.1 inches.

January 22–24, 2016
Snowzilla: East Coast Blizzard

A record warm December 2015 in the Northeast, in a winter with minor snowfalls through the first three weeks of January, changed abruptly when a snowstorm lumbered up the Atlantic Coast on January 22-23, 2016, delivering historic snowfalls.

A disturbance that came ashore in the Pacific northwest on January 19 traveled southeast on the eastern edge of a high-pressure ridge. A new area of low pressure formed in central Texas on January 21 that turned east across the Gulf states to Georgia, gathering moisture before yielding to a burgeoning coastal storm on January 22.

The nor'easter exploded east of Cape Hatteras, North Carolina, drawing energy from the unusually warm water. Wind-blown snow pummeled the Mid-Atlantic region on January 22–23, 2016, falling at the rate of two to four inches an hour for many hours as the storm slowed to a crawl off the Virginia coast. The wind increased to 35 mph creating blizzard conditions along the I-95 corridor.

Snowfall totals were maximized along the Interstate 78 corridor in Pennsylvania as far north as Schuylkill and Monroe counties. Accumulations ranged from 18 to 24 inches in the north between Pottsville and Stroudsburg, rising to around 30 inches in a band from Staten Island, New York, southwestward across central New Jersey and southern Pennsylvania.

Some of the outstanding snowfalls were record-breakers: Somerset 35.5 inches; Allentown 31.9 inches; New York City (JFK Airport) 30.6 inches, and a record depth of 28 inches; Harrisburg 30.2 inches; Baltimore 29.2 inches; New York City (LaGuardia) 28.2 inches; New York City (Central Park) 27.5 inches.

At the top of the list: Glengary, West Virginia (42 inches) and North Potomac, Maryland (38.5 inches), followed by Greencastle, Pennsylvania (38.3 inches), Gainesboro, Virginia (38 inches), Jackson Heights, New York (34 inches), Mount Mitchell, North Carolina (36 inches) and Morris Plains, New Jersey (33 inches).

The Mid-Atlantic cities hauled in substantial falls, notably 29.3 inches at Dulles Airport near Washington, D.C., and 17.8 inches in the nation's capital. Wilmington, Delaware had 17.8 inches.

Philadelphia was hit with the city's fourth greatest snowstorm (22.4 inches). Amounts grew significantly to the west and north: 32.4 inches at Spring Grove (York County); 32 inches at Shippensburg (Cumberland County) and Stowe (Montgomery County); 31.5 inches at South York (York County); 31 inches at Forks Township (Northampton County); 30.1 inches at Malvern (Chester County); 30 inches at Perkasie (Bucks County). In Reading, CoCoRaHS spotters reported totals ranging from 26.5 inches to 32 inches within a 10-mile radius of the city.

Impressive snowfalls piled up over northern Schuylkill County (24 to 30 inches—heaviest at Lake Wynonah) and southern Monroe County (26.5 inches atop Godfrey Ridge, with five inches falling between noon and 1:00 p.m.). Unofficial snow reports ranged as high as 28 to 32 inches near Kunkletown.

Fig. 4.6 January 23, 2016. Deep snowfall blankets the southern Poconos under 18 to 26 inches of snow. (Photo by Ben Gelber)

Farther north, 20.5 inches fell in Jackson Township to the west of Tannersville. Along the Interstate 80 corridor, East Stroudsburg received 18.2 inches and 14 inches blanketed Tobyhanna. Amounts tapered off dramatically just to the north, with merely one to three inches falling between Hawley and Scranton.

Fig. 4.7 January 23, 2016. Cars buried in the wake of an 18-inch snowstorm in East Stroudsburg. (Photo by Ben Gelber)

The January 2016 snowstorm ensnared upwards of 500 vehicles on the Pennsylvania Turnpike in the southwestern part of the state, trapping some motorists for part of the weekend. Rescue teams fanned out to provide food and water for those spending the night on the highway. Along the coast from Virginia to New Jersey, winds gusted to between 60 and 75 mph, and tides 10 feet higher than normal sent water pouring into low points along the southern Jersey Shore.

The death toll rose to 55, including heart attacks and traffic accidents. The number of power outages (480,000) and more than 10,000 cancelled airline flights attested to the magnitude of the blizzard that had a billion-dollar price tag.

March 13–14, 2017
Record Northeast Snowstorm

All the ingredients came together at the tail end of a warm winter to produce a snowstorm of historic proportions in northeastern Pennsylvania and southeastern New York less than a week before the official arrival of spring.

A favorable jet structure (coupled jet streaks) provided the upper-level support (lift) for a massive late-season nor'easter. A trio of upper-level disturbances in the arctic, polar and subtropical airstreams merged into a large storm along the coast of North Carolina and Virginia overnight on March 13–14, 2017. Mid-level frontogenesis, moisture convergence and upglide over a dome of arctic chill contributed to prodigious snowfall totals.

The storm passed east of Atlantic City around dawn on March 14, 2017, later moving past Long Island. Banding of snow in the northwestern quadrant of the storm brought record-breaking snowfall totals from the Lackawanna Valley northward to the Finger Lakes region of New York.

The largest snowfalls were reported in Wayne and Susquehanna counties, close to the New York border, and southward through northern portions of Lackawanna and Pike counties: Oakland 34.5 inches; both Honesdale and Ransom 31.1 inches; Damascus 30 inches; Towanda 27 inches; Lords Valley 26.3 inches; Hawley 25.9 inches.

Scranton's total of 22.4 inches edged out the March 1993 blizzard (21.3 inches) to establish an all-time single snowfall record. The ridges north and east of the Lackawanna Valley received up to 30 inches of fluffy snow, just north of where a zone of sleet reduced totals along and south of the Interstate 80 corridor.

The southern Poconos received between 18 inches at East Stroudsburg to 25 inches at Mount Pocono and Bushkill. In the upper Susquehanna Valley, Williamsport received 18.3 inches. Amounts diminished to 10 to 15 inches in the Lehigh Valley and suburbs of Philadelphia due to an extended period of sleet and icy rain.

Fig. 4.8 March 15, 2017. Late winter snowstorm deposits two feet of snow on the higher elevations of northeastern Pennsylvania, at Resica Falls. (Photo by Ben Gelber)

Binghamton, New York, received a record 31.3 inches, beating the February 1961 intensity mark for a 24-hour snowfall (23 inches) by a wide margin, and ultimately piled up 34.9 inches through the next two days, augmented by lake-enhanced snow streamers. The heaviest snowfall in New York was reported at Hartwick 1 N (48.4 inches).

March 2, 2018
Bomb Cyclone

A rapidly intensifying storm south of Long Island early on March 2, 2018, deepened to 28.76 inches (974 millibars), packing hurricane-force winds in southeastern New England, with gusts to 93 mph at Barnstable, Massachusetts.

The powerful storm system performed a loop east of New Jersey, generating wind gusts of 71 mph at Cape May, New Jersey, and 60 mph at Mount Pocono and Allentown. Blizzard conditions developed across the northeastern corner of Pennsylvania as winds reached 50 to 60 mph in Wayne, Pike and Monroe counties.

A driving rain (one to two inches) on the first day of March changed to snow around or shortly after daybreak on March 2, 2018, with snow falling at the rate of one to two inches an hour in the eastern part of the state. Accumulations averaged six to 12 inches north and west of Philadelphia. In the Poconos, 20 inches fell at Tobyhanna and 17 inches piled up at Mount Pocono.

Trees and power lines weighed down by heavy wet snow and rooted in soggy soils snapped and collapsed during hours of high winds, blocking roads for days. Vehicles were stranded on snow-clogged and tree-strewn roadways. Homeowners in rural parts of northeastern Pennsylvania were unable to leave their homes, which were obstructed by so many fallen trees. PECO Energy Company reported more than 616,000 customers lost power at the height of the storm in the greater Philadelphia area, and PPL Electric Utilities noted more than 60,000 were without power in the Pocono and Lehigh Valley to the north.

Governor Tom Wolfe activated the National Guard to rescue motorists and clear a path through downed trees and power lines. Warming shelters were set up for residents without power for up to five days after the storm. More than two million homes and businesses lost power in the Mid-Atlantic and Northeast. The snowfall in eastern New York ranged as high as 40 inches at Sharon Springs, west of Albany, and Syracuse was blanketed by 16.7 inches.

On March 6–7, 2018, a low-pressure area in the upper Ohio Valley began to transfer energy off the Virginia Capes, setting the stage for another nor'easter, which deposited six to 12 inches of snow on eastern and central Pennsylvania, with locally heavier amounts northwest of Philadelphia.

Top snowfalls were 16 inches at Richboro (Bucks County) and 14 inches at Rosemont (Montgomery County), with mixed precipitation turning to snow at Philadelphia (6.1 inches).

Tens of thousands of Pennsylvanians were still without power from the March 2–3 snowstorm when the second storm arrived. In the northwest, lake-enhanced snows totaled 15.6 inches at Erie, March 7 9, 2018. One to two feet of snow buried central New Jersey, with up to 26.8 inches at Montville (Morris County).

A third coastal storm in 10 days largely skirted Pennsylvania on March 12-13, 2018, with several inches of wet snow, but dumped one to two feet on southeastern New England. Yet in the northwest, more lake moisture contributed to snowfalls of 19.5 inches in Amity Township and 15 inches at Edinboro.

The first day of spring brought a fourth nor'easter, as a secondary storm developed east of the Delmarva Peninsula on March 21–22, 2018. Snowfall totals averaged 10 to 15 inches across southern Pennsylvania, with less amounts north and west that came in two rounds. In the southwestern mountains, 19.5 inches piled up at Winber 7 SE in Somerset County, and 19 inches fell at East York.

The major cities received a respectable helping of wet snow: Allentown 12.3 inches; Philadelphia 11.9 inches; Harrisburg 10.4 inches. Northwest of Philadelphia, 16.5 inches accumulated at Lansdale in Montgomery County. Snow amounts reached 14.5 inches at Queens Village and 20.1 inches at Patchogue on Long Island.

At East Stroudsburg, the March 2018 snowfall (30.6 inches) was the second greatest, after 1967 (33.5 inches). In the northwestern corner of the state, Erie (36 inches) had its snowiest March, surpassing 2008 by more than two inches. The snowfall in 2017–18 reached 166.3 inches, including 3.3 inches in April 2018, establishing a new single season snowfall record at Erie International Airport.

A final coastal disturbance on April 2, 2018, added another four to eight inches, coupled with minor snows on April 6 and 20, during one of the coldest Aprils on record, capping off an unusually snowy season at East Stroudsburg (73.2 inches) in the upper Delaware Valley.

December 16-17, 2020
Record North-Central Snowfall

Low pressure that produced an area of snow in the southern Plains on December 14, 2020, tracked to Kentucky, while broadly merging with a secondary coastal storm in the Southeast.

The rapidly intensifying winter storm triggered a slew of snow emergencies in the Mid-Atlantic area. An expansive stripe of accumulations that exceeded a foot extended from northern Virginia to Maine.

A shield of heavy snow overspread the Keystone State from the southwest beginning in the morning of December 16, 2020, that expanded northeastward, mixing with sleet in the eastern counties during the evening. Low pressure moved from eastern North Carolina to Chesapeake Bay, then off the coast near Atlantic City to east of Long Island, leaving a legacy of historic snowfalls over the Northern Tier of Pennsylvania and southern New York.

The ingredients came together along the East Coast. An upper-level trough, amplified by a mid-level disturbance, teamed up with a jet streak to produce ample snowfall rates. Additional energy was available from a surge of warm air, promoting additional lift, colliding with a fresh supply of arctic air from blocking high pressure over Quebec, Canada. Steep lapse rates aloft (rapid cooling) accounted for snowfall rates of five to six inches an hour at Binghamton (20.5 inches in six hours).

Massive totals above two feet piled up in the central and northern mountains. In Bradford County, near the New York line, 36 inches fell at Alba 3 ENE (revised total reported by NOAA's State Climate Extremes Committee). Accumulations in the county ranged from 30 inches at Towanda to 36 inches at Troy. The Binghamton, New York, NWS office measured a storm total of 40 inches, including 20.5 inches in six hours, smashing the previous record snowfall (35.3 inches on March 13-14, 2017), and establishing record snow depth of 39 inches.

Massive snowfalls occurred throughout all of the Northern Tier counties. The heaviest storm totals were reported in the upper Susquehanna Valley. Top figures were measured at Blossburg (31 inches), Canton (29 inches), Covington and Towanda (28 inches), and Susquehanna (23.5 inches). Snowfalls of 20 inches were measured at East Benton and Harveys Lake northwest of the Scranton area.

In the north-central mountains, the snowfall at Stevenson Dam reached 29.4 inches and 28 inches at Renovo.

The snow at Williamsport piled up to a depth of 24.7 inches, breaking the long-term record of 24.1 inches in January 1964. A fall of 24 inches at Wellsboro was among the heaviest on record.

The remainder of the state reported six to 12 inches. A total of 61 of 67 Pennsylvania counties were under a winter warning. Ironically, the only county in the state that did not fall in the range of a winter weather alert was Erie County.

The snowfall at Philadelphia (6.6 inches) was noteworthy because it ended a snow drought in the city, after 0.3 inch fell in the winter of 2019-20.

January 31-February 2, 2021
Long Storm

Another in a series of cross-country storms that brought flooding rain and deep mountain snows in California and the Southwest charged east in the closing days of January 2021.

A classic Miller B winter storm, which featured an Ohio Valley upper-air low moving in from the west, interacted with a developing coastal storm, as energy translated east of North Carolina. High pressure ensconced in the Northeast slowed the storm to a crawl (4 mph). Snow fell at Allentown for 50 consecutive observation hours.

As the coastal low deepened under a strengthening jet stream, fueled by Atlantic moisture and a cold conveyor belt (easterly winds), intense snowfall pivoted southwest across northern New Jersey to the Lehigh Valley on the afternoon of February 1, leading to copious snowfalls.

Spotter totals reached 36.1 inches at Nazareth, 31.2 inches at Springtown, 31 inches at Coopersburg, 30.5 inches at Bangor and 27.3 inches at Allentown—the second heaviest fall after January 2016 (31.9 inches). Heavy snow bands extended into northeastern Pennsylvania, where snowfall totals were a whopping 36 inches at Hawley, 33.2 inches at Spring Brook, 27 inches at Pocono Pines, 25 inches at Milford, 23.5 inches at Mount Pocono, and 19.1 inches at Scranton.

In the lower elevations on the edge of the southern Poconos, 27.3 inches fell four miles east of Saylorsburg, 26.1 inches at Kunkletown and 21 inches at Stroudsburg. In neighboring northwestern New Jersey, 36.2 inches fell southeast of Lake Mohawk, exceeding the state record (34 inches at Cape May in February 1899), 33 inches at Montague and 32 inches at Andover and Newton.

Farther west, 24.5 inches covered Albrightsville on the western edge of the Poconos, 20.5 inches at Schuylkill Haven and 20.5 inches in Franklin County in south central Pennsylvania. In Philadelphia and the southeastern part of the state to the lower Susquehanna Valley, 12 to 16 hours of sleet held totals to six to 12 inches. In the central mountains, accumulations ranged as high as 14 to 18 inches.

More Noteworthy Snowstorms

November

1809 November 24: "Strange to tell to future generations, snow about one foot, and tolerable good sleighing, a circumstance not known for many years if ever, in this land," in *Hazard's Register of Pennsylvania* (December 3, 1831).

1820 November 18–20: In the *United States Gazette*: a snowfall of 12 inches accumulated at Poughkeepsie, New York, with reports of heavy snow over the interior Northeast.

1831 November 21–22: Lewistown had a "severe snow storm" that left a depth of 10 inches (*Hazard's Register of Pennsylvania*, December 3, 1831). From the *Columbia Spy*: "It began to snow quite fast, the storm raging with great fury." The *Perry Forester* described extensive damage to trees, some twisted and toppled, as "the wind blew a hurricane." The heaviest fall, "about a foot," fell at Towanda. Several inches of snow and a coating of ice blanketed western and northeastern sections.

1862 November 7–8: The *Philadelphia Inquirer* reported that a "storm that commenced daylight on the seventh and continued until 2 o'clock [November 8] … has no parallel in any November." The *Easton Sentinel* stated that snow "reached a depth of about 14 inches," larger than any November snowfall in the Lehigh Valley. The Smithsonian station at North Whitehall caught 12 inches. In the southeast, 9.5 inches fell at Mount Joy in Lancaster County.

A correspondent from Delaware Water Gap reported 18 inches in the *Cultivator and Country Gentleman* (Albany, New York), which exceeds the modern November local snowfall record at Stroudsburg (14 inches on November 24, 1938).

1873 November 11–12: The first of two heavy early snowfalls blanketed Dyberry with 10 inches. The lower elevations had a few inches before a changeover to rain.

A second major storm on November 17–18, 1873, deposited 11 inches at Dyberry and Blooming Grove. The *Milford Herald* (May 5, 1874) cited S. D. Van Etten's diary that noted a 6-inch fall, and a Chester County site recorded 8 inches. The mean temperature at Philadelphia of 38.0°F is the lowest on record for any November.

1882 November 26–27 and 28–29: Back-to-back snowstorms affected the northeastern areas. On November 26–27, Blooming Grove had a total of 11.5 inches. Three to six inches fell in the valleys.

Heavy snow fell again on November 28–29, 1882, averaging four to six inches in eastern Pennsylvania, with 9 inches at New York City (14 inches in November 1882 ranks as the second snowiest November). Paterson, New Jersey, had a record snowy November (21 inches).

1885 November 22–26: A long storm brought alternating rain and snow in the lower elevations but deposited 24 inches at Tobyhanna, 19 inches at Canadensis (*Jeffersonian*), and 17 inches at Dyberry, where the observer confirmed "two feet" in the hills. The greatest snowfall was 27.5 inches at Drifton in Luzerne County. The central mountains had prolific falls. A press notice from Lock Haven stated: "Three feet deep in the hills."

Farther south, 10 inches fell at South Bethlehem, but only 3.5 inches at Easton closer to the Delaware River. Six to eight inches covered the hilly areas of eastern Pennsylvania (*Philadelphia Inquirer*).

1886 November 6–7, 12–13, and 25: The first of three major snowstorms to affect Pennsylvania deposited 16 inches at Warren in the northwest and 11 inches at Glen Summit in the northeast (*Jeffersonian*) on November 6–7, 1886. The elevated northern sections generally received three to six inches.

Heavy snow fell again on November 12–13, 1886: 18 inches at Dyberry and 13 inches at Warren, with a wintry mix and light falls in the lower elevations in the east.

A big Thanksgiving Day snowstorm on November 25, 1886, brought 12 inches at Scranton (*Jeffersonian*) and 10 inches at Dyberry. The three-storm barrage brought a total of 34 inches at Dyberry, a November record (1865–1903).

1892 November 9–10: An early season snowfall brought substantial totals in the east: Quakertown 7 inches; Harrisburg 4.8 inches; Easton 4 inches; New York City 2.3 inches.

A second storm on November 28–29, 1892, put down three inches at Philadelphia, six to nine inches in the southern Poconos (*Jeffersonian*) and up to 14 inches at Dyberry.

1898 November 26–27: A moderate snowfall on Thanksgiving Day, November 24, 1898, left an early blanket of snow up to four inches deep across eastern Pennsylvania, followed by strong winds and frigid temperatures.

A second, far more powerful coastal disturbance arrived on November 26–27, 1898, that was responsible for "some of the severest northeast gales and the heaviest falls of snow in New York and New England," according to the *Monthly Weather Review*. The steamer *Portland* made an ill-fated voyage and sunk off the coast of Cape Cod, Massachusetts, claiming 150 lives.

Low pressure over Michigan tracked southeast to Pittsburgh, merging with a deepening coastal low moving north from Hatteras, North Carolina. The consolidated storm center was located off the New Jersey coast by 8:00 p.m. on the 26th.

Strong winds developed as the storm "bombed" off the coast, creating deep drifts of up to 16 feet in New York City. The *New York Times* reported that "All the lines—horse, cable, and electric—suffered more or less." The barometric pressure reading at Boston dipped to 28.38 inches, accompanied by heavy gales. At least 50 persons died in the Portland Gale on land and 200 perished at sea.

The early-season East Coast blizzard brought record November snowstorm totals at New York City (10 inches) and Philadelphia (9.2 inches). The snowfall at South Bethlehem totaled 12 inches—heavier than any modern November snowfall. To the north, the Stroudsburg *Jeffersonian* reported 10 inches, with a "terrific gale of wind."

A windy, frigid early-winter day in the wake of the storm was followed by a third snowfall that brought moderate snow and sleet on November 29–30, 1898, ending the snowiest November: South Bethlehem 21 inches; New York City and Belvidere, New Jersey 19 inches; Philadelphia 9.2 inches.

1904 November 13: A heavy snowstorm extended from inland Virginia and North Carolina to Washington, D.C., and eastern Pennsylvania. A total of 6.1 inches accumulated at West Chester. Ten inches fell at Kennett Square and up north at Mount Pocono. The *Monroe Record* reported 10 inches at Bushkill. The Honesdale *Wayne Independent* (November 16, 1906) recalled falls of 12 to 18 inches in Wayne County.

1906 November 15–16: Another in a series of heavy mid-November snowstorms around the turn of the century deposited 12 inches at Scranton and Mount Pocono. A foot of snow blanketed Stroudsburg, recalled in the East Stroudsburg *Monroe Record* (November 20, 1910). At Easton, the fall was 4.5 inches.

1910 November 3–4: Rain changed to snow over the eastern highlands and accumulated to historic depths for so early in the season. The Milford observer Alla K. Doughty measured 4 inches, and also reported in her *Cooperative Observer's Meteorological Record* that 20 inches covered the high elevations of Pike County: "A remarkable fall of snow … varying in depth from an inch to 20 inches in various parts of the county. Back roads were blocked with snow and fallen trees. The rural mail carrier could not get through to deliver the mail."

The East Stroudsburg *Monroe Record* (November 20, 1910) mentioned 10 inches, the heaviest so early in the season. Other mountain reports included 15 inches at Tobyhanna and 20 inches at Porters Lake in southern Pike County. The Pocono Lake observer wrote down 12 inches (26 inches in total for the month). The Scranton total was 13.7 inches. Two to four inches covered the ground northwest of Philadelphia.

1935 November 17–18: A major pre-winter coastal storm slammed the interior Northeast, with hefty snowfalls and strong winds heaviest in the elevations. Amounts in Pennsylvania varied from 3.9 inches at Stroudsburg, with mixed precipitation, to 17.5 inches at Gouldsboro on the high Pocono Plateau. A foot of snow fell at Scranton.

1938 November 24: A great Thanksgiving Day snowstorm on November 24, 1938, snarled travel in the east: Gouldsboro 18.5 inches; Stroudsburg 14 inches (November record); State College 9.5 inches; Allentown 8 inches. Temperatures tumbled to record monthly lows on the 26th: Somerset -13°F; Quakertown -6°F; Gouldsboro -4°F; York -1°F; Stroudsburg 2°F.

A second storm on November 26–27, 1938, deposited seven inches at Allentown and four inches at Stroudsburg, establishing November snowfall marks of 15 and 18 inches, respectively. The greatest monthly fall was 21.2 inches at Philadelphia (Shawmont).

1939 November 5–6: An early season snowstorm blanketed interior Pennsylvania: 10.5 inches at Coudersport and 7.6 inches at State College. The Stroudsburg *Record* reported one to four inches covered the higher elevations.

1944 November 20–21: A heavy snowfall in the northeastern mountains opened a severe winter. The Stroudsburg *Record* reported a fall of 20 inches at La Anna in southern Pike County. Totals of eight to 14 inches occurred over high terrain. Mixed precipitation left several inches of snow and slush in the valleys. The month ended with a moderate snowfall on November 29–30, 1944, with up to six inches in the northern highlands.

1953 November 6–7: A massive and record early-season nor'easter brought wind gusts of 98 mph at Block Island, Rhode Island, and 78 mph at Brookhaven, Long Island. The powerful storm storm deposited 3 inches on Richmond, Virginia, and 6.6 inches at Washington D.C., with the low spinning northwest into southeastern New York. The snowfall totals became huge in Pennsylvania: Lock Haven 29.8 inches; Middleburg 27.5 inches; York 17.5 inches; Gouldsboro 17.3 inches; Harrisburg 15.4 inches; State College 13 inches; Mount Pocono 9 inches; Philadelphia 8.8 inches; Allentown 6.1 inches.

1954 November 2–3: For the second November in a row, heavy snow fell in the early part of the month, accumulating six to 12 inches across the northern highlands. The heaviest falls were 15 inches at Montrose and 13 inches at Wellsboro.

1962: November 3: Moderate to heavy snow blanketed much of the state, ranging as high as 11 inches at Montrose. Moderate falls occurred in the eastern valleys from Stroudsburg (2 inches) to Harrisburg (4 inches).

1968 November 12: Two early snowfalls kicked off winter in 1968–69. Wet snow totaling three to five inches blanketed northeastern Pennsylvania on November 10, 1968. Two days later, a full-fledged coastal system blasted the interior Mid-Atlantic with high winds and heavy snow.

State College was buried under 22.5 inches, with snow falling at the rate of three inches an hour. East Stroudsburg had 8 inches and Allentown received 6.4 inches. Three inches fell at Harrisburg and Baltimore before the snow turned to rain.

1971 November 24–25: The greatest Thanksgiving Day snowstorm in eastern Pennsylvania commenced in the early evening hours the day before the holiday, reaching a formidable depth. Although the snow changed to rain over the southeast, heavy snow continued to fall across the north.

Historic November falls occurred at Scranton (20.5 inches); Mount Pocono (24 inches); Freeland (27 inches); and Pleasant Mount (29 inches—43.1 inches set a November mark). The eastern valleys north of Lancaster, from Harrisburg to Allentown, received a little more than six inches before rain mixed in. Farther north, totals soared to 12.5 inches at East Stroudsburg and 18 inches at the high-elevation Zerbey Airport in Schuylkill County.

1980 November 17–18: Widespread heavy snowfall deposited four to 12 inches across the northern two-thirds of Pennsylvania: Ford City 12 inches; Pleasant Mount 11 inches; Pittsburgh 8.7 inches; State College 8.4 inches; Scranton and East Stroudsburg 8 inches.

1987 November 10–11: The Veterans Day Snowstorm of 1987 snarled travel on the holiday while bringing impressive storm totals: Stroudsburg 10.9 inches; Harrisburg 9.7 inches; Scranton 6.4 inches. The nation's capital had an all-time November record fall of 11.5 inches and up to 15 inches in the suburbs.

2007 November 18–19: A general snowfall of three to seven inches blanketed the lower elevations of eastern Pennsylvania north of Interstate 78. Farther north, 10 inches accumulated at Dushore and Laporte.

2014 November 26–27: At the end of the coldest 10-day period in late November since 1933 in Pennsylvania, a heavy snowfall commenced on the eve of Thanksgiving, dropping four to eight inches of snow on the eastern and central counties north of the Philadelphia area, and up to 12.3 inches in Lackawanna County at Newsom Ransom.

2016 November 19–21: A secondary coastal storm tapped into cold air pouring south behind a cold front to produce three to seven inches of snow over interior northeastern and east-central Pennsylvania. Totals ranged from 5.7 inches at Scranton to 12.8 inches at Pleasant Mount and 16.2 inches at Susquehanna. The snowfall at Binghamton (27.6 inches) and Syracuse (25.1 inches) continued through the 22nd to establish November records. A prolonged lake-enhanced contribution delivered 18 inches at Corry, Pennsylvania.

2018 November 15–16: A rapidly intensifying coastal storm interacted with a cold air feed from eastern Canada to bring a major snowfall to the Keystone State. The northern two-thirds of Pennsylvania received four to eight inches of snow. Rare early double-digit snowfall totals were reported in Centre, Franklin, Union, Somerset and Monroe counties. The top falls were: Windber 7 SE 14 inches; Coudersport 7 SE 13 inches; Bloomsburg 12 inches; State College 11.5 inches; Tyrone 11 inches; Mount Pocono 10.3 inches; East Stroudsburg 9.4 inches; Allentown 8.1 inches—a November record.

A delayed transition to rain resulted in a surprise heavy snowfall that coated the Interstate 95 corridor at New York City (6.4 inches—heaviest so early in the season), Philadelphia (3.6 inches) and Washington, D.C. (1.4 inches). The snow fell so heavily during the early afternoon on the easternmost leg of Interstate 78 in the Lehigh Valley that drivers were trapped in their vehicles for hours.

December

1840 December 4–6: An early heavy snowfall deposited 15 inches at Philadelphia (Peirce 1847). *Hazard's Register of Pennsylvania* reported 18 to 20 inches inland.

1841 December 16–17: In *Hazard's United States Commercial and Statistical Register* (January 5, 1842), an entry from the *Pennsylvania Inquirer and Courier* reported "near 18 inches." Over the interior, snow fell "to the depth of two feet" at Harrisburg and Pottsville—the greatest accumulation in the region so early in the season.

1887 December 17–18: The winter of 1887–88 was most famous for a great March blizzard, but the historic storm was actually not the heaviest snowfall of the season.

An early snowstorm on December 17–18, 1887, dumped up to two feet on eastern Pennsylvania. The Stroudsburg *Jeffersonian* put the fall at 26 inches in less than 24 hours, exceeding the modern

December record (18.8 inches on Christmas Eve 1966). A Reading dispatch in the same newspaper reported 18 inches, with snow still falling.

Professor Selden Coffin's measurement at Lafayette College in Easton came in at a respectable 22 inches, falling in 18 hours, starting at 5:00 p.m. on the 17th and ending by 11:00 a.m. on the 18th. This equaled the post-Christmas storm in December 1872 at nearby Egypt.

Areas south of the Lehigh Valley received much less, totaling 7 inches at Philadelphia and 10.6 inches in Baltimore. In northern New Jersey, 13.5 inches fell at Dover, dropping down to 6 inches at New York City. However, the *New York Times* reported much heavier falls at Poughkeepsie (12 inches) and Albany (16 inches).

1944 December 11–12: A deep snowstorm that deposited 12 to 18 inches in the western counties brought the greatest falls in the Monongahela River basin. From the *Climatological Data: Pennsylvania Section*: "Estimates of losses and expenses, incidental to the storm, range to 2,000,000 dollars in the Pittsburgh district."

The top 24-hour snowfall total was 26.5 inches at Erie—30.2 inches for the period of December 11–14, setting a city record that stood until December 2017. Recurrent lake-enhanced snows raised the December snow total at Corry to 76.5 inches.

1945 December 18–19: A nor'easter delivered a heavy snowfall in the east: Hawley 18.3 inches; 11.7 inches at Mount Pocono; 10 inches at Stroudsburg and Portland.

1997 December 29–30: An intense winter storm hugged the coast in late December 1997. Four to eight inches blanketed areas north of the Lehigh Valley, with a shift to freezing rain, but mostly snow persisted to the northwest: Long Pond 16 inches; Tobyhanna 15.2 inches; Honesdale 14 inches. Up to two feet accumulated in central New York.

January

1805 January 27–28: "Snowing been at it near 48 hours," noted Joseph Price in his diary at Wynnewood, northwest of Philadelphia, totaling 15 or 16 inches. Multiple heavy snows fell throughout the month.

1821 January 6–7: A major coastal disturbance brought rain and sleet to Charleston, South Carolina, on January 5, 1821, which became mostly snow over eastern Virginia. Secretary of State John Quincy Adams described a snowfall of 12 to 18 inches around the Washington area. At Philadelphia, the snow began at dusk on the 6th, piling up to an impressive depth of 18 inches (*Franklin Gazette*).

1936 January 17–19: Heavy snow, mixed with sleet and rain at the start, piled up to impressive depths in interior central and northeastern Pennsylvania: Gouldsboro 27 inches; Hawley 22 inches; Catawissa 22 inches; Scranton 20 inches; Stroudsburg 15.9 inches. Mixed precipitation in the southeast brought considerably less snow.

1945 January 15–17: Snow began falling in the late afternoon on January 15, 1945, continuing heavy at times on the 16th, mixed with sleet and freezing rain in the southeast. Hefty totals were recorded over the interior: Montrose 25.1 inches; Gouldsboro 22 inches; Hawley 21.3 inches; Harrisburg 21 inches; Stroudsburg 19.6 inches.

Gouldsboro had a total depth of 34 inches on the 29th. Montrose ended up with a record monthly snowfall of 57.7 inches and a substantial season total of 123.7 inches.

February

1829 February 20–21: The *Newport* (R.I.) *Mercury* had a citation of a big Philadelphia area snowstorm of "about 18 inches" that was judged to be "more severe than any within the memory of man" (*Connecticut Courant*, March 3, 1829).

1882 January 31–February 5: Back-to-back snowstorms on January 31 and February 4–5, 1882, left deep accumulations in the north. Dyberry received 15 and 17 inches, respectively, in the back-to-back storms. The total depth averaged three feet on Moosic Mountain northeast of Scranton (*Philadelphia Inquirer*).

1893 February 21–22: Snowy February, with moderate storms on February 12–13 and February 17–18, 1893, followed by a larger event on February 21–22, 1893: 20 inches at Dyberry and 15 inches at Stroudsburg (*Stroudsburg Times*). The February 1893 snowfall of 57 inches at Dyberry was a record for any month (1865–1903).

1894 February 25–26: Two heavy snowfalls, with wind, hit on February 12–14, 1894, and a heavier fall followed on February 25–26: Bedford 20 inches; Reading 16 inches (*Philadelphia Inquirer*); Stroudsburg 16 inches (*Jeffersonian*, March 11, 1894). New York City's Battery Park had 37.9 inches in February 1894—a record for any month in the city.

1906 February 8–9: Heavy snow in the interior northeastern counties: 22 inches at South Eaton and 15 inches at Scranton. Lesser amounts fell farther east with mixed precipitation.

1920 February 5–6: A massive coastal storm delivered a historic 33-inch snowfall at Gordon, atop the highest elevations of Schuylkill County. Gouldsboro reported 20.9 inches, Bethlehem 18 inches, Stroudsburg 15 inches, and New York City 17.5 inches.

Two moderate snowfalls at Gordon, in the higher elevations of Schuylkill County, including nine inches on the 15th, raised the depth to a record 48 inches, ending in a historic February monthly fall of 56.5 inches.

1921 February 19–20: Another heavy February snowstorm in the east: Gouldsboro 24 inches; Mount Pocono 21 inches; Bethlehem 17 inches; Stroudsburg 15 inches.

1926 February 3–5: A powerful snowstorm struck the eastern counties hardest again: Freeland 31 inches (all-time record snowfall, and a monthly total of 47.6 inches); Wilkes-Barre and Gouldsboro 20 inches; Tannersville 18 inches ("Very Bad Drifts Roads Closed," noted the observer); Scranton 15.5 inches. The East Stroudsburg *Morning Press* reported 16 to 20 inches and up to 25 inches in the Poconos. The *Morning Sun* gave a Stroudsburg total of 17 inches.

1940 February 13–15: A huge Valentine's Day snowstorm snarled travel for days: Towanda 27 inches (24 inches on the 14th); Hollisterville 23 inches; Scranton 17.3 inches; Gouldsboro 17 inches; Pittsburgh 16.3 inches.

1946 February 19–20: The Stroudsburg *Record* reported 15 to 18 inches in the hills and up to two feet on the ridges. Upwards of a foot of snow blanketed most of eastern Pennsylvania.

1947 February 20–21: A heavy eastern snow: 18.8 inches at Ephrata and 14 inches at Portland.

1972 February 18–19: An explosive storm centered off Atlantic City early on February 19, 1972, brought deep interior snows: Jersey Shore 24 inches; Williamsport 22.8 inches; Gouldsboro 16 inches; East Stroudsburg 12.2 inches.

1979 February 18–19: The Presidents' Day Blizzard struck a heavy blow along the southern fringe of Pennsylvania and the Mid-Atlantic. Temperatures at the start of the storm were in the single digits following a historic cold wave.

Philadelphia was blanketed with 14.5 inches; Harrisburg 14.2 inches; Somerset 14 inches. The snow depth at Cresson reached 42 inches on February 20. The heaviest fall was 25.7 inches at Rockville, Maryland, where the total reported snow depth on the airways observation reached 42 inches—a state record at the time.

1987 February 22–23: Intense snow bands lingered over Chester County for hours, producing historic snow intensity rates—18 inches fell at Coatesville between midnight and 4:00 a.m. on the 23rd—for a storm total of 23.5 inches. Most areas in southeastern Pennsylvania received six to 10 inches of snow.

March

1823 March 30–31: The Great Easter Blizzard of 1823 wreaked havoc along the Eastern Seaboard, bringing coastal rain and a deep inland snowstorm. Powerful winds toppled trees in eastern Pennsylvania and New Jersey. Two feet of snow piled up at Carlisle. The Stroudsburg *Jeffersonian* (April 20, 1854) recalled 24 inches locally.

1841 March 16–18: The *United States Gazette* (April 2, 1841) reported a blanket of snow up to three feet deep in the interior, extending to the Virginia Tidewater.

1892 March 1–2: A heavy inland snowfall was summarized in the *Philadelphia Inquirer*: Huntingdon 28 inches; Williamsport 18 inches; and Pocono Summit 14 inches. Government observers reported 17.5 inches at Dyberry and 11 inches at Quakertown. The weight of the snow knocked down railroad, telegraph and telephone lines.

A two-pronged storm began on March 16, with a heavier wave of snow on March 17–18, 1892. Philadelphia had a slushy 2.5 inches in the first round early on the 16th. The second system hit on St. Patrick's Day, totaling 15.4 inches at New York City and up to two feet in Connecticut (*New York Times*). Quakertown had 11 inches.

1916 March 15–16: A heavy northern snowstorm left 16 inches at Honesdale and 14 inches at Gouldsboro. An exceptional chill set in following the storm, with maximums at Honesdale on March 17–18 of 12°F and 15°F and a minimum of -11°F on the 18th. The lower elevations had mixed snow and ice with much less accumulation.

Snowfall totals in March 1916 were record-worthy at Honesdale (45 inches), Scranton (38 inches) and Williamsport (35 inches), with substantial snowfalls on March 6–7, 8–9 and 21–22, 1916.

1932 March 27–28: A late wet snowstorm deposited 28 inches at Wellsboro, 19 inches at Lock Haven, 16 inches at Morris Run, 15 inches at Gouldsboro and 12.5 inches at State College. Much of northern and western Pennsylvania received four to 12 inches. Wind gusts on the Allegheny Plateau reached 40 mph, according to the Morris Run observer, and blizzard conditions were reported by the Wellsboro observer.

1941 March 7–8: The Stroudsburg *Record* reported 15 to 20 inches in the Poconos; nearly 18 inches fell at Stroudsburg. New York City had 18.1 inches—the second greatest March snowfall (after 1888) since records commenced in 1869–70.

1947 March 2–4: A heavy snowstorm accompanied by strong winds lashed the northern highlands, blockading roads: Pleasant Mount 31 inches (25 inches in 24 hours March 2–3); Montrose 23.5 inches; Corry 20.5 inches; State College 12.5 inches. Eastern areas north of Philadelphia had a rain/snow mix, with generally six to 12 inches.

1950 March 20–24: A Long Storm brought substantial totals to elevated sections in northern Pennsylvania: south of Renovo 32 inches; Pleasant Mount 25 inches; Mount Pocono 15.1 inches; Scranton 9 inches, with mixed precipitation.

1956 March 16–19: Two waves of low pressures rippled northward along a nearly stationary coastal front. The first storm traveled from Alabama to Nantucket, Massachusetts, followed by a second system that formed off the Virginia Capes.

Cumulative snowfalls were as high as 17.4 inches at Stroudsburg and 23.9 inches at Newark, New Jersey (18.9 inches fell on March 18–19).

1962 March 6–7: The Great Atlantic Coastal Storm caused more than $200 million in damage from New England to Florida and rearranged the coastal plain in some areas, taking 33 lives. The huge storm churned up waves of 40 feet offshore, driven by 70 mph winds. A band of one to two feet of snow fell in south-central Pennsylvania, with a maximum of 26 inches at Shippensburg; Slippery Rock received 14 inches.

1969 March 1-2: Remarkable snow banding developed with a coastal storm that brought 13.5 inches at East Stroudsburg. A hefty 24 inches fell in the upper Delaware Valley at Matamoras, yet places only 25 miles on either side of the intense band received a few inches, while across the river at Port Jervis, New York, 33 inches piled up.

1970 March 29: A big Easter Sunday nor'easter deposited 16 inches at Gouldsboro and 12.5 inches at East Stroudsburg. Several inches of snow fell on the 31st.

1984 March 28–30: Rain turned to sleet and then heavy wet snow on the closing days of an unusually cold March, making many roads in the north impassable. Snow totals ranged as high as 24.5 inches at Montrose, 18.4 inches at Tobyhanna, and 18 inches at Pleasant Mount. Areas from Williamsport to Scranton and East Stroudsburg, and south to Shippensburg, received 10 to 12 inches, where the precipitation mixed with sleet and freezing rain for a long duration.

1997 March 31–April 1: An impressive late-season snowstorm that began as rain cranked up in the late afternoon hours of March 31, 1997, falling at the rate of two inches per hour. Storm totals early on April 1 ranged from 14.5 inches at East Stroudsburg and 16.5 inches at Mount Pocono to 26 inches four miles northeast of Gouldsboro (elevation 2,042 ft.). Deep drifts briefly trapped vehicles on Interstate 380.

April

1772 April 27: "In 1772, on the 27th of April, the snow fell nearly three feet on the level in New England," stated the *United States Gazette* (May 2, 1837), adding that it "snowed violently for six hours" at Philadelphia.

1803 April 17: The Stroudsburg *Jeffersonian* (April 22, 1875) recalled a "furious snow storm."

1818 April 21–22: *Niles' Weekly Register* (May 2, 1818): "Snow, in the Catskill mountains, was 18 inches deep, on the 22nd ult." The Hemstead *New York Chronicle Advertiser* (March 18, 1907) remembered "snow and ice" in New York City.

1820 April 1–2: "One of the heaviest snowfalls on record … drifts ten feet high" was recalled in the *Bethlehem Daily Times* (March 14, 1888), though the year was incorrectly listed as 1819. *Niles' Weekly Register* (April 15, 1843) confirmed an Easter Sunday snowstorm on April 2, 1820, extending east to Maryland's Eastern Shore.

1821 April 17–18: The *United States Gazette* noted three to four inches fell at Philadelphia. The *National Observer* in Washington, D.C., reported one inch in the city and 12 to 15 inches to the south and west over high terrain.

1826 April 6: Twelve inches of snow piled up in the eastern Pennsylvania hills (*Bethlehem Daily Times*, March 14, 1888).

1828 April 13–14: Nine inches fell at Pottsville (*Miners' Journal*). The *Register of Pennsylvania* (April 19, 1828; April 11, 1829) mentioned seven inches amassed at Harrisburg and "several inches" at Philadelphia.

1834 April 24 and 26: *Hazard's Register of Pennsylvania* noted light snowfalls at Bloomsburg and in surrounding sections (Bloomsburg Register; *Mauch Chunk Courier*). Snow fell as far south as Philadelphia, without any notable accumulation.

1837 April 18 and 23-24: The *Bethlehem Daily Times* (March 14, 1888) remembered five inches on April 18, 1837.

A second late snowfall on April 23–24 deposited six inches at Pottsville (*Miners' Journal*) and nine inches at Hanover (*Hanover Gazette*), mentioned in the *United States Gazette* (April 30, 1837). At Philadelphia, "it snowed violently for six hours" (*United States Gazette*, May 2, 1837). The Easton *Democrat and Argus* (April 27, 1837) added: "Sixty miles up the river the snow is four feet deep."

The following spring, snow fell again on April 23–24, 1838, totaling "several inches" at Pottsville (*Miners' Journal*), with snow extending down to Easton.

1841 April 10–14: Two separate storms combined to produce an historic April snowfall in New York City and near-record inland April snows. The first wave came on April 10–11, 1841, leaving 10 inches at New York City (Ludlum 1960).

A second coastal low on April 12–14, 1841, brought about a foot to New York City, described in the *Sussex Register* as "A snowstorm of great violence." The Erasmus Hall observer in Brooklyn logged an observational depth of 18 inches from both storms. W. I. Clark, writing in the *New York Times* (May 12, 1904), recalled that the two storms left a hefty snowfall "about two feet deep."

The helpful records of Charles Peirce (1847) stated that a foot of snow fell at Philadelphia on April 12–13, 1841, and "several inches" on the 14th. The *United States Gazette* (May 1, 1841) confirmed that 12 to 18 inches fell in eastern Pennsylvania in the second event.

1849 April 18–19: A blast of cold air brought light snow to the Mid-Atlantic on April 14–15, 1849, followed by a hard freeze as far south as Georgia on the 16th. A disturbance brought snow on April 18–19, 1849, that accumulated up to eight inches north of Towanda, noted in the *Bradford Reporter*.

The Belvidere, New Jersey, *Intelligencer* described the magical scene: "To see the green fields, in the freshness of maidenhood, donning the white veil." Two inches of snow fell at the Delaware County Institute of Science in Philadelphia, and the *National Intelligencer* remarked on "a blustering snowstorm" in the nation's capital at a record late date.

1857 April 19–21: Snow fell frequently in April 1857, with each storm bringing heavier totals deeper in April—the coldest fourth month on record in the region.

Heavy snow fell from the Deep South to the Ohio Valley on April 6-7, 1857, and again on April 13–15. The *Easton Argus* reported "several inches" on April 6–7, 1857. A light accumulation was observed in the mid-April storm. A third snowstorm on April 19–21 capped off a record snowy April in the interior Northeast.

Press reports generally listed falls of six to 12 inches in northwestern New Jersey and interior eastern Pennsylvania in a three-day storm that commenced on April 19, 1857, and did not let up until midday on April 21.

An item in the *New York Herald* on April 19, 1887, recalled "over two feet" fell at Wilkes-Barre on April 19–21. The Montrose *Independent Republican* said "the snow was about three feet deep." The reliable Pottsville *Miners' Journal* listed 18 inches, with up to three feet on Broad Mountain to the north. The *Easton Argus* reported "five or six inches." The Smithsonian observer at North Whitehall commented that "the snow was two feet and upwards deep" on the ridges to the north.

The *Monroe Record* (May 6, 1909) consulted the diary of Joseph Smith in Reeders, west of Tannersville, which noted an 18-inch accumulation. The *Wayne Independent* in Honesdale turned to the diary of C. F. Rockwell in an article on April 21, 1887. The top accumulations came on April 13–15 (8 inches) and April 19–21, 1857 (18 inches). Rockwell's journal added that South Canaan had a snow cover "nearly four feet in depth." A notice in the *Belvidere* (New Jersey) *Intelligencer* said the total snow depth reached a stunning 48 inches in the Beech Woods (Pocono Plateau).

1862 April 8–10: Snow came in two waves, starting on April 8, 1862, with a second burst April 9–10. In western Pennsylvania, 8 inches covered Indiana on April 8–9.

The first wave totaled nine inches at Pottsville, according to the *Miner's Journal*, with a total fall "to the depth of eighteen inches," and over the hills to the north "from two to three feet deep" on Broad Mountain.

The *Easton Sentinel* stated: "The depth of the snow was about 12 inches." The snow was even heavier across southern Pennsylvania. The *York Gazette* declared that "not much less than two feet in all fell." The Harrisburg *Patriot* put the accumulation at 18 inches, the greatest since 20 inches piled up from April 14–18, 1854. The Philadelphia *Public Ledger* mentioned 12 to 14 inches. Smithsonian station observers in Norristown (16.8 inches) and Mount Joy (17 inches) confirmed the historic April snowfalls.

Totals tapered off farther north to seven inches at Nazareth and eight inches at Reeders (*Monroe Record*, May 6, 1909). The *New York Times* reported six inches fell in the city.

1863 April 5: A *Washington Post* story (April 2, 2014) on late snows in the District cited the records of Reverend C. B. Mackee in Georgetown, who recorded 12 inches of snow—more than twice as much as any April snowfall since. The storm extended northward to New England. This followed a heavy snowfall on March 31, 1863.

1868 April 4–10: The culmination of the long, snowy winter of 1867–68 in Pennsylvania included a seemingly never-ending series of April snowfalls. Four inches of snow fell at Blooming Grove and Whitehall on April 4–5, 1868. A rare freezing day in April was observed at Bethlehem on April 5, 1868, as the temperature plunged to 25°F at 7:00 a.m., recovering to only 31°F at 2:00 p.m.

Back-to-back storms on April 7–8 and 9–10, 1868, deposited 12 inches and nine inches, respectively, at Dyberry, north of Honesdale. Stroudsburg historian Luther S. Hoffman, in his unpublished notes, recorded falls of 3 inches on April 7–8 and 8 inches on April 9–10. The diary of Joseph P. Smith, of Reeders, quoted in the *Monroe Record* (May 6, 1909), listed 10 inches on April 9–10, 1868.

In the Lehigh Valley, the Whitehall observer measured an inch of snow on April 7–8, 1868, in mixed precipitation, and 8.3 inches on April 9–10, 1868. Five inches fell at Ephrata. The Bethlehem observer recorded another late freezing day in the wake of the storm on April 10, 1868 (32°F/27°F).

A fourth spring snowstorm in a little over a week blanketed the southeastern counties on April 12, 1868, dropping 4 inches at Ephrata and Whitehall, but very little to the north. The monthly catch in April 1868 at Whitehall of 17.3 inches exceeds all modern April snowfall records in the Lehigh Valley.

Winter provided one more wallop on April 25, 1868. Little accumulation was reported over southeastern Pennsylvania, where rain mixed with wet snow, but the hilly areas to the north and west received another three to six inches.

The Stroudsburg *Jeffersonian* commented, "We had a most refreshing snow-storm in the neighborhood on Saturday last [25th], which lasted nearly the whole day, and made things look quite winterish." The snowfall was given by Hoffman as four inches, raising the monthly total to at least 15 inches, which is greater than the modern snowiest April (14.3 inches in 1982 at East Stroudsburg). Dyberry observer Theodore Day measured 25.8 inches of snow in April 1868, the snowiest April at his site (1865–1903).

1873 April 12: A heavy snowfall in the northeast deposited 9 inches at Blooming Grove at the end of a long winter. Two inches fell at Dyberry, and 1.2 inches covered the ground at Parkersville in Chester County. Light snow brought accumulations in the northern part of the state on April 17, 21–22 and 25, 1873, with several inches in the mountains.

1879 April 17–18: The Stroudsburg *Jeffersonian* noted eight inches on Blue Mountain to the south and 15 to 24 inches on the Pocono Plateau, but mostly rain in the valleys. The Honesdale *Wayne Independent* said, "Snow fell … to the depth of three feet." The *Philadelphia Inquirer* reported six to 10 inches of snow blanketed the higher elevations, and totals of 5 inches at Port Jervis and 12 inches at Monticello, New York, were impressive so late in April.

1883 April 23–24: "Rain, hail, and snow," reported the *Jeffersonian*, with up to six inches on the Pocono Plateau. The Dyberry observer noted three inches. A rain/snow mixture was evident in the valleys. New York City had 0.5 inch on the 24th.

1884 April 8–9: The first of two heavy snowfalls in northern Pennsylvania deposited six inches at Dyberry on April 2, 1884. The Stroudsburg *Jeffersonian* reported "A fall of eight inches, high winds, good sleighing" in the higher elevations.

A second storm on April 8–9, 1884, delivered a bigger blow: Hazleton 19 inches; Tobyhanna 18 inches; Moscow 15 inches (*Jeffersonian*, April 17). The Drifton observer (Luzerne County) measured 22 inches. The lower elevations generally received four to eight inches. Baltimore had a record April fall of 8 inches.

1887 April 17–18: Low pressure traveling across Kentucky and West Virginia brought a major late-season snowfall: Dyberry 14 inches; Scranton 12 inches; Stroudsburg 7 inches, mixed with sleet (*Jeffersonian*); Bethlehem 3.2 inches. A *New York Times* notice mentioned 10 inches at Lock Haven.

1889 April 6 7: A strong nor'easter brought heavy snow to the interior. Wind damage was heavy near the coast. Hurricane-force wind gusts were common, and a peak gust estimated near 120 mph occurred at Cape Henry, Virginia, before the anemometer was lost, according to the *Monthly Weather Review*. Snow totals ranged up to 12 inches at McConnellsville in Fulton County and 12.5 inches at Somerset.

1898 April 28: A "Storm of Snow Sleet & Rain," noted the Philadelphia Centennial Avenue observer. The nation's capital had its latest measurable snowfall (0.5 inch).

1902 April 7–8: A second consecutive April with a heavy snowstorm in southwestern Pennsylvania deposited 11.2 inches at Pittsburgh and 20 inches at Uniontown, which, combined with a 4.5-inch fall on April 1–2 set a monthly record.

1906 April 22–23: An area of moderate to heavy snow blanketed the higher elevations of northeastern Pennsylvania: Montrose 8 inches; Scranton 6.2 inches; LeRoy 6 inches.

1908 April 30–May 1: Snow blanketed the northwestern highlands: 3 inches fell at Saegertown and 1 inch at St. Marys.

1909 April 29: The latest significant low-elevation snowfall in Pennsylvania weather records occurred on the closing days of April 1909: South Sterling 7 inches; Scranton 6.8 inches; Milford 6.5 inches; Stroudsburg 5 inches (*Monroe Record*).

1915 April 3: Philadelphia endured one of its biggest snowstorms for any season in early April 1915 with 19.4 inches. A heavy band of snow 16 to 21 inches deep fell in a narrow corridor from Philadelphia to Trenton and across central New Jersey. A fall of 10.2 inches at New York City set a modern April record; Bethlehem had 5 inches.

1916 April 8–9: A heavy interior snowfall in eastern Pennsylvania brought substantial late-season totals in the eastern counties: Coatesville 15.4 inches; Lancaster 12 inches; Stroudsburg 11 inches; Altoona 10 inches; Bethlehem, Reading 7 inches; Philadelphia 4.1 inches.

1917: April 8–9: Another in a series of April snowfalls whitened the lawns and coated trees in the southeast: 7 inches fell at Philadelphia and 6 inches at West Chester.

1918 April 10–13: A double-barreled coastal system brought two periods of heavy wet snow to the Mid-Atlantic: April 9–11 and 12–13, 1918.

Maximum falls were well inland: Lock Haven 26 inches; Altoona 23 inches; Gouldsboro 22 inches; Freeland 21 inches; Towanda 20.3 inches; Scranton 12.7 inches; Bethlehem 9.5 inches. The Stroudsburg *Morning Press* reported 9.5 inches (the official observer collected 5 inches). Coastal areas caught two to five inches in the late phase of the storm. Washington, D.C., received a 2-inch snowfall on April 12.

1919 April 25–26: A light-coating snow blanketed the northern half of the state, generally less than an inch, and 1.9 inches at Gouldsboro. The East Stroudsburg *Morning Press* reported "intermittent snow and sleet" on April 25, and a frigid low of 18°F at Hazleton. West Bingham had a record late subfreezing day (30°F/18°F) in the northwest.

1923 April 14: Heavy snow blanketed the southern portion of the state, with totals as high as 16 inches at Cresson and 14 inches at Altoona. The southeast had a late snowfall of one to four inches. Philadelphia recorded 1.8 inches.

1924 April 1: A nor'easter brought a blanket of snow from the coastal plain to the Appalachians. Snow accumulations ranged as high as 10 to 15 inches in eastern Pennsylvania, with record April totals at Bethlehem (15 inches) and Allentown (12 inches). Large falls were observed at Quakertown (10.8 inches) and East Stroudsburg (9 inches, stated by the *Morning Sun*). Washington, D.C., (5.5 inches) and Baltimore (9.4 inches) had record April accumulations, and Philadelphia received 6.6 inches.

1928 April 27–29: A Gulf low and a secondary storm over western North Carolina joined forces east of the Virginia coast and continued northward. On April 27, 1928, rain changed to heavy wet snow across the western and central highlands. Eventually, the eastern counties north and west of Allentown received a slushy one to two inches.

The interior took the brunt of the heavy wet snowstorm. Nine inches fell at Mount Pocono, and up to foot covered the Pocono Plateau. At Freeland, the storm left 18 inches. Centre Hall received 24.5 inches; State College 17.3 inches.

Farther south, Carlisle had an impressive total of 26 inches and Altoona notched 25 inches. In the mountains, Somerset received 31 inches, which established an April state record of 38.5 inches, surpassing Salem Corners in 1894 (38.2 inches).

April 15-16, 1929: Ice and snow blanketed the higher elevations of northern Pennsylvania: nine inches in Towanda and 6.5 inches in Freeland.

1931 April 6–7: An elevation storm in the north left 6 inches at Freeland and 16 inches at Towanda.

1938 April 6–7: A heavy nor'easter deposited 11 inches at Scranton and Retreat, 9 inches at Gouldsboro, and 8 inches at East Stroudsburg (Pauline Learn observations).

1940 April 12–13: A winter storm that formed along the Carolina coast brought a rain that turned into heavy snow, as temperatures plummeted into the teens and low 20s in the north, accompanied by wind gusts of 40 to 60 mph. The heaviest falls were: 8.8 inches at Kregar, Westmoreland County; 6 inches at Portland, 4.1 inches at Bethlehem. The average snowfall was two to four inches. At Mount Pocono, the low on April 14 was a frigid 9°F.

1942 April 10–11: Another April snowstorm deposited 11 inches at Mount Pocono and 10 inches at Stroudsburg in the northeast. Up to an inch of glaze coated mountain sections, where snow alternated with sleet and freezing rain.

1943 April 13–14: The coldest April on record in Pennsylvania brought frequent snow showers and light coatings of snow. A cold surge brought a dusting of snow in the northern valleys and six inches at Gouldsboro.

1944 April 4–5: A coastal disturbance belted eastern Pennsylvania on April 4–5, 1944. Top falls were 9 inches at New Tripoli, 8 inches at Mount Pocono, and farther east as much as 12.6 inches at Layton, New Jersey. New York City had an unusually heavy April snowfall of 6.5 inches. Wellsboro had a remarkable late-season low of 6°F on April 6, 1944.

1946 April 26–27: A low-pressure wave developed along an advancing cold front, causing rain to change to snow across the northern and western sections of Pennsylvania, leaving one to three inches of snow on the higher terrain. The heaviest falls were 4.8 inches at Eagles Mere (elevation 2,020 ft.) in Sullivan County and 3.4 inches at Freeland (elevation 1,970 ft.) in Luzerne County, 3 inches at Hawley, and 1 inch at Scranton. The following morning at Kane, the reading of 14°F was the lowest so late in the season at the time.

1953 April 17–19: Intervals of snow accumulated to a depth of several inches across the northern and western highlands. Heavy totals include 9.5 inches at New Castle and 7.9 inches at Philipsburg. Flurries reached the southeast, including Philadelphia on April 20.

1956 April 7–8: A very cold, backward April brought frequent snows and a powerful nor'easter on April 7–8: Montrose 14.1 inches; Gouldsboro 14 inches; Scranton 7.4 inches; Stroudsburg 7.1 inches. New York City had a substantial fall of 4.2 inches.

On April 23–24, 1956, a late snowstorm coated the northeastern mountains with 9.5 inches at Montrose, 5 inches atop Blue Mountain south of Stroudsburg (William Hagerty notes), and 3 inches at Scranton, with a dusting in some valleys.

A reinforcing shot of cold air sent the thermometer to the lowest levels so late in the season on April 25 at Stroudsburg and Mount Pocono (20°F). In the northwest, Ridgway, in Elk County, had a January-like low of 11°F. Snow showers left 0.4 inch at Stroudsburg on April 26, and flurries were observed as far south as Reading.

1957 April 4: Large falls occurred at Philipsburg (13.6 inches), Mount Pocono (10 inches) and Stroudsburg (9.5 inches), as a slow-moving storm approached from the west. Thunderstorms with damaging wind gusts (75 mph) occurred on April 5–6.

1961 April 9–10 and 12–13: Gouldsboro had back-to-back snowfalls of 12 and 10 inches, plus two inches on the 18th, contributing to a record April total of 27 inches. Montrose had successive snowfalls of 14 and 4 inches. Stroudsburg picked up 5 inches on April 9–10 and 3.5 inches on April 12–13. Pittsburgh had 4 inches in both storms.

1982 April 5–6: A full-fledged April blizzard developed east of the Virginia Capes as energy translated to the coast from the Ohio Valley. The storm began in eastern Pennsylvania in the early morning hours of the 6th, creating blizzard conditions by afternoon to much of Pennsylvania eastward to New York City.

Snowfalls reached near-record proportions for April at East Stroudsburg (13 inches), Newark, New Jersey (12.8 inches—April record), Allentown (11.4 inches) and New York City (9.6 inches). Near the coast, snow, thunder and lightning were accompanied by wind gusts of 70 to 80 mph. Ice coated the streets of Philadelphia after 3.5 inches of snow and sleet accumulated.

1983 April 19–20: A blast of cold air brought a coating of snow across much of northern Pennsylvania on April 16–17, 1983, marking the start of an extraordinary week of winter in mid-April. Top falls were 7.5 inches at Montrose and 3.9 inches at Scranton, with mostly rain and a few flakes in the valleys.

A secondary shot of arctic air spilled into the state on April 18, with a quick 0.9 inch of snow at Allentown and 5.5 inches at Scranton. Meanwhile, a developing coastal system deposited a record late-season snowfall (1.8 inches) at Raleigh, North Carolina.

The nor'easter delivered the real blow on April 19–20: Montrose 17.5 inches; Tobyhanna 17.4 inches; Scranton 14.1 inches (April record); Stroudsburg 6.1 inches. Record late snows, with sleet, blanketed Newark, New Jersey (4.1 inches), Philadelphia (1.9 inches), Atlantic City (0.7 inch), and Wilmington, Delaware (0.5 inch).

A final shot of winter brought a transition from rain to snow on April 25, 1983, in northern Pennsylvania totaling 7 inches at Montrose, 5 inches at Gouldsboro, and 3 inches at Scranton. Record April snow totals were tallied at Montrose (34 inches) and Scranton (26.7 inches).

1986 April 22–23: Rain switched over to heavy snow across much of Pennsylvania at a late day that was historic. The largest totals were: Gouldsboro 24 inches (April snowfall 26.6 inches); Tobyhanna 21 inches; East Stroudsburg 14 inches (April storm record); Scranton 7.6 inches. An inch fell in the Lehigh Valley.

CHAPTER 4: WINTER STORMS AND OUT-OF-SEASON SNOWFALLS   155

Fig. 4.9 April 23, 1986. Historic late April snowstorm near Saylorsburg on Godfrey Ridge. (Photo by Michael Pontrelli)

1993 April 22–23: Heavy snow piled up across the elevated portions of the central Pennsylvania counties: Clarence 10.5 inches; Montrose 9.5 inches; State College 4.6 inches. Valley sections had a coating: Scranton 1.9 inches; Williamsport 0.8 inch; Harrisburg 0.6 inch; Stroudsburg 0.2 inch.

2007 April 15–16: A deep coastal storm pounded Pennsylvania and the Mid-Atlantic with torrential rain—in some places up to four inches—that changed to wet snow overnight, accumulating to as much as 14 inches at Francis E. Walter Dam in Luzerne County (elevation 1,509 ft.) and seven inches at Tobyhanna, with winds gusting to 52 mph at Pocono Mountains Municipal Airport. Binghamton, New York, had a substantial fall of 13.6 inches, an April record at the time.

Interior southeastern Pennsylvania received more than an inch of wet snow about 50 miles northwest of Philadelphia. The barometer at East Stroudsburg dipped to 28.85 inches on the 16th.

2012 April 22–24: A deep coastal disturbance drifted inland across eastern Pennsylvania, bringing rain and wind and pockets of heavy wet snow over the highest elevations as far east as the Poconos. The greatest totals fell over the western mountains. Laurel Summit reported 23.7 inches and a foot of snow blanketed areas around Penfield around 2,100 feet. Sylvania had a snow total of 14 inches, Bradford received 10 inches, and Montrose had 8 inches. Treacherous road conditions developed in the mountains that resulted in many accidents.

2020 April 17–18: Two days after a few inches of snow coated the high terrain, a wave of low pressure rippling along a frontal boundary over southern Pennsylvania was accompanied by a swath of heavier snow in the northern highlands, with a maximum fall of 9 inches at Coudersport and Leonard Harrison State Park (Tioga County). Northern portions of Bradford, Susquehanna, Wayne and Pike counties were blanketed with three to five inches of wet snow, and 7 inches fell at Warren Center in Bradford County. A dusting (0.3 inch) was observed at Scranton and Blakeslee, with a brief wintry mix in the lower elevations of the southern Poconos.

2022 April 18–19: For the second time in three years, a mid-April snowstorm coated portions of Pennsylvania. A wave moving up the Eastern Seaboard tapped into marginal cold air, changing rain to snow over the interior sections in an elevation storm.

Accumulations reached 14 inches at Laporte, 12 inches at Montrose, 10 inches at Athens in the Northern Tier, and 9.6 inches Blue Knob State Park in the Laurel Highlands. Much of western and northern Pennsylvania, including the valleys, received two to four inches of wet snow by daybreak on April 19. Even heavier totals were reported in south-central New York at Virgil (18 inches) and Binghamton (14.1 inches).

August and September Snows

Snow flurries occasionally skipped through the chilly late summer air in the 19th century in Pennsylvania. Frost in August was not uncommon near the end of the Little Ice Age in the interior Northeast. A press entry in the Stroudsburg *Jeffersonian* mentioned a "slight frost was observed, in many places" on August 26, 1858.

An anomalous early visitation of snowflakes occurred on August 26, 1885, in Pennsylvania. The *Jeffersonian* reported that "snowflakes were seen flying about in the air in this vicinity." Notices mentioned snowflakes as far south as the hills north of Reading. From Selinsgrove: "A snow squall passed over the mountains northwest … The snow fell so thick and fast that it was impossible for men on the road to see the mountain, a quarter mile distant." Snow flurries were spotted at Harveys Lake in the Back Mountain region of Luzerne County: "Snow fell this morning and continued until noon with slight intervals."

The Philadelphia *Public Ledger and Transcript* had an item regarding a light snowfall in western Pennsylvania on August 15, 1889: "Snow fell in sufficient quantity to cover the ground" in Greensburg, Westmoreland County. The site at Wellsboro recorded a trace of frozen precipitation. A "heavy frost" damaged fruit trees and garden crops following a "light fall of snow," reported in the *Harrisburg Evening Gazette*.

Another remarkably early appearance of frozen precipitation was noted on August 23, 1890. The Harrisburg Signal Service observer recorded a trace of snow, according to the *Monthly Weather Review*, and snowflakes extended to Northumberland and Northampton counties, followed by a light frost on the 24th.

The *Bethlehem Daily Times* had a dispatch from Milton that described a quarter-inch snowfall around 3:00 a.m. on August 23, 1890, at the end of thunderstorm. The *Indiana Democrat* in western Pennsylvania cited a report of "a quarter of an inch of snow on the awnings of the stores in the business part of the town." The story went on to say that freight cars on the Philadelphia and Erie and Philadelphia and Reading railroad lines were coated with 0.5 inch of snow. The *Democrat* carried a comment from a mail carrier, who encountered a "blinding snow squall, that lasted from five to ten minutes" in Lower Saucon Township in Northampton County, in the hills along his route to Easton. A month later, Blue Knob (elevation 2,500 ft.) had flurries on September 20, 1890.

The greatest September snowstorm in Pennsylvania weather history blanketed the northern mountains on September 28–29, 1844, remembered as a "remarkable snow-fall, in some parts of the county [Bradford] being 28 inches deep" (Heverly 1926). The *United States Gazette* reported "several inches in the interior of Pennsylvania."

Historical weather records show a few instances of snow in late September in the northern mountains of Pennsylvania. Frozen precipitation occurred at Dyberry on September 22, 1875, and sleet fell at Blooming Grove on September 23, 1885. Flurries were seen on September 30, 1888, in the northern highlands as far south as State College.

The *Easton Daily Argus* noted a mix of rain and snow on the evening of September 29, 1895, which was confirmed by a smattering of reports of snowflakes by local observers, including at one of the official Philadelphia sites and at Perth Amboy, New Jersey. The *Jeffersonian* carried an Altoona dispatch that mentioned a snow shower on September 30, 1895. A government observer at Edinboro in Erie

County recorded one inch, with daytime temperatures hovering between 37°F and 39°F.

In the 20th century, the earliest snow flurries were observed on September 10, 1924, at Freeland, with a few flakes seen the following year on September 30. A few early flurries were observed at Mount Pocono on September 17, 1928, and at a number of northern and western mountain sites on September 24–25, with a trace as far southeast as Colebrook (Lebanon County). Maximum temperatures were in the 50s to low 60s from September 24–October 1, 1928, the longest continuous spell of such chilly weather so early in the season.

Flakes mixed in with rain at Warren on September 25–26, 1940. A wintry low of 21°F at Kane on September 26 equaled the state record previously reached on September 22, 1904, at Saegertown (Crawford County) and Smethport (McKean County), and on September 28, 1914, at West Bingham (Potter County).

A chilly morning on August 24, 1940, brought a low of 30°F at Coudersport (Potter County), and a frosty 32°F at Stroudsburg the next morning, an August record (tied in 1965).

Flurries returned to the Pennsylvania highlands on September 28, 1942. A premature cold snap brought flurries across the northern counties of Pennsylvania on September 30, 1946. Berwick and Somerset recorded traces of snow. On the Pocono Plateau, flurries were flying at Gouldsboro, and at Mount Pocono on October 1. East Stroudsburg resident Pauline Learn, who kept a weather diary, jotted down "snow squalls" in her October 1, 1946, entry, the earliest incidence of snowflakes in the lower elevations in modern times.

An early snow appearance at Kane on September 22, 1947, preceded a state September record minimum of 17°F on September 28, 1947, at Hawley. Gouldsboro experienced a few flurries on September 30, 1947. Sightings of snowflakes occurred on September 20, 1956, at Gouldsboro and Scranton and across northern parts of the state. The Kane observer measured 0.2 inch—the earliest accumulation in state history.

An early freeze nipped crops in the Poconos and parts of northern Pennsylvania on August 30–31, 1965, with a minimum of 27°F at Coudersport, and upper 20s were observed on the Pocono Plateau. Readings dipped to freezing at Stroudsburg. Early autumn snowflakes next made an appearance at Corry on September 22, 1974, and Tobyhanna on September 23, 1981.

A remarkable late summer freeze affected the western highlands on the morning of August 29, 1982, when the mercury dipped as low as 23°F at Clermont 4 NW—a state August record low. Frost occurred as far south as Landisville in Lancaster County (35°F). The western mountains and northeast had frost again on August 29–30, 1986, with a minimum of 25°F at Bradford FAA. Philadelphia set an all-time August low of 44°F, edging out the previous record from August 31, 1965 (45°F).

Scranton had a trace of snow on September 29, 1983. The following autumn, the longtime observer Dixon Miller in Blakeslee, in western Monroe County, recorded that snow began falling just before midnight on September 27, 1984, with an accumulation of 0.2 inch on the grass at 2:45 a.m. on the 28th. The Freeland observer measured 0.1 inch. A WGAL-TV meteorologist reported sleet in Lancaster County at a record early date.

A sharp cold front carried a few historically early snowflakes at Pittsburgh, Bradford and Montrose on September 23, 1989. A significant early cold snap on September 29, 1993, included a trace of snow at Scranton and a rare September accumulation (1 inch) at Strongstown (elevation 1,880 ft.) in Indiana County—tying the state September mark set at Edinboro (1895). Flurries were observed in the Poconos just after midnight on October 1, 1993.

October Snowstorms

The earliest known nor'easter to bring snow to the coastal plain occurred on October 8, 1703, using the New Style calendar. At Philadelphia, according to the diary of Isaac Norris, quoted in Watson (1868): "We have had a snow, and now the northwest blows very hard. The cold is great, so that at the falling of the wind the river (at Philadelphia) was filled with ice."

A journal mentioned a light snowfall at Philadelphia on October 3, 1769 (Nese and Schwartz 2005).

An *Easton Argus* item on October 14, 1835, listed a light 0.5-inch snowfall witnessed at Montrose on October 1, 1835, in what turned out to be a harbinger of a long and snowy winter in 1835–36. However, the most impressive of all early-autumn snowfalls occurred in the cold autumn of the following year.

Snow flurries were reported at a number of locations in the northern Appalachians in late September 1836, cited in *Hazard's Register of Pennsylvania*. The *New York Evening Post* (October 4, 1836) noted an early snowfall on September 28–29: "Catskill mountains are covered with snow."

The first of two spectacular early snowstorms buried the highlands of Pennsylvania and New York on October 4–6, 1836. John Linn's *Annals of Buffalo Valley, Pennsylvania, 1755–1855*, had this to say: "A heavy snow-storm; snow one and a half feet deep in Penn's valley and on Buffalo and White Deer mountains" near Jersey Shore in Clinton County.

The Pottsville *Miners' Journal* reported 15 inches on Broad Mountain, which required many men to clear the railroad tracks. In the Wilkes-Barre area, the snowfall reached 11 inches (Harvey 1909). A fall of 20 inches occurred at Hollidaysburg (*New Hampshire Sentinel*, October 27, 1836).

The remarkable early autumn storm in Bradford County was recalled by Heverly (1926) in his county history: "Snow fell to the depth of nearly two feet." The *New York Evening Post* noted 10 inches at Binghamton. The *United States Gazette* mentioned "several inches" on ridge tops north of Lebanon.

A second great early snowstorm on October 11–12, 1836, brought several inches of wet snow to parts of eastern Pennsylvania and northern New Jersey, and sleet was observed at Philadelphia. The *Sussex Register* in Newton, New Jersey, described the snowfall as a "furious northeaster" that left a "sheet of snow that would not have disgraced the colder latitudes of Russia."

The *Easton Argus* carried this Pottsville dispatch on October 20, 1836:

> On Tuesday night [October 11] it commenced snowing, and continued until Wednesday morning, when it turned to rain at this place. On the Broad Mountain, the snow fell to the depth of 18 inches. ... At some places on the mountain we learn the snow is two feet deep.

On October 11, 1838, "part of the Allegheny mountain was covered with snow" (Peirce 1847). A heavy snowstorm occurred on October 31, 1838, when "10 inches fell on the interior of Pennsylvania and New York." The *United States Gazette* added that there were "several smart snow squalls" at Philadelphia. Rain changed to a wintry mix at New York City on October 3–4, 1841 (Ludlum 1961).

The Smethport *Settler and Pennon* expressed surprise upon waking up to snow that fell "to the depth of from eight to ten inches" on October 26-27, 1843, adding "as though we had indulged ourselves in a Rip Vanwinkle [sic] slumber." The *United States Gazette* noted "good sleighing" at Montrose. A Goshen, Connecticut, observer measured 4.5 inches.

A decade later, a heavy snowfall on October 24, 1853, reached the valleys from Virginia to Pennsylvania. The *Philadelphia North American and United States Gazette* (October 31, 1853) carried a notice from Sunbury about a 12-inch snowfall four miles northeast of the Susquehanna River community, following a rain/snow mix. Trees splintered under the weight of the snow and roads were nearly impassable. Six inches accumulated at Bedford.

The *Belvidere Intelligencer* in northwestern New Jersey reported, "Snow commenced about nine a.m. and continued for several hours … not less than four inches fell." The *Easton Argus* called it a "quite a respectable fall of snow" that left the hills white for three days.

Snow fell in late October 1859 at least twice. Bedford received one inch on October 20, which was a prelude to a heavy eastern snowfall down to the coastal plain on October 26–27, 1859. The heaviest total was 5.5 inches at North Whitehall, and 4 inches fell at Nazareth. Three to four inches blanketed New York City (Ludlum 1983).

A light accumulation occurred in the interior of Pennsylvania on October 14–15, 1860. On October 8–9, 1864, heavy snow fell in the northwest, where 8 inches fell at Indiana, followed by 4 inches on the 12th.

A few flakes were visible at Philadelphia on October 5, 1865, at a record early date, according to a *New York Times* notice. A trace was reported at Blooming Grove.

A coastal disturbance brought heavy snow on the night of October 6–7, 1873, in the northeastern counties, depositing 3 inches at Dyberry and six inches over the ridges, according to the observer Theodore Day. The Stroudsburg *Jeffersonian* reported three inches fell at Tobyhanna Mills in a storm that "lasted nearly all night" at Mountainhome. An item in the *New York Times* reported "one or two inches" at Hazleton, with the mercury at 32°F. Eight inches fell at Delaware, New York.

A little bit of snow whitened the interior elevations on October 11–12, 1875. A coastal disturbance on October 15, 1876, brought a little snow, and a record early snowfall at New York City (0.5 inch). On October 21–22, 1887, four inches fell at Greenville, in Mercer County, with an inch at Scranton and a dusting in the Poconos (*Jeffersonian*).

Snow flurries were noticed in northern Pennsylvania on October 2–3, 1888, as far south as Scranton, and in the northern sections on October 12 and 20–21, 1888. The *Jeffersonian* mentioned a high-elevation four-inch fall at Mount Pocono, although which event date is unclear.

Early flurries were widely observed on October 5, 1892, at lower elevations as far south as Harrisburg, and at a record early date at Washington, D.C.

A substantial northern elevation snowfall on October 10–11, 1906, totaled 10 inches at Warren, 7 inches at Saegertown, 4 inches at Montrose, and 1.5 inches at Dushore. Flakes reached down to the valley floor, and a dusting powdered Milford, as readings across the north remained in the 30s all day. Dushore tallied 5.1 inches on October 30–31. Pocono Lake dipped to 13°F on the 31st.

October 1917 brought several strong surges of cold air accompanied by snow or snow showers. Somerset received 13.5 inches on October 24 (monthly record) and 1.3 inches on the 30th. Warren (13.5 inches) and West Bingham (12 inches) had multiple snowfalls, including several inches that fell in the northwestern highlands on October 12, 1917. Pittsburgh had an early three-inch accumulation.

A powerful cold front crossed Pennsylvania on October 9, 1925, accompanied by an early dusting in the highlands and sleet at Harrisburg. The *Philadelphia Inquirer* and various press sources had notices of snowflakes flying from Wilmington, Delaware, northward to the suburbs of Philadelphia, Trenton and New York City.

There were reports of a coating up to an inch on grassy lawns in elevated areas around Scranton and Allentown, and a coating was even mentioned at West Chester. The Stroudsburg *Morning Press* stated that "grassy hills … were white" during "a showing of snow for 36 hours." Tannersville had a dusting and a high of 36°F. The fall at Gouldsboro totaled 2 inches. However, the local *Morning Press* mentioned 3.5 inches on the high plateau between Gouldsboro and Tobyhanna.

Snow was frequent over the high terrain in the northwest. Warren collected a hefty 7.3 inches on October 20–22, 1925, Saegertown had 4 inches and Pittsburgh received 2.2 inches. A cold front brought accumulating snow again on October 28, 1925, depositing 3.5 inches at Saegertown. In the southern Poconos, Tannersville received an inch, and three inches fell on the western Pocono Plateau, according to the *Morning Press*. The valleys had a light wintry mix.

But the highlight of this historically cold October in 1925 was a nor'easter at the end of the month. On October 30, 1925, with temperatures hovering in the low 30s, a storm walloped the state. The eastern counties had the greatest totals: 5 inches at Freeland; 4.3 inches at Quakertown; 4 inches at Tannersville and Phoenixville; 3 inches at Bethlehem and Gouldsboro. Erie had the greatest October snowfall in the state (14.7 inches)—a local record. Mount Pocono had a record October low of 12°F on the 31st.

An early snowfall on October 5–6, 1935, deposited one inch at Elk Lick and Somerset. The mercury dipped to 16°F at Somerset on the 7th. An unseasonably chilly coastal disturbance on October 19–20, 1940, brought an October record 2.2 inches of snow at Philadelphia, 2.4 inches at Lancaster and 4.2 inches at Johnstown. Three inches fell as far south as Dover, Delaware, and the nation's capital had a record October fall of 1.5 inches. Amounts tapered off to 0.5 inch at Allentown and Mount Pocono, and lawns were lightly snow-covered at Stroudsburg (*Daily Record*).

An elevation snow on October 18, 1948, brought 4.8 inches at Freeland and 1.5 inches at Hawley, with a rain/snow mix in the lower elevations.

Historic early snowflakes reached the southeastern lowlands on October 3, 1952, with a record early-season trace event at Coatesville, Landisville and Gettysburg. Accumulating snow fell across the northern half of the state on October 20, depositing two inches at Stroudsburg, 3.1 inches at Bradford and 3.5 inches at Mount Pocono. Measurable snow even reached the coastal plain at New York City (0.5 inch).

With a fresh snowpack, the mercury sunk to 7°F at Coudersport on October 21, 1952, setting an official state October record. The monthly *Climatological Data* report also noted a low of 5°F at a site 0.6 mile east-northeast of Pimple Hill "on a properly exposed minimum thermometer at an experimental station." Measurable snow fell again, though only at higher elevations, on October 28–29, 1952, totaling 3 inches at Mount Pocono (6.5 inches for the month) and 3.1 inches at Bradford, with flakes in the valleys.

A snowfall on October 25–26, 1962, set records at the time at Stroudsburg (3.5 inches), Scranton (4.4 inches) and Gouldsboro (6 inches). During a snowy period, 19 inches of snow fell October 24–27, 1962, at Corry, to establish a state monthly snowfall record.

A coastal disturbance on October 18–19, 1972, deposited snow across most of the state down to the coastal plain. Heavier totals in the eastern counties included: Clausville 5.5 inches; Montrose 4.6 inches; Allentown 1.4 inches. The Pocono region totals ranged from 2.5 inches at Stroudsburg to 4.4 inches at Long Pond on the high plateau. In the southeast, Devault received 4 inches, and 3 inches fell at Morgantown. Snow accumulated in western Pennsylvania; up to 4.5 inches at Derry and 1.8 inches at Pittsburgh.

Snowflakes and ice pellets fell across the Allegheny Plateau and central mountains on October 2, 1974, reaching State College at a record early date. Overnight snow showers dusted the Poconos on the 3rd: 1 inch at Hollisterville, and 0.2 inch at Scranton and Tobyhanna.

An eastward-moving system near the Mason–Dixon Line on October 16–17, 1977, tapped into unseasonably cold air, causing a cold rain to switch to heavy wet snow overnight in the mountains. An inch of snow fell at State College, Williamsport and Scranton, but above 1,500 feet it was a different story. A crippling, record-worthy 17 inches of wet snow piled up at Freeland, and more than a foot of snow fell at Hazleton and Jersey Shore in Clinton County. Eagles Mere had 11.5 inches, and Wellsboro tallied 7.5 inches. Five inches fell in the western Poconos at Gouldsboro and at Zerbey Airport (Schuylkill County). The heavy moisture content of the snow toppled thousands of trees in the mountainous sections, knocking out power and halting travel in some rural areas until plows cleared roads the next day.

An early-season coastal disturbance of historic significance brushed southeastern Pennsylvania on October 10, 1979. In the mountains of West Virginia, up to a foot of snow was reported, and 17 inches were measured in Big Meadows, Virginia (elevation 3,535 ft.). Philadelphia received a stunning record early snowfall of 2.1 inches, nearly matching the monthly mark (2.2 inches) of October, 19. 1940.

Both Washington, D.C., and Baltimore recorded their earliest measurable snowfalls—0.3 inch—in two centuries of record-keeping. One inch fell at Allentown and Tobyhanna, 3 inches at Nazareth, and 5 inches at Long Valley, New Jersey. In New England, Worcester, Massachusetts, received 7.5 inches, and 0.2 inch was noted at Boson's sea level Logan Airport.

An early-season nor'easter October 3–4, 1987, brought a predawn changeover to snow in the Northeast, resulting in historic early falls of 6.5 inches at Albany, New York, and more than 20 inches in the Catskills, causing trees still in mostly full leaf and power lines to collapse. Flurries descended into the valleys, setting an early-season record at East Stroudsburg, and 2 inches covered the high terrain at Gouldsboro. Flurries swirled around the northern suburbs of New York City at a record early date.

The greatest October snowstorm in western Pennsylvania fell on Halloween weekend in 1993. A total fall of 8.5 inches blanketed Pittsburgh on October 30–31, plus 1.7 inches on November 1, which added to the spooky backdrop for trick-or-treaters on October 31 (6.6 inches). The biggest storm total was 14 inches at Bradford.

October Snowfall Bonanza: 2000–2014

On October 9, 2000, snowflakes were observed at East Stroudsburg and ice pellets were reported at Bethlehem. Flurries occurred at Mount Pocono on October 7, 2001.

A dusting of snow October 25, 2002, in the Poconos at Tobyhanna was a prelude to a widespread snowfall on the 29th across northern Pennsylvania: State College 3.1 inches; Tannersville 2 inches; Saylorsburg 1.5 inches; East Stroudsburg 0.3 inch. Over higher terrain, 6 inches fell at Mahanoy City in Carbon County.

Snow flurries came especially early again in the following autumn. On October 1, 2003, Erie had a record early fall (0.2 inch). Flurries were observed at a record early date at Williamsport, Saylorsburg and State College on the 2nd. Another autumn snow on October 25–26, 2005, deposited 6 inches at Philipsburg and 1.8 inches at Tobyhanna, with a few flakes in the valleys.

On October 27–28, 2008, a massive nor'easter brought a wind-driven rain/snow mix to eastern Pennsylvania. Sleet and wet flakes mixed with rain during Game 5 of the World Series in Philadelphia with the Tampa Bay Rays, which had to be suspended in the bottom of the sixth inning. A brief changeover to all snow briefly brought slushy accumulations to parts of the interior eastern counties. In the Poconos, it was an all-out snowstorm: 20 inches fell at Gouldsboro 4 NE (21 inches for the month), 17 inches at Tobyhanna and 15 inches at Pocono Peak Lake—unprecedented for October.

Another early snow on October 15–16, 2009, dropped 5 inches of snow at Mount Pocono and Tobyhanna, 6 inches at Long Pond, and as much as 10 inches at Coudersport. A fall of 1.6 inches at East Stroudsburg was the heaviest so early in the valley. Early-season snow records were also set at State College (4.8 inches) and Saylorsburg (3.7 inches).

On October 2, 2011, State College saw flurries, and Philipsburg (1.5 inches) and Laurel Summit (2.2 inches) received their earliest measurable snowfalls. But that would only be a prelude to Snowtober on October 29, 2011, which put down five to 10 inches north and west of Philadelphia and 12 to 16 inches over the high terrain, shattering all October records in interior eastern Pennsylvania.

A year later, a post-tropical super storm named Sandy brought rain changing to snow in the western elevations on October 30–31, 2012. In the Laurel Highlands, 8.2 inches plastered Somerset, atop Laurel Summit. Additionally, 13.3 inches fell by daybreak on October 31, raising the storm total of 17.5 inches by the morning of November 1. An unofficial total of 28 inches was measured at the Seven Springs resort.

On October 4, 2014, a few flurries reached the western mountains and the Pittsburgh area.

May Snowfalls

Today, snowfalls in May in the Allegheny Highlands are not overly unusual, but snow reaching the lower elevations is rare. However, in olden times it was a different story.

On the evening of May 3, 1774, rain turned to snow from Virginia to New England. A seven-inch snowfall blanketed Lancaster on May 3, 1774, according to an item in the *Bethlehem Daily Times* (March 14, 1888). In *Historical and Pictorial Lititz* by Zook (1905), the Lancaster County accumulation ranged from six to eight inches.

Ludlum (1966) assembled an array of historical accounts of May snowstorms that affected the Mid-Atlantic region down to the coastal plain, based on regional journals. David Schultze's *Journals in Goshenhoppen* on May 4 stated: "Snow, four to six inches, icicles eighteen inches long formed and continued several days and nights." In *Extracts from the Diary of Jacob Hiltzheimer, of Philadelphia*: "The houses this morning are all covered in snow." Up to four inches of snow fell from northern Virginia to southern New England. George Washington's diary mentioned "spits of snow" at Mount Vernon.

Snow fell twice in early May 1803 in Pennsylvania. Six inches covered the ground in Bradford County on May 4–5 (Heverly 1926), and another six inches fell on May 7–8, 1803, in a widespread snowstorm, which also blanketed parts of eastern Ohio at a record late date. However, the snowfall in the lower elevations was shocking.

On April 20, 1854, the retrospective in the Stroudsburg *Jeffersonian* recalled: "There was a fall of snow to the depth of six inches." The late snowfall was accompanied by thunder and lightning at Philadelphia as rain changed to snow in the early morning hours of May 8, signaling the arrival of cold air drawn into the circulation.

Joseph Price's diary entry, in Wynnewood stated: "wind N.E. and Snowing most violent" with broken tree limbs. The glorious shade trees, "which lined many of the main streets of Philadelphia," were felled by the weight of the snow (Ludlum 1966). The *United States Gazette* (May 1, 1841) recalled an accumulation of "several inches," adding that "sleighs ran as in mid winter around Philadelphia." There is no modern precedent for such a May snowfall in the city.

Another rare May snowstorm reached the coastal plain on May 3–4, 1812, depositing nine inches at Bethlehem, remembered in the *Bethlehem Daily Times* (May 3, 1875). The *Newark Centinel of Freedom* reported, "Gardens and fields covered with snow at noon [on May 4]."

There was a noteworthy snowfall on May 10, 1829, to the depth of four inches, on the shaded north side of Blue Mountain, according to the *Lehigh Herald*, in an entry from the *Register of Pennsylvania* (May 23, 1829). A killing freeze followed on May 13.

A rare late May snowfall on May 14–15, 1834, left "a considerable fall of snow" in the northern mountains of Pennsylvania, according to papers from "Pottsville, Wilkes-Barre, and other places north and east" (*Hazard's Register of Pennsylvania*). Snow showers reached New York City, where the temperature fell from 40°F at 10:00 p.m. on May 14 to 28°F the next morning—a record late freeze in the city. The ice was a quarter-inch thick (*New York Commercial Advertiser*).

At Philadelphia, the mercury plummeted to 33°F on May 14, 1834, and 32°F on the morning of the 15th—a record late freeze in the city (*Hazard's Register of Pennsylvania*). The temperature plummeted to 21°F at Pottsville (*Miners' Journal*) on May 15, 1834. The *Register* mentioned a low of 30°F at Germantown, and the *Easton Centinel* mentioned a "severe freeze" on May 15.

The *Register* said, "We were visited by a degree of cold far more congenial to the rough, uncomely days of February." Afternoon readings were generally confined to the mid-40s from May 14–16, 1834. The historic late freeze extended as far south as North Carolina (Ludlum 1968).

The following spring, a heavy wet snowfall on May 20–21, 1835, accumulated 15 to 24 inches deep in Bradford County (Heverly 1926). *Hazard's Register of Pennsylvania* (July 4, 1835) carried a notice of "severe frosts" around Pottsville on June 22, 1835. Snow fell at Philadelphia on May 1, 1837, with up to two inches in the interior (*United States Gazette*).

Another late snowfall blanketed the northern highlands on May 24–25, 1838. The *Genessee Farmer* (Rochester, New York) carried a notice of a 10-inch snowfall at Bradford. Heverly (1926) recalled that snow fell all night in the north-central highlands, "over a foot deep" (incorrectly listed as 1839). The Pottsville *Miners' Journal* reported, "A considerable quantity of snow fell on Broad Mountain" north of Pottsville. A coating was observed at Pittsburgh (Ludlum 1968).

Snow fell down to the coastal plain on May 2, 1841. The *Easton Sentinel* reported, "A cold rain, together with several snow storms, of but short duration." The *Jeffersonian Republican* on June 1, 1842, had a notice of snow at Pittsburgh that fell on May 15.

A remarkable snow occurrence on May 24–25, 1845, was conveyed by the *United States Gazette* (June 12, 1845), sufficient to cover the ground in Berks County, at New York City and Providence, Rhode Island, at a record late date.

A May snowfall of historic proportions covered the high elevations in northeastern Pennsylvania on May 4–5, 1851. The *Easton Democrat and Argus* reported that six inches covered the high Pocono Plateau. "Snow fell at at Hazleton, Luzerne County, to the depth of eighteen inches," according to the Stroudsburg *Jeffersonian*, which exceeds all storms in May in Keystone State history.

A *New York Herald* dispatch from Washington mentioned that "a snow storm is prevailing here," indicating a coastal disturbance. The *Boston Daily Evening Transcript* confirmed that snow mixed in with rain at Philadelphia and New York City.

The spring of 1855 was abnormally cold. Local historian Luther Hoffman quoted the docket of a Squire Transue, reported in the local *Morning Sun* (April 30, 1928): "On the Tenth of May there fell a shnay {snow}, and on the eleventh it went away." The location would have been in in the southern Poconos near Stroudsburg. Although the year was not mentioned, this May snowfall almost certainly occurred early on May 9, 1855.

A rare May snowstorm brought accumulations to parts of northeastern Ohio and adjacent areas of Pennsylvania on May 8, 1855, confirmed in the records of a Crawford County weather observer. The Pottsville *Miners' Journal* stated: "Snow fell at Hazleton to the depth of 4 inches on Wednesday day morning [May 9]." Another dispatch, 12 miles from Pottsville, reported 5 inches at and that "the sleighing was excellent." The Pottsville observer noticed a reading of 36°F at 7:00 a.m. and 46°F at 2:00 p.m. on May 9, 1855.

The following year, a record late-season frost was reported on May 31, 1856, after the passage of a strong cold front. In Chester County, the Smithsonian observer recorded 34°F near sunrise. A notice in the *New York Herald* reported 38°F at

Brooklyn Heights on May 31, 1856. (The article recalled the same low reading in the city on May 21, 1850, when snow blanketed western New York, totaling two inches at Rochester.)

The Belvidere, New Jersey, *Warren Journal* reported a heavy snowfall on the Pocono Plateau on May 19–20, 1857: "Snow near Beech [Woods] is six or seven inches deep." Three inches blanketed Broad Mountain north of Pottsville (*Miners' Journal*). The North Whitehall observer noted sleet mixed with the rain. The *Warren Journal* described a "very cold rain" requiring "thick overcoats, large fires."

Measurable snow has not been observed southeast of the mountains in Pennsylvania since May 3–4, 1861. The Mount Joy observer in Lancaster County recorded a rare May accumulation (2 inches), and the *Central Press* in Bellefonte mentioned 3 inches. The heaviest recorded fall was 7 inches at Bedford, midway between Harrisburg and Pittsburgh. The Pottsville *Miners' Journal* reported that "we had quite a snowstorm." Observers northwest of Philadelphia noticed a rain/snow mix at North Whitehall, Norristown, Morrisville and New York City.

On May 1–2, 1862, the Indiana weather station reported 1.5 inches. Rain turned to snow on May 2–3, 1864, accumulating up to six inches at Tobyhanna (*Jeffersonian*). The *Belvidere Intelligencer* in northwestern New Jersey reported that "the hills around Belvidere were white with snow." The Blooming Grove observer measured 2 inches on May 2, 1866. Bradford County had "several inches" of snow May 8, 1867 (Heverly 1926). New York City's Central Park records listed 0.4 inch—the latest instance of measurable snow in May. The *New York Times* confirmed a "slight fall of snow" in the early morning.

Blooming Grove received two inches of snow on May 4, 1875, with flurries at Stroudsburg (*Jeffersonian*). Flurries were seen at Dyberry on May 12. Heavy frost occurred in the interior on June 14, 1875, with ice one quarter-inch thick in Bradford County (Heverly 1926).

A departing coastal storm on May 24, 1877, brought a wintry report in the Stroudsburg *Jeffersonian*: "Snowflakes were numerously seen." Several inches accumulated in the northern New Jersey highlands at a record late date (Ludlum 1983).

A widespread May snow occurrence caught the attention of the press on May 6, 1882. The *Jeffersonian* reported one inch—the only instance of measurable snow in May in the past 150 years in the Stroudsburg area in a valley location. A three-inch total was reported at Frackville in Schuylkill County in the higher elevations, and five inches fell at Hazleton. Snowflakes were witnessed at Easton for about an hour "but melted fast."

A historic cold snap on May 29–30, 1884, brought snowflakes at Catawissa, Wellsboro and Dyberry on the 30th. Heverly (1926) listed a snowfall of two inches at Barclay in Bradford County. Press notices mentioned a late damaging frost at Pittsburgh, Lancaster and Reading. Flurries were flying again late in the season at Dyberry on May 11–12, 1885.

A record late appearance of flurries at Pittsburgh happened on May 31, 1893. Merely two years later, flurries were widely seen in northern Pennsylvania on May 14, 1895. On May 8, 1898, Philadelphia had a record late trace of frozen precipitation. On May 4, 1900, Wellsboro (0.5 inch) and Renovo (0.3 inch) had a dusting of snow. Flurries were evident on May 29, 1902, at Wellsboro, Towanda, Dushore, Wilkes-Barre and elsewhere in the northern mountains, accompanied by a late freeze.

May snowflakes were unusually common in the first decade of the 20th century in northern Pennsylvania. The most noteworthy occurrence arrived on May 10, 1906, when Pleasant Mount (elevation 1,800 ft.) recorded three inches (*Wayne Independent*). Also the Stroudsburg *Daily Times* reported, "All day the white covering came down in Mount Pocono" and "a squall of short duration" was observed in Stroudsburg. Flurries were seen at a record late date at Trenton and Jersey City, New Jersey, and Washington, D.C. The mercury plunged to 23°F at Wellsboro on May 23, 1906, with a "killing frost."

Another late cold snap brought snow showers to Pennsylvania on May 11, 1907, totaling one inch at Scranton and Carlisle and 3 inches at Montrose—the heaviest so late in the season. Two to three inches were measured on the high Allegheny Plateau; Lawrenceville had 2.4 inches and Towanda 0.4 inch. A record late trace of snow was observed at Cape May, New Jersey. The mercury at Montrose dipped to a frigid 19°F in a widespread hard freeze that followed on May 12.

On May 1–2, 1908, "Snow fell on the Poconos" (*Monroe Record*). The Milford observer caught a rare low-elevation May snowfall (0.5 inch). The storm that began in the northwestern counties on April 30 brought 3 inches at St. Marys on May 1–2.

Snowflakes were observed on May 1–3, 1909, at State College and in the northern mountains. Milford had a few flakes on May 3, 1911. Mount Pocono recorded intermittent flurries on May 4–6, 1917, and 1 inch fell at Culvers Lake, New Jersey.

A dusting of snow blanketed mountainous sections of Pennsylvania on May 8–9, 1923, leaving 2 inches at Somerset, 1.8 inches at Gouldsboro, and one inch at Scranton and Carlisle. The *East Stroudsburg Press* noted rain and sleet. Baltimore, Washington, D.C., Lynchburg, Virginia, and Asheville, North Carolina, witnessed rare May snowflakes.

The latest widespread measurable snow in Pennsylvania came on May 24–25, 1925, following the passage of a powerful cold front and squall line on the night of May 23 that brought hail and damaging wind gusts. Three to six inches of snow whitened the mountains in Tioga County the following day and night. The heaviest snowfall was six inches at Leolyn in the north-central mountains. Pittsburgh had a record late snowfall (0.5 inch). The East Stroudsburg *Morning Press* mentioned traces of snow at Pottsville, Fern Ridge and Mount Pocono.

The following year, Freeland had a late trace of snow on May 24, 1926. On May 23–24, 1931, two to three inches of snow blanketed the Alleghenies.

The greatest May snowfall in modern Pennsylvania history occurred on May 1, 1945, at Monroeton (9.5 inches) in Bradford County. Four inches fell at Wellsboro, 2 inches at Towanda and 1 inch at Mount Pocono. Two to four inches of snow coated the high elevations again on May 3, 1945. Light snow was observed at Gouldsboro and few high spots on May 10. Monroeton set a state snowfall record for the month of May (16.5 inches).

During May 7–9, 1947, snow again blanketed the northern highlands, with a maximum of 3.4 inches at Meadville. In the northeastern region, 1 inch fell at Paupack, and flurries whisked across some eastern valleys, reaching to New York City.

A heavy, wet snowfall on May 10, 1954, dropped 4.8 inches at Freeland, 4 inches at Long Pond, and 2 inches at Mount Pocono. A light fall extended to Schuylkill County. Snowplows were summoned to hilly sections surrounding Scranton. A portion of Route 309 between Wilkes-Barre and

Hazleton was closed for several hours due to the heavy three-hour snowfall. Oak Ridge Reservoir, New Jersey, had 0.5 inch.

The backward spring of 1956, with sharply lower temperature departures from March to May, brought snowflakes at a record late date at Harrisburg on May 16, and flakes at Stroudsburg in the valley. On May 18, 1956, Pleasant Mount received one inch, and late flurries were again observed at Stroudsburg. Historic chill on May 24–25 brought consecutive subfreezing lows at Mount Pocono (29°F/25°F) and Stroudsburg (30°F/24°F).

Philadelphia caught a rare spectacle of snowflakes on May 1, 1963. One to two inches of snow blanketed the northern mountains, with amounts in the west ranging from 1.8 inches at Pittsburgh to 4.1 inches fell at Meadville. Scattered flurries whisked through northern areas on May 22–23, flying at Erie, Montrose and Towanda, followed by a late freeze in the north on May 24. Readings dipped to 18°F at Clermont 4 NW and Titusville, freezing plants and damaging crops. Frost reached the interior southeastern counties of Pennsylvania.

Snowflakes were widely observed on May 9–10, 1966, including in the eastern valleys, with light accumulations in the northern and western highlands. Scranton, Allentown and Harrisburg reported a trace on May 9. Pittsburgh had a record May snowfall of 3.1 inches, eclipsing the monthly record set three years earlier. Four inches fell at Lawrenceburg and 6 inches piled up at Galeton in the northwestern mountains.

Snow fell in the higher elevations of the state the following year across the north on May 6–7, 1967, measuring 1.5 inches at Hawley and 5 inches at Lakeville in the northern Poconos.

On May 17–18, 1973, a record heavy late-season snowfall deposited 4.5 inches at Montrose and Pleasant Mount; Gouldsboro had 2 inches, and Scranton recorded 0.4 inch. The Hazleton *Standard-Speaker* reported about one inch.

The high terrain was whitened again on May 18–19, 1976, in the northeastern part of the state. Gouldsboro received 1 inch; but more interesting, valley flurries were observed at East Stroudsburg University.

A true late-winter coastal storm brought a substantial snowfall on May 9, 1977, after rain the evening before changed to snow by morning as far southeast as Atlantic City, New Jersey, at a record late date. The heaviest totals were in the Poconos, above 1,900 feet. Tobyhanna received 8.2 inches and Gouldsboro had 6 inches. Ten inches fell atop Pimple Hill (elevation 2,215 ft.) and at High Point Park, New Jersey (elevation 1,410 ft.). Scranton had a record May snowfall of 2.4 inches. A light coating extended west to the central mountains and State College.

Sleet was observed at Stroudsburg on May 16, 1984, at a late date in the lower elevations. An upper-level pool of cold air descended on the Great Lakes and Ohio Valley on May 6–8, 1989, resulting in record late snows in Ohio and a substantial fall of 8.5 inches at Bradford FAA. The more elevated sections of northwestern Pennsylvania received two to five inches, and 0.4 inch fell at Erie.

In the following decade, a Lehigh Township observer four miles northeast of Gouldsboro (elevation 2,140 ft.) saw a few flakes as late as May 27, 1994. The same weather site experienced 0.2 inch on Mother's Day 1996 (May 12), with a few flurries reaching Honesdale, Blakeslee, Moutainhome and Sussex, New Jersey.

On May 18, 2002, wet snow fell over higher terrain in northern Pennsylvania, depositing several inches in the mountains. Weather service spotter totals ranged up to 3 inches at Sabinsville 3 SE, 2 inches at Montrose, and 1 inch at Wellsboro 4 SW. Flurries were observed at Hazleton and Honesdale. In New York, a record late-season storm blanketed Albany (2.2 inches) and Binghamton (1.7 inches), with flurries at Poughkeepsie.

A secondary surge of arctic air brought scattered snow showers on May 20–21, 2002, leaving a dusting atop Laurel Summit (0.3 inch). Flurries reached Williamsport and Pittsburgh. Freezing temperatures were observed in the northwest on consecutive mornings from May 18–22, and in the northeast on May 19–22, with the coldest spots dipping into the mid-20s.

Snow flurries were spotted at Mount Pocono on May 21–22, 2006. A slushy wet snow whitened the central mountains of Pennsylvania on May 12, 2008, and a dusting was witnessed in the western mountains on May 9, 2010, with flurries across the northern counties. On May 14–15, 2016, snow and sleet showers moved through the northern and western highlands, with traces at Williamsport and East Stroudsburg. A total of 1.2 inches fell at Clermont in the north-central mountains.

A blast of polar air on Mother's Day weekend 2020 was accompanied by a substantial snowfall in the Northeast and around the Great Lakes. Low pressure traveled from southern Illinois to southern Pennsylvania along a cold front on May 8, 2020, then exploded over the East Coast, dumping up to a foot of snow in northeastern New York and 15 inches in northern Maine on May 8–10. Up to six inches blanketed the hills of south-central New York.

Rain turned to snow across northern and western Pennsylvania on May 8, 2020, and NWS spotters reported accumulations of 2 inches at Corry 5 NE and 2.3 inches near New Albany 4 NW in Bradford County. Mountain Top in Luzerne County picked up 1.9 inches; 0.5 inch fell at Moscow 10 SE and Lords Valley. Flurries reached the lower elevations on May 8–9 at Tamaqua and East Stroudsburg, and dusted Long Pond (0.4 inch) on the Pocono Plateau.

More surprisingly, Lititz, in Lancaster County, had a slight fall (0.2 inch). Flurries were even observed at all the New York City metropolitan airports on May 9, 2020, which tied the latest date at Central Park (1977). A dusting of snow covered the hills of northwestern New Jersey, and 0.9 inch fell at Highland Lakes.

Mount Pocono logged maximum/minimum readings of 37°F/22°F on May 9, and a January wind chill of 8°F near sunrise—coldest so late since May 4-5, 1917 (38°F/20°F, 37°F/24°F). Johnstown (35°F/23°F) had its chilliest May day, and Altoona (40°F/26°F) reported a record cold monthly high. Bradford's morning low of 22°F eclipsed the daily record set on May 8, 1966 (34°F/24°F), and State College tied its May 1966 minimum of 27°F. The coldest site was 12°F at Laporte on May 10, 2020.

An elevation snowfall on May 2, 2022, set a state record, with 11 inches at Laurel Summit, surpassing the May 1945 single snowfall mark.

Summer Snowflakes and Freezes

The tail end of the Little Ice Age in North America brought a number of historic late snowfalls in the interior Northeast, particularly in the late 1820s and early 1840s.

Snow fell on June 5, 1825, at an undisclosed location in Pennsylvania, according to an item in *Freeman's Journal* (Philadelphia), also noted in Washington's *National Journal* (June 14, 1825). The

Register of Pennsylvania (February 2, 1828) had a notice that snow fell on June 11, 1827, in the high Alleghenies. The *Aurora and Pennsylvania Gazette* in Philadelphia (July 3, 1829) noted "some flakes" in northern New Jersey "a short distance from the ferry" on June 28, 1829, and a "slight fall of snow" at Saratoga, New York. Correspondents reported "severe frosts" at Pottsville on June 21–22, 1835, in *Hazard's Register of Pennsylvania* (July 4, 1835).

The latest measurable snowfall in the Keystone State happened on June 11, 1842. The Pottsville *Miners' Journal* reported that snow "covered the foliage with its virgin drapery, and giving to the summer livery rather a strange appearance." Snowflakes were observed at Harrisburg, easily the latest such occurrence south of the mountains.

The backward spring of 1843 also brought winter in June. Historian Luther Hoffman, who culled family weather records in Smithfield Township, north of East Stroudsburg, was quoted in the *Morning Sun* (April 30, 1928): "Snow fell two inches deep and the blossoming grain and rye were covered."

The *Ohio State Journal* in Columbus (June 14, 1843) mentioned snow at Buffalo and Rochester, New York. A half-inch dusted the ground a few miles from the lakeshore at Cleveland, according to the *Herald*. Frost extended south to areas around Baltimore and Annapolis, Maryland, the next morning.

Even more shocking, a snow shower was observed at Philadelphia, according to the records of Charles Peirce, with a temperature of 44°F. The Pottsville *Miners' Journal* confirmed the visitation of snow in the mountains of eastern Pennsylvania. Remarkably, the mercury plunged to 24°F at Pottsville the next morning, doing great harm to the vegetables and fruits.

The cold summer of 1843 continued to create news. An item in the *Ohio State Journal* on August 1, 1843, mentioned that "flakes of snow" mingled with rain on July 11, 1843, in Maine on "Allegheny Mountain," and frost killed corn in the low spots on July 25 in Stark County, Ohio, during six weeks of dry weather around Canton *(Ohio Repository)*.

An item in the Stroudsburg *Jeffersonian* on July 16, 1885, recalled "The Memorable July Frost" on the Fourth of July 1845, in a story from the *New York Sun*: "A heavy frost fell on the hills overlooking the Delaware valley" in eastern Pike County along Shohola Creek.

The Great June Frost of 1859, documented in detail decades later in the U.S. Weather Bureau's *Monthly Weather Review* June 1907 publication, scarred fields and memories of a growing season short-circuited. Entire fields of corn were wiped out in northern Pennsylvania and much of Ohio. A Pittsburgh dispatch in the *New York Times* reported, "A severe frost occurred in this neighborhood on Saturday night [June 4–5], and caused great damage to the wheat, corn and potato crops, and also to some of the fruit trees."

In *Old Time Tales of Warren County*, Bristow (1932) recounted the late-winter day on Saturday, June 4, 1859, as told by Civil War veteran James Clark, who was from Spartansburg.

Not only was there frost, but Clark recollected a rare June snowfall:

> There was a good crowd at the log rolling and us boys were kept busy with water pails … Along about nine o'clock it clouded up with a cold wind blowing and about eleven o'clock it began snowing, not a flurry but a real winter storm … By four o'clock in the afternoon there was a good two inches of

snow on the ground ... It just seemed as if calamity was in the air.

The dependence on farming raised great fear of hardship in a freeze that befell the northern and western Pennsylvania highlands as dawn approached on June 5, 1859. Clark's account noted: "No grass, not corn, no feed for cattle. The leaves began falling off the trees ... some of the woods ... as bare as winter." Young birds and rabbits were found frozen, and the following winter presented untold hardships due to a lack of feed for cattle. Heverly (1926) mentioned "a flurry of snow on the Fourth of July [1859] and so cold that persons wore overcoats at the celebrations" in Bradford County, though this late instance of snow likely occurred on June 4, 1859.

On June 5, 1882, at the end of a remarkably cold spring, an item in the Stroudsburg *Jeffersonian* said, "snow fell at Tobyhanna Mills." Snow flurries were also observed at Lansing, Michigan, on June 4 and Wytheville, Virginia, on the 5th.

The latest occurrence of snow or sleet in official Pennsylvania records occurred on June 23, 1902, at State College and in Washington County. The spring and summer of 1903 were historically chilly, too. Cassandra (elevation 2,100 ft.), in Cambria County, had traces of snow on May 27–28 and June 13, 1903.

The spring of 1907 was uncommonly snowy and wintry—the coldest in state history (52.0°F) for the months of April–June, a substantial 4.8°F below the long-term normal. Snow fell in April and May, and in early June in northern locations.

The average temperature at Morrisville was compared to the previous cold springs of 1816, 1857 and 1859 in the *Monthly Weather Review* (Henry 1907). The average temperatures in 1907 were taken at nearby Trenton, New Jersey, where records commenced in 1866 (Table 4.4).

Table 4.5
Average Winter Temperatures (°F)
Morrisville, Pennsylvania

Year	Apr.	May	June	July	Aug.
1816	47.0	57.0	64.0	68.0	66.0
1857	43.0	57.7	66.2	71.5	70.0
1859	47.4	61.0	66.9	71.6	71.3
1907	47.4	56.4	65.8	74.0	70.9
Coldest	1874	1917	1903	1914	1927
Trenton, NJ	44.4	54.6	66.4	72.0	66.8

Government observers recorded snowflakes at Hanover on June 1, 1907, and Wilkes-Barre on June 2, which was confirmed by press notices in the *Bethlehem Daily Times*, *New York Times* and *Philadelphia Inquirer*. Press notices mentioned wind-whipped flurries swirling around the tops of skyscrapers mingled with the mist. On June 2, the maximum/minimum readings at Pocono Lake were 46°F/32°F, with an absolute minimum of 30°F on June 4.

A memorable hard freeze occurred in some western highland locations on the mornings of June 9–10, 1913, with extreme minimums of 20°F and 25°F at Somerset. Northern and western highland sites reported frost, with readings of 28°F to 35°F.

The following spring brought a killing freeze on June 6, 1914, on the Pocono Plateau and a record June low of 25°F at Pocono Pines four days after a previous morning minimum of the freezing mark. After another June cool snap, local government observer Frank Bullock wrote: "The frost on the 17th killed local vegetation within a vicinity of ten miles."

A few snowflakes were observed at Mount Pocono on June 23, 1918, the latest date in state weather records. An unseasonably cold blast on June 14, 1933, followed a heat wave and highs in the 90s a week earlier. Afternoon readings stayed in the 50s in northern and western Pennsylvania. Frost followed early on the 15th, with lows of 28°F at Somerset, 29°F at Kane, and 33°F at Mount Pocono and Stroudsburg. The *Climatological Data: Pennsylvania Section* stated: "considerable damage was done to corn and truck crops."

On June 1, 1945, flurries were seen at Scranton and Gouldsboro. The temperature plunged to 32°F at Mount Pocono and 35°F at Stroudsburg. Local observer William Hagerty recalled in the Stroudsburg *Daily Record* (May 24, 1948) that 0.2 inch fell atop the high plateau above 2,000 feet.

A damaging freeze affected northern and western Pennsylvania and portions of Ohio on June 11, 1972, wiping out some crops and flowers in the valleys. The coldest readings in the state included: 21°F at Clermont 4 NW; 25°F at Bradford FAA and Kane; 27°F at Francis E. Walter Dam (Luzerne County); 29°F at Tobyhanna and English Center. Frost nipped buds as far south as Pittsburgh and Phoenixville (34°F) and in outlying communities northwest of Philadelphia.

On a few occasions, the mercury has dipped into the upper 20s in July in the cold hollows of the Allegheny Plateau. On July 5, 1929, the West Bingham observer recorded a minimum of 29°F. At Clermont 4 NW, a low of 28°F occurred on July 9, 1963.

Chapter Five

Cold Waves

'Tis rare to want the wholesome north-western seven days together.
—William Penn (1683), in a letter to the Free Society of Traders

AN ARCTIC OUTBREAK travels thousands of miles before reaching Pennsylvania. The delivery mode is a plunge of the polar westerlies, sending relatively unmodified frigid air across the snow-covered Canadian Prairies into the northern United States.

Although the shortest day of the year in the Northern Hemisphere is around December 21, most of the great polar blasts in the East have occurred in January and early February, attesting to the adage: "As the day lengthens, so the cold strengthens." One reason for the climatological lag is that expanding snow cover in early winter reflects sunlight and radiates warmth at night, counteracting the gradual increase in insolation as the sun's angle rises and daylight lengthens in January.

In the winter months, large areas of high pressure over the Arctic periodically break off and push polar air southward. Snow cover keeps the air refrigerated over greater distances from northern Canada to the northern Plains and upper Great Lakes. A bone-chilling blast of air borne on the wings of a northwesterly gale can send the temperature tumbling more than 25°F in a few hours.

Many historic cold waves that reached the eastern United States came by way of a cross-polar flow from Siberia through the Canadian Arctic, a trajectory that provides a perfect setting for a gelid air mass. Sparkling nighttime skies over an icy tundra and minimal daylight prevent the bitterly cold air from modifying during the long journey.

An infrequent source of brutally cold air in the central and eastern United States is the high-level whirlpool near the North Pole. A deep buckle in the jet stream in response to an expansive ridge of high pressure over the North Atlantic known as the "Greenland block" can dislodge bitterly cold air harbored by the polar vortex when stratospheric winds weaken. Pieces or lobes of the polar vortex rotate southward, unleashing an intrusion of brutally cold air in an amplified flow.

Cold air outbreaks go beyond simply tracking air temperatures and minimum values during the early morning hours. The wind-chill temperature is an apparent measure (not a real temperature) of how cold the air feels to exposed skin, which is calculated by the rate of heat loss. On a windy day, heat we radiate is drawn away. At first, the skin temperature lowers as the heart pumps blood faster to protect our extremities to compensate for body temperature. Prolonged exposure to extreme cold without adequate protection can cause a person's core temperature to fall to dangerously low levels, raising the risk of frostbite and hypothermia.

January 1780
Coldest Month

The winter of 1779–80 was characterized by deep snowfalls and unrelenting cold. The Delaware River at Philadelphia was shut down to navigation from December 21, 1779, until March 4, 1780 (Hazard 1828).

The Hard Winter of 1779–80 was the only winter in American history when all the bodies of water surrounding New York City were frozen solid for five consecutive weeks. The weather was intensely cold across Pennsylvania on January 6–7, 13–16 and 19–29, 1780.

Philadelphia physician Dr. Benjamin Rush reviewed the extreme conditions in January 1780, and in 1789 wrote: "During the whole of that month [January 1780]—except one day—it never rose, in the city of Philadelphia, to the freezing point." The records of David Rittenhouse of Philadelphia showed the temperature edging above freezing only once during the month, on January 31, 1780, when the eaves of houses were reported to drip melting snow and ice.

A notice in the *New York Packet* (Fishkill) listed a sunrise observation of -16°F at the British headquarters on Manhattan Island on January 29, 1780, which is comparable to the modern record low at Central Park of -15°F on February 15, 1934.

Preeminent weather historian David Ludlum, in his *Early American Winters: 1604–1820*, compared January 1780 to the coldest month of the 19th century in the region—January 1857:

> Due to the imprecise nature of the readings and varying methods of compiling means, it would be reasonable to place January 1780 and January 1857 in a class by themselves, as co-holders of the title of the coldest month in history. January 1857 probably produced greater extremes, but January 1780 stands preeminent for duration of cold without a break.

February 2, 1789
Bitterly Cold Morning

An exceptionally frigid morning on February 2, 1789, was well-documented in press notices at the time. A widespread snow cover of eight to 10 inches created a ideal conditions for bitterly cold air to settle in over eastern Pennsylvania, described in later years in *Hazard's Register of Pennsylvania* (December 28, 1828).

A thermometer owned by David Rittenhouse in downtown Philadelphia read -5°F at 6:00 a.m., nearly matching the low of -6°F reached the previous winter on Cold Tuesday, February 5, 1788. A location 20 miles west of Philadelphia dipped to -12°F. Peter Legaux in Spring Mill, 13 miles northwest of Philadelphia, observed -17°F, the lowest reading in his span of records (1786–1828).

The diary of Elizabeth Kieffer, deposited at the Lancaster Historical Society, listed a reading of -21°F at 7:00 a.m. on February 2, 1789, adding, "Colder than anyone living can recall" (Ludlum 1966).

January 1792
Cold January

The turnaround winter of 1790–91 brought a bitter reading of -6°F in February: "The preceding season [January 1790] was exceedingly mild—boys bathed in the river on the 2nd of January" (Hazard 1828).

A measure of severe winters is the duration of subfreezing conditions and the closure of rivers to navigation, a common thread in the late 1700s.

The next winter resumed the tendency toward frigid winters in the late 18th century in eastern Pennsylvania. By the end of December 1791, the Delaware River at Philadelphia was filled with ice. A brief thaw in early January opened the river for transportation, before even colder air arrived.

A thermometer belonging to longtime New York City weather-watcher Henry Laight never registered a day above 30°F from January 7–30, 1792, a remarkable spell of 24 days. The Delaware River at Philadelphia froze on January 7 and remained closed to navigation through March 6 (Hazard 1828).

The prolonged arctic chill intensified after a heavy snowstorm on January 18–19, 1792 (Ludlum 1966). The records of David Rittenhouse in Philadelphia noted a reading of -5°F near sunrise on January 23.

December 1796–January 1797
Bitterly Cold Early Winter

The winter of 1796–97 opened with an icy chill. A bitter Christmas Eve in 1796 brought a reading of -13°F at Northumberland, noted in the records of Joseph Priestly (Ludlum 1966). A few weeks later, an observation at Philadelphia descended to -13°F on January 9, 1797, located near the bridge over the Schuylkill River at the head of Market Street (Peirce 1847). Temperatures of -10°F occurred on successive mornings in Philadelphia on January 10–11, 1797, among the coldest readings ever known in the Quaker City.

February 9–11, 1818
Severe Cold

An intense winter storm tracking "probably across Tennessee and North Carolina" turned northeastward and redeveloped near the Virginia coast at the beginning of February 1818, ushering in bitterly cold air (Ludlum 1966).

Dr. Samuel Hildreth, from Marietta, Ohio, reported that "26 inches fell on the level" in 12 hours. The next week, from February 9-11, 1818, exceptionally cold weather settled in over western Pennsylvania, where readings plunged to -24°F at Meadville and -26°F in Fayette County, according to the *Huntingdon Gazette* (March 5, 1818).

January 1821
Cold January

The winter of 1820–21 was cold and snowy, including a leadoff snowstorm on November 11–12, 1820, that deposited six inches at New York City (Brickner 2006).

Pennsylvania Magazine of History and Biography (1891) published a detailed "Pennsylvania Weather Records, 1644–1835" that focused on accounts of extreme weather events. The journal summary noted that a "heavy north-east snowstorm, eighteen to twenty-four inches deep" blanketed eastern

Pennsylvania on January 6–7, 1821. Bouts of harsh chill followed, with lows of -3°F on January 19 and -7°F on January 25 at Philadelphia.

On January 25, the mercury plunged to -14°F at New York City, and thousands crossed the frozen Hudson River on foot (Ludlum 1968). A few days later at Pittsburgh "the thermometer marked 13°F to 14°F below zero" during the cold wave from January 27–29, 1821, with a comment, "Cattle frozen to death."

January 1835 Cold Week

A heavy snowfall of 10 to 20 inches blanketed much of Pennsylvania on December 29–30, 1834, setting the stage for the advent of the Cold Week in early January 1835.

Samuel Hazard, the venerable editor of *Hazard's Register of Pennsylvania*, collected weather dispatches from around the state. The lowest reading reported in his journal was -32°F at Pine Grove in rural Schuylkill County. Farther west, along the Susquehanna River, a reading of -31°F was observed at Milton.

The Pottsville *Miners' Journal* reported lows of -20°F on January 4 and -24°F on the 5th. The *Easton Centennial and Argus* noted daily extremes of 8°F/-12°F on the 4th and 10°F/-16°F on the 5th. An account from New York City gave readings of -5°F on January 4 and -7°F the next morning, comparable to the city's official January minimum of -6°F (1882).

The editor of the *Sussex Register* in Newton, New Jersey, noted that temperatures "varying one to ten, even to 32 degrees below zero" occurred "beneath a sky, too, undimmed during the whole time by a single cloud," and that the chill "exceeds the experience of our most venerable citizens."

In *Hazard's Register of Pennsylvania* (March 7, 1835), excerpts from correspondents around the state confirmed the frigid term of January 4–9, 1835. The records of the Reverend E. D. Barrett at Glade Run, about 18 miles east of Kittanning, listed sunrise values of -10°F to -15°F, which were reported in the *Kittanning Gazette*.

At Philadelphia, the lowest reading during the Cold Week occurred on January 5 (-6°F) at the Pennsylvania Hospital. In Edward Garriott's *Cold Waves and Frosts in the United States*, he cited *Hazard's Register of Pennsylvania* for the list of minimums on January 4–5, 1835, which appeared in an April 11, 1835, article (Table 5.1).

Table 5.1
Low temperatures (°F)
January 4–5, 1835

Poughkeepsie, NY	-35
Pottsville, PA	-24
Albany, NY	-32
Lancaster, PA	-16
Hartford, CT	-27
Washington, DC	-16
Providence, RI	-26
Baltimore, MD	-10

At Newark, New Jersey, a reading of -13°F (Ludlum 1983) is comparable to the modern record of -14°F (February 9, 1934). A Boston reading of -15°F would not be surpassed until February 9, 1934 (-18°F). In Upstate New York and northern New England, readings of -40°F or lower were reported at New Lebanon, New York; Bangor, Maine; Montpelier, Vermont; and Franconia, New Hampshire.

After several weeks of seasonable January chill, harsh winter weather returned in late January and early February 1835. A severe cold wave plunged southward across the Midwest on February 5. The temperature remained below 0°F day and night at Fort Dearborn (Chicago) on February 6–8, 1835, with a nadir of -22°F on the 8th at 7:00 a.m. Historic chill reached the Deep South, bringing exceptional chill at Charleston, South Carolina (0°F), and Jacksonville, Florida (8°F), which destroyed citrus crops.

The core of the cold air engulfed Pennsylvania on February 8–9, 1835, when the Pottsville temperature dipped to -6°F/-4°F, according to the *Miners' Journal*. The *Pittsburg Gazette* listed local extremes of -8°F/-13°F on February 8, noting that "the rivers are fast closed with ice." The lowest reading (-21°F) was observed at "Sewickley bottom," a community located about 14 miles northwest of Pittsburgh, according to *Hazard's Register of Pennsylvania* (March 7, 1835).

A snowfall on February 26–27, 1835, was followed by another period of bitter chill in early March. A notice published in the *Register of Pennsylvania* (March 14, 1835) mentioned a frigid sunrise reading of 4°F on March 2 at Philadelphia, comparable to the city's lowest monthly value in official records (5°F on March 5, 1872).

A powerful late-winter storm brought rain on March 10, 1835, that quickly turned to heavy snow: "The snow fell in greater quantities, and more rapidly than we ever recollect to have seen … in about an hour the snow had fallen to a depth of three to four inches." The snow depth at Philadelphia reportedly averaged eight to nine inches around sunset.

1835–36
Frequent Cold Waves

The winter of 1835–36 was outstanding for snow and persistent cold weather in the East and Midwest.

Early temperature records maintained at Pennsylvania Hospital in downtown Philadelphia place the winter (December–February) of 1835–36 (27.7°F) on a level with the coldest three months ever experienced in the Delaware Valley. The bitterest modern winter occurred in 1976–77 (28.0°F). The contemporary mean winter temperature at Philadelphia International Airport is 36.1°F.

Samuel Hazard, in his *Register of Pennsylvania*, took notice of a harsh early cold snap following a winter storm on November 23, 1835. A week later, on November 29, the temperature hovered between 20°F and 25°F at Philadelphia. The icy weather continued in early December. A low of 12°F was noticed on December 3, 1835. On December 12, 1835, Hazard wrote, "The weather for the last week or ten days has been severely cold."

An arctic outbreak of historic proportions descended on the Northeast during the daylight hours of December 16, 1835. The network of New York Academy thermometers dipped to -40°F in northern New York, at which point the mercury congealed (Ludlum 1968). A mid-afternoon value of -14°F at Norfolk, Connecticut, 80 miles northeast of New York City, attested to the severity of the polar chill.

In eastern Pennsylvania, bitter temperature tumbled to the single digits on the afternoon of December 16, 1835, and the Schuylkill River was soon covered with ice. The temperature skidded to 8°F at Philadelphia and -7°F at Pottsville.

A terrible fire broke out in New York City overnight on December 16–17, 1835, consuming hundreds of city blocks, which was made much worse by hydrants freezing in the near-zero chill (Ludlum 1961). New York City readings dipped to 3°F in Jamaica, Queens, and -2°F at East Hampton on December 17, 1835.

The weather moderated toward the end of December 1835, but a series of winter storms in January 1836 brought return bouts of frigid weather. The Pottsville *Miners' Journal* posted readings of -16°F, -22°F and -20°F on successive mornings from January 27–29, 1836.

During a five-day cold wave, on February 2, 1836, the mercury plunged to -4°F at Philadelphia, only peaking at 10°F, followed by a frigid morning on the 3rd (-4°F). Peirce (1847) wrote that February 1836 had a dozen "intensely cold days" and eight snowfalls that provided "good sleighing throughout the month."

The average temperature at Philadelphia in February 1836 was 24.0°F, fourth among the coldest Februarys: 1934 (22.2°F); 1979 (23.0°F); 1838 (23.6°F). The mean temperature at New York City's Governors Island in February 1836 (21.5°F) ranks as the second coldest February in New York City history, after 1934 (19.9°F).

February 5–7, 1855
A February Cold Wave

A true measure of the intensity of an arctic air mass is the depression of the temperature in the middle of the afternoon.

On February 5, 1855, the mercury at the academy station in Gouverneur, New York, stood at -20°F at 2:00 p.m. Afternoon readings on February 6–7 were brutally cold (-15°F and -10°F). On the morning of February 6, the temperature slumped to -40°F, at which point the mercury congealed "in the absence of a spirit thermometer" (Ludlum 1968).

On February 6, 1855, the Smithsonian observer at Pottsville wrote down a sunrise temperature of -3°F. The readings at 2:00 p.m. (-3°F) and 9:00 p.m. (-6°F) offered little improvement. Subzero readings were also noted on February 7 at 7:00 a.m. (-6°F) and 2:00 p.m. (-2°F). The temperature rose to 6°F in the evening, as snow commenced. A total of 14 inches of snow buried Pottsville on February 8–9, 1855.

At Norristown, the thermometer slipped from 17°F at 9:00 p.m. on February 5, 1855, to a frosty 4.3°F at sunrise. Readings continued to sag on February 6, drifting down to 2.8°F at 2:00 p.m. and -2°F at sunrise the next morning, before edging up to 2.5°F at 2:00 p.m. during a snowstorm.

Along the coast, readings plunged to -8°F at both Newark and New York City on the morning of February 6, 1855. The *Boston Evening Traveller* mentioned -44°F in West Randolph, Vermont, on the 7th. The *Traveller* thermometer hovered around -10°F, the bitterest daylight ever known at Boston.

1855–56
Siberian Winter

The winter of 1855–56 started early and rarely let up, aptly described by Schmidlin (1996) as a "Siberian Winter" in the Midwest.

The average temperature at Philadelphia in January 1856 was 24.2°F, one of the coldest Januarys on record. More impressively, the mean winter temperature for the months of January through March 1856 (27.8°F) nearly matched the coldest winter (December–February) ever recorded in the in Delaware Valley in 1835–36 (27.7°F).

The first snowfall of the season dusted the Pennsylvania highlands on October 13–14, 1855. The Smithsonian observer at Pottsville noted 0.3 inch, one of the earliest snowfalls measured locally. Flurries were observed on October 28, 1855, leaving another early coating (0.5 inch) at Pottsville, and also noticed by the Stroudsburg *Jeffersonian*.

Pottsville had snowfalls on November 21 (four inches), December 13 (six inches) and December 29 (six inches) in late 1855. A steady cold settled in around Christmastime, after a period of mixed precipitation, and temperatures rarely inched above freezing.

The temperature at Mount Joy in Lancaster County stayed below freezing at the prescribed 2:00 p.m. observation time from December 31, 1855, through January 26, 1856. The mean temperature of 17.3°F at Mount Joy in January 1856 is lower than in any month ever known in the region.

In the wake of a snowstorm January 5-6, 1856, the floodgates to polar air were opened. Chief Weather Bureau forecaster Edward Garriott (1906) wrote, "Snow remained in large quantities at Washington from the first of January to the middle of March [1856]."

The *Pittsburgh Gazette* thermometer "hanging on the easterly side of a column" reached a bitter -18.5°F at 7:00 a.m. on January 9, 1856, and went on to note conditions as "clear and exceedingly cold, with a high wind" and "rivers are frozen in place." A low of -20°F was obtained at New Brighton.

In Crawford County at Moss Grove (elevation 1,700 ft.), the Smithsonian observer reported -6°F early on January 8, 1856, and the same reading at 2:00 p.m., bottoming out the next morning at -22°F. A minimum of -18°F at Pittsburgh on January 9, 1856, stood the test of time until January 19, 1994 (-22°F).

The thermometer at Norristown slumped to -9°F at 7:00 a.m. on January 9, 1856, and merely recovered to 1°F at 2:00 p.m., ranking among the coldest daylight periods on record, comparable to January 18, 1857, and January 17, 1982. The *Easton Express* noted a morning value of -6°F. Pottsville experienced a rare subzero day on January 9 (-2°F/-11°F), recorded in the *Miners' Journal*, with a minimal rebound to 10°F the next afternoon.

The Nazareth 7:00 a.m. reading on January 9, 1856, was a frigid -16°F, rising to only 1.75°F at 2:00 p.m. The weather was still bitterly cold on January 10, when Nazareth had extremes of 7°F/-7°F. The highest reading at Nazareth for the month was a chilly 37°F. The temperature nudged past the freezing point at Norristown on just five days in January 1856 at the 2:00 p.m. observation time.

More snow blanketed Pennsylvania on January 12–13 and January 27–28, 1856. Longtime Newark, New Jersey, observer William Whitehead measured 33 inches in January 1856—snowier than any January at Newark until 2011 (37.4 inches).

Bouts of frigid weather returned in February and early March. The *Easton Express* on March 5, 1856, citing a Pottsville *Miners' Journal* dispatch, reported snow depths of 12 to 18 inches. Northern Pennsylvania thermometer readings dipped below zero on March 10, 1856, after a final significant snowfall.

January 1857
Coldest January of the 19th Century

After the frigid winter of 1855–56, bitter cold returned the following winter. The average temperature at Philadelphia in January 1857 (22.4°F) registered two degrees below that of January 1856. The only colder month in the city's history was January 1977 (20.0°F).

The average temperature at Easton in January 1857 (16.2°F) was comparable to February 1934 (16.4°F) at Allentown. The mean temperatures at New York City (19.6°F) and Flatbush (19.3°F) have only been approached in February 1934 (19.9°F) in the modern New York City records at Central Park.

At North Whitehall, northwest of Allentown, the temperature did not surpass freezing at the 2:00 p.m. observation time from January 6–26, 1857 (21 days), which exceeds the Lehigh Valley record at Easton in January 1893 (20 days).

The first of two exceptionally cold blasts arrived in Pennsylvania on January 17, 1857, when the mercury at Mount Joy in Lancaster County showed a steady fall, from 18°F at daybreak to 11°F at 2:00 p.m. to -6°F at 9:00 p.m. The morning of January 18 brought a reading of -12°F at Mount Joy. The afternoon temperature struggled to reach 3°F.

In Bucks County, along the Delaware River, the Morrisville observer noted 0°F at 2:00 p.m. and 9:00 p.m., a rare occurrence in southeastern Pennsylvania. The New York Academy thermometer at Copenhagen registered a bitter -42°F on January 18, 1857. The 2:00 p.m. observation at Cortland, New York, was an icy -10°F, a true measure of the depth of the chill.

While a cold dome of arctic high pressure hovered over the Northeast, a great blizzard developed on the southern margin of the polar front, depositing one to two feet of snow on the Mid-Atlantic region on January 18–19, 1857. A deep snowpack extended from northern Georgia to southeastern New England, creating optimal conditions for extreme chill.

A brief moderating trend on January 21, 1857, was accompanied by a coastal disturbance, paving the way for a second surge of brutally cold polar air that arrived in the Northeast on January 22. In northern New England, the gelid air caused the mercury to sink to -34°F at Craftsbury, Vermont, at 7:00 a.m. on the 23rd, rising to just -23°F at 2:00 p.m. A Smithsonian thermometer at Norwich, Vermont, recorded -44°F. Near the coast, a frigid low of -9°F was noted at Jamaica Academy in Queens. William Whitehead's Newark minimum observation was a bitter -12°F.

In Pennsylvania, the thermometer at North Whitehall registered 11°F at 2:00 p.m. on January 22, 1857, before falling to -7°F at sunrise on the 23rd. The afternoon reading recovered slightly to 8°F. The next morning, the mercury plummeted to -22°F at 7:00 a.m. on January 24, the lowest ever recorded in the region. In the afternoon, the temperature struggled to make it back to 10°F.

At Easton, the sunrise reading on January 24, 1857, at Lafayette College was an icy -16°F. A similar reading was obtained at Norristown, 15 miles northwest of Philadelphia. The Morrisville observer noted -17°F on January 24, and -10°F on January 26. The minimums at Stroudsburg were -29°F on January 24 and -26°F on the 26th, recalled by the *Jeffersonian* (January 19, 1893). The *Wayne County Herald* in Honesdale reported a minimum of -33°F on the 24th, which is the lowest on record in the region.

In hilly northeastern Pennsylvania, the Berwick Smithsonian observer recorded -24°F. The lowest unofficial reading in the state was -36°F at Rockport, south of White Haven, which was mentioned by the Belvidere, New Jersey, *Warren Journal* in an article on the last day of the month.

The Lafayette College records at Easton give the nod to January 1856 (15.66°F) over January

1857 (16.22°F). The records at Nazareth, however, rate January 1857 slightly colder overall (Table 5.2), based on the three daily observation times.

Table 5.2
Average Winter Temperatures (°F)
Nazareth, Pennsylvania

	7 a.m.	2 p.m.	9 p.m.	Average
Jan. 1856	12	25	15	17.3
Jan. 1857	12.4	22.2	14.4	16.4

An inspection of long-term records at Washington published in the *Monthly Weather Review* (March 1891) handed the award for the coldest month to January 1856 (21.3°F), outdoing January 1857 (21.5°F) by a frozen whisker. The mean temperature in the nation's capital in February 1856 (26.8°F) guaranteed that January/February 1856 stand out as the coldest two months in succession (24.1°F).

January 10, 1859
Coldest Daylight

The bitterest daylight of the past two centuries in eastern Pennsylvania occurred on January 10, 1859. A cold wave arrived on January 8 and reached its nadir on the 10th. William Whitehead's thermometer at Newark, New Jersey, plunged to -12.5°F early on the 10th, only rising to a maximum of -0.5°F later in the day—the coldest daylight in his records that commenced in 1843.

The *Sussex Register* in Newton, New Jersey, remarked that the chill on January 10, 1859, was "the severest cold ever remembered." The temperature dipped to -12°F and inched up to only -5°F in the afternoon—the coldest daylight in local history.

A check of the observations taken at Easton's Lafayette College show uniformly low readings on January 10, 1859: 7:00 a.m. (-9°F); 2:00 p.m. (-1.5°F); 9:00 p.m. (-7°F). Equally bitter values were observed at North Whitehall: -10°F at 7:00 a.m.; -2°F at 2:00 p.m.; -2°F at 9:00 p.m. This is the only instance of a true calendar subzero day in the Lehigh Valley.

Family records maintained by Charles and Alvin Hoffman at Franklin Hill, on the north side of East Stroudsburg, showed -14°F on January 10, 1859. The highest reading at Central High School in Philadelphia on January 10, 1859, was 4°F. The observational readings at Norristown varied from 1°F at 2:00 p.m. to 3°F at 9:00 p.m.

January 8, 1866
Cold Term of '66

The Cold Term of January 1866 sent thermometers plunging to the lowest levels ever recorded in January at Philadelphia and New York City. The weather was intensely cold from January 4–10, 1866, with maximum readings generally not exceeding the teens.

On the evening of January 7, 1866, a 7:00 p.m. reading of -3°F was noted at a police station in Upper Manhattan. At 6:00 a.m. on the 8th, the temperature had already plunged to between -12°F and -14°F. The *New York Times* reported, "During the afternoon it rose to zero, where it remained up to a late hour."

William Whitehead's thermometer at Newark registered -12.75°F, comparable to January 10,

1859 (-12.5°F), and January 24, 1857 (-12.0°F), and a reading from another city location on January 5, 1835 (-13°F)—comparable to the modern record of -14°F on February 9, 1934 (Ludlum 1983).

It was the same story at Philadelphia, where the minimum reading downtown on January 8, 1866, was -9.5°F, though the press reported temperatures in the suburbs as low as -14°F to -18°F, "which is said to be the coldest weather ever known there." At Stroudsburg, the *Jeffersonian* noted morning readings of -15°F to -20°F.

A northern Pocono weather station at Dyberry had a reading of -19°F at 9:00 p.m. on January 7, 1866, lowering to -22°F at sunrise on the 8th. The afternoon reading on January 8 recovered to just -4°F, before slipping back to -11°F at 11:00 p.m. The North Whitehall observer recorded -13°F at 7:00 a.m. on January 8, 1866, rising to 4°F at 2:00 p.m. The Norristown observer noticed -13°F around sunrise on the 8th, and an afternoon temperature of 6°F.

1867–68
Long, Cold Winter

The winter of 1867–68 was uniformly cold and snowy in Pennsylvania. Temperatures were substantially below normal every month from December 1867 through April 1868.

The first significant snowfall of the winter of 1867–68 arrived on December 10, 1867. A major snowfall on December 12, 1868, was accompanied by the coldest weather recorded so early in the season. The Stroudsburg *Jeffersonian* reported a frigid reading of 6°F at 5:00 p.m., after a nine-inch snowfall. On December 14, the thermometer descended to -22°F at Stroudsburg and -27°F at Dyberry. The weather station at Whitehall, near Allentown, registered -7°F at 7:00 a.m., in the wake of an eight-inch snowfall.

During the first 20 days of December 1867, the mercury climbed above freezing only three times at Whitehall. The cold, while not severe, continued unabated through January 1868. The thermometer at Dyberry failed to rise above freezing for more than two weeks from January 5–22, and several snowfalls powdered the region.

The steady cold from January 30–February 12 in Pennsylvania brought many subzero mornings. The temperature at Whitehall plunged to -6°F at 7:00 a.m. on February 2, and fell below 0°F again on February 4–5, 8 and 12, 1868. In the northern Poconos at Dyberry, the coldest morning came on February 8 (-28°F), as highs struggled to top 20°F.

The Stroudsburg *Jeffersonian* editor took stock of the season on February 20, 1868: "The weather has been uniformly cold, with the thermometer, for several days, ranging from zero to 22 degrees below that point."

Another cold blast on February 23, 1868, brought a 7:00 a.m. reading of -6°F at Whitehall, and afternoon temperatures stayed in the teens February 23–24, 1868. One of the lowest readings was -17°F at Dyberry on the 23rd. A heavy snowstorm set in on February 24–25, 1868, leaving 10.5 inches at Ephrata in Lancaster County, and more snow fell on the 27th.

The mean temperature in February 1868 at Dyberry was 13.6°F, lower than any other February (1865–1903) except 1875 (13.4°F). During the late winter cold term, the mercury at Harrisburg did not rise above freezing from February 22–March 5, according to the *Daily Patriot Union*.

March 1868 came in like a frozen lion across Pennsylvania. The low on March 1 at Dyberry was a nippy -13°F. A major winter storm swept up the East Coast a few days later, described by the *Jeffersonian* as "the heaviest snow-storm of the season."

Accumulations were greatest in the Pocono region, totaling 15 inches at Dyberry and 13.5 inches at Blooming Grove. The snowfall at Harrisburg was four inches, with less falling to the west. At the end of the storm on March 2, the 2:00 p.m. temperature at Dyberry slumped to 5°F and hovered in the teens in the southern counties.

On March 3, 1868, the temperature at Harrisburg only rose from 6°F to 12°F. At Whitehall the thermometer varied from 0°F at 7:00 a.m. to 9°F at noon. A few mornings later, the mercury tumbled to -22°F at Dyberry on March 5, 1868, the lowest reading ever recorded in March (1865-1903).

March 5, 1872
Coldest March Morning

The latest subzero daylight on record in Pennsylvania weather history occurred on March 5, 1872, when Blooming Grove registered extremes of -1°F/-6°F at 7:00 a.m. and 2:00 p.m. New York City had its coldest March Day (10°F/3°F). A Chester County site had daily extremes of 14°F/3°F.

Boston (-8°F) recorded its coldest March morning. In the nation's capital, March 5, 1872, was the coldest Inauguration Day on record before January 20, 1985, with a dawn reading of 4°F rising only to 16°F in the early afternoon. The remarkably low average temperature in Philadelphia in March 1872 (31.7°F) was only exceeded in 1843 (30.0°F) and 1885 (30.8°F).

1872–73
Two Bitter Outbreaks

A massive dome of arctic air settled over the Midwest in December 1872 during the week before Christmas. On the afternoon of December 22, 1872, the temperature at Dyberry stood at 2°F. At Pocopson, in Chester County, the thermometer peaked at a frosty 7°F.

The intense cold moderated briefly on the 23rd but was reinforced on Christmas Eve. Afternoon temperatures at Dyberry during the holiday period from December 24–27, 1872, ranged from 8°F–10°F, and at Pocopson from 12°F–13°F. The *Easton Express* reported maximum readings between 11°F and 19°F.

Morning temperatures on Christmas Day were near or below 0°F across the state, hitting a frosty -10°F at Blooming Grove in the northeast. In the central counties, a Union County observer jotted down -22°F at Lewisburg. The local government observer in Chester County recorded 0°F.

A massive snowstorm followed on the heels of the historic chill, beginning on Christmas evening 1872, which left one to two feet of snow. A surge of arctic air swept into Pennsylvania on December 29, 1872, sending the mercury into a downward spiral.

The records of John N. Stokes in Stroudsburg, routinely published in the *Jeffersonian*, gave a low of -18°F on December 30, 1872, which is still the benchmark for December. (The modern December record minimum is -14°F in 1963.) The low at Dyberry plunged to -14°F. The station at Egypt, northwest of Allentown, reached -10°F at 7:00 a.m.

More rounds of bitterly cold air followed in January and February 1873. On January 18, 1873, a temperature of -43°F was registered at La Crosse, Wisconsin, indicating a mound of cold air was in place, waiting for a shift in the winds to deliver the frigid air farther southeast. A heavy snowfall on January 27–28, 1873, accomplished that feat.

The mercury on January 29, 1873, at Dyberry fell from 4°F at 2:00 p.m. to -19°F at 9:00 p.m., before bottoming out at -31°F at 7:00 a.m. on the 30th—equaled once on February 10, 1888, during the period of record (1865-1903). The *Jeffersonian* reported an extraordinary low of -32°F at Stroudsburg, surpassed only in January 1912 (-35°F).

The observer at Egypt recorded -20°F at 7:00 a.m. In Chester County the temperature dipped to -18°F, a record to this day. In the south-central part of the state, a frigid -28°F was logged at Carlisle Barracks; there was a press report of -33°F at Lewisburg.

1874–75
Another Frigid Winter

The winter of 1874–75 would turn out to be even colder than 1867–68 and 1872–73. The average temperature at Philadelphia in January/February 1875 (25.3°F) ranks as the third lowest for consecutive months, after January–February 1856 (25.2°F) and December 1976–January 1977 (25.15°F).

The string of exceptionally cold winters in the late 1860s and early 1870s is evident in the data series at Dyberry (Table 5.3).

Table 5.3
Average Winter Temperatures (°F)
Dyberry, Pennsylvania

	Nov.	Dec.	Jan.	Feb.	March
1867–68	37.2	21.3	17.6	13.6	32.0
1872–73	31.9	17.9	18.3	19.1	26.3
1874–75	32.5	25.3	15.4	13.4	24.2
Average					
(1865–1900)	35.2	25.9	21.1	22.2	30.1

The autumn of 1874 was quite dry in Pennsylvania, though not unusually cold. Pittsburgh received just .06 inch of rain in October 1874, the driest month ever recorded in the city. Snow flurries were observed on October 13, 1874, in the northern mountains. The first real snowfall left several inches on the high ground on November 19, 1874.

A major snowstorm blanketed Pennsylvania on December 20, 1874, dropping 16 inches at Egypt and 12 inches at Stroudsburg (*Jeffersonian*). The air was moderately cold in December 1874, then periods of intense cold prevailed in January and February 1875. The 2:00 p.m. observation temperature at Dyberry remained below freezing from January 5–21, 1875 (17 days).

On January 9, 1875, the thermometer reading in New Castle plunged from 15°F at 7:00 a.m. to 0°F by 2:00 p.m., and -10°F by 9:00 p.m. The sunrise reading on January 10, 1875, stood at -18°F, rising to just 4°F in the afternoon. In the Lehigh Valley, Egypt recorded a sunrise low of -10°F on January 10, 1875.

Another bitter morning (-8°F) occurred on the 11th. A press dispatch reported a low of -18°F at Susquehanna in northern Bradford County on

January 10, 1875. The temperature at Dyberry plummeted to -12°F around sunrise on January 10, 1875, and recovered to merely 5°F at 2:00 p.m. before falling back to -15°F on the 11th.

A frigid outbreak in Pennsylvania extended from January 15–21, 1875, when readings at Dyberry never rose above the teens. The lowest sunrise temperature during the cold week was -16°F on the 19th. The mean temperature for the month of 15.4°F was the coldest January in the period of record (1865–1903).

January 1875 averaged an icy 22.6°F at Pittsburgh, with an absolute minimum of -12°F on January 10. February 1875 (20.8°F) was even colder, and only three Februarys since have been chillier. Daytime readings were generally in the 10s to lower 20s in northern and western Pennsylvania from February 4–8, 1875, and that was before the core of the coldest air settled in over the state.

A subzero day occurred on February 9, 1875, in the northwest. At New Castle, the mercury slumped to -12°F at 7:00 a.m., and from there slowly climbed to -6°F at 2:00 p.m. and -3°F at 9:00 p.m. In the northeast, the comparative readings at Blooming Grove were -7°F at 7:00 a.m., -5°F at 2:00 p.m. and -2°F at 9:00 p.m.

The Stroudsburg *Jeffersonian* reported lows of -6°F at Stroudsburg and -18°F at Tobyhanna. In the northwest, the value at Erie on February 9, 1875, was -16°F, which was the coldest figure on record locally until February 1979. New York City had its coldest daylight (2°F) since January 1859.

The thermometer at Blooming Grove never rose above freezing at any observation time from February 4–19, 1875 (16 days). Moderate snows fell over much of the state on February 11 and 19. The monthly mean temperature at Dyberry (13.4°F) in February 1875 was the lowest for any month (1865–1903).

A barrage of snowstorms hit the region on March 1, 3, 5 and 7–8, 1875, preserving the snowpack. The first day of spring was bitter across the state. Daily extremes at the Dyberry station on March 22 were exceptionally cold for early spring: 20°F/-6°F, followed by -13°F on the 23rd. On March 31, 1875, the observer commented: "Sleighing from January 8 to March 31, 1875," and "Snow two and a half to three feet deep in the woods."

Winter lingered through much of April 1875. A heavy snowstorm swept up the Eastern Seaboard on April 12–13, 1875, dumping six to 12 inches of snow north and west of Philadelphia. An arctic cold front arrived on April 16, 1875, sending temperatures plummeting to freezing by midnight in most areas, accompanied by snow showers.

The extraordinary chill is evident by reviewing the daily extremes at New York City on April 17 (37°F/26°F) and 18 (32°F/25°F), historically low readings so late in the spring (*New York Times*). The Harrisburg *Daily Patriot* confirmed the unprecedented chill on April 18 (30°F/20°F). A reading of 24°F was reported at Pittsburgh.

The Williamsport *Daily Gazette and Bulletin* described the exceptional late-season conditions as "very cold and disagreeable," which were accompanied by "two or three inches of snow" on April 16–17 and "three or four inches" on April 17–18 in the northern highlands, with a reading of 15°F at midnight on April 17. Additional snow showers dusted northern areas on April 19, and minimums were January-like along the coastal plain at Washington (22°F) and Baltimore (24°F) and Philadelphia (27°F).

December 1876
Deep and Persistent Early Cold

December 1876 opened with a blast of cold air, setting the tone for a consistently frigid month. On December 9, 1876, the 2:00 p.m. observation at Egypt registered 9°F following a fresh snowfall of 12 inches. On December 10, the morning reading plunged to -2°F and only recovered to 14°F in the afternoon. Philadelphia had its coldest day so early in the season, when the maximum/minimum temperatures were 17°F/4°F. The Stroudsburg *Jeffersonian* mentioned a low of -12°F. A later article in the *Daily Record* (December 10, 1948) noted a local minimum of -16°F.

Temperatures remained below freezing around the state except in the southeast from December 16, 1876, until a few days past New Year's Day. A snowstorm on December 18, 1876, freshened the snowpack and kept the air chilled. Philadelphia recorded its second coldest December (25.4°F) after 1831 (25.0°F).

Heavy snow fell on New Year's Day 1877 and into the next day, amassing a total of four inches at Philadelphia (*Philadelphia Inquirer*) and generally 12 inches more from Washington, D.C., to New York City and farther west. The *Jeffersonian* reported drifts of up to 20 feet in Wayne County, in the northeastern highlands.

January 3, 1879
Subzero Daylight

Bitterly cold air reached Pennsylvania on January 2, 1879, sending the temperatures plunging at New Castle, from 20°F at sunrise to 7°F at 2:00 p.m. and -6°F at 9:00 p.m.

The next morning found the mercury hovering around -16°F at New Castle. The daytime temperature inched upward to a frosty -5°F at 2:00 p.m. and -4°F at 9:00 p.m.

In the east, the corresponding 2:00 p.m. and 7:00 a.m. readings at Egypt on January 3, 1879, were 7°F/-7°F. The Stroudsburg *Jeffersonian* reported extremes of 6°F/-8°F.

In northwestern Pennsylvania, afternoon observations at New Castle on January 4–5, 1879, were limited to the single digits, culminating in a couple of subzero mornings. The wintry week was topped off by an eight-inch snowfall on January 8–9, 1879.

Another frigid blast arrived in the Northeast on January 18, 1879, in the wake of a heavy snowfall. The *Jeffersonian* in Stroudsburg listed minimums of -10°F on the 19th and -17°F on the 21st.

December 30–31, 1880
Coldest December Days

A series of bitterly cold arctic air masses invaded the eastern and central states in the winter of 1880–81, causing great hardships. The first blast arrived the third week of November 1880 and brought historically low readings to the northern portion of the country east of the Rocky Mountains.

Pennsylvania experienced January-like conditions beginning on November 19, 1880, and continuing through the remainder of the month. As far south as the Lehigh Valley, the weather observer at Egypt recorded subfreezing readings day and night for a week from November 21–28, 1880—the longest such period so early in the season.

The core of the cold air arrived on November 22, 1880, sending the mercury at Dyberry, in the northern Poconos, tumbling to 2°F, rising to only

16°F at 2 p.m. The afternoon reading in Egypt was 24°F. Another bitter day occurred on December 10, when the temperature at 2:00 p.m. read 5°F, before falling below 0°F later in the afternoon.

The morning of the 23rd brought a sunrise reading of -3°F—a record early subzero occurrence. There were also subzero values in the northern mountains on the 24th. The coldest morning occurred on November 26, 1880, when the mercury dipped to -6°F. The Egypt observer logged 3°F—lower than any modern November minimum in the Lehigh Valley.

Another cold blast on December 10–11, 1880, was a prelude to a consistently cold December, with periodic bitter surges of frigid air. An arctic front south of Pennsylvania became the focus for a string of disturbances that brought snow from the Deep South to the Mid-Atlantic region.

On December 28, 1880, Brownsville, Texas, received a rare two-inch blanket of snow that spread across the Deep South. Five inches fell at Montgomery, Alabama. As low pressure turned up the Eastern Seaboard the following day, heavy snow reached southeastern Pennsylvania, with totals surpassing 10 inches, which boosted the monthly total at Philadelphia to 22 inches.

The fresh snow cover paved the way for a blast of bitterly cold air of historic proportions. The observer at Egypt recorded a reading of -15°F at 7:00 a.m. on December 30, 1880, and only recovered to 0°F at 2:00 p.m. In the Poconos, the temperature at Dyberry ranged from -18°F at 7:00 a.m. to 0°F at 2:00 p.m. on December 30, 1880.

The Stroudsburg *Jeffersonian* reported a low of -10°F on December 30, 1880, with a 3:00 p.m. reading of 3°F on a sheltered thermometer on the south side of Main Street. The mercury tumbled to -10°F on December 31 and -17°F on January 1, 1881.

Farther north, press reports of low readings came from Milford (-15°F) and Dingmans Ferry (-18°F) to usher in a frigid New Year's Day 1881. Mount Ararat, 16 miles west of Carbondale, had a frosty -26°F on December 31, 1880, followed by a New Year's Day morning reading of -22°F.

Philadelphia experienced its coldest December day in history on December 30, 1880, with extremes of 5°F/-5°F (Table 5.3), and Pittsburgh bottomed out at -9°F. The New York City range of 4°F/-6°F was the coldest day in any month until December 1917. Freehold, New Jersey, had extraordinary chill (-0.5°F/-11°F). In northwestern New Jersey, Dodge Mines (elevation 1,100 ft.) had values of -7°F at 2:00 p.m. and -5°F at 9:00 p.m. on the 30th.

Table 5.4
Maximum/Minimum Temperatures (°F)
December 29–31, 1880

	Dec. 29	Dec. 30	Dec. 31
Dyberry	10/5	0/-18	6/-19
Philadelphia	14/4	5/-5	14/0

A minimum reading of -13°F was reported at Washington, D.C., on December 31, 1880, in the core of the gelid air mass. New Year's Day dawned equally frigid, with a low of -14°F that marked consecutive monthly records for the District.

A heavy snowstorm followed on January 4–5, 1881, totaling 12 inches at Egypt and Dyberry, establishing a deep snowpack that prolonged the frigid conditions. The mean winter temperature at Philadelphia (December–February) in 1880–1881 averaged 28.8°F, more than seven degrees below the contemporary normal. A cold spring into

April 1881 marked seven consecutive months of below-normal temperatures in the region.

Winter of 1884–85
Multiple Cold Blasts

The cold snowy winter of 1884–85 brought repeated shots of arctic air, starting a little before Christmastime.

A cold wave on December 19–20, 1884, sent the mercury tumbling well below zero in northern Pennsylvania on the 20th. The maximum/minimums at Dyberry were remarkably low for so early in the season on the 19–20th: -1°F/-10°F and 6°F/-15°F. The lowest reading at Blooming Grove was -12°F. Snow arrived on December 20–21, 1884, and subzero weather followed on the 22nd.

The cold week of February 11–17, 1885, ranks among the chilliest in mid-February records. On February 11, 1885: "The *Jeffersonian* instrument refused to go farther than 2 degrees above zero" in Stroudsburg, which, if official, would be the lowest February maximum on record. Dyberry had a subzero day (-1°F) on the 11th, with another icy day on the 17th (3°F). The lowest reading of the cold term was -15°F on the 12th, among the 14 subzero mornings in the month.

Frequent snowfalls in February 1885 sustained a respectable snow cover. The Stroudsburg *Jeffersonian* reported four inches fell on February 16, with an inch of melted precipitation, and up to 14 inches in the nearby higher elevations. Five inches fell on February 16–17 and six inches accumulated on February 24–25. Philadelphia amassed a record snow depth of 26 inches that stood until January 1996.

A historic cold week occurred on March 17–23, 1885. Dyberry had minimums of -16°F and -18°F on March 17–18 and -14°F on the 22nd—the lowest on record after the start of astronomical spring. Maximums ranged from 13°F to 23°F. Daytime maximums at Philadelphia from March 18–23, 1885, held in the 20s. A late-season record low of 6°F occurred on March 21. The mean monthly temperature at Philadelphia (30.8°F) falls nearly 13 degrees below the contemporary normal (43.6°F) and ranks as the second coldest behind March 1843 (30.0°F).

January 1893
Record Subfreezing Streak

Arctic air was a frequent visitor to Pennsylvania in January 1893, and Philadelphia endured two consecutive weeks of subfreezing weather. January 1893 (24.0°F) ranks only behind January 1977 (20.0°F) and January 1857 (22.4°F) as the coldest in the city's history.

Philadelphia experienced subfreezing weather on 18 of 19 days (January 4–22, 1893), with readings exceeding 20°F only once between January 10–18, 1893. Easton's Lafayette College weather station had below-freezing readings during a record span of 20 consecutive days (January 3–22, 1893), and the thermometer failed to rise above 20°F from January 10–18, 1893. The mean January temperature at Easton (16.5°F) was only a fraction of a degree higher than that of January 1857 (16.2°F).

Some places in northern Pennsylvania were below freezing for 22 straight days, through January 24, 1893. The lowest official temperature in the state was -25°F at Saegertown. The Stroudsburg *Jeffersonian* reported -18°F to -25°F on January 16, -22°F to -27°F on the 17th and -18°F to -22°F on the 18th. Temperatures dipped to -22°F at Dyberry and -14°F at Easton, Quakertown and York.

The *Philadelphia Inquirer* carried a summary of the icy readings as low as -24°F at Bear Gap, near Shamokin, and Spring Grove. Binghamton, New York, established an all-time historic low of -28°F. Historically low readings were observed of -20°F at Somerville, New Jersey, and -17°F at Millsboro, Delaware.

February 1894–1896
Trio of Frigid Februarys

The first of two arctic surges reached the Northeast on February 16–17, 1894, when the mercury tumbled to -14°F at Stroudsburg and -22°F at Tobyhanna, according to the *Jeffersonian*. The minimum at New York City's Central Park of -6°F is the lowest reading so late in the season.

Bitterly cold air infiltrated the region again on February 24, 1894, when the *Jeffersonian* reported a maximum of 8°F (lowest so late in the season), followed by a minimum of -14°F early on the 25th. The observer in Honesdale had a high of 3°F on February 24, 1894, and a nadir of -20°F the next morning.

February 1895 got off to a frigid start during a five-day stretch from February 5–9. Easton had maximums of 9°F and 8°F on February 6 and 8, with a minimum of -9°F on the 6th. The highs at Dyberry ranged from 2°F on February 6 to a maximum of 6°F during February 5–8, 1895, contributing to a frigid monthly mean of 14.8°F. A general snowstorm on February 8-9, 1895, deposited up to 10 inches at Easton.

An exceptionally cold day on February 17, 1896, marked the third consecutive frigid February in Pennsylvania. Maximum/minimum readings ranged from 1°F/-16°F at Dyberry to 10°F/-8°F at Easton. The *Jeffersonian* reported extreme lows of -14°F at Stroudsburg and -22°F at Tobyhanna. The *Philadelphia Inquirer* noted lows of -22°F near Hazleton and Wilkes-Barre.

February 9–15, 1899
Great American Cold Wave

Three moderate Mid-Atlantic snowfalls during the first week of February 1899 provided the requisite conditions for the greatest cold wave in the eastern United States. In northern Minnesota, the Leech Dam thermometer registered -59°F on February 9, 1899, as bitter air descended upon the northern Plains.

The following days brought subzero cold, worsened by snow and wind across the Midwest and Northeast. During the coldest stretch from February 8–13, 1899, the mercury generally remained below 10°F across most of Pennsylvania, except the southeast. The Somerset reading in the Laurel Highlands never rose above 0°F from February 8–10, 1899.

The coldest morning in the state occurred as the sky cleared on February 10, 1899, with a rare day of temperatures of 0°F or lower through a calendar-day at Dyberry, Harrisburg and South Bethlehem. The extreme low of -39°F at Lawrenceville, Tioga County, was the coldest official February reading ever known in the state—a mark that still stands. The maximum/minimum readings at Philadelphia (5°F/-6°F) yielded the coldest daily average in city records.

The *Stroudsburg Daily Times* reported lows of -10°F, -16°F and -14°F on February 9–11, 1899.

The morning readings at Tobyhanna dipped to -17°F, -25°F and -26°F. State College had its all-time bitterest day on February 10 (-5°F/-20°F).

In the west, Pittsburgh readings were subzero both day and night on February 9-10, 1899, with maximums of -3°F and -2°F—the longest subzero streak in history that included an all-time February minimum of -20°F on the 10th. The city shivered through a record seven consecutive days with subzero lows from February 8–14, 1899. The extremes on February 10, 1899, at Washington, D.C., of 5°F/-15°F were the lowest ever recorded in the nation's capital. Baltimore endured its coldest day (3°F/-7°F).

On the margin of the historic cold wave, the Great Blizzard of 1899 swept up the East Coast on February 11–14, 1899, dumping one to three feet of snow from the Mid-Atlantic to southeastern New England. A frigid morning on February 15, 1899, marked the end of an exceptionally cold week, with northern Pennsylvania minimums of -10°F to -15°F.

There would be a complete reversal of fortune on February 20–22, 1899, when the temperature soared into the 50s.

January 5, 1904
Coldest Pennsylvania Morning

A heavy snowstorm on January 2–3, 1904, paved the way for an arctic outbreak that would bring the lowest temperature ever recorded at a government weather station in Pennsylvania, causing hardships in the midst of unrelenting arctic cold waves.

In the southeast, the temperature rose to only 13°F in Philadelphia on January 4, 1904, and readings were barely above 0°F in the north. On January 5, the bottom fell out under a clear sky (see Table 5.5). Northwest of Philadelphia, readings plunged to between -10°F and -30°F.

An all-time state-record low temperature of -42°F was recorded at Smethport in McKean County, which exceeded the state extreme reached in February 1899. The next lowest reading was reported at Lawrenceville (-38°F) in Tioga County.

The *Stroudsburg Daily Times* reported lows ranging from -26°F to -31°F in the southern Poconos. A follow-up story (January 13, 1904) mentioned even lower readings: -34°F on Samuel Eschenbach's farm near Pocono Summit; -32°F at Blakeslee; -28°F two miles at the Stroudsburg home of local "weather prophet" Reverend Ira H. Hicks (Table 5.6).

Table 5.5
Minimum Temperatures (°F)
January 5, 1904

East		Central		West	
Towanda	-31	Lewisburg	-29	Coudersport	-33
Pocono Lake	-27	Wellsboro	-25	Grampian	-32
Milford	24	Selinsgrove	-24	Franklin	-30
Wilkes-Barre	-18	Montrose	-21	Emporium	-29

Table 5.6
Minimum Temperatures (°F)
January 5, 1904

Greentown	-34	Tannersville	-28
Bushkill	-33	Marshalls Creek	-28
Pocono Summit	-30	Brodheadsville	-20

Source: *Stroudsburg Daily Times*

Another cold wave arrived on January 18, 1904, driving readings far below zero in the northern highlands. On January 21, 1904, the *Times* reported: "Furnaces heaped with coal failed to keep a house warm and water pipes in many places were frozen shut."

The press reported extreme lows of -36°F on Transue Farm in Echo Valley, north of Shawnee, on January 19, 1904. Other noteworthy readings were observed at Analomink (-30°F) and Bushkill (-28°F). In the Stroudsburg area, local thermometers ranged from -16°F to -22°F, accompanied by a "bitter, cutting wind." Frigid minimums dipped to -15°F at Scranton, -21°F at Milford, -24°F at Pocono Lake and -25°F at Coudersport.

Frequent light to moderate snowfalls conditioned the arctic chill. Smethport had another extreme low of -27°F on February 2, 1904. A shot of brutally cold air on February 16–18, 1904, sent the mercury tumbling to exceptionally low levels so late in the season. Erie had a rare subzero day on February 16 (-2°F/-8°F), and single-digit maximums were common across the north.

The average three-month winter temperature of 28.4°F in the historic Philadelphia data series is the fourth chilliest since record-keeping started in 1825, exceeded only in 1835-36 (27.7°F), 1977-78 (28.0°F) and 1962-63 (28.3°F).

State records since 1895 reveal that the harsh season of 1903–04 (19.8°F) is the second coldest in modern history. The winter of 1904–05 (20.7°F) ranks fourth, the only instance of consecutive top 10 coldest winters in Pennsylvania.

A similar occurrence of back-to back extreme winters happened in 1976–77 and 1977–78, which was notable for headlines about the next ice age.

January 14, 1912
Bitter Cold

Snow cover and clear skies were responsible for another extremely cold January morning, eight years after the historic low temperatures in January 1904 (see Table 5.7).

The core of bitterly cold air was still west of Pennsylvania on January 12, 1912. A low temperature of -47°F was observed at Washta, Iowa. Des Moines endured its coldest day of the century (-14°F/-29°F) on the 12th.

Bitter cold air spilled east into Pennsylvania on January 13. The Pocono Lake observer (elevation 1,660 ft.) noted a maximum temperature of 3°F before the mercury plunged to -16°F around 9:00 p.m. Readings as low as -35°F were recorded at Stroudsburg and Pocono Lake on the 14th.

An all-time minimum reading of -27°F was notched at Lancaster on January 14, 1912. The minimum of -16°F at Bethlehem still exceeds the modern record of -15°F reached on January 21, 1994, at Lehigh Valley International Airport. In central Pennsylvania, a temperature of -17°F at the Pennsylvania State University weather station was the lowest reading at State College since February 1899 (-20°F).

Urban warmth afforded limited protection from the extreme cold. The minimum at the Weather Bureau in Philadelphia dipped to 0°F. However, at Washington, D.C., the urban heat island did not prevent the temperature from plummeting to -13°F on January 14, the lowest reading since January 1, 1881 (-14°F).

The *East Stroudsburg Press* listed numbing lows of -38°F at Bushkill and -32°F at Cherry Valley. Stroudsburg also hit -32°F on the 16th.

Table 5.7
Minimum Temperatures (°F)
January 14, 1912

Stroudsburg	-35	Lancaster	-27
Pocono Lake	-35	Lebanon	-17
Mifflintown	-31	State College	-16
Emporium	-31	Bethlehem	-16

1917–18
Frigid War Winter

Serious fuel shortages developed during the frigid War Winter of 1917–18 that caused great hardships for soldiers, who suffered in stateside barracks that were insubstantial.

A 1928 issue of *PennWays*, a Pennsylvania Department of Highways publication, traced the development of the state's snow removal policy to the cold, snowy winter of 1917–18 (Newton 2017). The War Department requested that the Department of Highways ensure that the Lincoln Highway across southern Pennsylvania (300 miles) would remain open for the war-time transport of troops and raw materials.

A chilly autumn in 1917 brought a historic early killing freeze in the northern highlands on September 11–12 that destroyed about 30 percent of the crops, with readings as low as 25°F to 30°F. The freeze ended the shortest growing season on record (109 days) in the interior valleys, where a late freeze nipped flowering plants on May 24, 1917.

Several waves of frigid air descended on Pennsylvania following heavy snowfalls on December 8 and December 13–14, 1917, the latter depositing up to 15 inches of snow in northeastern Pennsylvania.

The bitterest air mass arrived during the daylight hours of December 28, 1917, sending the temperature plummeting into the single digits and below zero by sunset in the north. The cold wave commenced one of the bitterest weeks in Pennsylvania history, lasting into the New Year (Table 5.8).

Table 5.8
Maximum/Minimum Temperatures (°F)
December 29, 1917–January 4, 1918

	Stroudsburg	Scranton	Mt. Pocono	Harrisburg
Dec. 29	10/-5	2/-10	5/-8	5/-2
Dec. 30	5/-12	3/-13	-7/-22	7/-3
Dec. 31	9/-13	5/-7	-2/-16	8/0
Jan. 11	4/-10	12/-3	0/-15	14/4
Jan. 1	9/-10	10/0	-2/-12	10/5
Jan. 3	12/-8	6/-5	8/-5	12/4
Jan. 4	18/-6	16/-5	6/-18	19/3

Scranton recorded its lowest December maximum (2°F) on December 29, 1917. At 5:00 p.m. on December 29, 1917, the Stroudsburg observer read -5°F, and at Mount Pocono (elevation 1,720 ft.) the mercury tumbled to -8°F after a maximum of 0°F. The mercury did not rise above 0°F at Mount Pocono until the afternoon of January 3, 1918—the longest subzero stretch in state history.

The break of dawn on December 30, 1917, brought some of the lowest December readings ever known, which felt even more brutal because of the gusty winds. Mount Pocono endured its coldest December day on record (-7°F/-22°F). Scranton had bitter extremes of 3°F/-13°F, setting a monthly mark for extreme chill. The lowest temperature in the state was reached at Ebensburg (-28°F) in Cambria County, which equaled the state December minimum reached in 1894 and 1914. New York City experienced its coldest day of the century (4°F/-13°F) at the Central Park Weather Bureau site.

Maximum readings in Pennsylvania remained below 20°F for eight straight days from December 28, 1917, through January 4, 1918, except in the extreme southeastern counties. December 1917 (19.3°F) ranks as the second coldest December in state history.

January 1918 (13.4°F) was even colder—the coldest first month until 1977. The lowest mean temperature (9.6°F) was observed at Drifton (elevation 1,633 ft.). After a bitter start, there was a quick thaw on January 12, 1918, when readings soared to 50°F or higher in the mild sector of a great blizzard that raged in the Ohio Valley.

The cold resumed mid-month with little respite. Snow fell frequently, keeping the air refrigerated. Moderate snowfalls occurred on January 12, 14–15, 18 and 21–22, 1918. Temperatures plunged to -10°F or lower on January 20-21 in the northern counties. The heaviest snowfall came on January 27–28, 1918, depositing 15 inches at Bloserville in Cumberland County, which raised the monthly total to 45 inches. The snow depth at Altoona was a hefty 40 inches by the end of the month.

A final arctic blast came on February 5, 1918: Mount Pocono 0°F/-21°F; Scranton 1°F/-10°F. In the upper Delaware Valley, longtime observer Luther Hoffman penned -18°F in his Smithfield Township records. The Stroudsburg site, a little to the south, noted extremes of 4°F/-11° F. The *Stroudsburg Times* on February 8, 1918, published a list of exceptional area lows: South Sterling -40°F; Mount Pocono -30°F; Bushkill and Cherry Valley -28°F; Henryville -24°F.

The winter of 1917–18 (December–February) stands out as the coldest in Pennsylvania weather history since records commenced in 1888, with an average temperature of 19.5°F—a frigid 7.5 degrees below the long-term mean. The mean winter temperature at Philadelphia of 28.5°F tied 1874-75 for the fifth coldest since 1825.

April 1, 1923
Latest Subzero Freeze

An extraordinary cold blast overspread the Great Lakes and Northeast on the closing days of March 1923. The East Stroudsburg *Morning Press* reported at midnight on March 28, 1923, the thermometer registered 6°F, dipping to 2°F at daybreak on the 29th. At Pocono Lake, the temperature plummeted to -4°F.

On April 1, 1923, the thermometer plunged below zero across northern Pennsylvania, and virtually the entire state established all-time April minimums on April Fools' Day. In the northern mountains, the mercury plunged to -5°F at Saegertown and West Bingham, -4°F at Bradford, -2°F at Freeland and 0°F at Mount Pocono. In the southeast, the "warmest" minimum readings were 14°F at Philadelphia and Shawmont. The lowest reported temperature in the nation was -34°F at Bergland, Michigan.

December 29, 1933
Frigid Eastern Daylight

Arctic air poured into Pennsylvania on December 28–29, 1933, following a general snowfall of six to nine inches on December 26–27.

Temperatures dipped below 0°F in the north on the morning of the 29th, with a minimum of -22°F at Montrose, and Wellsboro had a subzero day (-1°F/-13°F). The mountain region suffered through a subzero daylight. On December 29, 1933, the nightfall observation reached a brutal -15°F at Gouldsboro.

Early temperatures on December 30, 1933, ranged from -10°F to -20°F across the northern highlands. The East Stroudsburg *Morning Press* listed minimums of -36°F at Tobyhanna; -32°F at Mount Pocono; -22°F at Pocono Lake; -20°F in Cherry Valley.

The Stroudsburg area had its only true subzero calendar day of the century on December 30, 1933, with daily readings hovering between -1°F and -5°F on Academy Hill in East Stroudsburg. The maximum at Allentown was 4°F—the lowest in December since 1880.

February 1934
Coldest February

Bitterly cold arctic air plunged into Pennsylvania on February 8, 1934, holding maximums to the single digits in the north. Late-afternoon readings were already well below 0°F in the northern mountains at dusk (-15°F at Gouldsboro at 5:00 p.m.) and plunged to -22°F at Skytop around midnight, according to the Stroudsburg *Record*.

Historic chill greeted Pennsylvania early risers on February 9, 1934. In the central highlands, Bellefonte dipped to -31°F. In the north, Wellsboro checked in at -30°F. Record-breaking cold reached the coastal plain: Philadelphia (-11°F); Newark (-14°F); New York City (-15°F); Boston (-18°F).

These values are the lowest on record along the Interstate 95 corridor to this day.

The northeastern counties also experienced historic chill on February 9, 1934. In the high elevations, both Freeland (-7°F/-24°F) and Scranton notched their coldest days of the century (0°F/-19°F). Stroudsburg had a record cold February day (5°F/-16°F). The 5:00 p.m. reading at Mount Pocono (-8°F) followed a record February minimum (-25°F).

The Stroudsburg *Record* noted readings ranging from -18°F to -24°F and -30°F in the West End of Monroe County. Other cold spots in the northern Pennsylvania highlands included: -30°F at Corry; -29°F at Coudersport; -27°F at Warren; -26°F at Lawrenceville; -24°F at Brookville.

Temperatures around the state were generally below freezing from January 29, 1934, until February 10, 1934, and again from February 23–28, 1934. The late-month cold spell clinched an all-time February cold monthly minimum record for the state of 15.2°F, a stunning 11.3°F below the 20th century normal. The lowest observation of the month was logged at Lawrenceville in Tioga County (-34°F) on February 28.

Many Pennsylvania cities observed the coldest February in 1934. The lowest average temperature was 6.8°F at Ridgway (state record) and 7.4°F at Mount Pocono. All-time monthly low means were charted at Gouldsboro (8.2°F) and Stroudsburg (14.6°F).

The mean temperature at Allentown (16.4°F) was marginally colder than February 1893 at Easton's Lafayette College (16.5°F). The average temperature at Erie (14.0°F) set a February record. The monthly mean at Philadelphia (22.2°F) edged out January 1857 (22.4°F) for the coldest month, until January 1977.

1935–36
Frigid January-February

The winter of 1935–36 brought a series of cold waves to the eastern United States. A wintry December, with temperatures generally below freezing from December 20-31, 1935, gave way to moderating conditions in early January.

Yet frigid air continued to build in western Canada, destined to reload and reach Pennsylvania a few weeks later. On January 11, 1936, the thermometer at Langdon, North Dakota, failed to rise above 0°F, commencing a 42-day streak of subzero weather that lasted into early March.

A heavy snowstorm on January 18–19, 1936, buried the Keystone State under one to two feet of snow. After a few wintry days, with subzero mornings across much of the state, a brief thaw on January 22 saw the thermometer rebound to near 40°F in southeastern Pennsylvania.

Several hundred miles to the west, at Columbus, Ohio, the thermometer plunged on January 22 from 33°F at 8:00 a.m. to -5°F by 5:00 p.m. and -16°F at 10:00 p.m., heralding brutally cold air destined to lock the East in an unrelenting grip of winter.

We pick up the progress of the Polar Express in western Pennsylvania, where readings plunged to the single digits. Shortly before midnight, the temperature dipped to -14°F at Pittsburgh. The lowest minimum was reported at Ebensburg (-30°F) in Cambria County on January 23, 1936.

The maximum temperature on January 23, 1936, at Pittsburgh (-2°F) marked a rare zero day, after a morning low of -16°F. As the core of the cold air shifted east, the observer at Pennsylvania State University noted -2°F at 3:00 p.m. Gouldsboro had a zero day (-3°F/-18°F). The Harrisburg (9°F/-6°F) and Allentown (10°F/-5°F) high/low readings were frigid.

Morning lows on January 24, 1936, ranged from 5°F at Philadelphia to -5°F at Allentown and -10°F at Mount Pocono. The mercury dipped to -11°F in Pittsburgh, followed by a frosty maximum of 7°F.

The 12-day period from January 23–February 3, 1936, was one of the coldest in Pennsylvania weather history. The thermometer failed to rise above 20°F at Pittsburgh and Scranton from January 23–29, 1936. Pittsburgh shivered through six consecutive subzero mornings, the second longest streak in city weather records.

A reinforcing blast of arctic air arrived on January 27, 1936. Gouldsboro registered a daylight maximum of -1°F, and most locations in the north held in the single digits. Maximum temperatures stayed in the 10s and low 20s across Pennsylvania from January 31–February 3, 1936.

Subzero readings were recorded on the mornings of February 1 and 3, 1936, with the lowest reading reported at Gouldsboro (-22°F) on the 3rd. Gouldsboro had 32 consecutive sub-freezing days from January 16–February 16, 1936. The high temperature at Allentown was between 10°F and 26°F during the period from January 23–February 3, 1936, before edging up to 34°F on the 4th.

The cold moderated in the middle of February 1936, though several snowfalls beefed up the already deep snowpack. As much as 12 inches fell at Bedford on February 13–14. Another arctic high-pressure area drove the temperature down to a historic -60°F at Parshall, North Dakota, on February 15, 1936. Within a few days, the temperature would plunge well below zero across northern Pennsylvania.

A cold daylight on February 19, 1936, brought maximums in the single digits in the northwest and near 20°F in the southeast. The lowest minimum reading of -27°F occurred at Brookville. Minimum readings at Gouldsboro on February 18–22, 1936, were uniformly low: -14°F, -18°F, -13°F, -7°F and -13°F.

1942–43
Two Frigid Outbreaks

The winter of 1942–43 was not exceptionally cold, but it featured two major cold outbreaks.

Arctic air plunged southeast across Pennsylvania on December 19, 1942, following a moderate snowfall that left up to 10 inches in the central mountains. At 6:00 p.m. on December 19, the mercury slumped to 8°F at Freeland. The next morning brought subzero readings to all but the southeastern counties.

Daytime readings on December 20, 1942, were hard-pressed to reach 0°F in the northern mountains and barely edged above 10°F in the southeast. Montrose observed maximum/minimum values of -6°F/-17°F. Mount Pocono recorded -2°F/-17°F, for both sites the earliest subzero days on record. The 21st brought a low of -7°F at Mount Pocono. Somerset dipped to -25°F, lowest in the state.

Following a relatively mild January 1943, the middle of February was marked by a second polar surge of historic proportions that produced the lowest daily highs so late in the winter at many sites. On February 15, 1943, Scranton recorded a maximum of 0°F, after sinking to -13°F. Mount Pocono had a record low maximum for any month (-7°F) and a minimum of -19°F, yielding the coldest daily mean (-13°F) in history. Montrose also had a daily average of -13°F, with a range of -6°F/-20°F.

The station at Palmerton, in the east-central coal region, registered its coldest day on record (2°F/-16°F) on February 15, 1943. Near the Delaware River, Stroudsburg had extremes of 7°F/-16°F. The Stroudsburg *Record* listed these frigid numbers in the outlying Pocono districts: Scotrun -32°F; Sciota -31°F; Skytop -30°F; Saylorsburg -26°F; Swiftwater -25°F.

Along the coast, New York City had high/low values of 6°F/-8°F, the coldest daily mean temperature (-1°F) so late in the winter. The February 1943 cold wave was especially severe in New England. All-time records were set at Concord, New Hampshire (-43°F); Portland, Maine (-39°F); and Falls Village, Connecticut (-32°F—a state record).

Surprisingly, the mercury recovered to the 60s in Pennsylvania on February 20, 1943.

March 1960
Coldest March

The temperatures during the first 13 days of March 1960 were subfreezing across much of Pennsylvania—a record streak for so late in the season, with many days registering highs in the teens across the north.

After a major snowstorm on March 3–5, with an increasing snowpack, the mercury plunged to -20°F at DuBois 7 E on March 9 and Ridgeway 3 W on March 11, 1960. Coudersport had a total snow depth of 43 inches on the late date of March 27.

The coldest monthly mean temperature was 18.5°F at Pleasant Mount in northern Wayne County. All-time March chill was registered at Mount Pocono (22.1°F) and Stroudsburg (25.7°F). The statewide average temperature (24.5°F) was 11.3°F below the long-term mean.

January 1961
Coldest Since 1934

A major nor'easter brought one to two feet of snow across Pennsylvania and the Mid-Atlantic states on January 19–20, 1961, commencing a historic sub-freezing run.

The thermometer at Philadelphia did not exceed freezing for two full weeks during a frigid stretch from January 22–February 2, 1961, breaking the previous record of 13 days in January 1893. Stroudsburg had 17 consecutive days of subfreezing weather from January 19–February 4. During this remarkable span, a record 14 consecutive subzero mornings (January 21–February 3) were observed at Stroudsburg and York. The maximum temperature at Scranton failed to reach 20°F for nine straight days from January 20–28, 1961.

The coldest morning in the long arctic siege occurred on January 22, 1961, when the temperature plunged below 0°F statewide, except at Erie (2°F). The lowest reading was turned in at Mercer (-32°F) in the northwestern part of the state. In the Susquehanna Valley, Berwick established a new low mark of -20°F.

A reading of -25°F at Stroudsburg was the lowest since January 14, 1912. The Hawley observer recorded -27°F, the lowest reading observed at the site until January 1994 (-31°F). In the southeast, minimum readings of -19°F were recorded at Quakertown and Phoenixville. Philadelphia endured its first subzero morning (-4°F) since January 23, 1936. Allentown dipped to -12°F, a station record that lasted until 1994 (-15°F).

Exceptionally low readings persisted on the opening days of February 1961. The morning on February 2 was -4°F at Philadelphia, -14°F at Quakertown and -25°F at Hawley. Erie recorded an icy -6°F. Frigid readings of -26°F were reported at Clarion and Indiana. The lowest temperature in the state was -35°F, seven miles east of DuBois and at Ridgway. An unofficial thermometer registered -40°F at St. Marys in Elk County.

A heavy snowstorm on February 3–4, 1961, brought an end to the spectacularly cold pattern, as readings moderated into the 30s and 40s. During the third week of February 1961, a genuine thaw arrived, with highs in the 60s to lower 70s.

January 24–25, 1963
Polar Express

The winter of 1962–63 was punctuated by harsh cold and frequent snowfalls.

The mean winter temperature (December–February) in Pennsylvania of 20.9°F was more than seven degrees below normal and the sixth coldest winter in state records since 1895. The average winter temperature at Philadelphia (28.3°F) in 1962–63 is the fourth lowest in history, which is impressive because records date back to 1825.

A frigid air mass settled far to the south in the eastern part of the country in December 1962, bringing a freeze of historic proportions in Florida, where readings fell into the 20s early on December 13. Tampa recorded a stunning low of 18°F. A high-amplitude flow out of the northwest extended from Alaska to Florida.

Even colder air reached the Northeast near the end of the month. The lowest reading was -22°F at Clermont 4 NW on December 27. Tobyhanna observer Julia Grady reported a minimum of -15°F on the 28th. Frequent lake-effect snow showers and clipper systems added to the snow cover, which reached 41 inches at Rew.

A reinforcing surge of polar air on December 30, 1962, was accompanied by wind gusts of 40 to 60 mph, with a peak gust of 73 mph at Allentown–Bethlehem–Easton (Lehigh Valley International) Airport. The temperature plummeted to the single digits in northern Pennsylvania and then below 0°F overnight in places, compounded by bitter wind chill values below -25°F.

On December 31, 1962, afternoon temperatures in northern and eastern Pennsylvania struggled to reach 10°F, with extremes at Tobyhanna of 1°F/-11°F in the Poconos. The state experienced its coldest New Year's Eve since 1917, with the thermometer straddling 0°F in the northeastern highlands, and not much above that level elsewhere.

After some moderation in early January 1963, another historic blast of arctic air swept across Pennsylvania early on January 24, 1963. Frigid maximums were observed at Pittsburgh (-2°F), State College (0°F) and Williamsport (7°F). The following morning ushered in subzero lows; a low of -16°F at State College on the 25th was the coldest morning at the Pennsylvania State University campus since February 9, 1934.

Later in the month, Philadelphia had an icy low on January 29 of -5°F, which was the lowest temperature in the city since February 9, 1934. Bitterly cold values were observed at Lancaster (-11°F) and Quakertown (-18°F), and in the Pocono region, readings plunged to -18°F at Stroudsburg and -20°F at Tobyhanna.

Unrelenting cold set long-term snow cover records—106 consecutive days at Stroudsburg with an inch or more on the ground, a Keystone State record for an eastern valley site. The mean winter (December–February) temperature at Philadelphia (28.3°F) ranks as the third coldest in the city's history since records began in 1825.

January 1977 Coldest Month

The winter of 1976–77 is synonymous with unrelenting frigidity and the only visitation of snowflakes in South Florida.

A sinuous jet stream dove exceptionally far south across the eastern half of the country, driving a succession of arctic air masses southward all the way to the Gulf Coast. Early on January 19, 1977, flurries were seen for the first time in history at Miami Beach and Freeport, on Grand Bahama Island. Two inches of snow fell at Plant City, Florida.

The mean temperature in January 1977 in Pennsylvania was an icy 12.7°F, surpassing the frigid War Winter month of January 1918 (13.4°F). The coldest periods occurred on January 11–13, 16–18 and 28–31, 1977.

During January 1977, Bradford FAA (elevation 2,188 ft.) notched a state record low mean temperature for any month (4.3°F). A residential site four miles west of town recorded the state's nadir on January 13 (-36°F). Average maximum/minimum readings for the month at Bradford FAA (12.0°F/-3.5°F) earned the enduring recognition for the coldest month (4.3°F) in state weather history.

On January 17, 1977, Pittsburgh had a subzero day (-2°F/-17°F). The lowest reading that morning was turned in at Bradford FAA (-21°F), only rising to -10°F later in the day. Williamsport tied its all-time cold mark (-17°F) on the 18th, equaling February 1899 and January 1912, though this figure would be eclipsed in January 1994 (-20°F). In the east, daylight maximums only reached 4°F at Reading and -2°F at Wellsboro.

On January 27, 1977, Governor Milton Shapp ordered the closure of all public and private schools in Pennsylvania until February 1 to conserve natural gas. The measure was triggered by the effects of a 10-day shortage. Rapidly rising residential demand during the intensely cold winter forced dozens of utilities in the eastern half of the country to shut off service to industries to keep homes warm.

The fiercest wintry weather of the month in terms of travel arrived on the afternoon of January 28, 1977, as an arctic front crossed the state, accompanied by whiteout conditions. Interestingly, this was the first day since January 10 that most Pennsylvania stations climbed above freezing.

The weather change was dramatic. Shortly before 7:00 p.m., the temperature rose to 44°F in Philadelphia ahead of the arctic front, even as readings tumbled to -10°F at Pittsburgh. In the few hours following the passage of the arctic squall line, the temperature at Scranton plunged from 37°F to 4°F.

On January 29, 1977, temperatures slumped to very low levels: -35°F at Bradford 4 W, while the airport site had a bitter maximum of -6°F. The Bradford FAA confirmed the city's distinction as the Icebox of Pennsylvania, notching 51 consecutive days with readings at or below freezing (December 21, 1976, to February 9, 1977), beating out the previous 34-day record (December 13, 1969, to January 15, 1970). Montrose also recorded 51 consecutive subfreezing days through February 10, 1977, including a record minimum of -29°F on January 17, 1977.

The mean temperature at Pittsburgh in January 1977 was an icy 11.4°F, six degrees lower than the previous coldest January (1940). At the opposite end of the state, the average temperature at Philadelphia (20.0°F) easily bested February 1934 (22.2°F) and January 1857 (22.4°F) to become the coldest month in the city's weather history.

Other stations recording record coldest months in January 1977 were Erie 12.5°F; Hawley 13.2°F (nipping the February 1934 mark of 13.3°F); Williamsport 14.9°F; Scranton 15.0°F; Reading 15.9°F; and Harrisburg 20.1°F.

For a long-term affirmation of the cold, the average winter temperature at Philadelphia of 28°F established a modern record for consistent chill, only exceeded in the Pennsylvania Hospital records in 1835–36 (27.7°F).

February 1979
Prolonged Deep Chill

The winter of 1978–79 was severe in the Midwest, but not extreme in the East until February 1979, when arctic air eventually found its way into the Northeast.

As the core of the cold weather settled farther east in late January 1979, some northern mountain locations recorded subfreezing temperatures for 20 consecutive days, from January 31–February 19, 1979.

The coldest period of February 1979 commenced on February 9 and continued for 10 straight days, during which time the maximum readings at Scranton remained below 20°F. Daily highs at Philadelphia edged above 20°F only twice (February 15–16), with rare urban subzero mornings on February 10–11 (-1°F and -2°F) and 0°F on the 18th.

The lowest temperature in Pennsylvania was observed on February 18, 1979, reported at Warren and Clermont 4 NW (-34°F). The lowest monthly mean occurred at Bradford (10.0°F). Pittsburgh endured its coldest February (18.0°F), and Philadelphia (23.0°F) shivered through its coldest second month since 1934 (22.2°F).

The average statewide temperature in February 1979 (16.4°F) was the third lowest in February, after 1934 (15.2°F) and 2015 (16.1°F). A minimum of -52°F at Old Forge, New York, tied the all-time lowest temperature record in the eastern United States, previously observed at Stillwater Lake, New York, on February 9, 1934.

December 25, 1980
Frigid Christmas

An icy polar air mass descended on the Northeast on Christmas Eve 1980, bringing the lowest December temperature ever recorded in Pennsylvania, observed at Clermont 4 NW (-29°F) on Christmas morning 1980.

Gusty northwesterly winds buffeted the state on Christmas Eve as temperatures plunged through the teens into the single digits. A peak gust atop the fire tower near Long Pond on Pimple Hill (elevation 2,215 ft.) was clocked at 61 mph, noted by observer Dixon Miller.

Christmas Day dawned bitterly cold, with subzero readings throughout Pennsylvania, except in the extreme southern counties. In the northern mountains, the low temperature ranged from -10°F to -20°F. A Pike County National Park Service site at Loch Loman registered -15°F, which was equaled at Pimple Hill in the records of Dixon Miller.

Daylight readings on December 25, 1980, were exceedingly low. Montrose reached only 0°F; Pleasant Mount and Freeland peaked at 1°F; East Stroudsburg managed a high of 5°F. Pittsburgh recorded a maximum of 11°F on Christmas Day, after dipping to 0°F. Philadelphia endured its coldest Christmas daylight, with a daytime high of 9°F, after a low of 1°F. Boston had its lowest daylight maximum (-5°F) on record.

January 1982
Two Bitterly Cold Sundays

A vast polar high-pressure system over Alberta, Canada, on January 9, 1982, with a central atmospheric pressure of 31.10 inches indicative of a huge mound of frigid air, would soon be unleashed on the Midwest and Northeast.

The morning of January 10, 1982, proved to be the coldest ever recorded at Chicago (-26°F) at the time, though this record would be surpassed just three years later. As the Arctic Express chugged east, in western Pennsylvania the communities of DuBois (-1°F/-10°F) and Bradford (-2°F/-14°F) had subzero days. In the northeastern part of the state, the mercury at Tobyhanna registered -3°F at the 5:00 p.m. observation time.

A heavy burst of lake-effect snowfall at Buffalo, New York (28 inches—25.3 inches in 24 hours), got the snowball rolling. The temperature plunged to -5°F at Atlanta on the 11th. The next morning, the citrus crop sustained heavy damage in central Florida. A mixture of snow and ice fell along the Gulf Coast on January 12, 1982.

Meanwhile, a second surge of frigid air was building in the Upper Midwest. On January 16, 1982, readings bottomed out at -52°F in northern Minnesota, -25°F at Chicago and -26°F at Milwaukee. The bitterly cold air pressed eastward to the Atlantic Coast on January 17, 1982, on

Cold Sunday II, bringing even lower temperatures than the previous weekend. Temperatures nosedived to the lowest levels since February 1899 at Pittsburgh and February 1934 at Philadelphia.

Bradford FAA (-16°F/-27°F) shivered in the coldest observational day in Pennsylvania weather history on January 17, 1982. In the Poconos, a historic low daytime maximum of -8°F was recorded at Pimple Hill (elevation 2,215 ft.), near Long Pond, matching the Tobyhanna 5:00 p.m. observation, with a morning minimum of -18°F. The daytime high at Binghamton, New York, registered an all-time nadir (-7°F).

The maximum reading at Pittsburgh on January 17, 1982, was a bitter -3°F. The daylight maximum of -5°F equaled February 9, 1899 (Table 5.9), and the minimum of -8°F set an all-time cold mark (equaled in 1985 and exceeded in 1994). Erie (-3°F/-15°F), DuBois (-8°F/-21°F) and Altoona (-8°F/-20°F) had historic chill on January 17, 1982. State College registered extreme daily readings of -2°F/-17°F. Daytime highs at Williamsport and Allentown of -3°F marked the coldest daylight period for both cities. The daytime maximums at East Stroudsburg (0°F), West Chester (1°F), Reading (2°F) and Harrisburg (2°F) established January cold records.

Philadelphia endured its coldest day ever known in two centuries of weather history. The daylight maximum at Philadelphia International Airport (0°F) at 2:48 p.m. and morning minimum (-7°F) established January marks for frigidity (the minimum was tied in 1984). The daytime highs at Atlantic City (1°F), Baltimore (2°F), New York City (4°F) and Washington, D.C. (7°F) equaled or exceeded 20th century records for frigidity.

Table 5.9
Daytime Maximum Temperatures (°F)
January 17, 1982

Pittsburgh	-5
Allentown	-3
Williamsport	-3
Scranton	-2
Philadelphia	0

Another frigid morning followed on January 18, 1982, when readings were -10°F at Wilmington, Delaware, -8°F at Scranton, -7°F at Baltimore (January record), -5°F at Washington, D.C. (lowest since February 1934), and -4°F at Philadelphia.

The temperature remained below freezing in mountainous areas of northern Pennsylvania from January 8–28, 1982 (21 days). The monthly mean temperature at Philadelphia in January 1982 (24.7°F) ranks as the seventh coldest January since 1825.

1983–84
Frigid Christmas and a Polar Blast

A year after the record warm Christmas season in 1982, a massive mound of frigid air invaded the northern Plains in early December 1983, even as temperatures remained comfortably above normal in the East. Gradually, the bitterly cold air mass that originated over Alaska began to spread its wings from the interior Pacific Northwest across the Rockies to the Mississippi Valley; the Southwest was the only region outside the grip of polar air.

Low pressure developed along the polar boundary across Montana and North Dakota on December 14–15, forcing the polar air deeper into the midsection of the country, commencing one of the great cold waves in United States history. Snow covered 74 percent of the nation, keeping the air deeply chilled. The temperature dipped below 0°F in Omaha, Nebraska, at 3:00 a.m. on December 17, and subzero conditions prevailed for a record 202 consecutive hours, until 1:00 p.m. on the 26th.

The cold seeped into Pennsylvania by the 20th, though the core of the arctic air mass remained over the Midwest, until a small low-pressure area tracked across the Great Lakes on December 21–22. More than an inch of mixed precipitation fell on Pennsylvania on the opening days of winter. More critically, upper-level winds turned sharply northwesterly in the wake of the system, drawing bitterly cold air into the Northeast.

The central barometric pressure beneath a dome of frigid air over the Yukon reached 31.45 inches early on December 22, 1983. The mercury plunged to -50°F at Havre, Montana, on Christmas Eve. A vast high-pressure zone of dense cold air stretched from the shores of the Arctic Ocean to the Gulf of Mexico. A record high pressure for the lower 48 states crested near Miles City, Montana, at 31.42 inches. Houston, Texas, commenced a record stretch of continuous subfreezing temperatures that lasted 91 hours.

On December 24, 1983, the mercury bottomed out at -25°F at Chicago's O'Hare Airport and only recovered to -11°F in the early afternoon, setting a record low daily maximum. The arctic air was accompanied by powerful winds gusting past 40 mph, triggering blinding snow squalls that deposited one to three feet of snow downwind of Lakes Erie and Ontario. The snow drifted to incredible depths of 10 to 25 feet.

The mercury at Pittsburgh slipped below 0°F by 6:00 a.m. on Christmas Eve and did not climb above that frigid point for 51 straight hours, until 9:00 a.m. on the 26th (a record that stood until January 1994). The highest temperature at Pittsburgh on Christmas Day was 0°F, after a record low of -12°F.

In the eastern part of the state, temperatures plunged through the teens into the single digits on Christmas Eve. Christmas Day was exceptionally cold in northeastern Pennsylvania. Freeland (elevation 1,970 ft.) had high/low readings of -5°F/-15°F. Single-digit maximums were the rule, rivaling Christmas Day 1980: Scranton 4°F; East Stroudsburg 6°F; Allentown 8°F; Philadelphia 10°F. Washington, D.C., had its coldest Christmas Day (14°F/3°F) on record.

The Southeast was subjected to a nasty combination of snow and freezing rain, culminating in a damaging freeze on December 24–26, 1983. Christmas Day brought record lows at Atlanta (1°F), Jacksonville (10°F) and Orlando (19°F). The hard freeze extended to South Florida, and Miami dipped to 33°F on the 26th.

As many as 125 cities in two dozen states reported record minimum temperatures during the Christmas holiday. Temperatures moderated after Christmas, but another brutal surge of arctic air infiltrated the country in mid-January 1984.

The polar vortex shifted south to near Hudson Bay in southeastern Canada, funneling arctic air southward again. Attesting to the severity of the chill, absolute record lows were observed on January 21, 1984, at Toledo, Ohio (-20°F), and Elkins, West Virginia (-24°F).

The morning of January 22 was exceptionally cold in eastern Pennsylvania. A modern January record low was registered at Philadelphia (-7°F). In the Pocono region, the temperature plunged to -18°F in East Stroudsburg and -23°F at Tobyhanna.

A thaw followed the cold wave in January 1984, sending temperatures into the 40s and 50s on January 25–27, 1984. February was one of the mildest on record, ending with a "heat wave" on February 23–24, when readings soared to near 70°F in Pennsylvania.

However, like a recurring theme, winter was still not through in March 1984. A snowstorm on March 8–9, 1984, brought another blast of polar air, sending the thermometer below 0°F in many sections of the state on the 10th. March 1984 was the coldest third month in Pennsylvania since 1960.

January 19–21, 1985
Great Arctic Outbreak

A warm December 1984 may have given the impression that the winter of 1984–85 would be a mild one. The warmth was misleading, of course.

A large Canadian high-pressure area migrated southward into the north-central United States on January 19, 1985. Maximum/minimum readings at International Falls, Minnesota, plunged to historic levels—from a low of -38°F to a high of -28°F on the 19th. The following day brought brutal chill to the remainder of the Midwest and Ohio Valley. A record low of -27°F at Chicago edged out the previous all-time mark, and the daytime high of -11°F reached an all-time nadir (equaled on January 6, 2013).

The Arctic Express rolled into Pennsylvania on the night of January 20–21, 1985, with temperatures falling to between -10°F and -20°F and wind chills near -40°F. Pittsburgh tied its all-time minimum temperature on January 20–21 (-18°F), registering a maximum of 1°F on the 20th. Bradford had frigid extremes of -2°F/-17°F on January 20. Most of northwestern Pennsylvania endured temperatures below zero all day.

The following morning, temperatures across northern Pennsylvania generally bottomed out between -10°F and -20°F. The chilliest reading in the state was -29°F at Derry. Farther to the east, Williamsport dipped to -14°F. Tobyhanna reported -20°F.

Philadelphia and Baltimore both reported lows of -6°F, one degree shy of the January 1984 record. Even more impressive, Wilmington, Delaware, established a new January mark (-14°F), tying February 1934 for the city's coldest morning on record.

Maximum/minimum readings were uniformly frigid on January 21, 1985: State College 3°F/-17°F; Scranton 3°F/-14°F; Lebanon 9°F/-21°F; Philadelphia 8°F/-6°F. On that day, an icy noon reading of 7°F at Washington, D.C., forced the presidential inauguration indoors for the first time in American history.

The morning of January 21 brought historic chill in the Deep South. All-time lows for frigidity were shattered at Knoxville (-24°F), Nashville (-17°F) and Birmingham (-10°F). Historic minimums were also recorded at Savannah, Georgia (3°F) and Charleston, South Carolina (6°F). A state minimum record (-22°F) was notched at Hogshead Mountain, South Carolina. South Florida suffered through a severe freeze, posting lows of 5°F in Pensacola; 16°F in Daytona Beach; 21°F at Tampa; and 31°F at Miami. The cold that gripped the Southeast wiped out two-thirds of the vegetable crops.

The lake-effect snow machine was in full gear in late January 1985. A total of 30 to 42 inches of snow fell around Buffalo, and up to five feet piled up in favored locations downwind of Lake Ontario. Drifting snow shut down the New York State Thruway from Rochester, New York, to the Pennsylvania line. Wind chill readings dipped to -50°F in northwestern Pennsylvania.

December 1989
Coldest December

When winter arrived in early 1989 with a Thanksgiving Day snowstorm along the Eastern Seaboard on November 22–23, it decided to stay through Christmas Day.

Frequent snow showers and squalls visited the northwestern part of the state in December 1989. Measurable snow fell at Erie every day from December 11–23, with the greatest contribution during the four-day period of December 19–22 (26.6 inches), producing a record snow depth of 39 inches on the 21st. The monthly snowfall at Erie (66.9 inches) set a record that stood until December 2017. The heaviest monthly snowfall in the state in December 1989 was 71.3 inches at Union City Filtration Plant.

Maximum temperatures at Bradford were stuck in the teens for 13 days (December 15–27) during the cold term. Pittsburgh registered subfreezing conditions for 22 consecutive days from December 7–28. Pittsburgh had its coldest day on December 22, 1989, with extremes of 7°F/-12°F.

On December 23, the mercury plunged to -17°F at Bradford 5 SW. The core of the cold air reached eastern Pennsylvania on Christmas Eve, setting numerous daily record lows: -22°F at Frances E. Walter Dam, in Luzerne County, -20°F at Tobyhanna; -12°F at East Stroudsburg; -5°F at Scranton.

Tobyhanna endured 24 consecutive subfreezing days from December 7–30, 1989. East Stroudsburg counted an early-season record 17 straight subfreezing days between December 14 and 30. During an eight-day stretch through Christmas Day, minimums were 0°F or lower, and highs were limited to 15°F to 24°F from December 19–25.

December 1989 was the coldest in state history (17.4°F). The frigid pattern set local records for December chill at Bradford (12.0°F), East Stroudsburg (17.9°F), Scranton (18.6°F) and Erie (21.7°F).

January 19–21, 1994
Historic Cold Wave

A heavy snowfall over the Ohio Valley and Middle Atlantic region on January 16–17, 1994, laid the groundwork for the arrival of a gelid air mass. The extreme cold claimed the lives of at least 150 in the United States.

On January 18, 1994, St. Cloud, Minnesota, had a maximum reading of -17°F and a minimum of -34°F, before bottoming out at an icy -40°F on the 19th. Daytime maximums on January 18, 1994, were historically low in Ohio at Delaware (-12°F), Dayton and Findlay (-11°F). Temperatures would remain at or below 0°F for a record 56 consecutive hours at Cleveland and Columbus.

The frigid air spilled into Pennsylvania on the morning of January 18, when the mercury dipped below 0°F at about 6:00 a.m. in Pittsburgh. The temperature remained at or below 0°F at Pittsburgh for a record 52 consecutive hours, until 10:00 a.m. on the 20th, exceeding December 1983 by one hour. The calendar-day extremes on January 19,

1994, at Pittsburgh (-3°F/-22°F) yielded the lowest daily average on record and equaled the lowest maximum temperature in the city's weather history (February 1899).

On January 19, 1994, an all-time record minimum of -18°F was observed at Erie. The low temperature at State College also reached -18°F. Maximums ranged from 0°F to -9°F (DuBois) in the western half of the state to only slightly above 0°F in the east. Altoona had an exceptionally frigid day (-3°F/-25°F).

Bitter subzero days at Harrisburg (-1°F/-14°F) and Coatesville (-1°F/-11°F) were historically cold. Scranton (-2°F/-13°F), Palmerton (2°F/-12°F) and Allentown (2°F/-11°F) had record daylight chill. At Philadelphia (6°F/-5°F) and Washington, D.C. (8°F/-4°F), the maximums set January cold marks. Kane (-6°F/-35°F) and Bradford FAA (-6°F/-28°F) experienced a memorably cold day.

In the Poconos, minimums dipped to -21°F at Tobyhanna and -25°F at Gouldsboro on January 19, 1994. A Mount Pocono weather station reported extremes of 1°F/18°F, and another private Canadensis site had a daytime high of -5°F. Scranton residents who ventured outdoors braved the city's coldest daylight (-2°F) since 1982.

Exceptional radiative cooling under a clear sky with a snow cover brought all-time record low readings on January 21, 1994: Harrisburg -22°F; Scranton -21°F; Williamsport -20°F; Allentown -15°F. Glacial minimums included: -25°F at East Stroudsburg (lowest since January 22, 1961); -31°F at Tobyhanna (lowest since January 14, 1912); and -31°F at Hawley (all-time record). A National Park Service thermometer near Bushkill in the middle Delaware Valley (elevation 420 ft.) recorded the impressive consecutive minimums of -30°F, -21°F and -33°F on January 19–21, 1994.

January 2014
Polar Vortex Arrives

The polar vortex became a household term in January 2014. A vast swirl of brutally cold air descended on the Northeast and Midwest, passing over Hudson Bay. The news media embraced the term "polar vortex" as if it were a new concept, even though the description of the core of the coldest air in the lower stratosphere above the Northern Hemisphere first appeared in meteorological literature in the late 19th century.

A cold wave arrived on January 6–7, 2014, sending the mercury into a free fall, dipping more than 50 degrees in 24 hours across the Ohio Valley and Northeast. The thermometer at East Stroudsburg tumbled from 52°F at 7:00 a.m. on January 6 to -2°F at daybreak on the 7th.

Extremes on January 7, 2014, included: Pittsburgh 4°F/-9°F; DuBois 0°F/-13°F; Mount Pocono 1°F/-9°F; East Stroudsburg 7°F/-3°F. The readings felt 30 degrees colder with the wind chill running as low as -30°F to -40°F.

Fig. 5.1 January 27, 2014. Snow rollers formed in western Pennsylvania at Clarion, following a surge of arctic air in the wake of an Alberta clipper system and brief thaw. (Photo by Ben Gelber)

Alberta clipper systems reinforced the bitter chill into February 2014. Another frigid day on January 27 brought extremes of 7°F/-8°F at Pittsburgh and 3°F/-11°F at Dubois, and western areas reached -20°F at Portersville, -19°F at Rural Valley and -16°F at Clarion—coldest of the winter. The absolute minimum in the state was -23°F at Oswayo in Potter County on the 29th.

In the snowy winter of 2013–14, East Stroudsburg had continuous snow cover from January 18 to March 29 and nearly matched that string in the frigid winter of 2014–15 (January 24 to March 27), reminiscent of the long winters of the 1960s. Erie received 138.4 inches of snow for the 2013–14 season, the sixth greatest total on record.

February 2015
Second Coldest February

A broad area stretching from the Great Lakes to the Northeast averaged more than 10°F below normal in February 2015, a level not experienced since January 1977.

The mean temperature in Pennsylvania in February 2015 (16.1°F) took second place, trailing only February 1934 (15.2°F). The arctic chill was caused by a persistent trough of low pressure over Hudson Bay in eastern Canada that fed arctic air southward. Each wave of cold air spilled across the snow-covered landscape with little modification.

Bradford FAA recorded a frigid day on February 15, 2015 (-2°F/-14°F), followed by a bitter low of -25°F the next morning. Coudersport had a maximum of 0°F on the 15th. The coldest readings in the state were reported in the interior northwest: -32°F at Chandlers Valley (Warren County) and Oswayo (Potter County). The thermometer at Erie dipped to -18°F, which set a February mark and tied the all-time mark reached on January 19, 1994.

On February 20, 2015, Pittsburgh notched the city's latest reading of -10°F or lower. Pleasant Mount had the lowest reading on February 24 (-20°F). On February 28, lows in the northwest plunged to -19°F at Bradford FAA, -21°F at Clarence and -25°F at Chandlers Valley.

The monthly mean temperature at Hawley (12.0°F) exceeded the all-time monthly mark set in January 1977 (13.2°F) and February 1934 (13.3°F). Pleasant Mount (9.2°F) broke the January 1977 station record (10.7°F). Chandlers Valley (6.8°F) tied Ridgway (February 1934) for the coldest February mean temperature in state records.

March 2015 came in like a lion across Pennsylvania. On March 6, the mercury slumped to a record daily low of -5°F at Pittsburgh. Harrisburg broke monthly minimum records on consecutive mornings (March 6–7), with lows of 0°F and -1°F.

Chapter Six

Rainstorms and Floods

If it thunders on All Fools' Day, it brings good crops of corn and hay.
—Weather folklore

RAIN PROVIDES VITAL nourishment for our lawns, vegetable gardens and field crops. But sometimes when it rains it pours, and trouble ensues if the ground is thoroughly saturated. The situation becomes dire when torrential rain falls on impervious surfaces or sodden hillsides, resulting in urban and small stream flooding.

A mere one inch of rain falling in a single acre of ground or farmland equals about 27,154 gallons of water, weighing about 133 tons, according to the U.S. Geological Survey (USGS). A cubic foot of water weighs 62.4 pounds.

Water flowing at 4 mph exerts a force per unit area which is comparable to the power of an EF2 tornado (111-135 mph). A current with 6 inches of moving water can knock a person down. A foot of water will float small vehicles, and 18 to 24 inches can carry away large vehicles. A car displaces 1,500 pounds for every foot of rising water, which weighs 1,500 pounds less for every foot of water pressing on the side of the automobile. This is why it is very dangerous to drive through floodwaters.

Flash floods strike with little or no warning. Torrential downpours overwhelm storm-sewer systems and turn roads into raging rivers in minutes. In the past decade (2011–2020), more than half of all flooding fatalities in the United States were in vehicles swept away by high water.

Flooding killed an average of 85 people in the United States in the most recent 30-year period of NOAA records (1991–2020), more than in any other type of severe weather event. Annual losses between 1987 and 2016 totaled $8 billion.

Historic flooding in the Midwest and Plains during the late winter and spring of 2019 was catastrophic for farmers, inundating fields and crop warehouses. Barge traffic that carried shipments of grain, fertilizer and supplies was halted for months on bloated rivers. Nearly 20 million unplanted acres in the Corn Belt sharply reduced the annual yields, as crops planted early languished in flooded fields from northwestern Ohio to South Dakota.

Two of the most intense documented rainfall events in Pennsylvania history occurred in the southern Pocono region. On August 1, 1915, the Stroudsburg government observer J. Clyde LeBar measured 7.50 inches in a three-hour deluge.

A similar scenario played out in the West End of Monroe County on August 16, 2003, when 7.25

inches of rain fell at Kunkletown. The NWS office at Mount Holly, New Jersey, confirmed that an estimated six to eight inches of rain fell near Gilbert (*Storm Data*). Yards were under water after a stream overflowed at Brodheadsville. Firefighters employed "makeshift dikes of ladders and tarp to steer flood waters away from homes."

High-resolution forecast models sometimes struggle to capture small-scale, or mesoscale, features that produce excessive rainfall, and pinpointing the location of flash flooding often occurs after a storm is in progress. The most dangerous flash flooding situations arise when repetitive thunderstorms carry drenching downpours over the same area for several hours when the flow is parallel to a focusing boundary. This is known as "training" because cells behave like train cars unloading tanks of water on a single track.

A drenching overnight rain totaling nearly seven inches on June 10–11, 2018, affected parts of Berks, Bucks, Montgomery and Philadelphia counties, shutting down portions of Interstate 76 (Schuylkill Expressway) for nearly 10 hours, causing long delays at the start of the morning commute. A clogged drainpipe in Montgomery County exacerbated the flooding.

On the evening of June 20, 2018, a warm front became a pathway for slow-moving storms southwest of Pittsburgh, dumping totals exceeding three inches of rain in a little more than an hour. Flash flooding resulted in 30 swift water rescues, and there was a fatality in Bridgeville.

A year later to the date on the night of June 19–20, 2019, a narrow complex of regenerating thunderstorms moved northeast from Maryland and Delaware, aligned with a stationary front. In two to three hours, three to five inches of rain pelted southeastern Pennsylvania. The heaviest rainfall totals were observed south of Reading and in the northern suburbs of Philadelphia. The storm total at Philadelphia International Airport was 4.63 inches.

A motorist was rescued after becoming stranded in high water in Lansdowne. Residents were evacuated in Wyomissing, in Berks County, when their apartment building filled with six feet of water.

A cold front moving through eastern Pennsylvania on the late afternoon of July 11, 2019, unloaded four to six inches of rain on southeastern Berks County into northwest Montgomery County, triggering a rare flash flood emergency just before 5:00 p.m. at the beginning of rush hour. Numerous water rescues followed. Three people were killed, including a pregnant woman and her nine-year-old son in Berks County near Boyertown. A tragedy unfolded on Manatawny Creek, which flows into Schuylkill River, where a five-year-old reportedly drowned at a Berks County farm.

Heavy Rain and Flash Floods

Hydrology determines where water flows and piles up on the landscape. Potential flooding is mitigated by dams that regulate the flow of water and control the frequency of flood events. Levees constructed in floodplains usually prevent rivers from overflowing.

In flash floods, rivulets exceed the capacity of the landscape to accept runoff that quickly overwhelms storm sewers and drainage systems, and pools in the low-lying spots. Runoff on an acre of paved surface is 10 to 20 times greater compared to grassy areas.

The foothills and Appalachian valleys are prone to flooding when heavy runoff water drains downhill and fills the small tributaries downstream that enter the main stem of a river. The steeply sloped terrain funnels torrents of water into narrow channels that empty into an extensive watershed.

Excessive rainfall often occurs in slow-moving thunderstorms where the mean cloud layer flow is parallel to a stalled front. A plume of moisture transported by a low-level jet is intercepted by a shallow boundary, triggering multiple rounds of storms. Mid-level energy produces a surface low that pumps more moisture into the frontal zone. The warm conveyor belt process feeds the uplift, occasionally tapping an atmospheric river, which brings drenching downpours atop sodden soil.

Historic floods in the Mid-Atlantic and Northeast are usually the result of low pressure hemmed in by blocking high pressure off the Atlantic Coast, which directs a stream of moisture from the Gulf of Mexico, Atlantic Ocean, and less frequently from the Caribbean Sea, in an atmospheric squeeze play. Upper-level energy associated with a trough of low pressure and a southeasterly flow enhance upslope circulations, wringing out the sopping air.

A moisture-laden tropical system tracking up the Eastern Seaboard or from the Gulf states is a recipe for flooding. Warm-layer clouds building in a corridor of moisture are efficient rainmakers. A zone of moisture convergence dispenses copious rainfall rates that cause soil erosion.

River floods evolve more slowly in response to a prolonged discharge of water falling on saturated soil, where the flow rate exceeds the capacity of the channel. Large watersheds are more susceptible to extensive flooding, receiving input from numerous tributaries in the watershed that enters the main stem of the river. Unlike flash floods, which typically occur in the spring and summer, river floods are more likely to occur during the late winter and spring following several days of rain and melting snow that run off frozen ground.

During a thaw following a persistent cold regime, chunks of river ice tend to lodge against embankments and barriers as the ice breaks up, creating ice jams that form impoundments. When an ice dam gives way, a surge of water overtops natural barriers and floods lowland areas beyond the natural floodplain.

Automated Flood Warning System

Systematic streamflow data account for the velocity and volume of runoff, which have been recorded for a little more than a century (Kury 2011). The flow rate of water is measured in cubic feet per second (cfs), which is monitored by a network of automated gauging stations positioned by the United States Geological Survey (USGS).

Hydrographs of river levels provide critical information on the flow rate and discharge, enabling meteorologists at 13 regional NWS River Forecast Center (RFC) locations to issue flood warnings.

The NWS Integrated Flood Observing and Warning System (IFLOWS) was developed in 1979 in the Central Appalachians, where flooding poses a significant risk due to the elevated terrain, with steep slopes prone to channeled runoff. The initial system was completed in 1981 and quickly expanded to include most of Pennsylvania.

The network continued to grow with the installation of new equipment in the early 1980s, a joint effort to provide state and federal agencies with reliable warnings and crest forecasts. The current Automated Flood Warning Systems (AFWS) collects data from 1,700 sensors in a dozen states.

Major Pennsylvania Rivers

The eastern boundary of Pennsylvania is formed by the Delaware River, which rises up in the Catskill Mountains in southeastern New York, where two branches come together and trace a southeasterly course to form the border of New York and Pennsylvania.

The silvery Delaware River then turns to the southwest for a time before again flowing southeast—a pattern of alternating jogs along a 330-mile course to the Atlantic Ocean by way of the mouth of Delaware Bay. The Delaware is joined by the Lehigh and Schuylkill Rivers in southeastern Pennsylvania.

The broad Susquehanna River drainage system encompasses 27,510 square miles of New York, Pennsylvania and Maryland, including 21,038 square miles in Pennsylvania. The North Branch Susquehanna River rises near Cooperstown, New York, and threads a sinuous 464-mile course through the rugged Allegheny Plateau. The main stem is joined by the West Branch just north of Sunbury and continues south, emptying into the mouth of Chesapeake Bay at Havre De Grace, Maryland, and the Atlantic Ocean.

The Juniata River, one of the main tributaries of the Susquehanna River, flows about 104 miles, slicing between several prominent ridges, meeting the Raystown Branch Juniata River near Huntington before merging with the Susquehanna 15 miles northwest of Harrisburg.

In western Pennsylvania, the Allegheny River flows 325 miles from southwestern New York to Pittsburgh. Point State Park marks the confluence of the Allegheny, Monongahela and Ohio Rivers. The origin of the 130-mile Monongahela River lies in the mountains of north-central West Virginia.

Early Pennsylvania Floods

A sudden rise in river levels was referred to as "freshet" in the early American press. Pennsylvania pioneers were warned by Native Americans against building in a natural floodplain along the Delaware River, but the admonition was generally ignored. Some settlers believed that the high water provided cover to navigate the rocky shores and facilitate moving crops to the marketplace. The misjudgment was inevitably disastrous.

The earliest documented flood in the Delaware Valley occurred in 1687 (Ludlum 1983). A report by the United States Geological Survey (USGS) recounted a great inundation on February 27, 1692.

Historic Tales of Olden Time, Concerning the Early Settlement and Progress of Philadelphia and Pennsylvania, a letter written by Phineas Pemberton on February 27, 1692, described floodwaters 12 feet above the normal high-water mark at Delaware Falls, near Trenton, New Jersey, that reached the upper stories of homes.

There are scant historical references to major flooding that befell settlers living along the Lehigh River in 1739 and the Schuylkill River on July 15, 1757 (Kury 2011).

In western Pennsylvania, trader James Kenny documented a flood at Fort Pitt that occurred on January 7–8, 1762 (Schafer and Sajna 1992):

This Morning ye flood increasing still we had ye Bato up to ye Door, by Noon ye Street fronting our door under water; many People brought Goods to us for preservation; got Going with Canoes between ye Houses & Batoes, I set to work and got all our peltry up stairs & ye Wollings &cc up about Dusk ye water got to power into our Cellar increasing with ye …

Another great flood on March 8, 1763, forced Kenny to flee his trading post and home. Floodwaters crested at about 40.9 feet, or "just a few inches below the second-floor gunports of The Blockhouse in Point State Park" (Schafer and Sajna 1992).

Heverly (1926) described a destructive 1771 flood along the Susquehanna River:

> On May 28th [1771], the Susquehanna river rose to an unprecedented height, inundating both the towns of Sheshequannunk (Ulster) and Wyalusing. At the latter place great damage was done by the water sweeping off fences and stock. The inhabitants of Sheshequannunk were compelled to take to their canoes and retire to the wooded Heights back of the town.

March 15, 1784
Great Ice Flood

Another great flood affected the Susquehanna River Valley at the end of the Long Winter of 1783–84, when deep snow cover and intense cold caused ice to build as waterways froze over, only to collapse during a thaw and driving rainstorm (Heverly 1926):

> The breaking up of the Susquehanna river on the 15th of March 1784, greatly distressed the inhabitants who had built their houses on the lowlands near the banks of the river. The uncommon rain and large quantities of snow on the mountains together with the amazing quantity of ice in the river … swelled the streams to an unusual height—ten and in many places twenty feet higher than had ever been known since the settlement of the country.

Bradsby (1893) quoted Major James Moore, who was posted at Fort Wyoming and observed the inundation of the Wyoming Plains, now the area of Wilkes-Barre, writing on March 20, 1784:

> The people in this country have suffered exceedingly from the late freshet. Not less than 150 houses have been carried away. The grain is principally lost, and a very considerable part of the cattle drowned. The water rose thirty feet above low-water mark.

An unknown number of structures were swept away in the flood, recounted by Charles McCarthy (1972) in the *Wyoming Valley Observer*.

October 6, 1786
Pumpkin Flood

Floods that greatly impacted agricultural areas were referred to as "pumpkin floods" because they inundated the fertile bottomlands, where crops were most abundant.

The Pumpkin Flood on October 6, 1786, in the midst of a soggy autumn sent most eastern Pennsylvania rivers out of their banks—the Susquehanna, Delaware and Schuylkill Rivers and the tributaries.

Rupp's *History of Lehigh, Northampton, Monroe, Schuylkill and Carbon Counties* noted the widespread flooding on the Lehigh River. A Moravian mission station founded in 1746 Weissport was the site where the Trippey home floated downstream, eventually catching in a clump of trees. The parents survived by clinging to tree limbs and were later rescued, but their two children were swept away by the raging water.

Heverly (1926) described the high water that afflicted the Susquehanna lowlands:

> Crops were swept away, and the bosom of the river was covered with floating pumpkins. The loss was severely felt, and many cattle died the following winter for want of sustenance. For years this freshet was designated by the old inhabitants as the Pumpkin Flood.

The "Flats" surrounding Wilkes-Barre were subjected to moderate flooding in July 1809, January 1831 and May 1833, but nothing as severe as the events in 1784 and 1786.

February 9–10, 1832
Pittsburgh Flood

Following another period of high water in late winter, in March 1907, the meteorologist-in-charge at the city Weather Bureau in Pittsburgh, Henry Pennywitt, passed along historical clippings from the *Pittsburg Gazette* (February 14, 1832), reprinted in the *Monthly Weather Review*.

A period of severe early cold weather, accompanied by significant snowfalls, in December 1831 set the stage for trouble when the weather moderated in January 1832. The break-up of ice and melting snow left the ground saturated. Frequent rains fell from February 5–9, 1832, with only "slight interruption," triggering major flooding in southwestern Pennsylvania.

The water was said to be "higher than had been known by any living inhabitant of this city or neighborhood, inundating the low ground of the boroughs of the Northern Liberties and Allegheny and the greater part of the city of Pittsburg north of Liberty street."

May 9, 1833
Rivers Run High

Rain began falling in Pennsylvania on May 4, 1833, heaviest on May 8, before tapering off the next day. The Easton Sentinel account stated the Delaware River rose 16 feet in 24 hours, submerging the dam at the mouth of the Lehigh River and the canal system, blocking the route of the Philadelphia Stage. Lumbermen suffered an estimated $15,000 in losses as logs, boards and equipment floated away.

The *United States Gazette* (May 20, 1833) reported "freshets in our great rivers" caused widespread damage: "Along the Susquehanna, much loss has been sustained," and two lives were lost in the region. Five flood fatalities were reported at Troy, New York.

June 26, 1835
Urban Flood

The *United States Gazette* in Philadelphia described a storm "of almost unprecedented violence," declaring that the "the vividness of lightning and the noise of thunder were terrific." The account said that the rain poured "as if there were windows in heaven."

Torrential rain accompanied by intense thunder and lightning swamped city sewers and basements. Homes and businesses on Market Street had flooded cellars, destroying "merchandise, dry goods especially, contained in them." The *Pennsylvania Inquirer* deemed the event "one of the most violent storms that has occurred in this vicinity for years."

Storm accounts were summarized in *Hazard's Register of Pennsylvania* on July 4, 1835. Flooding was widespread on the west side of the city, where walls of buildings under construction collapsed. The *Philadelphia Gazette* said the storm began shortly after midnight on June 26, 1835, and continued for about three hours, as "the rain fell in torrents." The press account anticipated damage to mill dams and the wheat crop.

January 8, 1841
Bridges Freshet

A deep snowstorm blanketed eastern Pennsylvania on December 4–6, 1840, depositing 12 to 18 inches of snow. A thaw followed a frigid period during the opening days of January 1841, accompanied by a heavy rainfall. Runoff quickly swelled the Delaware River, as the heavy discharge encountered large blocks of ice downstream that choked the channel.

The Bridges Freshet of January 1841 earned that moniker because nine bridges were washed away along the Delaware River (Dale 1996). The river stage at Easton reached 35 feet, a remarkable 13 feet above flood stage, swamping low-lying areas and causing significant damage.

The Stroudsburg *Jefferson Republican* described the antecedent conditions:

> The weather which had been for three days previous intensely cold, began to moderate, and early on Tuesday morning [January 5, 1841], it commenced raining … It continued raining without intermission until eleven o'clock Thursday night [January 7] … The great body of snow which had fallen speedily dissolved, and together with the rain, produced the greatest freshet ever in this section.

Rupp (1845) wrote about the wreckage that swept away 90 percent of the bridges in Monroe County. Mills, tanneries, barns, shops, and a number of homes were damaged beyond repair. A young New Jersey man drowned while attempting to salvage his property along the Delaware River.

A paper presented at the Bucks County Historical Society in 1927, entitled "Floods and Freshets in the Delaware and Lehigh Rivers," brought forth information copied from the Bible of George Wyker, a longtime resident of Upper Tinicum Township in Bucks County (Ludlum 1983). Wyker's historical notes regarding the flood of January 8–9, 1841, which were read at the meeting, indicated that water levels were higher than at any previous time in 107 years (at least since June 1734).

His records recounted that four bridges on the Delaware River and five over the Lehigh River were washed away, causing "a great deal of damage to property of every description, and on the Lehigh, it carried away several houses, with all their furniture, and several lives were lost."

August 5, 1843
Delaware County Cloudburst

Y. S. Walter (1844) described "a hurricane of great violence" that exploded over Delaware County between 4:00 and 5:00 p.m. on August 5, 1843.

The widespread damage pattern and torrential rain would have been associated with a powerful thunderstorm complex passing over the region. Nineteen people died west of Philadelphia, mostly trapped by rising water (Ludlum 1970).

A Delaware County Institute of Science (1910) committee report described rapidly-moving clouds that "appeared to approach from different directions, and to concentrate at a point not very distant from the zenith of the beholder." The account remarked on "the immense quantity of water falling (which it carried with it in one continuous sheet) … rendered it impossible to see a distance of more than fifty yards."

Flash flooding triggered by the intense rainfall resulted in an "almost instantaneous rise" in the water, from five to 10 feet, and seven to eight feet on Crum Creek in 10 minutes. At Avondale, the water reportedly rose 19 feet in a little more than a half-hour.

March 14–15, 1846
Spring Floods

Sixty-two years to the date following the Great Ice Flood of March 1784, another serious spring flood affected eastern Pennsylvania. Heavy rain and melting snow had piled up to a considerable depth in February, which Heverly (1926) wrote led to a "great freshet in the Susquehanna; nearly all the river bridges and those on the larger creeks swept away."

In the southeast, we again turn to George Wyker's notes from March 15, 1846: "Quite a high flood in the river, but not by 2 or 3 feet as high as that of 1841. The above I have written on the 8th day of April 1846, being now in my 80th year" (Ludlum 1983).

July and September 1850
Great Floods

Two episodes of disastrous flooding afflicted the foothills in eastern Pennsylvania in the late summer of 1850. A tropical storm moving north from Virginia soaked the region during the night of July 18–19, 1850, which culminated in destructive flooding that killed at least 20 people along the Schuylkill River (Ludlum 1963).

An even greater flooding catastrophe occurred less than two months later. A light rain on the evening of September 1, 1850, turned into a torrential downpour, triggering deadly flooding along the Schuylkill and Lehigh Rivers. Runoff from the Appalachian foothills overwhelmed dams on September 2, 1850. The *Pennsylvania Inquirer* in Philadelphia reported a "heavy and incessant rain—the unusual amount of 3.80 inches—is recorded as having fallen in one day."

An Easton *Argus* dispatch in the *Jeffersonian Republican* (September 12, 1850) told the story:

> The heavy rain on Sunday and Monday [September 1–2, 1850] caused the Lehigh River to rise to a fearful height. At the mouth of the river, it was about twenty inches higher than July last [July 19, 1850]. On Monday afternoon between two and four o'clock the water rose at that place about five feet, filling many cellars in the lower part of the town, and destroying considerable property by its sudden and rapid movement.

The event was described as "one of the most destructive floods ever known upon the waters of the Lehigh and Schuylkill." Trestles became clogged with debris that formed dams, which eventually gave way, unleashing a deadly wall of water. A tragic dam break at Mauch Chunk (now Jim Thorpe) sent torrents of water downstream, sweeping away bridges and homes. No loss of life occurred at Mauch Chunk, but the death toll at Pottsville was put at "fifteen or twenty," and "30 to 40 houses were swept away."

Bradsby's *History of Luzerne County Pennsylvania* recounted the circumstances leading to 22 deaths along two tributaries of the Susquehanna River in the southwestern part of the county.

In Luzerne the loss of life and property was greatest on the small streams. Solomon's creek rushed down the mountain's side with fearful impetuosity, destroying the public highway and the improvements of the Lehigh & Susquehanna company at the foot of the plane. The Wapwallopen, with its increased volume, dashed madly over the country, sweeping away two of the powder-mills of Knapp & Parrish. The Nescopeck, undermining the dam above the forge of S. F. Headley, bore off to the Susquehanna on its turbulent flood the lifeless bodies of twenty-two men, women and children. These unfortunate people had assembled in one house near the forge. The house stood upon elevated ground, and was supposed to be the best place for safety. One man, fearing to trust to the stability of the house, took up his child in his arms, and calling to his wife, who refused to follow, rushed through the rising waters, and gained the hillside. When he turned to look behind him, house, wife and friends had disappeared.

Around Tamaqua, "thirty-three persons were drowned, sixteen being members of one family Forty homes were destroyed in Tamaqua along the Little Schuylkill River. The confluence of the Little Schuylkill River and main stem of the Schuylkill River at Port Clinton was the scene of a terrible tragedy, where 26 persons lost their lives (Bradsby 1893). Serfass (2013) counted the total loss of life in Schuylkill County of 62 persons—the greatest weather disaster the Coal Region has ever experienced.

The Reading *Gazette and Democrat*, on September 7, 1850, told the story of devastation to homes, property, bridges and railroads across the Schuylkill Valley, calling the historic flooding "the direst calamity which has ever befallen them." At least eleven deaths occurred in the Reading area. The Schuylkill River crested at 23.2 feet "above low-water mark at the foot of Penn Street."

Modern records reveal higher stages on only two occasions: June 22, 1972 (31.3 ft.) and during a week of heavy rain culminating on June 28, 2006 (23.75 ft.).

June 4–5, 1862
Great Flood of '62

In Joseph Levering's *History of Bethlehem, 1741–1892*, the Great Flood of '62 was deemed "the most disastrous flood on record in the valley."

Weiss (1863) prepared his "Incidents of the freshet on the Lehigh River," stating that the loss of life was about 150 persons, by far the worst natural disaster in Lehigh Valley history.

A pattern of moderate to heavy rain enveloped Pennsylvania on June 2–4, 1862. The Smithsonian weather station in Indiana, in the western part of the state, received eight inches of rain (16.38 inches for the month). Torrential rain fell over the eastern counties on June 3–4, setting the stage for major flooding.

The Lehigh River rose faster than had ever been known on the afternoon of June 4, 1862, and surged out of its banks after nightfall. Eighteen bridges were destroyed along the Lehigh River from White Haven to the junction with the Delaware River, leaving only three that survived the onslaught (Lehigh Gap, Bethlehem and Easton). The Delaware, Lackawanna and Western Railroad and Delaware Canal were under water. Dams that powered grist mills burst, sending swirling turbulent waters downstream.

Historian Thomas Eckhart, in the first of his three-volume series entitled *The History of Carbon County*, wrote extensively about the flooding that claimed the lives of more than 30 persons in the county. He described the network of 20 dams operated by the Lehigh Coal and Navigation Company that "supplied the canal locks and provided depots for logs to be sawed or transported."

The first dam break occurred at Dam No. 4 at White Haven, after pounding water caused booms supporting sawed lumber and logs to break apart. The rising upper Lehigh River became choked with thousands of logs and debris that acted as battering rams, destroying the dams and locks. Once a logjam burst, the crushing force of released water caused successive dam failures.

River communities were inundated, starting at Mauch Chunk (now Jim Thorpe), where more than 50 buildings were swept away, killing four persons. Two young women drowned when their boat capsized in the swollen Lehigh River. Another 19 people drowned at Burlington, and four lives were lost at Weissport on the east side of the river near Lehighton. Seven persons were killed at Penn Haven in the flooding.

Levering cited records maintained by the Bethlehem Bridge Company that noted the water gauge on the Lehigh reached 20.5 feet around noon on June 5, 1862, slightly exceeding the previous high mark in January 1841 (20 feet). The Delaware, fed by mountain tributaries up north, continued to rise as it churned southward.

A dispatch from Delaware Water Gap in the *New York Times*, 25 miles north of Easton, reported "quantities of furniture, store goods, bridges, house, cattle … all going down the river." The Delaware, Lackawanna and Western Railroad suffered severe damage when the tracks were submerged.

A mile above Easton at Glendon, the *Times* story reported that water reached the second story of homes:

> … all the bridges between this and Mauch Chunk are swept away. The Lehigh Bridge is partly gone and will probably be totally demolished. All the canals are under water. The iron works have stopped, and the railroads are submerged. Many people were drowned in their houses, so sudden was the rise.

The *Monroe Democrat* summarized the severe flooding in lower Stroudsburg after Pocono Creek rose rapidly early on June 5, 1862, reporting that rising waters were "carrying away many houses and bridges. The damage done was very great."

> It commenced raining on Tuesday evening [June 3], and kept steadily on through the whole of Wednesday day and night. The water poured down without intermission, and everybody expected that there would be a respectable freshet in our streams, but no

one dreamed that such a terrific and destructive flood was about to sweep over the land. It came so sudden that many families in lower portions of our town were obliged to leave their houses in boats, and the water rushed into the lower stories of their dwellings before they could remove their furniture to the upper stories.

On June 19, 1862, the *Monroe Democrat* listed the damages along the Delaware River at Easton as totaling a hefty $10 million. Property losses in the Stroudsburg area were estimated at $75,000, according to historian and former chief burgess Dr. Jackson Lantz, in *Picturesque Monroe County, Pennsylvania*.

March 16–17, 1865
St. Patrick's Day Floods

Heverly (1926) wrote that the St. Patrick's Day Flood on March 17, 1865, brought "the highest water, twenty-eight feet, ever known in the Susquehanna" in Bradford County.

In northern Pennsylvania, the March 1865 flood came to be called the "Barrel Flood" because thousands of oil barrels waiting to be shipped were carried away in the rising waters. The authors of *The Allegheny River* wrote: "Oil City swept its banks clean of thousands of barrels, derricks, and drilling rigs and sent it all cascading down the river to knock out the Pennsylvania Canal aqueduct at Freeport."

A press dispatch from Franklin in the *Philadelphia Inquirer* added: "We are having the greatest flood ever known in this region. The bridge at Oil City and the French Creek bridge at Franklin, are swept away. Miles of the railroad track are gone, and the telegraph lines are washed away, houses, tanks and barrels, full and empty, cover the river. The loss is estimated by millions."

The Great Freshet caused considerable damage along the West Branch Susquehanna River. Raging waters took out all the bridges between Williamsport and Danville. The report stated that about half of Danville was under water. Runoff fed the main stem of the Susquehanna River, triggering "unprecedented" flooding at Harrisburg, when the river rose two and a half feet above the "destructive freshet of 1846" level.

A wire dispatch from Harrisburg reported that a "larger portion of the lower and eastern end of town" had been inundated. Paxton Creek, on the eastern edge of the city, "backed up" and submerged many homes.

On March 17, 1765, the floodwaters swept "thousands of timber logs, with millions of feet of sawed lumber downstream," and the lower portion of the city was under water. The story continued: "Much suffering has been created among many poor families who live at Middletown and in the villages along the shore, clear to Columbia." The Susquehanna River at Wilkes-Barre reached 33.1 feet, a level that was not equaled until March 20, 1936.

October 3–4, 1869
Saxby Gale Triggers Flooding

A tropical storm located south of Cape Hatteras arrived early on October 4, 1869, eventually making two landfalls, one on Cape Cod, and later near Eastport, Maine. A cold front and low-pressure trough near the Eastern Seaboard provided a lifting mechanism for the surge of tropical moisture (Ludlum 1963). The heaviest rainfall measured was 12.35 inches at Canton, Connecticut.

In eastern Pennsylvania and across New Jersey, a light rain commenced on the evening of October 2, 1869, turning into a steady rain that at times became torrential in the following evening. Rainfall totals at New Jersey weather stations averaged near five inches, which would send rivers into high flood in the north (Ludlum 1983).

In the northern Poconos, observer Theodore Day reported 4.50 inches of rain at Dyberry. Eight inches of rain filled up the gauge at Mount Joy in Lancaster County. In Montgomery County, 6 inches of rain fell at Plymouth Meeting. Marcus Corson commented in his "observational notes": "Disastrous freshet on the Schuylkill. ... Water higher than the great flood of January 26, 1839." The water reached the second stories, stated the Philadelphia *Evening Telegraph*.

The *Philadelphia Inquirer* on October 5, 1869, described the rainfall extremes of 1869, ranging from "unparalleled drought" for several months to a "freshet of unprecedented violence." The story stated, "bridges are carried away, factories, dwelling houses, ice houses, &c, are submerged, boats are swamped, and the river is swollen to three times its usual size."

Corson noted in his report that four people drowned along the Schuylkill River between Norristown and Philadelphia, with "much property damaged." On October 4, 1869, the Schuylkill River crested at an all-time high of 17 feet at Philadelphia, six feet above the modern flood stage. The crest at Reading of 21.6 feet was the highest since September 1850, causing major flooding on Conshohocken Creek (Kury 2011). Farther north, water broke through the cribbing of a dam below South Stroudsburg along McMichael Creek around 7:00 a.m. on October 4, 1869, churning wildly over lower Main Street and wiping out a bridge (Lantz 1897).

Fig. 6.1 October 4-5, 1869. The oldest known photograph of flood damage in Pennsylvania at Stroudsburg, following the Flood of October 4, 1869, which appeared in Picturesque Monroe County *(1897).*

In a flood retrospective published on February 13, 1896, the *Stroudsburg Times* summarized the calamitous scene:

> The great volume of dark, angry waters soon swept away the old grist mill, with all its machinery, swirled down on the woolen mill and Wallace's sawmill, past J. O. Saylor's building, carrying ruin with its fearful march.

Historian Thomas Eckhart (1996) compared the flood of 1869 in Carbon County to the flood of 1862: "Overall damage to the [Lehigh Valley Railroad] line as a result [of] this flood was comparable to losses sustained in the June 1862 flood."

The cause of the heavy rain was an offshore hurricane, presaged a year earlier by Lieutenant Stephen Saxby (1804–1883), a British officer in the Royal Navy. In November 1868, Saxby published in the *London Standard* (Schwartz 2007) that he prefigured the alignment of the Sun, Moon and Earth, and the proximity of the Moon.

After the storm's passage, Saxby's prediction attracted wide attention.

A perigean spring tide occurs three or four times a year, when the new or full moon is closest to Earth at the same time the Sun, Moon and Earth are most closely aligned. However, the perigean spring tide has little impact on our weather, though the celestial arrangement contributes to high tides that sometimes exacerbate coastal flooding.

August 12, 1873
Lehigh Valley Flash Flood

An interesting account of a storm in the *Allentown Chronicle* reprinted in the Stroudsburg *Jeffersonian* on August 28, 1873, described a torrential storm at Guthsville, a small community situated five miles to the southeast of Allentown, that evolved late in the afternoon of August 12, 1873.

Streets and cellars were swamped as water rose as high as the first floors of several homes. Roads were washed out, and property damage to homes and yards was considerable.

The press report continued: "Of all the severe rains known to the oldest inhabitants in Lehigh County that of Tuesday afternoon [August 12] from about a quarter past three o'clock till a quarter past five, was the most copious … It is somewhat remarkable that the very hard rain only visited a semicircular belt of territory about a mile and a half wide, from four to six miles to the southeast of this city, leaving Allentown with only a usual summer rain."

A singular occurrence was observed between 5:00 and 6:00 p.m. near the headwaters of Cedar Creek caught the attention of Hiram J. Schantz, Esq., who "noticed a body of water five- or six-feet high rushing towards the mill from above, coming from a direction where no creek ran."

The whirlwind was thought to be "a waterspout which it is said descended near Crackersport and rushed across the fields carrying fences before it." About one-half of Schantz's sawmill, along with the waterwheel, was wrecked, with damages of $5,000. The storm characteristics are covered in more detail in Chapter 7.

The trigger for the heavy downpours was a cold front that merged with a coastal disturbance during a soggy period from August 12–14, 1873. Copious rain, accompanied by a raw northeast wind, pelted the coastal plain. A chilly late-summer shower turned into a downpour in Philadelphia that totaled 5.21 inches, followed by 1.15 inches on the 14th. Measurable rain, totaling 9.96 inches, fell on 11 consecutive days from August 12–22, 1873.

July 26, 1874
Pittsburgh Flash Flood

The city of Pittsburgh (then Pittsburg) suffered through its worst weather disaster, in terms of loss of life, on a humid summer evening on July 26, 1874. Heavy downpours associated with a cluster of slow-moving thunderstorms pounded the north side of the city. The subsequent runoff in the hilly areas channeled torrents of water into unsuspecting neighborhoods.

As reported in the *New York Times*: "At about 8 o'clock on a Sunday evening a storm set in which, as far as the destruction of human life is concerned, was the greatest calamity which has ever visited our city, and, in some degree of horror, rivals the recent Mill-Run disaster in Massachusetts [in May 1874]."

The awful destruction was attributed to "the sudden down-pouring of immense volumes of water." The "copious showers during the earlier part of the day" saturated the ground before a heavy evening thunderstorm set in.

Damage in the city of Pittsburgh was judged to be "comparatively light" to streets, curbing stones and shade trees, which were washed out by heavy drainage racing down the hillsides. Major flash flooding unfolded two miles north of Allegheny City, along the Allegheny River on the north side of Pittsburgh.

The community of Allegheny City, considered then a sister city of Pittsburgh, was particularly vulnerable to flash floods because "the avenues ran through narrow valleys, hemmed in by great hills." The enormous runoff filled the lower floors of frame houses even before some residents realized what was happening.

Houses along formerly quiet, narrow streams were swept away by torrents of water. Severe losses occurred along Butcher's Run, Wood's Run and Plummer's Run. A stream that ran parallel to Saw Mill Run in the community of Charter Valley and emptied into the Ohio River "three miles below" was the scene of many deaths.

Heart-rending accounts of entire families swept away were common: "All the little runs, emptying into Charter's Creek, became rushing rivers. The loss of human life has been frightful."

A few miles downstream, an observer commented on a "huge, inky, black funnel-shaped cloud which overhung the city, the narrow end being lowest, while the dark parts emitted almost continuous flashes of lightning." No wind damage was reported.

Local newspapers gave widely varying death tolls, which was understandable considering the scope of the disaster and overlapping reports, and the sensational nature of reportage. Grazulis (1993) cited 134 fatalities that were incorrectly attributed to a tornado at Erie in some early government publications.

December 10–11, 1878
Early Winter Floods

Heavy rain overspread eastern Pennsylvania early on December 9, 1878, emptying into creeks and rivers that were higher than normal after a heavy rainfall at the beginning of the month. Torrential rain fell over the headwaters of the Delaware River in southeastern New York, causing the river to spill over its banks along the New York border southward to the Delaware Water Gap. Swollen creeks caused damage, washing away railroad tracks, bursting dams and causing several small bridges to crumble.

From the *New York Times*, on December 12, 1878:

> The heavy rain which commenced Monday morning [December 9, 1878] ceased at 10 o'clock last night, changing suddenly into a gale. The Delaware and Lackawaxen Rivers at noon to-day were higher than they have been since 1836 when millions of feet of lumber and other valuable property swept away. The Delaware, between this place and the Delaware Water Gap, overflowed its banks in many places, and thousands of acres of flat lands are submerged between Milford and Stroudsburg, a distance of 33 miles.

A Scranton dispatch reported on the flooding in low-lying areas along the Susquehanna River that engulfed several homes in the Kingston Flats near Wilkes-Barre. Water covered the Lackawanna and Bloomsburg railroad tracks in the lower Wyoming Valley. Harrisburg was surrounded by floodwaters in low-lying areas on the northwest side of town that swamped many homes.

The meteorological setting featured an intense early winter storm centered over Arkansas on Sunday morning, December 8, 1878. The storm tracked northeast, reaching the New York City area on the 10th. At 6:00 p.m., the barometer at New York City registered an extremely low reading of 28.80 inches (975 millibars), and the temperature rose to 56°F in the warm sector. A strong wind ensued, placing eastern Pennsylvania, southeastern New York and southern New England in the northwest semicircle of the storm, where the heaviest rain typically falls.

High winds battered the East Coast, and heavy rains resulted in serious flooding from northern New Jersey and eastern New York to New England. Major flooding occurred at Paterson, New Jersey, where the Passaic River was higher than it had been since the snowmelt flood of mid-April 1854.

February 6–7, 1884
Snowmelt Floods

A soggy February in 1884 followed a stormy January, triggering serious floods resulting from a combination of heavy rainfall and the breakup of ice. The Ohio River at Pittsburgh crested at 33.3 feet around 11:00 p.m. on February 6, 1884, the highest stage since February 1832 (35 feet), and well above the level reached on February 8, 1883 (27.5 feet).

The February 1884 flood was confined mostly to lower sections of Pittsburgh. Railroad communications and gas lines were cut off for several days until the water receded. More than 5,000 residents were left temporarily homeless. The damage total was put at $2 million in the *Monthly Weather Review*.

An ice jam broke along the Susquehanna River, which rose 17 feet in 12 hours on February 7, 1884, flooding low-lying areas around Wilkes-Barre and Kingston, which were buried under as much as 12 feet of caked ice. The monthly rainfall was quite substantial in February 1884, with totals of 7.40 inches at Millville Depot in Wayne County and 7.29 inches at West Chester.

June 4, 1885
Flash Flood Near Pittsburgh

The skies opened around 8:00 p.m. for two hours over the northeastern part of Allegheny City, on the north side of Pittsburgh. The deluge was "estimated at not less than ten inches," according to the *Monthly Weather Review*. The Allegheny River rose seven feet at Wood's Run in 15 minutes, inundating cellars and homes, with water covering the first floor of businesses in the area. Livestock suffered greatly.

The stormy pattern produced a flash flood at Erie on June 21–22, 1885, resulting in a landslide at the "water-works." Streams overflowed and "washed out the sidewalks and streets," and "great damage was done to the crops."

Damaging tornadoes were reported on June 21 in the evening at Ravenna, Ohio, injuring two persons, and a rare late night tornado occurred at 2:20 a.m. at Friendship, New York.

August 1–2, 1886, Wyoming Valley Flash Floods

The Wilkes-Barre area was visited by a severe thunderstorm overnight on August 1–2, 1886, resulting in flash flooding that killed a young man along Tobys Creek, a tributary of the Susquehanna River near Plymouth in Luzerne County.

High water tore through several mills, raising Harveys Creek 10 feet higher than was ever known before. Flash floods swept away "saw and grist mills … iron bridges" at Trucksville, Harveys Lake and Luzerne (*Jeffersonian*).

May 31, 1889 Great Johnstown Flood

The seeds of the Great Johnstown Flood tragedy on May 31, 1889, were sown by arrogant indifference regarding the integrity of an obsolete earth-fill dam built by the state to extend the old canal system.

The South Fork Dam was constructed between 1838 and 1853 to furnish water to the Pennsylvania Canal. The imposing structure stood 72 feet high, nestled against a ridge, at an elevation of 450 feet above the industrial city of Johnstown, with a vulnerable population of 28,000 situated 14 miles downstream from Lake Conemaugh.

To supply the Pennsylvania Canal system, numerous dams and reservoirs were constructed in the 19th century, before railroad systems became the primary mode of transportation for moving goods across the state. By the late 1800s, the canals were largely abandoned, leaving nature to take its toll on the reservoir.

A group of wealthy Pittsburgh investors that included Andrew Carnegie, Henry Clay Frick and financier Andrew Mellon purchased the 160-acre site in 1879. The exclusive South Fork Fishing and Hunting Club provided a secluded retreat for the magnates that was home to private residences.

The reservoir that spanned 2.5 miles and measured 1.5 miles in width was up to 100 feet deep. Modifications were made to the spillway by the South Fork Fishing and Hunting Club to shore up the dam by lowering the pressure. The earthen foundation was the only flood protection for the valley below the junction of South Fork Creek and the Little Conemaugh River, which joins the Stonycreek River downstream to form the Conemaugh River.

The absence of a solid central foundation and the addition of a roadway along an embankment of the dam affected the trestle supports, a condition that some residents were wary about a decade before the devastating Great Johnstown Flood.

The atmospheric scenario that triggered widespread flooding in western and north-central Pennsylvania from May 31–June 2, 1889, was traced to a Pacific storm that took shape over southwestern Texas (Ludlum 1989). A deep trough of low pressure provided a lifting mechanism for a conveyor belt of moisture from the Gulf of Mexico to the Great Lakes, funneling unseasonably cool air south that came into conflict with tropical air streaming north.

A fine mist turned into a gentle rain on the Thursday evening, May 30, 1889, which was also Memorial Day. After a lull, rain commenced falling before midnight, with much greater intensity. The following morning, the reservoir water level was rising at one inch per hour. As early as 11:30 a.m. on May 31, 1889, the president of the club, Elias Unger, had engineers work feverishly to open the sluiceway to release pent-up pressure on the floodwall by digging a secondary spillway as water began overtopping the South Fork Dam.

The young resident engineer John G. Parke Jr. feared the dam was going to give way and alerted the telegraph operator at South Fork, but the wires were damaged in the rainstorm the night before. Parke reportedly rode 18 miles, frantically sounding his whistle and "crying out that the dam was bound go to" (McCullough 1968). Yet few valley residents evacuated their homes or fled for the hills.

As the afternoon wore on, it became apparent that the dam was on the verge of a catastrophic collapse. Crews frantically attempted to shore up the crumbling structure to no avail, and the middle of the dam finally gave way at 3:15 p.m. Through this breach of 300 feet, water from the South Fork of the Little Conemaugh River swelled mightily and raced through the valley floor at 40 mph.

The thunderous torrent attained a height of 40 feet, sending 20 million tons of water pouring through the lowland, an event that was described in various press accounts as sounding "like a cannon ball." The merciless floodwaters were judged to be three times more powerful than Niagara Falls, carrying an inky mass of trees, houses and boulders.

Fig. 6.3 June 1, 1889. Stereoscopic views after the Great Johnstown Flood, from the Pennsylvania Railroad Depot (top), near the Club House and iron railroad bridge, and at Main and Clinton Streets (top). (Courtesy Pennsylvania State Archives)

Fig. 6.2 June 1, 1889. Stereoscopic views following the Great Johnstown Flood where the Woodvale prison had floated to from a mile away (top), and Main and Franklin Streets, near the junction of Conemaugh and Stony Creeks (bottom). (Courtesy Pennsylvania State Archives)

The rampaging wall of water first consumed the villages of South Fork, Mineral Point and East Conemaugh, wiping out everything in its path. Debris turned into "battering rams" and piled up to a height of 40 feet. Railroad engines weighing 20 tons parked on the Pennsylvania Railroad yard tracks "were tossed like chips of wood" for distances of thousands of feet.

The towns of Franklin and Woodvale were lost in the onslaught of pounding waters that were soon headed directly for lower Johnstown, a vital manufacturing hub that included the Cambria Iron Works. The force of the water against the hillside caused a backwash up the Stonycreek River, wiping out the center of town in 19 minutes (Watson 1993).

An article entitled "Johnstown Flood; Other Disasters," that appeared in the *New York Times* on October 1, 1911, recalled the horrific destruction:

The mass of debris borne by the flood was checked by the bridge of the Pennsylvania

Railroad, and an effective dam was made. The water recoiled upon the city, meeting there the wing that had been diverted and had flowed around the city. The result was a gigantic whirlpool which ground to pieces any building that escaped the first onset. The noise of the destruction was maddening to the survivors.

An ominous rumble was the only warning nature provided to Johnstown residents before a tall dark spray became visible above the deadly wave of water. By then it was too late to seek higher ground in the face of the impending doom.

Thirty acres of water-logged debris careened into the stone arch railroad bridge. Many of the desperate survivors clinging onto the old Stone Bridge structure at the intersection of the rivers died in a horrific manner. Trapped debris caught fire, after sparks ignited the spilled kerosene fuel that caused the bridge to become engulfed in a maelstrom of water and flames.

Fig. 6.4 June 1, 1889. Great Johnstown Flood aftermath near the center of the flood-ravaged community. (Courtesy Pennsylvania State Archives)

Fig. 6.5 June 1, 1889. Homes in Johnstown reduced to rubble following the Great Johnstown Flood. (Courtesy Pennsylvania State Archives)

Eyewitness accounts of the unfolding devastation were recounted by David McCullough in *The Johnstown Flood*:

> It began as a deep, steady rumble, they would say; then it grew louder and louder, until it had become an avalanche of sound, "a roar like thunder" was how they generally described it. But one man said he thought the sound was more like the rush of an oncoming train, while another said, "And the sound, I will never forget the sound of that. It sounded to me just like a lot of horses grinding oats."
>
> Everyone heard shouting and screaming, the earsplitting crash of buildings going down, glass shattering, and the sides of houses ripping apart. Some people would later swear they heard factory whistles screeching frantically and church bells ringing.

Hundreds of desperate souls were swept off rooftops and buildings in the lower part of the city. Ruptured oil tanks ignited and burned some of the refugees alive despite valiant rescue efforts that were hampered by the loss of fire-fighting equipment.

Historian John Bach McMaster (1852–1932) summarized the disastrous flooding that occurred in western and central Pennsylvania on May 31–June 1, 1889, which appeared in two installments in *The Pennsylvania Magazine of History and Biography* in July 1933, one year after his death (p. 220):

Between Renovo and Clearfield nine small bridges were carried off, six miles of roadbed washed away and fifteen miles of track overturned. Between Lewisburg and Tyrone four bridges were lost, four others hurt and one mile of roadbed washed out. At Keating and Linden, at Lewisburg, Williamsport and Montgomery one hundred and six freight cars which had been placed on the bridges to weigh them down went with them. At Lock Haven a lime car took fire, and thirteen cars were burned. Between South Fork and Johnstown three bridges, thirty-three locomotives, eighteen passenger cars and three hundred and fifteen freight cars were destroyed. Three more bridges went out between Johnstown and Pittsburgh, and three in Blair county between Altoona and McKees [now Mckees Rocks].

Of the rain, which, in Pennsylvania, fell on the western slopes of the mountains, much went down the tributaries of the Allegheny. Of these tributaries, three more notable for size are Clarion river, Red Bank creek and the Conemaugh river.

His account of the flood continued in the October 1933 journal installment (p. 316):

Twenty of the sixty-six counties in the state had been flooded. In each life had been lost, millions of dollars of property destroyed, hundreds of people stripped of homes and earthly possessions, whole villages devastated, and the health of a million inhabitants seriously threatened. From the mouth of the Susquehanna to the sources of the West Branch and the Juniata, and from the mouth of every large tributary of these rivers to the mountains was an unbroken succession of ruined and injured towns. In some the reservoirs had been swept away and the water supply cut off.

Six large bridges were washed away between Renovo and Harrisburg, and eight more were damaged. Railroad track and roadbed washouts were widespread, and dozens of miles of telegraph lines were toppled. Damage on the Juniata River included 14 bridges from Duncannon to Tyrone. An untold number of culverts and small county bridges were lost.

In the Johnstown area, more than 1,600 homes were destroyed by the wall of water 35 feet high. The catastrophic flooding left at least 2,209 dead, and many of the 979 missing likely perished (Shank 1988). Disease took even more lives in the aftermath of the calamity. The city would take more than five years to begin to recover from the immense damage to property, estimated at $17 million at the time (more than $5.6 billion in modern dollars).

We will never know precisely how much rain fell in Johnstown on May 30–31, 1889, because the Signal Service observer, Mrs. H. M. Ogle, died in the flood. An observer at Wellsborough (now Wellsboro) logged 1.70 inches at daybreak on May 31 and added 7.45 inches as of 4:20 a.m. on June 1—a three-day total of 9.80 inches. The official Harrisburg measurement was 7.56 inches. Very likely these figures were far less than the amount of rain that poured into tributaries of the Conemaugh River.

Major flooding also occurred along the West Branch Susquehanna River at Renovo and Clearfield, relayed by telegraph to the city of Williamsport on the morning of May 31, 1889. However, it would be another 24 hours before the full impact of the heavy rain upstream would be felt in the city. Around noontime on Saturday, June 1, 1889, floodwaters topped the Pennsylvania Railroad line in downtown Williamsport and swamped the area north of the railroad tracks.

Williamsport was then a "prosperous lumber town," and losses included "about 200 million feet of logs and about 40 million board feet of lumber" (Eckhart 1996). The Susquehanna River crested at Williamsport at 32.4 feet—12.4 feet above flood stage, which was almost seven feet higher than the record 1865 flood.

By the time the water reached its maximum depth around 9:00 p.m., about 75 percent of the city was under water that was six feet deep in the buildings on Market Square (Shank 1988). Many lives were lost when people were swept away by rising water.

Twenty people died in the Nittany Valley in flooding incidents and seven fatalities occurred in Wayne Township west of Lock Haven in Clinton County. Only one person drowned at Lock Haven because residents had wisely taken precautions after receiving the news that Renovo had been hit hard by flooding.

William Shank (1988), author of *Great Floods of Pennsylvania*, summarized the destruction in considerable detail:

> The millions of board-feet of logs and sawed lumber which had been swept down the Susquehanna and its tributaries by the great floods of 1889 did great damage as they went downstream. Like a tremendous army of battering rams, the heavy logs and lumber destroyed everything in their path—bridges, factories, houses. They razed the whole town of Milton and swept on. Eventually, they came to rest at tidewater in Chesapeake Bay.

Floodwaters tore through the Juniata River valley in south-central Pennsylvania between Tyrone and Lewistown, with the worst damage at the junction of the Raystown Branch Juniata River. Again, there was significant loss of life where the water was reportedly eight feet above the record 1846 flooding, reaching 35.9 feet at Newport. Only one bridge survived the flood in Huntingdon County, and the town of Huntingdon was completely isolated by floodwaters.

The American Red Cross attended to the homeless and injured survivors, under the direction of Clara Barton. The Pennsylvania Railroad reported 172 bridges were damaged or carried away by the rushing waters, and 50 miles of track were damaged or destroyed.

In the decade following the 1889 flooding, river gauges were constructed to create a systematic measurement of streamflow. The peak discharge occurred at the stone viaduct just above Johnstown, which gave way, unleashing a massive surge of water, according to a University of Pittsburgh at Johnstown study.

In "A Scientist's Autopsy of the Johnstown Flood," researchers Carrie David Todd and three colleagues, using modern hydrological mapping, calculated the rate of water and debris at 15,600 cubic yards per second, comparable to the mean flow of the Mississippi River. One of the findings in the reanalysis was that the dam actually was lowered three feet by the club, not one foot as claimed by the industrialists in court in the years following the devastating flood.

June 4–5, 1892
Oil Creek Flood and Great Fires

On the 125th anniversary of a disastrous flood and fires that swept through the communities of Spartansburg and Oil City, the NWS office in Cleveland, Ohio, published a description of the flood, culled from local newspapers and government weather station data.

The historical weather records in June 1892 reported "a total of 373 Weather Bureau Offices, 166 river stations, and 59 special rainfall stations across the United States." At the time, there were no government flood warnings, and forecasts were transmitted by telegraph to several hundred cities around the country, often received at railway stations.

A wave of heavy rain events soaked northwestern Pennsylvania in late May and early in the spring of 1892. On June 4, a warm front lifting north was accompanied by torrential downpours that brought an estimated rainfall of 10 inches at Spartansburg. The Oil Creek rose to a height of 30 feet, fanning out about three-quarters of a mile wide, causing death and destruction at Riceville and Centerville southward to Titusville and Oil City.

Fig 6.6 June 5, 1892. A view of South Franklin Street in Titusville, after the flood and devastating fires of June 4-5, 1892. (Courtesy Titusville Herald *[Oct. 7, 1897] and Crawford County Historical Society)*

A disaster unfolded a little after midnight, when an earthen dam crumbled seven miles above Titusville, which inundated portions of Titusville and Oil City. The water from Oil Creek spread out over a distance "half a mile wide" south of Titusville. Several benzene tankers in Oil City on both sides of the creek ignited, damaging or destroying 200 to 300 homes.

Storage tanks emptied oil and other poisonous discharges into the creek, which exploded into flames. The disaster, which was recounted in the Oil City *Derrick* on June 29, 1955, was triggered by the wicked combination of flooding and fires ignited by gas, as the water swamped the homes of unsuspecting residents in the creek hollow.

More than 130 were killed and otherwise missing and unaccounted for in the valley, including 54 in Oil City and 72 in Titusville. Damages exceeded $3 million at the time. The *Pittsburg Dispatch* (June 10, 1892) reported a toll of the deceased and missing of around 150 people, including nearly 100 fatalities in Crawford County and close to 60 in Venango County.

May 20–22, 1894
Eastern Pennsylvania Floods

Light rain that had been falling for about 24 hours over central and eastern Pennsylvania turned into a downpour on the night of May 19–20, 1894.

Saturated soils could no longer hold the runoff, causing creeks to overflow their banks. Serious flooding was reported at Williamsport, where the West Branch Susquehanna River overflowed, leaving city streets under up to eight feet of water. In the central mountains on May 20–21, 1894, water was reportedly higher than during the flood of 1889.

Major flooding occurred along the Conemaugh, Juniata and Susquehanna Rivers in the south-central mountains. At Harrisburg, the Susquehanna River crested at 25.7 feet, close to the record crest attained in 1889. In Wilkes-Barre, two children drowned.

On the Delaware River at Easton, the flooding was the worst since October 1869. Jordan Creek, a tributary of the Little Lehigh Creek, combined to raise the Lehigh River above flood stage near Allentown, before arriving at the Delaware. The rainfall was 8.66 inches at Mauch Chunk (now Jim Thorpe) in eastern Carbon County. This rain emptied into the Schuylkill River and neighboring tributaries, causing severe flooding. The average Pennsylvania rainfall in May 1894 of 8.88 inches set a state monthly record that stood until June 1972 (11.23 inches).

February 6–7, 1896
Winter Flood in Eastern Pennsylvania

February 1896 opened on a wintry note, with a widespread heavy snowfall on February 2–3 in Pennsylvania. A thaw ensued, followed by a soaking rain that began on February 5, 1896, and continued for several days, falling on frozen ground.

Rainfall totals in the range of two to three inches came on top of melting snow, contributing to heavy runoff on saturated soils. In Pike County, the Blooming Grove observer caught a record February rainfall of 3.31 inches in 24 hours.

The Delaware River topped its banks in southeastern Pennsylvania, swamping railroads and industrial areas. The Wissahickon Creek flooding was "the worst ever known to have occurred in the memory of the oldest inhabitant," reported the *Philadelphia Inquirer*. Two persons died in Pottstown and three others "made a miraculous escape" from rising waters along the Manatawny and Schuylkill that wiped out a bridge at the Warwick furnace plant.

The Jordan and Lehigh Creeks overflowed, and water filled the streets in low-lying sections of Allentown and Bethlehem, where the headline stated: "LEHIGH ON THE RAMPAGE." In the southern Poconos, two bridges were washed away at Bartonsville along Pocono Creek, including a New York, Susquehanna and Western Railroad structure. Many track washouts were reported.

The *Stroudsburg Times* (February 13, 1896) reported that dams were overrun by high water, causing flooding at Effort, Kellersville and South Stroudsburg.

Never in the recollection of the oldest inhabitant has such a steady volume of water poured down upon this county, as commenced about 1 o'clock of the above morning [February 7]. The center of the attraction seemed to be McMichael's Creek, which at the South Stroudsburg falls was a miniature Niagara. The water rushed over the dam with a boom-like cannon, dashing up a wave fully 40 feet high. Damage to the railroads running through the Stroudsburgs was estimated to be $100,000.

Flooded basements were common throughout the region, and roads resembled "a regular sea" as water poured through town.

August 24, 1901
Carbon County Flash Flood

Two heavy rainstorms in late August 1901 led to a flash flood that took four lives in Mauch Chunk (now Jim Thorpe).

Eastern Pennsylvania was swamped locally with three to six inches of rain on August 18–19, 1901, including especially large totals at Philadelphia (4.71 inches) and Hamburg (6.10 inches). Another wet period from August 22–24, 1901, deposited one to three inches in saturated areas. Hamlinton, Wayne County, had rain every day from August 17–24, 1901.

The Lehigh River, already swollen by the previous rainfall, raged through Mauch Chunk, pouring down Broadway with great force, flooding stores and basements. The boilers at the Mauch Chunk Electric Company went out, plunging the town into darkness (Eckhart 1996). The torrent of water created deep gullies in the middle of town and washed out many roads and trolley tracks.

December 13–14, 1901
Widespread Flooding

The *Philadelphia Inquirer* covered the breadth of flooding triggered by a powerful disturbance along the Eastern Seaboard, accompanied by high winds and torrential rainfall, on December 14–15, 1901.

Floods, such as have rarely devastated the State of Pennsylvania, are reported from all points between Altoona and Easton. West and East, and from the northern state line to as far south as Lancaster. Millions of dollars' damages have resulted from a storm of wind with little less violence than a tornado, and a downpour of rain characterized by some as almost a continuous cloudburst.

Heavy flooding occurred on the Susquehanna, Lackawanna, Lehigh and Schuylkill Rivers. There were numerous reports of railroad tracks and roads being washed out and bridges torn from their foundations as rivers and creeks overflowed their banks.

Houses along river basins were undermined by loosened soils from the high water. Damage along the Schuylkill River was estimated at not less than $1.5 million. High winds tore the roofs off an opera house and several hotels in Lancaster, and there was considerable property damage suggestive of a small tornado.

Fifteen miles north of Reading, country roads were submerged under six feet of water and homes were badly flooded. Residents could not recall a time when water had risen so rapidly. In the east-central hills, the Mahanoy Creek at Ashland was described as a "raging torrent" that ripped up railroad tracks and washed away outbuildings and bridges.

A tragic railway accident occurred when a portion of the Pennsylvania Railroad Bridge over Lycoming Creek between Williamsport and Newberry collapsed. The engine and nine freight cars of the westbound Oyster Express plunged into 20 feet of water, killing an engineer, a brakeman and a fireman, who were in the cab of the engine.

More than 20 township and railway bridges were destroyed in Schuylkill County alone. Large mudslides damaged sections of the Pennsylvania Railroad at Rock Station east of Auburn. In the Lehigh Valley communities of Easton, Allentown and Bethlehem, the December 1901 flood was considered to be the worst since June 1862. Extensive flooding occurred around Easton as the rising Delaware River backed up into swollen tributaries.

The Lehigh Valley Traction Company, which furnished electricity to Allentown, was flooded. The Lehigh River was so high at 10:00 p.m. on December 15, 1901, that the Allentown Gas Company turned off the gas, darkening much of the city. Backwater flooding also occurred along the Little Lehigh River and Jordan Creek, cutting off South Allentown from the rest of the city.

A dam along a privately owned two-mile lake at Naomi Pines in the Poconos burst, sending even more water spilling into the Little Lehigh River, which aggravated the situation. In neighboring Carbon County, the pounding water tore out a span of a new steel railroad bridge and destroyed another railroad bridge at Penn Haven. The Lehigh River rose five feet higher than it had during the August 1901 flood. The county was darkened again by the loss of electricity. Damage to the canal and railroad lines paralleling the river was extensive.

The town of Weissport, on the east side of the river from Lehighton, was covered by four feet of water (Eckhart 1996). The storm lashed Tamaqua for 24 hours before abating early on December 15, 1901, causing "awful damage." The Wabush Creek spilled over its banks, flooding the business district. At Pottsville, four feet of water poured down Railroad Street, flooding basements and extinguishing the town's steam-heating plant fires.

Farther north, the Susquehanna River inundated Monroeton in Bradford County, sweeping a woman to her death. In Wayne County near Honesdale, the Penwarden Pond Dam broke in Dyberry and a courier was dispatched to alert residents downstream of the imminent threat. Streets were flooded in Scranton and there was "a great deal of damage" around Wilkes-Barre.

The rains ushered in a cold wave, suggesting a powerful collision of polar and tropical air that contributed to the 24-hour deluge. Powerful winds were associated with a large temperature drop, accompanied by a vigorous cyclonic depression.

February 28–March 1, 1902 Ice Jam Flood

The latter half of February 1902 brought two major winter storms, followed by heavy rain and a thaw. In eastern Pennsylvania, eight to 12 inches of snow blanketed the region on February 16–17, 1902. A second storm brought a substantial load of snow and ice on February 21–22. Mixed precipitation at Philadelphia totaled 1.96 inches.

As the weather thawed, rain began to fall on February 25, 1902, saturating the heavy snowpack that covered Pennsylvania while loosening great masses of ice that were lodged in the waterways.

The *Philadelphia Inquirer* described the breakup of ice that unleashed a large quantity of water into eastern Pennsylvania rivers:

Bearing masses of ice, logs, debris of every description, and even tearing tugs and scows from their anchorage, the usually placid Schuylkill River tore its way over its course yesterday with terrific vehemence, wreaking ruin in many places and incapacitating works along its entire banks.

The ice freshet that followed in February 1902 was much like the one in February 1896. High water caused major flooding along the Susquehanna River and on some tributaries, leading to widespread flooding at Chambersburg, Shippensburg, Lancaster and York. Significant flooding occurred along the Schuylkill River at Pottstown and Norristown. A large dam owned by the Glasgow Iron Company was swept away in a rush of water and debris. In western Pennsylvania, the flooding was extensive around Waynesboro involving "all the streams in this vicinity."

Another heavy rainfall on February 28, 1902, deposited 1.40 inches at Philadelphia, aggravating the already serious high-water issue. The latest rainstorm combined with melting snow and ice to push rivers and creeks beyond their banks again in eastern Pennsylvania, causing flooding along the Lehigh River and Delaware River.

The *Wayne Independent* in Honesdale, in an account of the flood that appeared in the "Bicentennial Edition" (February 4, 1978), described how the Lackawaxen River overflowed its banks on February 28, 1902. Huge chunks of ice swept through Honesdale, causing Main Street to resemble an ice floe. Dynamite was used to break up the ice jam at Park Lake, but the effort was in vain.

The Main Street Bridge was swept away, and before long, one-third of Honesdale below Park Lake was under water. North of the Lackawaxen River, floodwaters backed up to High Street, filling cellars with water. More bridges were taken out over the ensuing days as the ice and water pressed on.

A week later, a weakened stone arch bridge over Wallenpaupack Creek went out with another round of high water. In neighboring communities, floodwaters reached almost to the second floor of homes in Milanville, and significant flood damage to homes and businesses occurred in Damascus and Cochecton, New York.

The new $250,000 Lehigh Valley Railroad Bridge at Bridgeport, Carbon County, amidst the onslaught of water came down, and two crewmen lost their lives as the bridge gave way (Eckhart 1996). The Lehigh River went into high flood, causing widespread damage at Mauch Chunk (now Jim Thorpe) and Weissport across the river from Lehighton.

The Lehigh River at Allentown was reportedly two feet higher than it was during the previous record flood in June 1862, and at Easton reached a height of 33 feet. The flood stage of 16 feet at Bethlehem was exceeded by 10 feet, reportedly nearly three feet above the June 1862 mark. The Schuylkill River at Reading attained a height of 21.5 feet, comparable to the 1869 flood.

Flooding affected parts of Pittsburgh, Reading, Philadelphia, Allentown and Wilkes-Barre. Along the Susequehanna River, the water achieved its highest level in Wilkes-Barre about 10:00 a.m. on March 2, 1902, when "all flood records were broken" after "three or four feet of water" swamped the "lower end of the city." West Nanticoke was flooded in a short time, and the electric railroad system in Wyoming County was washed out.

The *Climate and Crops: Pennsylvania Section* summary (February 1902) offered this account of the flooding:

In Pittsburg there has been no parallel for the flood at that point since 1832. The waters in the Susquehanna, Lehigh, and Delaware rivers on this side of the Alleghenies have not been so high before in the last forty years. The destruction of life has not been large, but the destruction of property was never greater.

October 9–10, 1903
Pumpkin Freshet

Weather historian David Ludlum (1983) described the storm of October 8–10, 1903, as the "greatest rainstorm in the history of the New York metropolitan and northern New Jersey areas."

The remnants of a tropical storm drifting up the Atlantic Coast caused light rain to overspread Pennsylvania on October 7–8, 1903, which soon turned into a driving rainstorm. The incessant rain tapered off to drizzle on October 10, but runoff swelled creeks and tributaries throughout eastern Pennsylvania and New Jersey.

Water levels along the Delaware River at Portland rose 8.7 feet above flood stage, exceeding the high mark reached during the Bridges Freshet of January 8, 1841. The Delaware, Lackawanna and Western Railroad tracks running parallel to the river were submerged under more than a foot of water, which entered the first floor of area businesses. A trolley bridge was swept away, and Slateford was inundated just to the north.

By the time the Delaware River crested in the early evening of October 10, 1903, a vast swath of the valley was immersed in muddy waters. Pumpkins bobbed up and down in the water, carried away by the roiling floodwaters that consumed cropland.

Bridges collapsed along the way, taking the lives of unsuspecting onlookers, starting with a 651-foot span between Matamoras and Port Jervis, New York. Severe flooding occurred in Easton at the junction of the Bushkill and Lehigh Rivers, which feed into the Delaware.

The aftermath of the flood was described in Dale (1996):

> The city lost hotels, trolleys, gas and electric plants, and its sewer plants, as well as most businesses. Homes along the Lehigh, Bushkill, and Delaware rivers lay underwater. The Mineral Springs and Salts Eddy hotels were swamped. Phillipsburg [New Jersey] received similar devastation.

In Pike County, Delaware River tributaries were in high flood. Several fatalities were reported along the Manunka Creek near Milford. Regional damage estimates reportedly topped $200,000, including $60,000 in the eastern Poconos.

Rainfall totals were exceptionally heavy in the northeastern corner of Pennsylvania. Stroudsburg drugstore proprietor J. Clyde LeBar, prior to becoming the official local government observer in 1910, kept daily weather observations. His measurement of eight inches established a long-standing record, recalled in the Stroudsburg *Record* (October 13, 1949).

Farther north along the Delaware River, Alla K. Doughty, the government observer in Milford, logged a remarkable 9.90 inches of rain from October 7–10, 1903, with a 24-hour total of 7.70 inches on October 8–9, 1903. On the Pocono Plateau, a weather station at Pocono Lake received a storm total of 6.90 inches. North of Honesdale, the measurement at Dyberry was 7.63 inches.

The rain was heavier along the coast closer to the track of the tropical storm. In northeastern New Jersey, a total of 15.04 inches at Paterson came close to the huge rainfall amounts of September 1882, which also occurred during the passage of an offshore tropical storm. The Passaic River at Little Falls, New Jersey, reached an all-time crest of 17.5 feet, and along with the Ramapo River sent floodwaters on the rampage, taking out virtually every dam and bridge. The New York City measurement of 11.17 inches established an all-time 24-hour rainfall record for the city.

March 14–15, 1907
Pittsburgh Flood

On March 10–11, 1907, a slushy four to eight inches of wet snow fell on much of western Pennsylvania.

Heavy rain and thawing conditions on March 12–14, 1907, set the stage for flooding in western Pennsylvania for the second time in the winter of 1906–07. The Ohio River rose to 35.5 feet on March 14, surpassing the record stage of 30 feet recorded on February 10, 1832. A summary of the watershed discharge appeared in the *Monthly Weather Review*:

> An inspection of the weather maps and special reports shows that the flood at Pittsburg can be attributed mainly to the enormous volumes of flood water caused by the excessive rains and melting snows from March 12–14 over the Kiskiminetas and Youghiogheny watersheds. The Monongahela, of course, contributed largely, but not so much as in the January [1907] flood, when stages above the mouth of the Youghiogheny were from 3 to 5 feet higher.

The Ohio River would eventually crest at 38.4 feet, supposedly the highest flood stage since that of March 9, 1763 (adjusted to 41.1 feet), noted by Shank (1988). Telegraph and telephone services were wiped out in the aftermath of heavy rains and electrical storms on March 13, 1907, inhibiting the transmission of warnings.

Ice jams broke free in various places, releasing huge quantities of water that coursed through the Kiskiminetas, Allegheny, Monongahela, and Youghiogheny Rivers toward Pittsburgh. The Ohio River surpassed flood stage between 6:00 p.m. and 7:00 p.m. on March 13, 1907, eventually reaching 31.1 feet (8.1 feet above flood stage) at 8:00 a.m. on March 14.

On the evening of March 13, 1907, the Conemaugh-Kiskiminetas River had already reached an elevation of 18 feet—11 feet above flood stage and the highest level since that of the Johnstown Flood on May 31, 1889. South of Pittsburgh, the Youghiogheny River crested at a record height of 28.2 feet at West Newton—5.2 feet above flood stage and 6.2 feet beyond all previous high-water marks.

Flood damage at Pittsburgh was estimated at $5.6 million, hitting the industrial and manufacturing districts the hardest. Nine lives were lost in the area, including three resulting from a railroad bridge collapse.

September 30, 1911
Austin Flood

In a soggy conclusion to the hot summer of 1911, the Philadelphia Weather Bureau rain gauge picked up 9.99 inches of rain from August 23–30, 1911. The maximum daily falls occurred on the 30th (2.32 inches) and 31st (3.41 inches).

Another week of heavy rain in late September 1911 in the northern mountains of Pennsylvania caused many streams to overflow. After a sopping day on September 29, the next morning offered the promise of dry conditions. However, the pleasant turn in the weather belied the impending disaster that would afflict the small town of Austin, between St. Marys and Coudersport in Potter County, on the fateful afternoon of September 30, 1911.

A huge dam had been built on a hill about a half-mile above Austin in 1909 that spanned the valley of Freeman Run. The dam was built to supply reserve power for the Bayless Pulp and Paper Mill that provided financial opportunities for the industrial town of 3,200 in the southwestern part of Potter County.

The concrete structural marvel was 530 feet long, 49 feet high and 32 feet wide at the base. The dam was constructed based on modern engineering principles. However, in the middle of January 1910, after a series of heavy snowfalls that were followed by a thaw and heavy rain, the Bayless Dam bulged out 36 feet. A small portion of Austin was soon under water.

The 1910 flood raised concerns and triggered protests about the physical integrity of the dam and concrete reinforcements. However, most residents were satisfied that the dam was safe, with the exception of a diligent resident named William Nelson, who continued to take daily readings and warn of the potential for disaster (Nuschke 1960).

On September 30, 1911, the reservoir was full for the first time and spilling over the top, while holding 4.5 million gallons of water. Suddenly, a small hole developed in the western end of the dam just past 2:30 p.m. and the pressure caused a failure.

Skies were sunny on the afternoon of September 30, 1911, so no one could have anticipated that the roar was an impending wall of water heading directly for Austin. Disaster struck quickly and without mercy, whisking away homes with residents inside. The wall of water continued past Austin, swallowing up the small communities of Costello and Wharton a short distance away.

The death toll was initially estimated in the hundreds, but apparently many were able to make a narrow escape, and that figure was reduced to 78 (Shank 1988). Marie Kathern Nuschke wrote about the catastrophe in *The Dam That Could Not Break*.

No person who witnessed the flood wants to remember the night that followed it. It continued to rain hard. A heavy mist settled over the Valley and through it streaked billowing black smoke while the bright red lights from the flames that came from burning debris pierced it now and then, giving a ghastly color to the faces of the people who continued to walk aimlessly on all the streets.

In the end, shoddy construction and excessive rain triggered the collapse of the Austin Dam, unleashing nearly 400 million gallons of water. Over a distance of eight miles there was much destruction, which became a worldwide news story. The collapse was the sixth greatest dam failure in the nation's history at the time, resulting in legislation being introduced to improve dam safety.

August 3, 1915
Mill Creek Flood

A tropical system came ashore along the northeastern coast of Florida on August 1, 1915, tracking to the north along the Eastern Seaboard. Remnants of the disturbance crossed central Pennsylvania early on August 4, 1915, before curving northeastward across southern New England.

Moisture streamed along the eastern flank of the Appalachians into Pennsylvania and New Jersey on August 3–4, 1915, generating heavy downpours that triggered serious flooding. Three waves of torrential rain pounded Erie and vicinity late on August 3, which caused Mill Creek to overflow its banks.

The tranquil tributary that flows through town to Lake Erie had a history of flooding, most recently in May 1893, when floodwaters washed homes off their foundations and wrecked many bridges in the county, as recounted in the *Erie Daily Times* (August 5, 1915).

The story had this to say about the flash flooding on August 3, 1915: "Yesterday the flood meant a magnificent spectacle for some; a hard search for others; and for many, a gruesome search through the morgue for missing friends or relatives." A two-block portion of the city was "devastated" by the flooding that left "several hundred families homeless."

Five feet of water covered the streets following 13 hours of relentless downpours. The Weather Bureau rain gauge in Erie caught 5.40 inches on August 3, 1915 (5.77 inches in 13 hours on August 3–4). Three inches fell between 4:00 p.m. and 7:00 p.m. Unofficial rainfall totals were as high as 20 inches, according to the *Monthly Weather Review*.

A lake formed in a blocked culvert at 26th and State Streets, at the present site of Veterans Memorial Stadium. Police and firemen were unable to clear it of debris before the pressure caused a break around 8:45 p.m., releasing a huge volume of water when the walls caved in a "great roar" (www.eriecountyhistory.org).

A narrow channel of water filled with flooding debris (trees, outbuildings and chicken coops) rushed through downtown Erie and surrounding neighborhoods at 25 mph. Residents were caught without warning in an area three miles long and six blocks wide. At least 33 persons were swept away to their deaths in the city's worst natural disaster—other sources list the death toll between 36 and 40.

Houses crumbled in the torrent and were washed away from their foundations as the waters reached the second-story dwellings. Damage to the heart of Erie was extensive: 225 homes and more than 300 other buildings, according to the Erie County Historical Society records. The twisted wreckage of structures and automobiles caked in mud, along with human remains, presented a macabre sight. The entire city was plunged into darkness as water swirled around power plants and lines of communications were cut off in all directions. Damages were estimated at $3 million.

Four days after the destructive Mill Creek flood, a tornado formed just north of nearby Harborcreek, whirling through the small community before heading out into Lake Erie. There were no serious injuries, and damage to crops and vineyards amounted to several thousand dollars.

August 22–23, 1933
Hurricane Floods

Heavy rains attended the passage of the Chesapeake-Potomac Hurricane on August 22–23, 1933, over southeastern Pennsylvania, causing serious flooding that was described to be the worst in the state since 1889. The Pennsylvania Highway Department estimated that the damage to bridges and roads alone was $800,000, according to a report in *Climatological Data—Pennsylvania*. Crop losses were heavy.

Forty-three weather stations in Pennsylvania reported an excess of six inches of rain on August 21–23, 1933. The heaviest rainfall was 13.82 inches at York (8.48 inches in 24 hours), establishing a rainfall record for the state and eclipsing the previous mark of 11 inches on June 25–26, 1884, which also occurred at York. The monthly total of 17.70 inches was also a state record at the time.

The Schuylkill River reached a stage of 19.7 feet at Reading, not far below the crest of 23.2 feet observed on September 2, 1850. Much of the south side of Reading was heavily flooded. Basements filled with water, and floodwaters rose as high as the first floor of dwellings.

Flooding was widespread over southeastern Pennsylvania, including the lower Delaware Valley. Mauch Chunk Creek in Carbon County spilled out of its banks on the afternoon of August 23, 1933, sending water pouring down Broadway in Mauch Chunk (now Jim Thorpe). The rushing water caused considerable damage through the center of town, isolating residents in some areas.

The gas company, which was owned by the Pennsylvania Power and Light Company, laid temporary lines while repairing the damage, estimated at $50,000. The Big Creek at Weissport lived up to its name, flooding numerous properties. A bridge at Nesquehoning was swept away, isolating New Columbus (Eckhart 1997).

Two weeks later, on September 3, 1933, another flash flood brought on by a cloudburst struck Carbon County and lasted about 12 hours. A dam broke in East Mauch Chunk that aggravated the flooding and damaged the Lehigh Valley Railroad, including two railroad bridges. A 17-year-old boy averted disaster by warning a train engineer along the Central Railroad of New Jersey shortly after midnight of a washout near Nesquehoning (Eckhart 1997).

Torrential rains sent the Codorus Creek roaring out of its banks, cresting at a record height of 24 feet at York, which exceeded the June 1884 flood record (Shank 1988). This is the second highest crest on record, after June 22, 1972 (26.36 feet).

July 9–10, 1935
Northeastern Floods

A period of wet weather commencing on July 7, 1935, turned into tropical downpours on July 8–10. In southern Wayne County, the weather station at Gouldsboro received 6.47 inches of rain on July 9–10; a storm total of 7.26 inches over four days caused widespread flooding.

Heavy rains also hit central and southern New York on July 7–8, 1935, overfilling the Susquehanna watershed, allowing excess water to work downstream. Ithaca, New York, received 7.90 inches of rain, triggering major flooding in the region that took 40 lives in New York.

In eastern Monroe County, 6.94 inches of rain fell at Stroudsburg, including a record 24-hour total of 4.70 inches for the month of July. The weather station at Bethlehem's Lehigh University recorded 5.10 inches on the 9th.

Farther south, the swollen Susquehanna River caused extensive property damage and flash flooding that claimed the lives of 12 people. Storm damage in the upper Susquehanna Valley was estimated around $25 million (Ludlum 1982).

Rounds of heavy rain on July 9, 1935, brought the Delaware and Schuylkill Rivers and tributaries out of the banks. Nine inches of rain fell in the Pottsville area near the headwaters of the Schuylkill River on July 8–9, 1935. The Perkiomen Creek, one of the tributaries entering the Schuylkill, reached a height of 18.26 feet, the third highest on record.

March 17–20, 1936
St. Patrick's Day Floods

The winter of 1935–36 brought a heavy accumulation of snow in January and February, coupled with periods of intensely cold weather. The snowpack began to recede by late February and early March, though a considerable quantity of snow remained over the northern mountains of Pennsylvania.

A series of heavy rainstorms falling on mostly frozen ground during the middle of March combined with melting snow over the higher elevations to bring historic flooding to much of Pennsylvania and the Ohio Valley, inflicting Pittsburgh's greatest weather disaster.

An anomalously wet weather pattern developed over Pennsylvania on March 9–10, 1936, and a heavier rain fell on March 11–12. A series of winter storms had left a considerable quantity of snow in the higher elevations, which, combined with a steady rain, resulted in unusually heavy runoff. Ice jams quickly developed along streams, causing heavy flooding in lowland areas. Many places in the state received an inch of rain or more on March 11–12, 1936, and 2.50 inches fell at Gouldsboro.

The Susquehanna River at Harrisburg flooded north of downtown and in the West Shore suburbs on March 13, 1936. Light rain commenced falling again in western Pennsylvania on March 15–16 and became heavy statewide on March 17–18, lasting intermittently through the 20th. In the northwest, heavy rain changed to wet snow on March 17, 1936, accumulating to a depth of 20 inches at Corry.

Fig. 6.7 March 19, 1936. St. Patrick's Day Flood along the swollen Susquehanna River. (Courtesy Pennsylvania Historical Society)

Fig. 6.8 March 19, 1936. St. Patrick's Day Flood inundated the center of Harrisburg. (Courtesy Pennsylvania Historical Society)

The greatest 24-hour total in March 1936 was 4.50 inches at Buffalo Mills on March 17–18.

There was no room in the soils and waterways for additional heavy rain. Shank (1988) summarized the terrible plight of Pittsburgh residents who lived in the Golden Triangle district:

> Unconfirmed rumors of a dam break at Johnstown had Pittsburgh citizens jittery on Tuesday, March 17, 1936. Water had been rising all day, reached 27.1 feet at noon, and by late afternoon was invading the Golden Triangle. A fire which broke out in the Crucible Steel Plant, on the lower north side, added to the confusion and apprehension as streets became clogged with traffic leaving the flood area.
>
> Forecasters were still predicting a maximum crest of only 33 feet. By Wednesday morning terror gripped the city as water swirled around office buildings downtown and homes on the north side. Power went off all over Pittsburgh. Phone lines went down. More fires broke out.

Floodwaters in downtown Pittsburgh crested at an all-time record level of 46 feet at 7:00 p.m. on March 18, 1936, which was 21 feet above flood stage, and 7.5 feet above the previous high-water mark on March 15, 1907. Fifteen feet of water covered the Triangle section, forcing thousands to flee for their lives.

Boats maneuvered through flooded city streets that resembled canals in an attempt to rescue trapped victims, as fires and gas explosions rocked the downtown area. The National Guard and state police struggled to maintain order and prevent looting in downtown Pittsburgh, which was sealed off to the public.

Fig. 6.9 March 19, 1936. St. Patrick's Day Flood damage at the Golden Triangle, at the confluence of the Allegheny and Monongahela Rivers near downtown Pittsburgh. (Pittsburgh Post-Gazette *Photo Archives, copyright 1999, all rights reserved*)

Fig. 6.10 March 19, 1936. St. Patrick's Day Flood covering Liberty Avenue in Pittsburgh. (Pittsburgh Post-Gazette *Photo Archives, copyright 1986, all rights reserved*)

Lorant (1964) reported that 3,000 residents were injured and more than 100,000 lost their homes in the Pittsburgh area alone. The March 1936 issue of *Climatological Data* called the losses in many Pennsylvania watersheds "appalling and unprecedented." The death toll at Pittsburgh was 67 people, and 22 were lost around Johnstown. The summary noted, "many other reported missing were almost certainly lost in the flood, which conservatively put loss of life in Pennsylvania at upwards of 100."

Johnstown was hit hard again on March 17–18, 1936, when the water reached a depth of 14 feet after several days of pouring rain, resulting in the deaths of 25 people. A total of 77 buildings were destroyed and 3,000 were substantially damaged, according to the history compiled by the Johnstown Area Heritage Association (JAHA). As many as 9,000 were left homeless in "The Flood City," with damages totaling $25 million (Lorditch 2007).

In the northern Poconos, 60 people were rescued from floodwaters on March 18, 1936, along the Lackawaxen River around Hawley. A little farther south, Stroudsburg was drenched with 6.92 inches of rain on March 17–18, 1936, much of it falling in 24 hours. The weather station at Mount Pocono reported 5.20 inches fell in 24 hours. Mountain tributaries feeding into the Delaware River raised the water level at Easton to the highest point since October 1903, where the river crested 10 feet above flood stage, sending water into downtown Easton and neighboring low-lying communities.

The Pittsburgh area suffered incalculable destruction in the St. Patrick's Day Flood, estimated by government engineers around $250 million (1936 dollars), raising the total losses in the state of Pennsylvania to $300 million (Shank 1988). It would be several months before the city began to recover from the terrible devastation. Property losses were estimated at about $9 million in Wilkes-Barre and throughout the Wyoming Valley, and more than $28 million downstream on the Susquehanna River.

Flooding along the Delaware River in March 1936 was not as severe as in October 1903, because the northern tributaries did not reach flood stage at the same time, as they did in 1903. Back-channel flooding slowed the discharge into the main stem but still submerged rail lines and roads. When the river peaked on the evening of March 18, 1936, many of the islands, including Island Park, were under water.

The Stroudsburg *Record* on March 19, 1936, described the flooding in the Poconos:

> The swirling, roaring waters, which last week pounded down the river, hurling huge cakes of ice, now poured the fullest vengeance against banks, homes, and bungalows and swept everything before them as they rushed along. Banks were eaten out, transportation stopped, homes were flooded, bungalows and garages were washed away, and lives were endangered, although there were no fatalities up until noon today.

Serious flooding was widespread in Pennsylvania along every major river. The West Branch Susquehanna River emptied into the streets of Williamsport, cresting 11 feet above flood stage on March 19, 1936, at a height of 33.57 feet, 1.2 feet higher than in 1889. Sunbury, at the confluence of the West and North Branches, was swamped under six to 12 feet of water, as the river crested at a record height of 34.65 feet, 11 feet above flood stage.

Along the Susquehanna River, Wilkes-Barre suffered major flood damage as the river rose 11 feet above flood stage, within inches of the record height attained on March 18, 1865 (Shank 1988). Some 4,000 residents were marooned by the swollen river, which caused extensive damage and much suffering. Above Wilkes-Barre, the river at Towanda crested at 25.03 feet, slightly above the record 1865 stage.

In south-central Pennsylvania, heavy flooding carried away bridges and homes at Mount Union, where the Raystown Branch flows to the Juniata River. Major property losses affected Huntingdon. Farther downstream at Newport, the gauge read 34.34 feet, only slightly below the historic crest of 35.9 feet in 1889 (Shank 1988).

Surging waters from the Juniata fed into the lower Susquehanna, adding to the rising waters around Harrisburg on March 17, 1936. Over the next two days, water from four to 15 feet deep covered almost a third of the city. Maximum flood stages were recorded at 6:00 p.m. on Thursday, March 19, reaching 29.23 feet at the official Nagle Street gauge and 30.33 feet upstream at Walnut Street (Shank 1988).

The worst flooding in Pennsylvania claimed 69 lives in the Pittsburgh area, and damages in the state added up to $300 million. More than 60,000 steelworkers were out of work following the St. Patrick's Day Floods that killed 107 statewide.

Reservoir, dam and levee projects were constructed in the years following the March 1936 disaster, authorized by an act of the U.S. Congress on June 22, 1936, to protect the state from future catastrophic floods. The modified Flood Control Act of 1938 gave Army engineers the right to obtain property for the sake of flood control and to construct high flood walls in areas deemed necessary.

May 22–23, 1942
Lackawaxen River Flood

A prolonged heavy rain drenched northeastern Pennsylvania from May 20–23, 1942, sending rivers and streams out of their banks. The 1942 flood was the worst in the history of the Lackawaxen River Valley and Wayne County, exceeding the previous record flood stages attained in February–March 1902 and July 1935.

Upstream, a weather observer at Carbondale reported seven inches of rain fell between 4:00 p.m. and midnight on May 22, 1942. A rain gauge at Tanners Falls, along the Dyberry Creek north of Honesdale, collected 4.96 inches between 6:00 p.m. and midnight, with a storm total in excess of seven inches, sending Dyberry Creek out of its banks.

Records maintained by the Department of Forests and Waters showed a rapid rise on the Lackawaxen River near Hawley between 10:00 p.m. on May 22 and 4:00 a.m. on May 23, reaching a height of 17.4 feet. This reading was 6.23 feet above the peak discharge observed in March 1936.

Honesdale took a devastating blast when a wall of water 10 feet high roared through town from the overflow of Seelyville Dam around 9:30 p.m. The floodwaters reportedly rose 15 feet in about 15 minutes. The gong of a fire alarm was sounded two hours later—the only warning that residents in Honesdale received of what became a horrific disaster.

The dark of night and loss of power prevented warnings from being adequately disseminated. Twenty-four people died at points along the Lackawaxen River between Prompton and Hawley. Six bridges were lost, 46 homes were destroyed, and 1,200 residences were damaged (www.poconohistory.com/MonroeCounty.htm).

Eight inches of rain fell at Hickory Run in eastern Carbon County on May 20–22, 1942. The observer at Mauch Chunk (now Jim Thorpe) measured 6.14 inches in 24 hours on May 22–23. A wooden bridge crossing Robinson's Creek between Coalport and East Mauch Chunk collapsed around midnight, stranding residents.

Fig. 6.11 May 23, 1942. Heavy downpours sent the Lackawaxen River out of its banks, triggering a deadly flash flood in Wayne County. Looking north from Willow Avenue, Honesdale. (Courtesy Elizabeth Korb Schuman)

Flooding extended southward along the Lackawanna, Lehigh and Schuylkill Rivers, where floodwaters rose 10 feet past flood stage. At Weissport, the Lehigh Canal was inundated and homes were submerged under eight feet of water from the flooded Lehigh. Damages to the Lehigh Coal and Navigation Company, between White Haven and Easton, were put at $5.7 million (Eckhart 1997).

The death toll in the state mounted to 33 persons, as high waters continued to sweep through the region until early on the 23rd. Flooding was also reported along the Delaware and Susquehanna Rivers. The *Climatological Data* report noted, "Property damage, conservatively estimated, amounted to $13,000,000."

July 17–18, 1942
North Central Cloudburst Flood

The summer of 1942 was unusually stormy. A series of cloudbursts associated with slow-moving thunderstorms drenched central Pennsylvania on July 17–18, 1942, taking 15 lives.

Excessive rainfall from a thunderstorm complex turned quiet mountain streams into raging torrents that were channeled through narrow mountain passes. Spectacular rainfall estimates ran as high as 15 inches in less than 24 hours in Cameron, Potter, McKean and Elk counties, reported in the *Annual Summary for Pennsylvania* in the annual *Climatological Data* publication.

Communication was cut off at Emporium, a town of 4,500, which was especially hard-hit by widespread flooding as dikes were overwhelmed. Flooding at Port Allegany in McKean County claimed six lives and left about 400 people homeless, according to reports in the *Bradford Era*.

Coudersport suffered heavy flooding, where the Allegheny River swamped Main Street in five feet of water. Dam breaches occurred above Austin and Johnsonburg. There were reports of waves forming atop the floodwaters, including on the East Branch of the Clarion River at Glen Hazel (Eisenlohr 1952).

On July 23, 1942, highway department chief engineer T. C. Frame reported that damage to Pennsylvania highways and bridges was estimated at $500,000. Frame said, "We lost about 16 bridges ranging from 75 to 250 feet." The July *Climatological Data* report listed a maximum official 24-hour rainfall total of 8.48 inches at Coudersport. The rain gauge at the Smethport Highway Shed registered 6.68 inches, before the gauge was lost in the flood.

A bucket survey undertaken after the storm in McKean County estimated that the area between Smethport and Port Allegany received a world-record six-hour rainfall of 30.60 inches on July 17, 1942. A USGS Water Supply Paper by Eisenlohr (1952) listed three extraordinary totals: 35 inches at Port Allegany, 34.50 inches at Turtlepoint, and 32.70 inches northeast of Coudersport, on July

17–18, 1942. The Port Allegany total has traditionally been accepted as the state record, adding up to a stunning monthly total rainfall of 40.90 inches (Jennings 1950).

The Port Allegany rain total was an estimate derived from a volume of 37 quarts of water in a 40-quart milk can, according to the U.S. Weather Bureau report (1942). Similar bucket measurements were obtained at the surrounding sites in excess of 30 inches of rain. However, the dimension of the mouth of the bucket could have resulted in a figure that was too high (Shein et al. 2013).

For that reason, the state July rainfall record properly belongs to Park Place—19.81 inches in July 1947.

July 1947
Record Soggy July

The spring and summer of 1947 were cool and unusually stormy. The United States Soil Conservation Service reported soil and crop losses in excess of $900 million in May and June 1947.

A persistent and unseasonably strong jet stream was the likely source of trouble, providing a pathway for frequent storms, from the Midwest across the Great Lakes into Pennsylvania. A weather modification experiment in New Mexico involving a silver-iodide generator used for rainmaking studies in 1947 was also implicated in the unusually stormy pattern in the Midwest (Petterssen 2001).

July 1947 was an unusually wet and cool month in Pennsylvania. Measurable rain fell on 16 days, and the month tied July 1920 for the coolest seventh month in more than 50 years of state records.

The heaviest monthly rainfall total (19.81 inches) was measured at Park Place, atop a ridge in northern Schuylkill County (elevation 1,900 ft.), five miles northwest of Mahanoy City. The whopping total is the heaviest monthly rainfall ever recorded in July in Pennsylvania, and the second greatest for any month, after Mount Pocono (23.66 inches) in August 1955. Record July rainfall figures were reported at Freeland (15.32 inches), Erie (13.27 inches) and Hawley (12.94 inches).

A slow-moving front on July 7–8, 1947, stagnated over the mountains of northern Pennsylvania, creating an upslope pattern that forced moist Atlantic air to ascend a shallow layer of cooler air. Waves of thunderstorms contributed to the torrential rainfall.

At Freeland, 5.86 inches of rain came down in 24 hours. The maximum rainfall was measured at a weather station at Park Place, where 7.83 inches fell in 24 hours. Five-inch rain totals were common in southern Wayne County.

Only one week later, an anomalous summer low-pressure system brought heavy downpours to the Keystone State on July 16–17, 1947, totaling 4.29 inches at Hawley, 3.29 inches at Park Place and 3.10 inches at Reading. However, the greatest rainfall event in the stormy month was yet to come.

On July 21–22, 1947, a complex of thunderstorms dumped 6.32 inches of rain on Erie in three hours on the 21st, for a storm total of 10.42 inches in 24 hours. As much as 20 inches were estimated to have fallen in other parts of the city, though the airport received a little less than four inches.

High water forced the evacuation of 66 Erie residents, and a state of emergency was declared. Basement flooding and damage to homes and businesses, primarily south and east of the downtown area, was conservatively estimated at $250,000.

August 1955
Hurricane Floods Bring Devastation

Segments of northeastern Pennsylvania suffered through the most destructive flood in the region's history in August 1955. Every tributary of the Delaware River from Honesdale to Philadelphia exceeded flood stage, as trillions of gallons of water were unleashed.

Prior to the arrival of twin hurricanes in August, the spring and summer of 1955 represented the driest growing season in the state since 1909. July 1955 was the hottest month in recorded Pennsylvania history. Corn and hay crops withered in the fields, and parched vegetables were puny by midsummer.

Ironically, drenching rainfall unloaded by the remnants of Connie was considered highly beneficial at the time, despite pockets of flooding. Six to 10 inches of rain pelted eastern Pennsylvania, associated with the passage of Hurricane Connie's remnants, on August 10–14, 1955.

Kresgeville reported 7.29 inches in Monroe County's West End. Stroudsburg received 6.82 inches, with the heaviest downpours coming in a 24-hour period on August 12–13 (6.17 inches). In the northern area of the county, Mount Pocono recorded 9.84 inches on August 11–14, 1955.

In the southeastern areas, Chester and Montgomery counties reported major flooding. A rain gauge at Neshaminy Falls filled with 7.29 inches. In York County, 9.24 inches accumulated at New Park (6.50 inches in 24 hours).

Two persons died—one in Lancaster County and a second near Coatesville, "presumably when automobiles were caught in rain-swollen creeks," according to *Climatological Data—Pennsylvania*. A minister from Princeton, New Jersey, drove off a dark road into the turbulent waters of Tobyhanna Creek, in the Poconos. His body was recovered in the car several days later.

Flooding closed roads along the Perkiomen Creek in Bucks County, and homes were flooded near the French and Brandywine Creeks in Chester County. Industrial plants near Philadelphia along the Neshaminy Creek were also hit by high water. In the west, basement flooding was widespread in southern Allegheny County near Pittsburgh, and fruit trees in Erie County sustained thousands of dollars in damage from high winds.

Rivers stayed within their banks because the streamflow was low during a prolonged summer drought, despite a non-tropical rainfall on August 7–8, 1955 (Schwartz 2007). Meanwhile, forecasters were warily eyeing a strengthening Hurricane Diane, which would make landfall near Wilmington, North Carolina, on August 17 as a minimal hurricane.

The northerly trajectory took Diane inland across Virginia and Maryland. Although some thought the storm threat had essentially ended by 11:00 p.m., writing off Diane as a "dying" storm, flood alerts were posted for inland areas.

On the night of August 17–18, 1955, a weakening Diane still had plenty of moisture to unload. Ten inches of rain soaked the Blue Ridge Mountains of Virginia, where upslope motions contributed to the torrential downpours. A trough of low pressure lingering over the eastern foothills created a confluent boundary that added lift.

Matters were made worse on August 18 because the circulation of Diane took an unexpected right turn across the southeastern corner of Pennsylvania during the early evening. Excessive rains caused tremendous runoff that cascaded off sodden hillsides into the watersheds of Brodhead and Paradise Creeks in the Poconos and the Lackawanna River near Scranton.

Fig. 6.12 August 19, 1955. The swollen Delaware River, after a floating house split the Northampton Free Bridge connecting Easton, Pennsylvania and Phillipsburg, New Jersey. At one point water covered the bridge deck. (Courtesy Easton Express-Times)

Fig. 6.13 August 19, 1955. Widespread flooding in downtown Easton, with the "World's Fastest Car Wash" living up to its advertisement. The Delaware River rose to a record height of 43.7 feet. Flood stage at Easton is 22 feet. (Courtesy Easton Express-Times)

Fig. 6.14 August 19, 1955. Flooding in Easton at Northampton and Front streets, where the water ignored the one-way sign and surrounded the city. (Courtesy Easton Express-Times)

The heaviest rainfalls occurred in southeastern Lackawanna, southwestern Wayne and in northern Monroe counties, funneling huge volumes of water into the mountain tributaries that flow through the scenic highland communities.

A prolific cloudburst unloaded eight inches of rain from 2:00 to 6:00 p.m. on Tamaqua, where one person died in the floodwaters (Shank 1988). The rainfall at Hazleton totaled 9.31 inches in 21 hours. The heaviest rain accumulation in northeastern Pennsylvania on August 17–19 totaled 11.11 inches at Pecks Pond.

In the Poconos, the airport weather station at Mount Pocono recorded 10.75 inches (10.63 inches in 24 hours). The former Weather Bureau observer Harry Greene measured 13.24 inches in the borough of Mount Pocono, according to the Stroudsburg *Daily Record*. Fifteen miles to the southeast, Stroudsburg received 6.15 inches in 24 hours, a figure nearly identical to Connie's deluge a week earlier.

In Paradise Valley, debris that accumulated in an arch culvert 50 feet below the railroad tracks at Devil's Hole Creek caused the embankment to collapse on the fateful evening of August 18, 1955. The event sent a wall of water and hundreds of tons of rock and dirt hurtling downstream through the narrow channel of Paradise Creek.

The failure of two culverts led to washouts on the Delaware, Lackawanna and Western Railroad. At one site, the track was "suspended 30 to 40 feet in the air" (*Daily Record*). A great tragedy was averted only because Train No. 5 came to a halt after passing the Cresco station around 7:00 p.m. The conductor left the train to assess the tracks ahead. A state fire inspector surveying the dire situation in his Jeep flagged down the engineer to warn him of the slide, who hurried back to the train and safely backed up the cars to the Cresco station.

Helicopters were called in to airlift the 235 railroad passengers safely to a nearby resort, a relatively new form of rescue at the time (Schwartz 2007). The train was removed several weeks later after track repairs were made.

In a *Pocono Record* flood history 50 years later on August 22, 2005, retired East Stroudsburg Fire Company chief Marvin Abel (1912–2010) recalled that the collapse of a hydroelectric power dam from the force of water filled with floating trees led to the subsequent failure of dozens of dams downstream, triggering an unfolding disaster.

The bucolic Paradise Creek, a renowned trout stream, where leading conservationists such as Theodore Roosevelt and Gifford Pinchot stayed at the Henryville House nearby and fished in the pristine mountain air, swelled from the cascade of water unleashed by the Devil's Hole Creek.

The unbridled fury of the rising wall of water caused more than 40 small dams across feeder streams to crumble like dominoes, powering an expanding volume of water coursing through the hillsides, isolating Henryville and swallowing up structures, which choked the creek with debris.

The surge of water obliterated every impoundment, aided by a drop in elevation of 1,200 feet on the escarpment. Dozens of small bridges, some privately maintained, were torn away, adding more debris to the deadly torrent. Large pine trees were uprooted and transported downstream with sundered small structures. The normally shallow, picturesque Brodhead Creek swelled 25 feet in 15 minutes, according to the *Daily Record*, raising the water to a devastating height of 30 feet that swept past Analomink.

An anxious group of 46 visitors, mostly women and children, who were staying at a religious retreat waited out the deluge in 14 small cottages that were known as Davis Cabins. As the situation became increasingly dire, three residents departed in the dank darkness, traipsing across a bridge to the Pinebrook Bible Conference retreat center that sat on higher ground (Shafer 2010).

The remaining 43 guests retreated to the upper floor of a two-story wooden-frame structure called the Big House. The surging waters had surrounded the building and created a horrifying clamor. Fleeing into the attic, the group huddled and prayed, singing songs for comfort. Disaster struck when the Clubhouse was swept off its foundation. People living downstream would never forget the terrified sounds of screams and cries for help over the sound of rushing waters.

Only six of the 43 residents survived the horror, clinging to tree limbs and debris (Shafer 2010). A special *Pocono Record* section, *Remembering the Flood of 1955*, published in 2014, listed the names of 38 people who died near the religious camp that included a fatality at Analomink. Eight vacationers from Philadelphia at Price's Cottages perished as the swirling waters upstream isolated Canadensis.

Residents of the twin boroughs of Stroudsburg and East Stroudsburg were largely unaware that a deadly wall of water was imminent until it was too late. The quiet creek that separated the two communities, normally about five feet deep, turned into a violent maelstrom filled with all manner of debris.

Low-lying areas were quickly inundated before many residents could escape.

A group of 100 women playing bingo in the Day Street Fire Hall were evacuated as rising water cascaded through the building. Despite the work of valiant rescuers, including local firefighters in boats, six people died in the angry floodwaters. Four bodies were recovered inside the decimated building.

Fig. 6.15 August 19, 1955. The rampaging Brodhead Creek floodwaters tore through the twin boroughs of Stroudsburg and East Stroudsburg, taking scores of lives. (Courtesy Monroe County Historical Association)

Fig. 6.16 Bridge at Canadensis into Brodhead Creek in the Poconos. (Courtesy Monroe County Historical Association)

Devastating flooding occurred near the junction of Brodhead Creek and McMichael Creek in South Stroudsburg. Homes along Brodhead Creek in the "Flats" section of East Stroudsburg were torn from their foundations and tossed into the abject blackness, bobbing in the churning water like bathtub toys.

The Maplehurst and Day Street Flats, in the lower section of East Stroudsburg, were submerged beneath as much as 10 feet of water when the creek expanded to 1,000 feet in width, extending into lower-lying areas on the Stroudsburg side of the creek. In all, 30 lives were lost in the Stroudsburg area; all but two drowned on the East Stroudsburg side. The State Bridge built in 1904, which connected the two boroughs, was destroyed. An emergency structure was completed two days later. In the meantime, helicopters flew in essential medical supplies, food and water to trapped residents who were isolated.

Farther downstream, where the Brodhead Creek emptied into the rising Delaware River, a catastrophe was averted that was vividly recounted in Shank's *Great Floods of Pennsylvania*:

> A miracle saved a number of campers and cottagers on Shawnee Island in the Delaware, who had watched the Delaware rise all day on August 18, until by ten o'clock that night it was too turbulent to escape to the mainland. Suddenly the rain stopped and the waters smoothed. The campers could hardly believe their eyes—the Delaware had reversed its direction and was placidly running *upstream*!

South of Shawnee Island, where the Brodhead surged into a narrow strait of the Delaware to the east, a "huge water barrier" had halted and reversed the flow of the main stem of the river long enough for island residents to escape by boat. In a scene reminiscent of the biblical parting of the Red Sea, the water resumed its destructive course just as the refugees made their way to safety.

A weather site two miles northwest of Paupack received 9.07 inches of rain in 24 hours. The East Branch Wallenpaupack Creek was met by Bridge Creek, before pouring into the main stem south of Greentown, where ten people died in homes and farmhouses that washed away. The rising waters cut off Newfoundland, and one death occurred at South Sterling. An engorged Wallenpaupack Creek would meet Middle Creek at Hawley, claiming two more lives in the dank night.

The reservoirs prevented a greater catastrophe by retaining enough water to slow the onslaught. On August 20, 1955, the Wallenpaupack Dam was overtopped, but rumors of its imminent collapse proved to be unfounded. The following year, the Brodhead Creek was channelized. In 1959–1960, the Dyberry Dam, 2.9 miles upstream from where the creek joins the Lackawaxen, and the Prompton Dam were completed by the U.S. Army Corps of Engineers, to protect Hondesdale and Hawley.

Serious flooding affected communities along the Lackawanna River. The river gauge down at Old Forge, five miles south of Scranton, reached 20.05 feet early on August 19, 1955, overflowing in portions of Wayne, Lackawanna and Luzerne counties.

Scranton and Dunmore suffered historic flooding that developed after a deluge of 4.58 inches in 24 hours. Bridges were destroyed in the Flats portion of South Scranton, where homes were awash in mud-soaked water, some torn from their foundations and carried 100 feet downstream, as Roaring Brook lived up to its name.

Thirty-two homes were lost around Richter Avenue in Scranton, and 18 were destroyed in the Little England section, where two persons died. Homes were damaged by flooding upstream at Dunmore. Floodwaters took out the Pennsylvania Water Company's distribution system. Scranton's water supply was lost for a day (Schwartz 2007). The mayor declared a state of emergency that would continue for five days, shuttering most businesses and industries.

The swollen Delaware River was choked with debris from raging tributaries toppled the historic Portland-Columbia covered bridge. Construction of the 725-foot-long wooden structure had commenced in 1831 and was completed in 1869. The longest covered bridge in the nation withstood the Pumpkin Freshet of 1903.

On August 19, 1955, the Delaware River rose three feet higher than ever known at the nearby Riegelsville, New Jersey, Warren County gauge. The force collapsed the four-span bridge, when three-quarters of the last covered bridge across the Delaware vanished (Ludlum 1983). Parts of the relic from another era rammed into the 60-year-old Phillipsburg-Easton free bridge, taking out the middle section, which plunged into the roiling waters. Along the Delaware, four other bridges were severely damaged or wiped out. The road over the Washington Crossing Bridge was flooded, and rising waters destroyed the Ewing Bridge near Philadelphia.

A record crest occurred on the Lehigh River at Bethlehem of 25.9 feet at Bethlehem, 10 feet above flood stage. At Allentown, the river hit a record 23.4 feet, exceeding May 1942 (21.7 feet). Homes and businesses were evacuated farther east, as the river joined the Delaware at Easton, causing heavy damage in the city's worst flood.

An unofficial water measurement put the crest at 43.7 feet, more than five feet above the previous October 1903 record (38.3 feet) at Easton. Fifteen industrial plants were destroyed in the flooding around Allentown, leaving 15,000 workers temporarily unemployed (Schwartz 2007).

The Wabash Creek and Little Schuylkill River, which empty into the main branch of the Schuylkill River, swamped low-lying sections of Tamaqua, where one person died. In Philadelphia, floodwaters poured down Broad Street, where cars floated downstream and homes were inundated. In New Hope, the Bucks County Playhouse took on lots of water. High water approached the tops of telephone poles at New Hope and Yardley, and hundreds were left homeless in Upper Black Eddy.

The summary in *Climatological Data—Pennsylvania* reported that the Hurricane Floods of 1955 affected 6,600 square miles, deemed "the most disastrous flood ever recorded in eastern Pennsylvania." Navy and Marine helicopters rescued children trapped at small camps. Government and volunteer pilots airlifted hundreds of trapped victims to safety, including youngsters at Pocono area camps. The Civil Air Patrol handled food drops.

There were many heroic tales of residents who risked their lives ferrying residents to safety using boats (Shafer 2010). The Army managed the task of distributing relief supplies and helped recover the bodies of victims. The American Red Cross joined other agencies to set up shelters at area churches, schools and fire stations.

Washed-out rail lines halted transportation for a couple of months along the Delaware, Lackawanna and Western Railroad Company. The repairs proved too costly, forcing the railroad to merge with Erie Railroad, ultimately leading to the creation of the Erie Lackawanna Railway five years later in the wake of ruinous losses totaling $16 million to track, rail cars and equipment. About 150 rail and road bridges were washed away, and 30 dams were damaged or destroyed by the raging waters. Damage statewide was put at more than $70 million.

The circulation center of Hurricane Diane passed over central sections of the Garden State before moving back out over water. The peak discharge along the Delaware River at Riegelsville, Warren County, was measured at 340,000 cubic feet per second, about 1.2 times greater than that of the Great Flood of 1903 (Ludlum 1983). In New Jersey, six deaths occurred in flood-related circumstances.

More than a foot of rain fell on parts of Connecticut, and nearly 19 inches swamped the Berkshires in Massachusetts, triggering historic floods in New England that killed 91 persons. President Eisenhower declared Pennsylvania and seven other states disaster areas.

The loss of life attributed to Diane was put at 191 people, but likely topped 200, including 101 victims in Pennsylvania. The total damage at the time, counting indirect economic losses, was about $1 billion—the costliest hurricane in United States history up to that time.

Not surprisingly, August 1955 was the wettest August in Pennsylvania, averaging 8.42 inches, surpassed only by May 1894 (8.88 inches) and June 1972 (11.23 inches). August 1955 was the wettest month on record at Mount Pocono (23.66 inches).

Civil Defense in Monroe County counted 78 deaths, which included two persons who died in accidents during rescue and recovery efforts. The remains of several victims were eventually discovered in 1959, 1960 and 1969.

The bodies of two youngsters from New York and New Jersey who perished in the high waters at Davis Cabins a few miles north of Stroudsburg were never found. David Pierce (2015) noted, "At least 62 dwellings in East Stroudsburg" were lost and 115 borough houses reported heavy damage, with another 190 receiving minor impacts.

In Monroe County, 59 bridges were destroyed in the flood, including a total of 42 highway and 17 rail bridges (Dale 1996). Virtually every bridge washed away or had damage on Route 209. The costs were estimated at $28.5 million across the eastern part of Pennsylvania.

After the devastating floods that struck Pocono communities in 1955, levees were installed in low-lying areas of Stroudsburg and East Stroudsburg along Brodhead Creek. The flood control project resulted in the construction of two new dams in northeastern Monroe County, in addition to dams finished in 1960 across the Lackawanna River and Dyberry Creek.

June 22–25, 1972
Agnes Delivers Devastating Floods

The costliest weather-related disaster in Pennsylvania history was wrought by an early-season tropical system moving up the Eastern Seaboard in June 1972.

Tropical Storm Agnes brought heavy rain and historic flooding that claimed 50 lives in central Pennsylvania and caused $2.12 billion in damage. The combination of flooding and fires damaged about 68,000 homes and 3,000 businesses. Nationally, the flood toll reached $3.1 billion in 12 states, taking 122 lives.

The minimal early-season Gulf hurricane was born east of the Yucatan Peninsula on June 16, 1972. Hurricane Agnes lashed Cuba with high winds and torrential rain before crashing into the Florida Panhandle on June 19, 1972, near Valparaiso.

On Monday, June 20, 1972, the first plume of moisture reached Pennsylvania, accompanied by heavy rain. The storm weakened as it tracked north over Georgia and the Carolinas, but regained energy over the Virginia Capes. An upper-level trough over the Eastern states worked in tandem with high pressure off the coast of New England to slow the forward progress of Agnes, which turned inland on June 21–22 near New York City.

Over the next five days, the remnants of Agnes looped across the highlands of north-central Pennsylvania and south-central New York. A prolonged heavy rain event soon overwhelmed the carrying capacity of rivers and streams, from southern New York to Virginia, pushing waters to all-time high levels that caused widespread flooding.

Pennsylvania Governor Milton Shapp declared a State of Emergency early on Thursday, June 22, 1972. The Susquehanna River continued to rise, threatening low-lying areas around the state capital. Locally, six to 12 inches of rain fell in 24 hours in the Harrisburg area on June 21–22. A gauge at York collected 13.50 inches.

A watershed research site operated by the U.S. Department of Agriculture (USDA) in extreme western Schuylkill County recorded 14.80 inches in 24 hours near Mahantango Creek (Bailey and Patterson 1975). The rain gauge recorded 19.50 inches during the passage of Agnes, mentioned in the monthly NOAA *Storm Report*.

Flooding along the main stem of the Susquehanna River was historic, the greatest at least since the 1784 spring flood. In an area from Harrisburg to Chesapeake Bay, peak flows exceeded 7.5 million gallons per second (one million cubic feet per second).

The daily rainfall totals at Harrisburg on June 21 (5.81 inches) and 22 (9.13 inches) resulted in a record 24-hour catch (12.55 inches), yielding a storm total of 15.26 inches through the 26th. The Schuylkill River at Reading reached a record stage of 31.3 feet on June 22, far above the flood stage of 15.5 feet.

The raging floodwaters cascaded over existing embankments, forcing nearly 250,000 Pennsylvanians to flee their homes, three to six feet higher than in March 1936. Early on June 23, high water rampaged through the state capital of Harrisburg in the Riverside section, covering the north side up to four blocks from the river (Shank 1988). Water surrounded the governor's mansion on North Front Street, pouring into the first floor. Nearly half of the city of Harrisburg was under water.

During Saturday morning, June 24, 1972, the Susquehanna River would eventually crest at 32.57 feet, some 3.34 feet higher than the March 1936 record flood stage. The stream flow at Harrisburg during peak discharge was a little over a million cubic feet. The river crested a mere foot below flood stage.

Dams prevented more extensive flooding in northern Pennsylvania, although Shank (1988) described some close calls: "The City of Williamsport owes its escape from the 1972 Flood completely to the work of the Corps of Engineers. The 1955 flood levees, plus the three major upstream dams ... effectively kept all major flood water out of the city." The flood stage reached 34.75 feet along the West Branch Susquehanna River at Williamsport, exceeding the 1936 mark by 1.2 feet. Farther downstream, Muncy, Milton and Lewisburg were hard-hit hard by raging floodwaters.

At the convergence of the North and West Branches, Sunbury, which had been the scene of major flooding in March 1936, suffered severely again in June 1972. However, a system of floodwalls, levees and five pumping stations erected by the Army Corps of Engineers, completed in 1948, saved Sunbury from a calamity.

In northeastern Pennsylvania, flooding was widely anticipated in the Wyoming Valley, but the scope and depth far exceeded what anyone could have imagined based on past experience. As early as 6:00 a.m. on Friday, June 23, 1972, Civil Defense reported that the river level had already exceeded the previous high-water mark of 33.07 feet attained on March 20, 1936 (Shank 1988).

Wyoming Valley residents listened to a steady stream of information on radio and television, warning of imminent flooding and urging them to go to local disaster shelters. Evacuations began around Wilkes-Barre and Kingston as residents hoped dikes constructed after the March 1936 floods would withstand water levels of 37 feet.

By mid-morning on June 23, 1972, forecasters realized rising water levels could reach 40 feet, requiring the evacuation of Wilkes-Barre residents. Ten thousand volunteers applied sandbags to weaker links in the dikes and openings in the bridges, in a desperate attempt to stave off the inevitable.

The bad news came at 1:12 p.m., as the first dike to give way occurred at Forty Fort near a cemetery. The National Guard was charged with the gruesome task of retrieving the bodies and caskets afterward. A succession of dike failures prompted an endless string of dire warnings for residents downstream along the Susquehanna River.

At 2:37 p.m., the large Wyoming Dike failed at the Midway Shopping Center. With catastrophic flooding imminent, 100,000 residents of Wilkes-Barre, Kingston and surrounding communities (50 towns in all) were forced to move to higher ground, congregating in designated public buildings and schools.

Around 7:00 p.m. on June 24, 1972, the river finally crested at Wilkes-Barre at a height of 40.91 feet—18.9 feet above flood stage and nearly eight feet above the previous high-water mark in March 1936. The water had topped the levees by four feet, ruining more than 25,000 homes and businesses and causing about $1 billion damage.

Another 2,278 homes were damaged by water nine feet deep in downtown Wilkes-Barre. In Kingston, nearly every home suffered water damage. Many dwellings were deemed uninhabitable after residents returned to survey the destruction days later.

Floodwaters along the North Branch Susquehanna River brought record high-water marks at Sunbury, Bloomsburg and Danville. The water at Sunbury reached 12 feet above flood stage at noon on Saturday, June 24, a foot higher than in March 1936.

Flooding was widespread on June 22, 1972, even before the Susquehanna swallowed up the central section of Danville under five feet of water (Shank 1988). The river crested at 32.32 feet at 7:00 p.m. on June 24, exceeding the record of 30.71 feet on March 9, 1904, and more than four feet above the March 1936 level.

Record flooding also hit Reading, where the Schuylkill River rose 8.5 feet above flood stage (31.5 feet), easily topping the previous historic height of 23.3 feet (September 2, 1850). Flood damage was extensive at Pottstown and Norristown. Three persons died in the flooding in the Philadelphia area.

Fig. 6.17 June 25, 1972. The Susquehanna River roared through York, after the passage of Tropical Storm Agnes. (Courtesy Pennsylvania State Archives, photograph from the York Daily Record)

Heavy flooding was also reported along the Codorus Creek at York following a round of torrential tropical downpours on June 22, 1972. An average storm total of 16 inches fell over the watershed, pushing the normally placid creek to a height of 26.4 feet in the late afternoon—2.4 feet above the 1933 record crest—"cutting the city in two" (Shank 1988). Nonetheless, the Indian Rock Dam prevented a greater disaster by holding back nine billion gallons of water.

Fig. 6.18 June 25, 1972. Downtown York swamped by the overflowing Susquehanna River, following several days of torrential rain totaling more than 12 inches. (Courtesy Pennsylvania State Archives)

Fig. 6.19 June 25, 1972. The Susquehanna River fanned out more than a mile wide near York. (Courtesy Pennsylvania State Archives)

In western Pennsylvania, Pittsburgh suffered $45 million of flood damage after the Ohio River crested 11 feet above flood stage (35.85 feet), though still nearly 10 feet below the record 1936 stage. Dams built by the Corps of Engineers on upstream tributaries saved the city of Johnstown from a repeat of previous disasters, which consisted of the enlargement of portions of the Little Conemaugh River and the construction of a reservoir south of the city.

The largest monthly rainfall total in Pennsylvania in June 1972 from an official reporting station occurred at York Pump Station 3 SSW (20.45 inches). Harrisburg received a record 18.55 inches of rain. The previous state June rainfall record was measured at Greensburg (13.74 inches) in June 1951.

On June 30, 1972, Pennsylvania was declared a federal disaster area, and eligible for assistance. The death toll mounted as floodwaters receded and more bodies were discovered. An estimated 220,000 residents in the region afflicted by high waters were forced to leave their homes as floodwaters ravaged scores of communities. Some would never be able to return to their demolished homes.

The loss of life in flooding caused by Agnes was greatest in Pennsylvania (50) and New York (25). Pennsylvania suffered the greatest economic losses, which would be about $14 billion in today's dollars. The death and property toll were heavy in northern Virginia, where at least 10 inches of rain caused $222 million of damage and floods took 13 lives.

September 25–26, 1975
Eloise Swamps Susquehanna Region

Merely three years and three months after being inundated by floodwaters unleashed by Agnes, the Susquehanna Valley was struck by another deluge, as the remnants of Hurricane Eloise headed north from the southern Appalachians through northern Virginia, crossing Philadelphia and New York City.

Eloise made landfall on September 23, 1975, near Fort Walton Beach, Florida, only 10 miles to the west of where Agnes came ashore in June 1972. Eloise took a different path, tracking west of the Appalachians and losing some moisture before reaching Pennsylvania. Unfortunately, a stationary front over the Mid-Atlantic region provided a substantial lifting mechanism in addition to the rising terrain, forcing warm moist tropical air to climb over cooler air.

As in June 1972, a blocking upper-air pattern over Pennsylvania greatly enhanced rainfall totals. At Harrisburg, 7.22 inches of rain came down in 24 hours on September 25–26, 1975, setting a September rainfall record of 14.97 inches. Extensive flooding developed along the Susquehanna River and many of its tributaries in central and northeastern Pennsylvania on September 26–27, 1975, forcing nearly 20,000 persons to evacuate their homes. In north-central Pennsylvania, floodwaters in Tioga County at Westfield and Lawrenceville exceeded June 1972 levels, after six to 12 inches of rain fell in the Susquehanna Basin

The earlier addition of the Raystown Dam northwest of Harrisburg controlled flooding along the Juniata River, one of the major tributaries of the Susquehanna River. The Susquehanna crested an average of 10 feet above flood stage along the West Branch at Milton, Lewisburg and Muncy downstream to Harrisburg. Five people lost their lives directly caused by flooding in Pennsylvania.

An area along the Susquehanna River, from about Williamsport to Sunbury extending south to the Wyoming Valley was protected from a disaster by levees and dikes that were built after the Agnes flood. Nonetheless, precautionary evacuations continued until the water began to recede.

July 19–20, 1977
Another Johnstown Flood Tragedy

The city of Johnstown has a long history of devastating floods, most notably the Great Johnstown Flood of May 1889, which took an estimated 2,200–3,000 lives, and the St. Patrick's Day Flood of March 1936.

History would be repeated when a small cluster of thunderstorms regenerated over the higher elevations west of Johnstown on the night of July 19, 1977. The stalled complex brought heavy rain, commencing around 8:30 p.m., that continued nearly non-stop until around 4:30 a.m. on the 20th. Up to a foot of rain fell in West Taylor Township (11.82 inches), northwest of Johnstown.

Torrents of water poured off the deep hillsides of Cambria County into the Little Conemaugh River, which meets the Conemaugh River just above Johnstown, filling the tributaries of Solomon's Run, Sam's Run and Peggy's Run, as immense pressure built up on the Laurel Run Dam.

The eventual dam break sent a wall of water 10 feet high rushing through the valley, wiping out five more dams along the way. More than 128 million gallons of water poured through the Conemaugh Valley into Johnstown with little warning, sweeping away unsuspecting residents in their homes and cars (Lorditch 2007). The ensuing disaster took at least 84 lives and resulted in damages totaling $240 million at the time.

The 1937 Johnstown Local Flood Protection Project (LFPP) grew out of the Flood Control Act of 1936, enacted by Congress to protect the region from a repeat of the 1936 flooding devastation. Re-engineered channels, and the addition of two floodwalls along an 8.8-mile stretch of the Congemaugh, Little Conemaugh and Stonycreek Rivers, were completed in 1943 by the Army Corps of Engineers.

Those flood controls in the Little Conemaugh Valley averted a disaster in June 1972 but were unable to handle the volume of rain that fell on the night of July 19–20, 1977, according to the U.S. Army Corps of Engineers report. In the 1990s, major repairs of the LFPP were undertaken at a cost of $40 million, and newer projects shored up the floodwall along the lower Conemaugh River (Lorditch 2007).

November 4–5, 1985
Election Day Flood

After Hurricane Juan made landfall near Pensacola, Florida, on October 31, 1985, a piece of the system drifted north to the Smoky Mountains, pushing rain north to Pennsylvania. A new low-pressure center formed in the Gulf a few days later and moved along the old frontal boundary draped east of the Appalachians to the Mid-Atlantic region.

Tropical moisture was already flowing along the Eastern Seaboard, even before the storm stalled for a time west of the nation's capital on November 5, 1985. Four days of heavy rain drenched southwestern Pennsylvania and West Virginia from November 2–6, averaging four to nine inches.

At the junction of the Cheat and Monongahela Rivers, which flow north from West Virginia into southwestern Pennsylvania, floodwaters crested 14 feet above flood stage at Point Marion in southern Fayette County. The flood gates on Lake Lynn in West Virginia were opened on November 4, sending water pouring into Port Marion that reached eight feet in depth.

The worst flooding in Pennsylvania and West Virginia occurred on Election Day, November 5, 1985. The Monongahela River crested at 42.7 feet at the Maxwell Lock and Dam in Brownsville, 10.7 feet above flood stage—the highest stage in 50 years. Thirty-five businesses were flooded in Brownsville. Making things worse, 23 barges broke from their moorings and became entrapped in the Maxwell Lock and Dam. This key link to local shipping would be out of commission for six weeks. The local economy suffered millions of dollars in production losses.

Thirty-four homes were destroyed and 355 suffered damage, which totaled $15–20 million in Fayette County (*Storm Data*). In Greene County, 300 people were evacuated at Greensboro, where the Monongahela rose 18 feet above flood stage. A total of 178 homes were ruined and 915 had minor damage, costing $6.15 million.

In Washington County, the river exceeded the 1967 record flood stage by six feet, with a peak discharge of 220,000 cubic feet at Charleroi as the Monongahela topped out at 44 feet. Forty barges broke free, 2,600 people were evacuated, and one person drowned in North Charleroi. California University reported $1 million in damages.

In southwestern Pennsylvania, 3,349 homes were damaged and 2,000 became homeless. Damage statewide totaled $83 million. Governor Dick Thornburgh called up 600 National Guardsmen to assist with the massive clean-up, and thousands of residents had no access to clean water.

In the four-state region of Virginia, West Virginia, Maryland and Pennsylvania, which were declared disaster areas, resources poured in to aid flood victims. Nationally, the storm damage totaled $1.4 billion in 1985 dollars. Sixty-two persons died, including 38 in West Virginia—the state's second greatest flood disaster. (The Buffalo Creek flood killed 125 on February 26, 1972).

November 1985 brought numerous rain events in Pennsylvania—the wettest November in state history since 1895. The greatest total precipitation (13.93 inches) was reported at Chalk Hall, establishing a state record for the month. Pittsburgh also set a November mark (11.05 inches).

The destructive flooding around Election Day November 1985 prompted Congress to move on public safety concerns by funding the expansion of IFLOWS in Pennsylvania, West Virginia and Virginia, along with parts of North Carolina and New York that shared a history of major flooding. By the mid-1990s, the advent of Doppler weather radar improved numerical forecast model output, centralized communications, and expedited flood warnings.

January 1996
Great Snowmelt Floods

Record snow depths of two to five feet during the first two weeks of January 1996 would soon turn into a secondary disaster after the middle of the month. A series of heavy rainstorms combined with sudden rapid snowmelt, and ice jams sent huge volumes of water pouring into Pennsylvania waterways. Significant flooding occurred in 57 of the state's 67 counties, and 19 residents perished.

Record flooding occurred along several tributaries of the Susquehanna River, particularly the West Branch on January 19–21, 1996. The main stem at Harrisburg was especially high and powerful, causing the Walnut Street Bridge to collapse. At Safe Harbor, the waterpower facility sustained major damage as did the hydroelectric plant in Holtwood.

The large water content of the snow and rapid thawing set the stage for catastrophic flooding. The first batch of heavy rain was accompanied by a cold front on January 18–19, 1996. A total of 3.03 inches of rain fell on Williamsport in 24 hours on the 19th, setting a January record. The wind gusted to 67 mph in a thunderstorm at Philadelphia.

The combination of melting snow and heavy rain falling into a dense snowpack set the stage for massive runoff into Pennsylvania creeks and streams that would result in near-record flooding in many areas. Rivers packed with chunks of ice backed up until the temporary blockage collapsed, cutting loose torrents of water.

The Loyalsock and Lycoming watersheds in north-central Pennsylvania were severely affected by flooding. A tragedy occurred when an ice jam broke loose along the Lycoming Creek, unleashing a wall of water that roared through several small communities and killed six residents. Hundreds were forced to flee their homes as water rose suddenly downstream.

At Wilkes-Barre, the Susquehanna River rose to an elevation of 34.4 feet (12.4 feet above flood stage). The river crested 7.7 feet above flood stage at Harrisburg, causing heavy damage. Ice floes along the Delaware River dammed up a large volume of water, inundating riverfront property at Shawnee and Price's Landing on January 19.

One to two inches of rain pelted the frozen surface and promptly drained into streams and creeks, inundating roads and highways. Heavy damage was done to the 90-year-old Shawnee Inn and Country Club and along the Delaware Water Gap National Recreation Area. Flood damage in eastern Pennsylvania was estimated at $50 million and totaled $25 million in Monroe County alone. Widespread flooding caused damage in the millions of dollars along the Delaware River, and smaller creeks throughout Monroe, Northampton, Lehigh, Berks, Bucks, Chester and Delaware counties.

Flooding along the Perkiomen Creek in Montgomery County resulted in two deaths, and another person died in a storm-related incident. As many as 5,000 residents were evacuated in Bucks County, mainly from Riegelsville to Yardley along the Delaware River, which reached its highest level since August 1955.

In western Pennsylvania, the moisture-loaded weather system triggered heavy rain totaling two to four inches in 24 hours, resulting in the worst flooding downstream from Pittsburgh since Agnes in June 1972. The Allegheny and Monongahela Rivers crested within five hours of each other. The Allegheny River peaked at 10 feet above flood stage on January 20, 1996, at a level of 34.6 feet, only slightly lower than the high-water mark recorded in June 1972 (35.8 feet). Downtown Pittsburgh and Point State Park were severely flooded.

Heavy rain returned on January 27–28, 1996, pushing rivers in the east to flood levels again. One to two inches of rain accompanied a cold front on the second weekend in a row, taking another life in Bucks County along the Delaware River. Water rescues were required after drivers attempted to go around barricades and became trapped in rising waters.

The streamflow of the Susquehanna River at Harrisburg in January 1996 reached 570,000 cfs, though this was not as high as the rate of 1 million cfs recorded in June 1972 during the Agnes floods. Total financial losses in Pennsylvania incurred during January 1996 from the blizzard and floods were estimated at $1 billion.

Floodwaters damaged 52,000 homes, and 11,000 suffered irreparable damage, resulting in 200,000 evacuations. Storm surveys reported that 2,000 businesses were damaged in the floods, plus 1,500 bridges and 78 parks. The cost of the water damage that impacted state highways and bridges totaled nearly $500 million.

The 1996 calendar year brought multiple flood events. Four suburban Philadelphia counties were hard-hit on June 10, 1996, and 10 western counties suffered flash flooding in July. Parts of central Pennsylvania dealt with major flooding in the wake of ex-hurricane Fran in September 1996.

More than $1 billion in flood-related damage occurred in the Keystone State in 1996 (Martin 1997), during which time there were a record five disaster declarations, summarized in the *Pennsylvania Almanac*: "Approximately 42,000 individuals and businesses received more than $355 million in federal and state aid."

August 20, 1999
Bradford Flood

A stalled upper-air storm and disturbances pivoting around low pressure instigated pockets of rain and thunderstorms that culminated in serious flooding in Bradford, McKean County, on August 20, 1999, the worst the area had seen since 1947.

A burst of heavy rain, mainly between 2:00 a.m. and 5:00 a.m., dropped 3.42 inches of rain at the Bradford Central Fire Station. A second round of slow-moving storms dumped torrential rain in the city of Bradford and areas around the region between 1:00 and 4:00 p.m. that added another 5.56 inches in the gauge at the fire station, for a storm total of 8.98 inches. Streets turned into rivers and water poured off the hillsides, with nowhere else to go.

The flooding situation became serious in the city of Bradford by 3:00 p.m. on August 20. At least 10 water rescues were required to pull trapped residents and motorists from dangerous situations, though only one injury was reported. As many as 211 homes and 53 businesses incurred water damage. Twenty-two homes reported major losses, and eight were destroyed. The damage toll in Bradford was put at $25 million.

Other areas hard-hit by heavy rain that caused localized flooding included Richfield, in the eastern part of Juniata County, where an estimated eight to nine inches of rain resulted in damage totaling $500,000. Thirty homes and four businesses had losses from high water. Flooding was reported in Union County at Lewisburg, with reported losses of $100.000.

September 17–18, 2004
Ivan Swamps Pennsylvania

The passage of Hurricane Ivan's remnants overnight on September 17–18, 2004, resulted in unexpected widespread flooding that afflicted 60 of the state's 67 counties (Schwartz 2007), after twice as much rain fell as had been predicted. Damages from the heavy flooding around the state were estimated at $264 million.

Low pressure traversed the central Appalachians and was centered in Virginia on September 18, 2004, having triggered incessant downpours for about 10 hours in Pennsylvania. Four to 10 inches of rain pelted the heart of the state, and unofficial totals were as high as 13 inches.

Millvale, Heidelberg and Etna were especially hard-hit in the Pittsburgh region, where streams overflowed, filling basements and the lower floor of homes that were left caked with mud after the water receded. Chartiers Creek poured into homes in Carnegie, where waters were more than four feet deep.

The Susquehanna River at Bloomsburg crested six to seven feet above flood stage from Bloomsburg to Harrisburg. The river at Wilkes-Barre peaked at 34.96 feet, far above the flood stage of 22 feet—the fourth highest on record. Flooding was extensive in the Scranton area, where the overflowing Lackawanna River swamped 150 homes.

Along the Delaware River, water levels were the highest witnessed since August 1955, reaching 33.45 feet at Easton, where the flood stage is 22 feet. Shawnee was engulfed in floodwaters that spread across Routes 611 and 209 in the Delaware Water Gap National Recreation Area.

Fig. 6.20 Widespread flooding swamps Shawnee Golf Course on September 18, 2004, in the wake of the remnants of Hurricane Ivan. (Photo by David Kidwell, Pocono Record*)*

The *Centre Daily Times* in State College covered water rescue at Milesburg, Philipsburg and Coburn following downpours of four to eight inches. High winds combined to leave 1,500 homes and businesses without electricity. Route 322 was closed by the Pennsylvania Department of Transportation (PennDOT) around Lewistown Narrows, and flooding closed eastbound lanes of Interstate 80 between Milesburg and Bellefonte. Flooding was severe along Bald Eagle and Penns Creeks.

The heaviest storm totals were observed in the southern and eastern counties on September 16–18, 2004: Hershey 7.90 inches; Saxton 7.36 inches; Williamsport 6.60 inches; Scranton 6.03 inches; Altoona 6.01 inches; Pittsburgh 5.95 inches (24-hour record); State College 5.17 inches.

Additional rainfall from the remnants of Jeanne on September 27–28, 2004, helped set a state September average rainfall record (8.87 inches) at the time, surpassing 1975 (7.95 inches), though the mark would be toppled in 2011 (10.10 inches).

October 7–8, 2005
Tammy's Remnants Cause Flooding

Tropical Storm Tammy developed just east of Florida on October 5, 2005, crossing the coast at Jacksonville with 50 mph winds. The tropical system drifted west to Alabama before being picked up by a non-tropical low.

Abundant moisture was drawn northward by the frontal system, causing heavy rain to break out across the Mid-Atlantic region on October 7–8. Excessive totals were triggered by the interaction with low pressure and a slow-moving cold front in the East. A westward-moving subtropical depression southeast of Bermuda on October 8 supplemented the stream of moisture absorbed by the wet weather system. The ensuing flooding in the Mid-Atlantic region claimed 10 lives.

Allentown had a record rainfall of 9.65 inches on October 7–8, 2005, including a 24-hour record on the 8th (8.71 inches). The official rain gauge at East Stroudsburg caught 9.88 inches, including a 24-hour station record for any month (8.26 inches). Philadelphia established a 24-hour October rainfall record (5.94 inches). Widespread urban and stream flooding occurred along the Schuylkill River. Substantial flooding in Lancaster County, following upwards of more than a foot of rain, resulted in 34 water rescues, and Route 322 was impassable from Ephrata to New Holland (*Storm Data*).

Early October 2005 was very wet in eastern Pennsylvania with so much tropical moisture around. The weather station at East Stroudsburg had 10 consecutive days with measurable rain from October 6 to 15, 2005, totaling 14.05 inches, which contributed to a state October rainfall record of 16.88 inches that overtook the previous rainfall mark observed in August 1955 (15.09 inches). Allentown's rainfall (13.16 inches) topped August 1955 (12.10 inches), the city's wettest month until August 2011 (13.47 inches). Philadelphia's October total of 8.68 inches eclipsed 1902 (6.66 inches).

June 23–29, 2006
June Monsoon

A plume of moist tropical air from the southeast circling a Bermuda High and colliding with a cold front over the interior Mid-Atlantic region beginning on June 23, 2006, brought historic rainfall totals to eastern Pennsylvania. Widespread flooding in Pennsylvania resulted in more than $100 million in damages to homes, businesses and infrastructure, forcing upwards of 300,000 residents to evacuate.

The influx of Atlantic moisture encountered very warm air near the surface, causing air parcels to rise rapidly, forming clusters of slow-moving showers and storms focused along a trough of low pressure. The steamy air underwent further ascent over the eastern Appalachians, increasing the convective rains.

As usual in these situations, upper-level winds were parallel to a stalled front over the Appalachians—a catalyst for tropical downpours and extensive flooding. A digging upper-air low swirling over the Great Lakes enhanced low-level convergence and instability. Finally, a subtropical low traveled north from near Jamaica in a strong southerly flow, edging into North Carolina on June 27, 2006, and enhancing the rain train until the upper-level low departed on the 29th.

The heaviest 24-hour rainfall in Pennsylvania fell on June 27–28, with a total of 7.10 inches at Equinunk 2 NW in Wayne County. The great cumulative rainfall (June 23–29) was a staggering 16.00 inches at Tamaqua, Schuylkill County, where the monthly total soared to a spectacular 20.11 inches.

In the southern Poconos, Mount Pocono set new daily rainfall records June 25–27 (2.76, 2.77, 3.14 inches). Cumulative rainfall amounts from June 22–29, 2006, ranged from 8.98 inches at East Stroudsburg to 10.96 inches near Bossardsville. A rain gauge at Blakeslee recorded 7.02 inches in 24 hours on June 26–27.

During the entire wet episode from June 23–29, 2006, more than a foot of rain fell in pockets from northwestern Wayne County southwestward across western Monroe, southern Luzerne, eastern Carbon and Schuylkill counties, south to Lancaster County. Additional historic totals included: 13.31 inches at Nescopeck, 13.00 inches at Holtwood, and 12.14 inches at Camelback Mountain.

Along the Delaware River, gauging stations reported the fourth highest crest on record at Belvidere, New Jersey, and at Washington Crossing, exceeded only in 1955, 1903 and 2005. At Tocks Island, near Delaware Water Gap, the river crested at 33.87 feet (12.87 feet above flood stage), several feet below the August 1955 benchmark (37.40 feet).

Fig. 6.22 Streets turned into rivers and homes flooded in Portland along the Delaware River on June 29, 2006. (Courtesy Pocono Record*)*

The Schyulkill River at Reading attained its second highest level on record (23.75 feet on June 28, 2006), more than eight feet above flood stage, which was the highest reading since Hurricane Agnes visited the region in June 1972 (31.30 feet).

On June 29, the North Branch Susquehanna River crested at 11.53 feet near Waverly, topping the March 1936 mark (10.40 feet) in the Upper Susquehanna region. Widespread flooding occurred downstream between Wilkes-Barre and Columbia, where the river reached its highest level in the past 300 years.

Streams and main stem rivers throughout eastern and central Pennsylvania rose to moderate or major flood stage. The NOAA *Storm Data* summary reported that Governor Ed Rendell signed a Declaration of Disaster Emergency that included 46 counties in Pennsylvania. A total of 21 counties were deemed eligible for federal aid.

There were more than 1,200 water rescues, and hundreds of bridges and roads were closed: "At least 65 bridges were damaged, with an estimated 23 requiring total replacement." Statewide, the American Red Cross estimated that 7,800 residences suffered damage or were destroyed by flooding.

Fig. 6.21 Water lapping up at the entrance of Shawnee Inn in following several days of intense rainfall on June 29, 2006. (Photo by Keith Stevenson, Pocono Record*)*

Fig. 6.23 The Delaware River overflowed in Delaware Water Gap, swamping homes on Broad Street, after nearly a week of torrential rain on June 29, 2006. (Photo by David Kidwell, Pocono Record*)*

Tobyhanna had its wettest June (12.10 inches) in 2006, topping 1972 (10.14 inches), and Allentown established a June rainfall mark (9.13 inches).

August 19, 2011
Pittsburgh Flash Flood

On August 19, 2011, a deadly flash flood hit Pittsburgh during rush hour, after two inches of rain fell in 37 minutes. Three persons died when they became trapped in their vehicle in nine feet of water on Washington Boulevard around 4:00 p.m. A fourth person drowned after being pulled by swirling water into a storm sewer.

September 7–8, 2011
Remnants of Lee

The wet pattern of 2011 continued in early fall. Tropical Storm Lee took shape over the Gulf of Mexico on September 2, 2011, with modest thunderstorm activity near the center, making landfall on September 4 along the southern coast of Louisiana as a subtropical storm.

The remnant low-pressure area linked up with a cold front that draped across the Appalachians. The frontal zone intercepted a southerly surge of rich moisture beneath an upper trough over the Mississippi Valley, triggering rounds of soaking rain that fell on ground saturated by the recent passage of Hurricane Irene on August 28.

Incredible rainfall amounts were observed, including 21 inches at Colonial Beach, Virginia. Binghamton, New York, had a record 24-hour rainfall of 7.49 inches on the 7th, and a storm amount of 9.02 inches in 40 hours. (The 24-hour record was established the year before from the remnants of Tropical Storm Nicole on September 30-October 1, 2010.) The Susquehanna River at Binghamton hit a historic flood stage of 25.18 feet, higher than at any time since records were first kept in 1847.

The heaviest downpours in Pennsylvania fell on September 6-7, 2011, in the Susquehanna region, ranging from six to 12 inches, causing widespread and flooding and heavy damage. Hershey received 6.98 inches on the 7th and a total of 12.18 inches from September 5–8. Swatara Creek near Hershey crested at 26.8 feet on the evening of September 8—an incredible 19 feet above flood stage and nine feet higher than the previous high-water mark on July 26, 2018, at the site.

Areas surrounding Hersheypark were underwater. A total of 125,000 residents were ordered to leave their homes due to rising waters in the south-central part of the state. Ten inches of rain fell in 24 hours at Myerstown. Local observers in Lancaster County reported up to 15 inches at Elizabethtown and Mount Joy, according to the Lancaster *Intelligencer Journal*. The Bloomsburg Fair was cancelled for the first time in its 157-year history due to flood damage to the fairgrounds.

In north-central Pennsylvania, there was widespread destruction at Sonestown in Sullivan County caused by the collapse of a dam on Birch Creek that flooded Route 220. Record flooding also occurred on Loyalsock Creek, a tributary of the West Branch Susquehanna River in eastern Lycoming County, where bridges were heavily damaged.

Portions of Wilkes-Barre devastated by Agnes in June 1972 were evacuated on September 8, 2011. The Susquehanna River crested at 42.66 feet in Wilkes-Barre on the 9th, a stunning 20.6 feet above flood stage, which surpassed the historic level reached in June 1972 (40.91 feet). This time the levee system did not fail as it did during the passage of Agnes. However, unprotected communities from Pittston to Nanticoke were hard-hit, resulting in 100,000 persons being forced from their homes, according to the Scranton *Times Tribune*.

The Governor's Residence in Harrisburg was evacuated, along with 10,000 residents in downtown areas in low-lying portions of the city. The Susquehanna River reached its fifth highest level on record in the city (25.17 feet) on September 9, 2011, more than 10 feet above flood stage. The river at Danville crested at 31.55 feet, greater than 11.5 feet above flood stage, and just shy of the all-time level in June 1972 (32.32 feet).

Heavy damage ran into the millions of dollars in Dauphin, Lebanon and York counties. Hundreds of buildings were destroyed and thousands more were affected.

NOAA reported a total of more than 6,000 homes and businesses were damaged by flooding in Pennsylvania and more than 1,000 were destroyed. Seven deaths were reported statewide, mostly people trapped in their vehicles on flooded roadways. High water carried away bridges and submerged roadways.

Fig. 6.24 Soccer fields and clubhouse in Minisink Hills, Smithfield Township, submerged after Hurricane Irene blew past northeastern Pennsylvania on August 28, 2011. (Photo by David Kidwell, Pocono Record)

The surface reflection of Lee faded, but a mid-level circulation persisted over the southern Appalachians, before spinning down at the end of an exceptionally soggy onslaught that came on top of a wet August in the Keystone State.

The September 2011 rainfall at Harrisburg totaled 18.43 inches, which was the second heaviest monthly tally after June 1972 (18.55 inches). Bear Gap reported 14.81 inches of rain during a six-day span from September 5–10, 2011, establishing a new state rainfall record (25.41 inches) for any month, topping the August 1955 total of 23.66 inches at Mount Pocono.

July 28–29, 2013
Philadelphia Summer Rainfall Record

Philadelphia was drenched with episodes of high water that triggered bouts of flash flooding during the sultry days of late July, in the midst of the wet summer of 2013 that brought an incredible 29.71 inches of rain (June–August). Record rainfalls in June (10.56 inches) and July 2013 (13.24 inches) followed the city's wettest August in 2011 (19.31 inches).

On the afternoon of July 28, 2013, a training line of torrential rain set up from the eastern edge of the Lehigh Valley south to Wilmington, Delaware. A stalled boundary acted as a convergence zone for moisture.

An amazing 7.35 inches fell in four hours at Philadelphia International Airport from 3:00 p.m. to 7:00 p.m. Heavy storm runoff quickly inundated highway ramps between Interstate 95 and Broad Street and extended onto Interstate 76 and across the river into New Jersey. The daily rainfall in the city (8.02 inches) on July 28, and 24-hour total of 8.26 inches July 28–29, 2013, eclipsed the previous record during the passage of ex-hurricane Floyd in September 1999 (6.77 inches).

October 20–21, 2016
Central Mountains Flooding

A slow-moving cold front aligned with a moist southwesterly flow acted as the focus for repeated thunderstorms on the night of October 20–21, 2016, in northern Pennsylvania.

Four to eight inches of rain on the Allegheny Plateau caused widespread flooding, leading to evacuations and water rescues. Roads were covered by water and debris in low-lying areas and dozens of bridges were damaged or destroyed. Several hundred homes reported moderate to heavy losses caused by high water. A ruptured Sunoco Logistics pipeline deposited more than 50,000 gallons of gasoline into Loyalsock Creek, which flows into the West Branch Susquehanna River, polluting the water supply.

One man, a retired local teacher, died in Clinton County when a tree landed on his home. Storm wind gusts were estimated at 100 mph, according to the NWS State College storm report. Barns were destroyed by high winds (80 mph) in Cambria County. About 100 residents in Bald Eagle Valley had to leave their homes due to the flooding and washouts. Two homes were carried away, and hundreds of other residences sustained damage in Centre County.

Torrential rain turned roads into rivers at Hillsgrove in Sullivan County, where Loyalsock Creek raged out of its banks. High winds downed power lines and rising water took swamped vehicles and railroad beds. A bridge over Loyalsock Creek in Montoursville was swept away. The damage to infrastructure calculated by Pennsylvania Emergency Management Agency (PEMA) and FEMA totaled around $2 million.

July and August 2018
East-Central Floods

Prolonged bouts of heavy, and at times non-stop, rain moving south to north in the Mid-Atlantic region in late July and early August 2018 triggered widespread flooding, when surface-based moisture was lifted by persistent mid-level low pressure. Up to two feet of rain fell between July 21 and August 14 in several locations in eastern Pennsylvania.

A blocking Bermuda High teamed up with upper-level low pressure rotating slowly across the Great Lakes and Northeast to funnel deep, subtropical moisture northward. Six to 12 inches of rain fell in a broad band from southeastern Virginia through east-central Pennsylvania and Upstate New York from July 21–25, 2018. Rainfall rates were one to two inches an hour.

Moderate flooding occurred in the Susquehanna Valley, after creeks feeding the main stem of the river northeast of Harrisburg spilled over their banks. Major flooding also occurred in Schuylkill County in the hilly east-central part of the state. Spring Creek, which flows through Hersheypark, overflowed, forcing the amusement park to close for several days, after muddy floodwaters rose beneath the Skyrush roller coaster. The only time the water was higher was during the passage of Tropical Storm Lee in September 2011.

The NWS Middle Atlantic River Forecast Center (MARFC) reported major flooding along the Swatara Creek east of Harrisburg at Harper Tavern and Hershey, forcing evacuations in parts of Swatara Township. Less significant flooding was reported along the Susquehanna River at Harrisburg. Rapidly rising water closed roads and forced evacuations in Derry Township at Hummelstown. Up to 10 inches of rain fell in Dauphin County in the four-day monsoonal pattern. All-time July rainfall records were notched at Harrisburg (12.09 inches) and Williamsport (11.99 inches).

A second wave of flooding on August 13–14, 2018, affected most of the same areas hard-hit in July. High water was reported in areas from around Harrisburg and Williamsport to southern New York, and extending eastward to Pottsville and the western suburbs of Philadelphia. A corridor of three to seven inches of rain fell on August 13. Scranton received 4.34 inches in 24 hours, second only to August 18, 1955 (4.40 inches).

Serious flash flooding occurred in Lancaster, Chester, Schuylkill, Columbia, Wyoming and Bradford counties. Homes in Tremont, north of Harrisburg, were flooded for the third time in less than three weeks, after a creek overflowed. A portion of the Schuylkill Expressway was shut down due to the high water. Water rescues and stranded vehicles were a common scene all day in Havertown, Darby, Port Clinton and Benton in Columbia County along Fishing Creek.

Drenching downpours wrung out from a saturated atmosphere coupled with strong storms rumbling across portions of Maryland and southern Pennsylvania sent creeks out of their banks, trapping residents in their homes and cars in Upper Darby and at Port Clinton. Floodwaters in Carbon County in the Pottsville area affected places cleaning up from heavy flooding a few weeks earlier that had affected hundreds of homes and businesses.

On August 31, 2018, a stationary front in southeastern Pennsylvania served as the focus for torrential rainstorms that triggered more flooding in Lancaster, Lebanon, Dauphin and York counties. Mount Joy received 10.72 inches of rain in less than 10 hours. A swath of Lebanon County between Myerstown and Jonestown received more than five inches. Pottstown tallied 6.04 inches, and Northampton had a sopping 5.16 inches, adding a soggy coda on an exceptionally wet month.

Floodwaters swamped highways and railroad tracks. Portions of Route 283, near Route 772, and Route 30 in York County were under many feet of water for a time. Flooding washed over Airport Road in Fredricksburg, Lebanon County, causing the asphalt to buckle. Oil Creek overtopped its banks. More than a half-dozen water rescues, including vehicles and homes, occurred between Manheim and Mount Joy.

August 2018 was the third wettest on record at Williamsport (9.10 inches) and Harrisburg (7.06 inches). Harrisburg recorded 21.37 inches from

June through August, the second soggiest behind 1972 (23.33 inches). The highest totals reported by WGAL News 8 Storm Team were around Lebanon (31.74 inches), Lancaster Airport (24.52 inches) and York-Thomasville (22.58 inches). The meteorological summer of 2018 was the wettest in state history, as far back as 1895, with 19.11 inches of rain, topping 1928 (17.78 inches).

Then came the remnants of Tropical Storm Gordon on September 7–10, 2018, which brought excessive 24-hour rainfall totals in the southwestern part of the state. Pittsburgh received 5.63 inches, with more falling at Allegheny County Airport (7.51 inches). A local meteorologist, Dennis Sims, measured 8.14 inches in Bridgeville. Rain gauges in Westmoreland and Washington counties approached seven inches, and portions of Armstrong, Beaver, Butler and Greene counties counted nearly six inches in 72 hours.

By the end of 2018, annual precipitation records were established at State College (63.97 inches), Scranton (61.08 inches) and Pittsburgh (57.83 inches), breaking previous records set in 1996, 2011 and 2004, respectively.

August 3-4, 2020
Tropical Storm Isaias Deluge

The northward trajectory of Hurricane Isaias after making landfall in eastern North Carolina late on August 3, 2020, took the weakening tropical storm through the Delaware Valley, resulting in flooding that caused millions of dollars in damages.

Hundreds of residents of Southwest Philadelphia and Darby City in Delaware County were forced from their homes when Cobbs Creek overflowed. Flooding caused significant damage in the Manayunk section of Philadelphia and along the Schuylkill River in Montgomery County, cresting at 20.55 feet early on August 5, which was more than 7.5 feet above flood level.

Serious flooding occurred throughout eastern Pennsylvania. During the height of the storm on the afternoon of August 4, a total of 8.59 inches of rain fell at Wynnewood, Montgomery County, and 7.25 inches at Center Valley, Lehigh County. Additional spotter reports included 7.48 inches at Saint Davids, Delaware County, and 7.43 inches at Sellersville, Bucks County. Allentown received 4.92 inches and Philadelphia had 4.16 inches.

In the southern Poconos, rain gauges recorded 7.00 inches at Canadensis and 6.95 inches at both Saylorsburg and Bossardsville. The maximum total was 7.71 inches at Hamilton Square, a little southwest of Stroudsburg. A weather station in East Stroudsburg received 6.43 inches, and 6.61 inches fell in Smithfield Township north of town. In Pike County, Milford totaled 4.72 inches. (An additional short burst of rain after 10:00 p.m. associated with a cold front was not included in the storm totals.)

July 12, 2021
Philadelphia Flash Flood

The sky opened over the Philadelphia area at the start of the evening rush hour on July 20, 2021. A hot, muggy day was broken by torrents of water unleashed on southern Bucks County and northern parts of Philadelphia, resulting in dozens of water rescues. Water poured into homes and flooded roads, stranding motorists, and forcing highways to be shut down.

The heaviest rain fell in lower Bucks County, with 10.32 inches at Croydon, 9.85 inches at Bristol Township and 8.69 inches at Bristol. Florence, New Jersey, received 7.63 inches. The totals were described as a 100-year flood. Up to five inches fell in northern parts of the city of Philadelphia.

The combination of a tropical air mass, daytime heating, and a frontal boundary draped across the Lehigh Valley and central New Jersey fueled the deluge in southeastern Pennsylvania, before weakening and moving off to the northeast of the Delaware Valley.

September 1-2, 2021
Ida Triggers Catastrophic Flooding

Hurricane Ida caught the attention of forecasters on August 26, 2021, as a tropical depression in the Caribbean Sea near Jamaica. Ida passed the edge of western Cuba the following day. On August 28, the storm intensified rapidly from a Category 1 to a Category 4 hurricane, when top sustained winds rose from 80 mph to 150 mph in 24 hours by 8:00 a.m. on August 29. The powerhouse storm headed northwest across anomalously warm Gulf waters.

Hurricane Ida made landfall on August 29 at 12:55 p.m. near Port Fourchan, Louisiana, blasting the southeastern part of the state with extreme winds that toppled trees and high-voltage transmission lines, leaving more than 902,000 without power, and thousands more in Mississippi. Maximum sustained winds averaged 150 mph (Category 4). Twenty-six storm-related deaths were reported in Louisiana.

After slamming into the northern Gulf Coast, Ida quickly weakened to a tropical depression as the storm moved inland and headed northeast through the Tennessee Valley, southeastern Kentucky and West Virginia by the morning of September 1, 2021. The situation worsened when the tropical rainstorm merged with a frontal boundary and interacted with an upper-air trough over the Ohio Valley, which when coupled with jet energy, enhanced the lift of moist tropical air.

Flash flooding ensued from the Pittsburgh suburbs to the Susquehanna Valley and southeastern Pennsylvania. Tremendous amounts of rain, falling at the rate of three to four inches an hour on saturated soil, pelted the region that was still soggy from the recent passage of two tropical systems (Fred and Henri).

Fifty-two deaths were reported from Virginia to Connecticut, mostly due to flash flooding. The greatest statewide tolls occurred in New York (18) and New Jersey (30), which was the deadliest weather-related disaster in the Garden State since Hurricane Sandy claimed 37 lives in October 2012. Eleven people died in flooded basement apartments in New York City.

New York City's Central Park weather station received a record one-hour rainfall of 3.15 inches between 9:00 and 10:00 p.m., with a storm accumulation of 7.19 inches. Water gushed through subway stations and submerged cars and buses up to the windows. Rescuers in boats assisted trapped flash flood victims. Newark Liberty International Airport was flooded on the ground floor, after a record rainfall of 8.44 inches. Eleven inches fell at Flemington, New Jersey, and 11.75 inches at Haines Falls, New York.

Flash flood emergencies were declared in southeastern Pennsylvania and New Jersey, and in New York City for the first time ever. Multiple water rescues were undertaken by boat, as numerous streams and creeks overflowed, stranding motorists. A total of 450 water rescues transpired in Montgomery County, northwest of Philadelphia, and more than a few thousand rescue operations were reported in southern Pennsylvania.

More than 200,000 customers lost power from Pennsylvania to Connecticut in the storm. Five people died in Pennsylvania from weather-related causes; three died in Montgomery County, including one death caused by a tree falling on a car, killing the driver. Another fatality occurred after a motorist drove into floodwaters in Bucks County.

Frequent drenching afternoon downpours with severe thunderstorms moved rapidly north through the region. Roads turned into raging rivers, and highways filled with water around Philadelphia. The Vine Street Expressway (I-676) flooded, and the muddy Schuylkill River lapped just beneath the 22nd Street overpass. The highway was closed from Broad Street to the Schuylkill. Seven large pumps were brought in to remove the water, leaving a large amount of mud and debris. Flooding forced PennDOT to shut down the Schuylkill Expressway (I-76). MLK Drive, and many more major thoroughfares that were now impassable. Heavy flooding was reported in the Manayunk section of the city.

The Schuylkill River crested at 16.35 feet at Philadelphia on September 2, more than five feet above flood stage and the highest level recorded since October 4, 1869. The Brandywine Creek at Chadds Ford crested at 21.04 feet, far above the flood stage of nine feet, breaking the September 1999 record of 17.15 feet during the passage of the remnants of Hurricane Floyd.

Swatara Creek at Harper Tavern in Lebanon County crested at 18 feet, nine feet above flood stage. The Lehigh River at Glendon reached five feet above flood stage. Record stages along the Perkiomen Creek at Graterford (24.2 feet), East Branch of Brandywine Creek near Downingtown, and Schuylkill River at Norristown (26.85 feet) hit more than 13 feet above flood stage.

The heaviest rain totals were measured in Chester County, reaching 9.43 inches at Coatesville and 8.17 inches at Phoenixville. Storm totals topped six inches in much of eastern Pennsylvania. Some of the largest rainfall reports came in from at Tower City, Schuylkill County (7.90 inches), and Manchester, York County (7.74 inches). Excessive totals were noted at Geistown, Cambria County (7.46 inches), Hershey, Dauphin County (7.29 inches), and Palmyra, Lebanon County (6.95 inches).

In the northeast, the heaviest rainfalls occurred at Gouldsboro, Wayne County (6.80 inches), and Beach Haven, Luzerne County (6.75 inches). Farther south, Bossardsville, Monroe County (6.67 inches), reflected the uniformity of the torrential amounts unleashed by Ida.

Four to six inches of rain doused southwestern Pennsylvania. At Jefferson Hills, below Pittsburgh, a total of 5.72 inches raised creeks and streams to levels past flood stage in Allegheny County. Roads were submerged. A school bus in Shaler Township stalled in rapidly rising waters. Forty students and the teacher had to be rescued. Another bus would become stranded in Stowe Township.

July 15, 2023
Bucks County Tragedy

Torrential rain falling on sodden ground in Bucks County around 5:30 p.m. on July 15 2023, caused a swollen creek to overrun roadways, sweeping at least 11 cars away, Seven people died in the flash flooding near Washington Crossing. A nearby rain gauge caught 5.31 inches on July 15–16.

Chapter Seven

Thunderstorms, Tornadoes and Whirlwinds

The rush of the cyclone was estimated to be 500 feet wide. Its appearance was that of a dense black cloud revolving at a terrific rate. In the heart of it the gloom was like the darkness of midnight …

—*Philadelphia Inquirer*, August 4, 1885

BILLOWING CUMULUS sprouting through a hazy sky in a cauliflower-shaped formation is the initial sign of a localized thunderstorm percolating on a muggy afternoon. A cooling breeze whooshes through swaying tree limbs amidst rumbles of thunder reverberating in the distance.

The charged atmosphere, redolent of moisture, reveals an ominous inky sky, peppered by flashes of lightning and flanked by low-flying clouds. The opening salvo of a cacophonous, show-stopping drama is announced by an onrushing downpour, accompanied by booming thunderclaps that rise to the level of a cannonade. Furious gusts of wind sweep rippling sheets of rain across the pavement and the steamy air chills precipitously.

The commotion gradually diminishes and the incessant rumbling recedes. The periwinkle sky is brightened by a glorious rainbow that fills the horizon and denotes the grand finale framed by a golden sunset, streaked with crimson and salmon hues.

Thunderstorms

At any given moment around the world, approximately 2,000 thunderstorms are in progress. In the United States, about 100,000 thunderstorms percolate annually, from garden-variety thundershowers to gusty storms rolling across the landscape.

The birth of a single cell thunderstorm begins uneventfully on a warm sticky day. The sun warms the ground, which heats the air. Invisible air parcels are launched like miniature hot air balloons, cooling by expansion as the pressure decreases. Once the parcel temperature lowers sufficiently to become saturated, water vapor condenses into tiny droplets.

Instability is defined as the amount of energy required to cause air to rise. A corridor of warm, humid air undergoing heating underneath chilly air aloft causes clouds to bubble up, creating an unstable atmosphere that fuels updrafts. Moist air parcels release latent heat upon condensation that enables a cloud to grow. With sufficient lift and the presence of ice particles at higher altitudes, a thunderstorm is born.

Electrification of Storms

Lightning strikes the surface of the earth an average of 100 times per second, which equals eight million bolts per day. In a storm, the separation and buildup of opposite electrical charges leads to a discharge of current, realized in a lightning flash that consists of a series of strokes lasting about 30 microseconds.

In the United States, about 25 million cloud-to-ground strikes occur annually, causing an average of 39 fatalities and about 400 injuries (1991–2020). Most lightning casualties are males (81 percent), usually caught in open areas such as golf courses or athletic fields.

Lightning is a massive spark of electricity that carries 100 million to one billion volts of electricity, according to NOAA's National Severe Storms Laboratory (NSSL). An average lightning bolt generates 30,000 amps of current, but a powerful stroke can produce more than 10 times as much. The discharge of electricity occurs when the insulating property of air finally breaks down. Lightning completes a circuit between the air, clouds and ground, equalizing regions of opposite charges in the atmosphere.

Fig. 7.1 Lightning strike on August 2, 2011, branching out near Lake Wallenpaupack, viewed from Newfoundland, Wayne County. (Photo by Keith Stevenson, Pocono Record*)*

High flash rates tend to be followed by heavy downpours. Storm electrification is a product of the interaction between positively charged ice crystals and negatively charged water droplets, which occurs more readily in tall storms with strong updrafts that create opportunities for precipitation particles to freely collide.

The sound of thunder is a consequence of energy released by a lightning stroke that heats the surrounding environment to a temperature around 50,000°F—five times hotter than the surface of the sun. A rapid expansion of the heated channel of air occurs faster than the speed of sound, producing a shock wave we hear as a thunderclap that results from the sharp increase in pressure.

Sound travels one million times slower (1,130 feet per second at sea level and a temperature of 68°F) compared to the speed of light. This is the reason we don't hear thunder until after the precise time of a lightning strike. Sound waves travel at about 0.2 mile per second, depending on the temperature, propagating faster in warm air and closer to the surface.

Sound waves are affected by the humidity, wind velocity and terrain features that reflect wave motions. A temperature inversion, which is more common at night and in a snowstorm, refracts sound waves back toward the ground, resulting in a prolonged rumbling over a greater distance. A jarringly close strike (flash-boom) rattles the walls and our nerves. In a typical warm weather scenario, if we hear thunder five seconds after seeing lightning, the storm is about a mile away.

How thunderstorms generate lightning is incompletely understood. Scientists are learning more about how storms become electrified by videotaping multiple flashes and analyzing high-speed camera images that capture thousands of frames per second. Lightning location systems (LLCs), both ground and satellite-based, provide a large network to monitor flashes.

Electric charge separation necessary for lightning begins with robust updrafts and downdrafts that distribute charge centers on precipitation particles. The collision of ice crystals within the clouds creates opposite charge centers in a thunderstorm. The popular theory postulates that the upper part of the towering storm cloud gains positive charge, and the lower portion gathers negative charge. The interaction of water droplets and ice particles leads to thunderstorm electrification.

In a strong updraft, water droplets and ice are lofted to high, cold altitudes. Ice crystals form by deposition (vapor to solid), aggregation (collision of ice crystals) and riming (liquid water freezing on contact). Ions gravitate toward the lighter, partially melted snowflakes and graupel (spongy ice or soft hail), transferring a positive charge to the frozen particles.

Accelerating downdrafts carry warmer, heavier hail and droplets to the mid-levels of the storm cloud (near 15,000 feet), which become negatively charged. The negative charge induces a flow of electrons to a positively charged electric field that shadows the storm close to the ground. Air is an insulator, and therefore a poor conductor of electricity, so a steep voltage gradient of about one million volts per meter is required between opposite charge centers, which are distributed upward, downward and sideways.

Once electrification exceeds the breakdown potential, electrons flow downward along an ionized channel and the path of least resistance. The discharge starts with a stepped leader advancing in discrete segments of 150 feet at a speed greater than 200,000 mph, all happening in a millionth of a second. Weakly charged electrons are lowered toward the ground, attracting positive charges or streamers that branch upwards along protruding objects such as trees, metal fences, buildings and power transformers.

The connection between downward moving negatively charged stepped leaders that branch out near the ground with a channel of positively charged particles moving upward along tall objects establishes a conductive pathway for the discharge of electricity. The highly luminous return stroke follows the leader channel upward at a speed of 200 million mph. As many as 10 to 20 strokes typically occur in a fraction of a second, until the charge is dissipated, short-circuiting the cloud.

The human eye cannot resolve the multiple discharges we perceive as a bright flash or bolt of lightning that is only as wide as a human finger. Dozens of stepped leaders branching toward the ground have been captured in a single flash by high-speed cameras. In recent years, researchers have identified lightning branching upward from the tops of tall trees, buildings, and even mountainsides, when the electric field is enhanced. Rarely, a storm exhibits an inverted electrical structure conducive to positive strikes near the core.

About 20 percent of all the recorded lightning flashes comprise potentially dangerous cloud-to-ground strokes. Intra-cloud lightning occurs when intervening layers of clouds and rain diffuse the narrow in-cloud flash, which appears as sheet lightning. Flashes within and between clouds account for 80 percent of the observed lightning bolts.

Positive lightning arising above a thunderhead is more intense and especially hazardous. A "bolt from the blue" usually catches a hapless victim without warning. Originating in the upper portion of a thunderstorm, a connection occurs between the positively charged cirrus anvil and negatively charged ground streamers more than 10 miles in the distance. Positive lightning is more powerful, with a maximum charge of 300,000 amperes and one billion volts, 10 times that of negative strikes.

Some lightning discharges last longer in a long-continuing current (LCC) succeeding the main bolt, creating a more likely scenario for an object struck by lightning and heating up to its combustion temperature. Lightning is responsible for igniting forest fires and damaging power lines, which is why electric utilities review lightning data to determine vulnerable sections of transmission lines.

A seldom-seen and spectacular positive discharge occurs in the upper atmosphere between 18 and 40 miles above the storm, in the low ionosphere, and has been given colorful names such as sprites, elves and jets.

A distant light show on a hot, muggy night is sometimes incorrectly referred to as "heat lightning." There is no such thing; the thunderstorm responsible for the flickering is too far away for the thunder to be heard by the observer.

A rare occurrence of a highly mysterious phenomenon was recorded on the evening of June 12, 1911. A spherical manifestation generally called "ball lightning" captured the attention of the Stroudsburg *Jeffersonian*, described by a mesmerized eyewitness as a "ball of fire" that fell on West Main Street. Family members stated the electricity "seemed to break and then there was a deafening crash that sort of dazed all the members of those households."

On April 2, 1947, the state *Climatological Data* report noted: "Electrical phenomena of a peculiar nature occurred in Buchannon Valley, near Cashtown, in Adams County." Various observers reported that "balls of fire floated inside and outside of houses."

An eyewitness, W.E. Cole, heard a "pistol shot" around 10:15 p.m., which he said was followed by "a red ball of fire more than a foot in diameter." Racing to his basement, Cole heard "another shot-like sound, and more glowing balls of fire appeared." He reported that tree trunks appeared "dynamited" and covered the ground "almost white with fragments" as superheated sap likely exploded.

Deadly Lightning Strikes in Pennsylvania

The annual average death toll for the United States from direct and indirect lightning strikes is 39 (1991–2020). Long-term records published in NOAA's *Storm Data* counts 140 lightning deaths in Pennsylvania (1959–2023), the ninth highest total in the nation. Between 2011 and 2020, nine Pennsylvanians were killed by lightning.

On June 13, 2019, two recent high school graduates were struck and killed by a lightning bolt on an island near the lake in Mammoth Park in Mount Pleasant. The *Pittsburgh Post-Gazette* noted that 35 years ago to the day, June 13, 1984, two teenagers fishing at a golf course in Hempfield, also in Westmoreland County, were hit by lightning, and one later died. Another youngster was seriously injured by lightning in Beaver County.

The lightning strike that caused the most casualties in the state occurred on July 24, 1908, at the Pennsylvania National Guard encampment in Gettysburg. Four people were killed and 26 injured, according to the NOAA monthly *Climatological Report: Pennsylvania Section*. Numerous storms struck the southern part of the state from July 22–25, 1908, causing property damage and localized flooding.

On September 4, 1937, two prominent businessmen and two caddies were struck and killed by lightning while taking refuge under an oak tree at the Long Vue Country Club in Penn Hills Township, Allegheny County. Several injuries were also reported.

The Franklin *News-Herald* covered a deadly lightning strike on August 4, 1939, that killed a six-year-old boy and seriously injured two women. A bolt of electricity shocked "scores of youngsters" standing along a wire fence during a field meet at the Wyoming Valley Playground and Recreational Association in Kingston, Pennsylvania. A farmer was killed the same day by a lightning strike in Perry County.

A deadly lightning strike on June 23, 1965, took the lives of four golfers and injured six people who had sought refuge under a wooden shelter at the Armco Country Club, located 12 miles from Butler. On August 24, 1968, lightning struck tent poles at the Crawford County Fair, killing two people and injuring 72.

A tragedy on September 18, 1984, occurred on a soccer field in Chester County, injuring 29 people, including two dozen players, several coaches and a bus driver. Three were seriously injured and one died five days later (*Storm Data*). The Allentown *Morning Call* stated that the injured players "were huddled at the center of the field when the bolt struck," and that several spectators on the sidelines were hurt.

Fans leaving Pocono Raceway in Long Pond on August 5, 2012, at the end of a rain-shortened race were caught in a nasty storm that produced multiple lightning strikes, killing a man and seriously injuring nine others behind the grandstands in the parking area.

Livestock were killed and some homes and barns were damaged in a severe electrical storm; one strike left a three-inch hole "as though it had been caused by a bullet" at Williams Stoddard's Bushkill farm. High winds left patterns of storm damage in parts of Wayne, Pike and Monroe counties into northwestern New Jersey, possibly indicative of embedded weak tornadoes.

An unexpected lightning strike in the middle of winter at Eynon, near Scranton, on January 13, 1993, struck the backyard of a house at 2:30 p.m. The NOAA *Storm Data* account said that the force was responsible for "breaking an underground water main and then traveling through the cellar and out the front of the house." The summary continued: "The foundation of the house and garage were damaged, and a wall was blown off the garage. The electric meter was found 100 feet from the house." A five-year-old boy suffered minor burns in a nearby garage.

On August 18, 2019, nine people (two adults and seven children) were hurt at the Dolphin Swim Club in Bucks County, about 20 miles northeast of Philadelphia, around 5:00 p.m. when a tree struck by lightning fell on a tent. Three of the victims suffered serious injuries.

Thunderstorm Frequency

Thunderstorms can be isolated in nature or develop in an organized pattern ahead of a frontal boundary that intercepts a flow of moist, unstable air.

The highest number of annual thunderstorm days in Pennsylvania—greater than 40—occurs in the western part of the state. Moist air is forced to ascend the Allegheny Plateau, coupled with proximity to unstable air in the Ohio Valley and lake-breeze boundaries, contributing to more frequent storms in western Pennsylvania compared to other areas of the state.

Central and eastern Pennsylvania communities average 30 to 35 thunderstorm days annually.

The fewest number of thunderstorms is observed in the northeastern corner of the state—25 thunderstorm days—in the cooler elevated regions farther removed from higher instability.

Most thunderstorms do not go on to cause wind damage, but about 10 percent are classified as severe across the United States, accompanied by wind gusts in excess of 57.5 mph and/or quarter-size or larger hail. A wind gust of 50 mph pushing against a tree in full leaf 50 feet high exerts a force of 1,000 pounds.

Thunderstorm Classification

Thunderstorms form singly and in clusters. Storm development is sparked by instability fueled by a flow of warm, humid air. Vertical development is enhanced by faster winds aloft accompanying jet stream energy (dynamics), in the vicinity of a frontal system and contrasting air masses. Lower pressure near an axis of moisture convergence is a focusing mechanism for storm updrafts.

A localized air mass storm on a soupy afternoon meanders in a sluggish flow and possesses a short life span of less than an hour, announced by streaks of lightning and sharp thunderclaps. A lonely storm hovering over a ridgetop that forms due to uneven heating soon rains itself out in place.

A pulse thunderstorm boils up on a very steamy afternoon and is announced by booming thunder and sharp cloud-to-ground lightning, framed by a charcoal gray sky. On days with plenty of available potential energy, daytime heating and leftover boundaries from recent showers encourage "popcorn storms" to blossom. Slow-moving pulse-type storms drifting along in a seasonally weak mid-level flow normally build quickly and then simmer down in an hour or two when the rain-cooled downdraft cancels the updraft.

Elevated convection forms in an environment where storm energy is released atop cool air near the ground. The surge of warmth forced above a layer of cooler air near the surface destabilizes the atmosphere. Elevated thunderstorm clusters often produce considerable lightning but are not likely to mix down damaging wind gusts. Storms that form along a warm front are capable of hail and frequent lightning.

Multicell clusters reflect a collection of thunderstorms in various stages of evolution and dissipation. Groups and lines of storms are triggered by an upper-level wave of energy—a cold pocket with low pressure, stronger mid-level winds and counterclockwise spin conducive to lift. Merging cells and intersecting boundaries are more likely to result in damaging wind and flash flooding.

Squall Lines and Thunderstorm Systems

An atmospheric profile that indicates the winds are increasing in speed and/or shifting direction with height, or vertical wind shear, is a favorable setup for the evolution of organized updrafts and strong to severe storms.

Storms firing on a humid day are energized by insolation. After a dry slot erodes the lingering low stratus cloud deck, buoyant air parcels ascend in a colder, dry environment with pockets of instability resulting in upscale storm growth.

A dip in the jet stream, or shortwave trough, is the source of deep-layer shear that instigates lift, spin and moisture confluence. The edge of a heat dome in the spring and summer offers an ideal overlap of shear and instability. Strong mid-level winds sustain updrafts that initiate storm growth. The ingredients come together at the interface of contrasting heat and humidity zones near low pressure, causing air to rise rapidly as storms bubble up.

Diverging air aloft with a jet stream disturbance promotes broad ascent, amplified by a wind speed maximum. The compensating low-level return flow impinges on a surface front, pumping in tropical air and acting as a lifting mechanism. Differential heat energy where skies are mostly clear intensifies the instability gradient along a ribbon of moisture.

Upper-air forcing and a convectively enhanced circulation serve to increase the counterclockwise spin, which is conducive to storm organization.

Along the southern edge of the westerlies, the buoyancy increases. A strengthening flow begins to tilt burgeoning updrafts away from developing rain-cooled downdrafts, which maintains a storm inflow that fosters updraft longevity. Interactions with neighboring outflows, pooling moisture and a surface trough promote sustained lift.

Microwave satellite imagery captures the fine details of the atmosphere slowly coming to a boil. A perturbation in the flow coaxes upward motion and the formation of billowing clouds. The pocket of low pressure chills the top of a layer faster than the lower portion warmed by heat released upon condensation, as rising parcels become saturated. The release of convective instability turns agitated cumulus into a cluster of storms.

A key severe weather parameter is convective available potential energy (CAPE), defined as the amount of positive buoyant energy an air parcel would have if lifted over a fixed distance between the cloud base and equilibrium level. CAPE values provide a good indication of potential updraft intensity by calculating the instability.

The storm-scale convection-allowing models (CAMs) capture short-term dynamics inside the thunderstorm, which require higher resolution to simulate updrafts and downdrafts. During the late spring and summer, the high sun angle heats the ground, which warms the layer near the surface, causing parcels to attain buoyancy and tap into pockets of stronger winds aloft.

Thunderstorm development can be delayed by a capping inversion in a relatively warm, dry layer between 3,000 and 7,000 feet. The elevated mixed layer originates from north-central Mexico or over the Desert Southwest. The pressure-cooker setup indicates trapped humid, unstable air below, until a wave or front blows the lid off. Storm inhibition is soon overcome, often explosively.

If the winds aloft are primarily unidirectional, the mean environmental flow becomes aligned with a frontal system, triggering line segments or multi-cell storm clusters that bring torrential rain, strong wind and small hail, often merging into a linear storm complex. In an environment with deep-layer shear, a self-sustaining mesoscale convective system, or MCS, evolves in an area of pooling moisture and convergence, which is driven by a piece of mid-level energy.

Clusters of thunderstorms that congeal into a small complex can develop a bowing structure resembling a backward C, where powerful wind gusts are concentrated. A quasi-linear convective system (QLCS) with kinks takes on the shape of a dogleg. Notches in the line reflect the transient circulations (mesovortices) capable of spawning spin-up tornadoes.

A QLCS in a high shear/low instability setting is often accompanied by powerful wind gusts. An increasing buoyancy or pressure gradient between low pressure (inflow) and high pressure pooling in the rain-cooled air (outflow) forms a storm-scale pressure gradient dipole. The resulting low-level rear inflow jet propels the MCS at greater speeds.

Rear inflow jets along a bowing segment are associated with downbursts. The interface between surging rain-cooled outflow and air flowing into the storm from the front end, or inflow notch, is conducive to brief spin-up tornadoes.

The merging rain-chilled downdrafts spawn a migrating outflow boundary that becomes a mini cold front. New back-building storms erupt along the tail, where southerly surface winds intercept a line of moisture convergence. New updrafts form a flanking line of storms on the consolidating cold pool in a synergistic process that leads to upscale storm growth. Sometimes, the interactions trigger a gravity wave, which acts like ripples on a pond that further disturbs the loaded atmosphere.

In a high-velocity flow, taller storms translate momentum down to the surface as blasts of wind, preceded by a striking shelf (arcus) cloud and ragged rain curtain.

Composite outflows collide and interact with boundaries and terrain circulations, causing localized wind shear that is usurped by a passing updraft, which sometimes causes rotation and the risk of an isolated tornado in the surge of outflow.

The release of latent heat with condensation combined with rotation imparted by the Coriolis force contributes to the formation of a mid-level circulation feature in a highly unstable setting with strong wind shear. A mesoscale convective vortex (MCV) takes on the appearance of a comma shape, with a sharpening bowing structure. Small circulations occasionally spin up brief tornadoes near the apex of the bow and around the head of the MCV. A counterclockwise (cyclonic) vortex on the northern end is infrequently accompanied by a "bookend vortex" on the southern edge that rotates clockwise.

In the summertime, a cluster of storms rolls over the Corn Belt during the heat of the day and absorbs a large quantity of water vapor released by plant transpiration, where dew points are very high. At night, an oval-shaped mesoscale thunderstorm complex (MCC) is strengthened by a low-level jet. Nighttime cooling stabilizes the lower atmosphere, which decouples from the stronger winds a little above the surface. The steepening pressure gradient between the High Plains and Mississippi Valley ramps up a nocturnal low-level southerly plume of moisture that collides with cooler, drier air.

A thunderstorm complex eventually becomes outflow dominant, as rain-cooled air undercuts the inflow into the storm system, which "gusts out" over time. Convergence weakens when storms outpace the instability or encounter drier air between 5,000 and 10,000 feet. Individual cells run out of steam, as exhibited by decreasing lightning flashes and warming (shrinking) cloud tops.

Sinking motions behind a mid-level shortwave and nighttime cooling ultimately suppress the updrafts as the instability wanes. A trailing impulse is unlikely to spawn additional storms once the air mass has been convectively overturned and thoroughly worked over by a multicellular system. However, the remnant MCV often outlives the decaying complex and triggers renewed storms the next day, as the atmosphere recovers and recharges near residual boundaries.

Microbursts and Derechos

Around nightfall on May 11, 2015, an intense low-pressure area tracked toward the Great Lakes, directing unseasonably warm sticky air northward ahead of a cold front. A line of severe thunderstorms moved from northeastern Ohio into northwestern Pennsylvania.

An intense thunderstorm struck the west side of Erie, damaging a mobile home park on Route 5 in Millcreek Township. Snapped trees ended up on dozens of homes and cut power, as lines toppled. Surges of wind wrecked four homes and damaged scores more, including from a subsequent fire.

Intense downdrafts form when falling raindrops evaporate in entrained drier air, chilling the column and accelerating the downward rush of air that hits the ground with force. Rain-cooled outflow fans out with an expanding cold pool, producing a gust front that generates new updrafts, while other storms fall apart, collapsing in a heap of rain. Melting hail also removes heat and intensifies the pulses of wind.

Momentum from the mid-level winds is pulled downward in a robust downdraft. A concentrated pattern of wind damage, up to 2.5 miles, splayed in a divergent path is a classified as a microburst; a broader swath is a macroburst. Wind gusts in a downburst typically reach 50 to 80 mph in a rain core, which can be augmented in the direction of the storm motion within an upright thunderstorm rolling in a generalized eastward direction.

A description of a rural microburst appeared in the *Allentown Chronicle*, which was witnessed five miles northwest of the city, resulted in damages in the thousands of dollars.

A descriptive account was reprinted later in the *Jeffersonian* (August 28, 1873):

> Of all the severe rains known to the oldest inhabitant, that of Tuesday afternoon [August 12, 1873] from about quarter past three o'clock till a quarter past five, was the most copious. … The streets [of Guthsville] were a stream of water two or three feet deep. … In the afternoon between five and six o'clock, Hiram J. Schantz, Esq. stood above his mill at the head of Cedar Creek, and without any apparent cause he was turned around twice by some invisible force. … A moment after he noticed a body of water five or six feet high rushing towards the mill from above coming from a direction where no creek ran, and which was the water of a waterspout which it is said to be descended near Crackersport, and rushed across the fields carrying fences before it.

The storm system must have been extensive. The Philadelphia weather office measured 5.21 inches of rain on August 13, 1873. During a soggy 11-day stretch from August 12–22, a total of 10.12 inches of rain drenched the city.

In a very muggy environment with light steering currents, water loading of downdrafts leads to a powerful rush of air when a storm collapses on itself—a wet microburst. Storm outflow fans out radially and reinforces a convergent boundary that migrates away from the gust front, forming new cells.

Downbursts are responsible for thousands of reports of damaging gusts each year that snap large tree limbs and knock down power poles. Large trees with shallow root systems already loosened by soggy soil are susceptible to being uprooted, often toppling on power lines, homes and vehicles.

The word *derecho*—a Spanish term for "direct" or "straight ahead"—first appeared in print in 1888, coined by Gustavus Hinrichs, founder of the Iowa State Weather Service. A derecho is a widespread, long-lived wind event, accompanied by damaging gusts reaching 58 mph, extending for at least 400 miles, with a path width of 60 miles or more that expands into a crescent-shaped storm complex.

About two-thirds of the derechos in the central and eastern United States occur between May and July, when the available energy and instability peak. A cluster of storms organizes with the support of a wave of mid-level energy in the flow in a moisture-rich environment. The Corn Belt is a favorite zone for a summer derecho because plants give off moisture. A serial derecho event develops when a storm complex is largely separated from the main inflow, while a newly evolving line of gusty storms propagates southeastward into an untapped steamy air mass.

On the late morning of June 29, 2012, a group of thunderstorms formed over eastern Iowa, signaling an active convective day. The advancing line of storms transited on the northern edge of an exceedingly hot and steamy air mass. By early afternoon, building thunderstorms exploded southeast of Chicago, where the temperature had risen to 100°F south of a quasi-stationary boundary. A few hours later, the cluster of storms was joined by a new batch in central Indiana that coalesced into a destructive squall line.

Wind damage reports initially poured in across northern Illinois. Indiana and Ohio in a developing massive bowing structure. The complex eventually reached all the way to southern Virginia in about 14 hours, racing southeast at an average speed of 59 mph (NOAA). A peak gust of 91 mph occurred at Fort Wayne, Indiana, and winds speeds topped 80 mph in central Ohio, felling thousands of trees and utility poles that left more than a million Ohio residents without power for more than a week in the midst of a sweltering heat.

On the blistering afternoon of June 29, 2012, Washington, D.C., reached a record high of 104°F, offering extreme convective available potential energy (CAPE). The southeastward-moving derecho weakened as the line crested the Appalachians but gained a second wind in the cauldron of heat and humidity east of the mountains. A peak wind gust of 81 mph was observed at Tuckerton, New Jersey, early on June 30, 2012. Damage was widespread south of the Mason-Dixon Line in Virginia.

The June 2012 derecho introduced the term to the public through media coverage. The infamous windstorm directly impacted 10 states, killing 22 people, mostly from falling trees, resulting in more than $2.9 billion in total damage. About 4.2 million customers lost power in the path of the storm system that traveled 800 miles before moving off the Atlantic Coast around 4:00 a.m. on June 30. More than 1,100 reports of damaging winds were listed by local NWS offices.

A damaging winter windstorm on February 11, 2009, occurred with intense low pressure that tracked northward through the Great Lakes into southeastern Canada. Along the trailing cold front, a line of thunderstorms formed in the evening hours that produced significant wind damage across much of Pennsylvania. A peak wind gust of 92 mph was recorded at Allegheny County Airport, near Pittsburgh, amidst intense wind gusts of 60 mph and higher overnight on February 11–12, 2009, associated with a steep pressure gradient between departing low pressure and Canadian high pressure.

In eastern Ohio, western Pennsylvania, and northern West Virginia more than 500,000 homes and businesses lost power during the relentless assault. An embedded microburst with a half-mile path was discovered north of Yellow Creek State Park in Indiana County that damaged several homes and toppled many trees. Maximum gusts were gauged around 80 mph.

The strong winds affected eastern Pennsylvania beginning shortly before dawn behind the cold front, knocking down susceptible trees and power lines and ripping off roofs. Many vehicles suffered damage caused by falling trees. Across the state, about 250,000 homes and businesses lost power, which was not fully restored for another two days (*Storm Data*). Stretches of Routes 309 and 611 were closed due to downed trees.

A peak wind gust of 64 mph in Forks Township, Northampton County, and 55 mph at Philadelphia International Airport left pockets of damage. In the city of Philadelphia, PECO said that 85,000 homes and businesses lost electricity, where roofs and billboards were damaged and snapped tree limbs fell on homes. The greatest damage occurred in Bucks and Delaware counties.

Hailstorms

Hail has a brighter signal on Doppler radar because ice reflects considerably greater microwave radiation than liquid water. Large hail develops in moist air at cold, high altitudes, sustained by strong updrafts of 50 to 100 mph. Ice particles are suspended, picking up water droplets that freeze on contact.

Towering thunderstorms building in an area of strong winds aloft develop tilted updrafts that last longer in the cold upper atmosphere, especially in the case of discrete cells. Lumps of ice that hit the ground, from quarter size (1 inch) to softball size (4 inches), fall when the tumbling ice is too heavy to be suspended.

On July 23, 2010, a hailstone that measured 8 inches across (18.62 inches in circumference) was found at Vivian, South Dakota.

Hailstorms result in an average of $1 billion in damages annually in the United States, flattening crops, tearing shingles and cracking windshields. A hailstreak can cover a considerable distance of 5 to 25 miles and extend around a mile wide. Leaves are stripped as the hail piles up in drifts. The terminal velocity of a large hailstone can reach 100 mph. Hail discharged by a tall thunderstorm falls at the fastest rate, traveling at an angle rather than straight down. Infrequently, injuries and even a rare fatality have been reported, when people are caught out in the open during a bombardment. The loss of farm animals caught outside in a hailstorm is less uncommon.

The physics of hail formation begins with suspended ice crystals and graupel collecting on aerosols that act as freezing nuclei for water droplets caught in robust updrafts. Supercooled droplets collide with ice, windborne dust, salt, clay particles and microscopic pollutants. The embryonic hailstone grows by accretion as liquid drops freeze on its edge, progressively encasing the ice crystals in alternating layers of clear and translucent ice that resemble the skin of an onion.

The chaotic currents in thunderstorms create myriad, irregularly shaped hailstones. When the hail becomes heavy enough so that gravity pulls balls of ice downward as the updraft weakens, a hailstorm ensues. The hail is tossed sideways as it exits the storm. Evaporative cooling serves to maintain the integrity of a hailstone, even with partial melting through warmer air in the lower levels, which reaches the ground mostly intact.

Early newspapers are replete with accounts of hailstreaks. A *Country Gentleman* correspondent in Philadelphia noted, "a very severe hailstorm, which did much damage to the crops as well as property" on July 20, 1859, after a warm spell.

In *Climate and Crop Service: Pennsylvania Section*, there was an account of a widespread hail and wind event on September 17, 1896, in eastern Pennsylvania:

> One of the most severe hail storms since that of 1871 swept over the eastern portion of the State on the afternoon of the 17th, from Lycoming County in a southeasterly course over portions of Columbia, Montour, Northumberland, Schuylkill, Lehigh, Berks, Bucks, Montgomery, Chester, and Delaware Counties. Several buildings were struck by lightning and burned, and the damage by wind and hail to buildings, window glass, trees, grape vines, orchards, garden truck, poultry, etc., amounted to many thousands of dollars. Many of the hailstones were reported as large as hens' eggs, and there were wagon loads the size of walnuts.

A *Philadelphia Inquirer* headline the next day blared, "HAIL AND WIND WROUGHT HAVOC" the previous afternoon. Bloomsburg was hit by "the most disastrous hail storm ever known in this vicinity." Property damage caused by lightning, hail and wind that unroofed buildings was put at $20,000. The hailstones in Shenandoah that were "the size of large plums" shattered 3,000 windows. A dispatch from Bucks County reported that "nearly all the telegraph and telephone poles between Bristol and Croydon are down."

In Sellarsville, a stone pillar was torn away from a barn, and a hundred oak trees were "shattered and twisted" by a tornado. That storm destroyed a stable, carried a large water tank 400 feet and tore a railroad platform from the foundation, which was "stood on end." Many businesses and homes suffered broken windows, especially on the northern exposures, and the skylights of the cigar factories "containing 150 panes of glass" were destroyed. A storm that hit Perkasie damaged small buildings and ruined hundreds of thousands of bushels of apples.

A severe hailstorm on August 9, 1915, struck Carlisle, breaking thousands of panes of glass in an hour, the result of hailstones at least 1.5 inches in diameter. Another storm on August 31, 1922, described in *Climatological Data: Pennsylvania Section*, put down enough hail around West Chester that the next day it was still two feet deep, and drifting.

A severe hailstorm on the afternoon of July 21, 1929, near Johnstown unloaded a barrage of hail that broke more than 2,000 windows, piling hailstones into drifts two to four feet deep.

The *Wayne Independent* described the damage caused by high winds and a violent hailstorm on June 23, 1941, that produced hail "as large as a fist," piling up three feet deep in drifts west of Hawley. Crops were destroyed, windows were smashed, and chickens were killed during the 45-minute storm at Maplewood.

Severe storms pounded parts of eastern Pennsylvania for hours from 7:00 to 11:00 p.m. on July 9, 1945. The *Climatological Data: Pennsylvania Section* report described "hail the size of hickory nuts" (up to two inches in diameter) that contributed to storm damage that totaled about $4 million. Excessive rainfalls of up to eight inches fell at Bangor and Nazareth; observer totals included 6.20 inches at Easton and 3.82 inches at Stroudsburg, causing flash flooding.

A severe weather outbreak in western Pennsylvania on September 25, 1994, brought funnel clouds and large hail in places in the early evening. Hail ranged from quarter-sized to as large as 3.80 inches in diameter seven miles northwest of Meadville near Conneaut Lake, where 40 homes sustained damage at Little Corners in Hayfield Township, Crawford County. Cars and windows had heavy damage.

On June 22, 1996, hail up to three inches in diameter accompanied powerful storms that developed along a warm front in southern Monroe County, affecting an area from Kresgeville to Saylorsburg between 4:00 and 5:00 p.m.

Hail battered portions of Cumberland, Northumberland and Carbon counties during May 26, 2011. Storms raked eastern Pennsylvania, accompanied by tornadoes and flooding rain. Hailstones in Carbon County were as large as three inches in diameter, breaking windshields and denting vehicles in Lansford and Summit Hill. PPL Electric Utilities reported that about 120,000 customers lost power in the storms, according to the Associated Press (AP).

The passage of a powerful cold front on May 22, 2014, was accompanied by supercell thunderstorms during the afternoon in eastern Pennsylvania. In Berks County, near Wyomissing, hail the size of tennis balls (2.5 inches) shattered windshields, slashed crops and siding, and damaged a local mall. A U.S. Airways jet had a cracked windshield when landing at Philadelphia International Airport. Ping-pong ball-sized hail (1.5 inches in diameter) was reported in Monroe County.

Tornadoes

A tornado is one of the most visually striking phenomena on Earth, a violently rotating column of air capable of inflicting a corridor of destruction.

Every year we see stunning and often tragic images of flattened houses, collapsed walls, roofs torn off, mangled mobile homes, twisted transmission lines, corrugated metal and aluminum siding ripped from structures, toppled utility poles and rows of trees snapped or sheared off—the handiwork of a writhing corkscrew-shaped vortex of low pressure that tore through a community.

Technology has raised the number of confirmed tornadoes in the United States annually, which are responsible for an average of 69 deaths and 1,500 injuries (1991-2020). Tornadoes rated EF3 or more cause greater than 80 percent of those fatalities.

During earlier times, tornado deaths were far greater, averaging 260 annually between 1912 and 1936. Nocturnal tornadoes are responsible for two-and-a-half times the number of fatalities compared daylight storms in the United States.

During the period of 1975–2000, the mean number of tornado deaths per year in the United States dipped to 54 (Brooks and Doswell 2001). A national record streak of 283 consecutive days without a tornado fatality ended on February 24, 2018, in a late-winter storm outbreak. Only 10 tornado deaths were reported in the country in 2018, the fewest since 1875. However, despite improved warning lead time, large tornadoes are capable of incurring a heavy loss of life.

Tornado Frequency in Pennsylvania

In an average year, 16 tornadoes are confirmed in Pennsylvania (1989–2022 data), with peak activity in May, June and July. Historically, an unusually high number of significant tornado swarms have caused widespread damage in parts of Pennsylvania between May 28 and June 2, abetted by seasonally strong wind shear and increasing humidity.

The greatest number of tornadoes reported in Pennsylvania since 1950 have occurred in the western counties of Westmoreland and Crawford. A second active zone encompasses York and Lancaster counties in the southeast. An isolated pocket with more tornadoes than most other counties is Lycoming County, partly due to the valley of the West Branch Susquehanna River to the north of the rugged Appalachian ridges.

The undulating Allegheny Plateau in western Pennsylvania has the highest incidence of tornadoes, with proximity to deeper moisture transported up the Ohio Valley. Fewer tornadoes are reported farther east because of the elevated topography and a decrease in instability. A zone of increased tornado activity southeast of the Appalachian ridges is the result of flatter terrain, unfettered access to humid air streaming north along the Eastern Seaboard and the presence of a lee trough and convergent boundary.

Prior to the 1970s, weak tornadoes were not always entered into the database, so care must be taken when evaluating storm frequency and distribution. Doppler radar, implemented in the 1990s, has markedly increased the number of small storms discovered in storm surveys based on radar indications.

Most Pennsylvania tornadoes are narrow and short-lived compared to the larger storms typical of the Midwestern and Southern states. In recent years, the average storm path length observed in the Commonwealth was about 2.5 miles, and the width was 150 yards. Occasionally, a stronger EF2 tornado or greater, with 111 mph winds or higher, causes significant damage. The longest recorded track was 69 miles on May 31, 1985 (NWS/SPC).

Tornado Season

In an average year, more than 1,225 confirmed tornadoes are reported in the United States. Some are isolated events, while others arrive in clusters over a few days, resulting in significant damage, with the high risk of injuries and loss of life.

Tornado watches are issued by the Storm Prediction Center (SPC) in Norman, Oklahoma, for areas of up to 50,000 square miles. The threat of hail, damaging wind gusts and tornadoes are assigned probabilities using predictive models that incorporate kinematic (wind) and thermodynamic (temperature, pressure, moisture) data.

The nation's heartland has been referred to as Tornado Alley, extending from central Texas north to eastern Nebraska. The western extent brushes the High Plains and the eastern edge includes the Mid-South and lower Ohio Valley. A second zone of tornado activity during much of the year is the Deep South, especially in the winter and early part of spring, with a secondary season in the autumn. Studies have shown an eastward shift in the peak tornado frequency, likely due to a more prevalent southerly flow tapping a warmer Gulf of Mexico.

The topography, geography and meteorology create a landscape for contrasting air masses that frequently clash. The relatively flat terrain in the midsection of the country is a battleground for a flow of moisture-rich air that interacts with fronts and energy dispensed by waves aloft, spawning more tornadoes than any place in the world.

Tornado season shifts from the Southern states in late winter and early spring to the southern and central Plains, Midwest and southern Great Lakes in May and June. In the northern tier of states, tornado season peaks in early summer.

The highest frequency of tornadic thunderstorms occurs in zones reflecting large contrasts in temperature and wind, both horizontally and vertically. In the summer, the jet stream retreats north in response to a diminishing temperature gradient across the Lower 48, sharply reducing the tornado threat.

A short second tornado season in late autumn reflects strong areas of low pressure tracking from the southern Plains to the Great Lakes. Cold air masses push farther south behind an invigorated jet stream. Deep-layer wind shear is sufficient to overcome modest instability that results from less intense daytime heating, a setup favorable for bands of low-topped thunderstorms capable of spawning tornadoes.

The passage of a tropical storm often brings relatively weak tornadoes after the system moves inland and encounters friction over rough terrain. Strong mid-level rotation associated with the circulation center sometimes spins up short-lived tornadoes in the spiraling rainbands.

Tornado Formation

Tornadoes form in an environment where winds are shifting direction and gaining speed with height. A storm updraft that usurps strong low-level shear along an axis of moisture is more likely to spawn a funnel cloud capable of reaching the ground.

Storm updrafts acquire spin at the interface of an instability gradient and outflow boundary. The atmosphere is primed where cool air lies adjacent to pockets of sunshine. The colocation of a moist southerly low-level jet at a few thousand feet and fast-moving southwesterly winds in the upper levels promotes deep-layer vertical wind shear. The intense wind field tilts updrafts downstream away from downdrafts, providing ventilation that sustains moisture-rich inflows.

A discrete storm off by itself is more likely to rotate. A thunderstorm moving off a frontal boundary at a sharp angle feeds off the relatively unimpeded inflow jet, farther removed from updrafts and downdrafts along a line segment. Right-splitting cells raise a red flag because those storms have the best access to an unfettered inflow of moisture and ambient wind shear.

The effective bulk shear represents the wind shear gradient from the cloud base to the mid-levels, which is the lower half of the storm. A mean shear vector directed at a sharp angle from a surface boundary favors rotating storms. Research shows that a bulk shear value difference greater than 30 mph significantly increases the risk of a tornado.

Tornadogenesis begins when strong winds aloft impart a horizontal rolling motion that breaks down into smaller vortices at a critical value. Robust updrafts that punch through turning tubes of air stretch the column in the vertical and ingest ambient spin (vorticity). As the inflow increases and the column tightens, the rotational velocity increases, much like how a figure skater gains speed by drawing in his/her arms to shorten the radius, based on the conservation of angular momentum.

Fewer than 20 percent of all supercells spawn tornadoes, though most strong tornadoes originate from a broader storm circulation. The mesocyclone slowly contracts from a diameter of up to 10 miles to about two to six miles, as the pressure drops.

The supercell resembles a spaceship or upside-down layer cake, with a lowering cloud base that is sometimes obscured or rain-wrapped. A funnel cloud is revealed with a skirt-like collar cloud that is the demarcation line between more stable air.

A lowering wall cloud reveals laminar striations, usually in the southwest quadrant of the storm, with a rain-free cloud base.

A mature rear flank downdraft (RFD) carries mid-level spin to the near-surface. Inside the rotating column, air is descending rapidly and warming by compression, accelerating a buoyant RFD as it wraps around the mesocyclone and meets up with a forward flank downdraft. Under the right conditions, a condensation funnel appears beneath the cloud base, surrounded by twisting filaments.

The classic rotation signature of a tornado correlates with a surge of rain pushing out in the shape of a mini bow. As air is turned inward and lofted, a "hook echo" sometimes appears on Doppler radar, where a tornado often hides in a mesocyclone.

A tornado can assume a variety of appearances, from a narrow cylindrical pendant with a whirling column of dirt (dust envelope) to a menacing wedge tornado as wide as the column is tall. The storm is sometimes eerily backlit by the sun as a funnel descends. Tornadoes that are either obscured by darkness or rain-wrapped create a particularly dangerous situation because they arrive with less visual warning signs.

The wind inside a tornado is unstable and sometimes breaks down into subvortices, primarily on the southern edge of the parent vortex, possessing rotational winds 50–100 mph faster than the main vortex. Such "suction vortices" are associated with near-complete destruction of some properties while adjacent ones are left relatively unscathed.

A long-track supercell poses a significant threat to spawn a family of tornadoes, as the storm system undergoes cycling, inflicting widespread damage. Eventually, rain-cooled air chokes off the inflow and a tornado will "rope out" like an elephant trunk and lift back into the clouds.

Landspouts and gustnadoes are not associated with a supercell. A skinny, weakly rotating column of air occasionally develops when a thunderstorm interacts with low-level wind shear. Storm outflow stretched vertically by an updraft can create a slim vortex. Landspouts are more common in the High Plains churning across farm fields and are counted as tornadoes, unlike gustnadoes that form along a rain-cooled gust front and create narrow eddies in a column of dust. Air forced to flow around a solid structure accelerates due to lower pressure on the opposite side resulting from the pressure gradient.

A dust devil occurs with intense surface heating that initiates updrafts stretched in the vertical. The column of hot air contracts with increasing spin on a sunny, breezy afternoon. A hard-baked field is the best expanse for a narrow whirl 10 to 1,000 feet tall that resembles a miniature tornado.

On April 21, 1963, a landspout at Reading descended from a clear sky around 5:30 p.m., tearing a brick veneer off the side of a school building, uprooting trees and knocking down power lines along a 0.2-mile path. Keen (1992) wrote that the storm appeared as a "black spiral, easily one-half mile high." The path width was 15 yards.

A feisty landspout on April 26, 1987, three miles northwest of Williamsport ripped the plastic off a greenhouse and lifted a pick-up truck and camper off its blocks, moving it 15 feet and scattering debris more than 300 yards.

Tornado Detection

Before the advent of Doppler weather radar in the 1990s, only a quarter of all confirmed tornadoes in the United States had warnings, compared to 75 percent in 2004. Weather Surveillance Radar technology classified as WSR-57 and WSR-74 replaced surplus military units, designated by the year of development (1957 and 1974, respectively). These conventional radar systems could not measure storm winds.

Although tested in the 1950s, the NWS did not deploy Doppler radar systems until the early 1990s, which came after the introduction of the newer WSR-88D in 1988. The Next Generation Weather Radar (NEXRAD) network was implemented in a $4.5 billion NWS modernization plan in the early 1990s, which included the installation of Doppler systems at 122 Weather Forecast Offices (WFOs).

Currently, 159 NWS Doppler radar systems are located in the United States and Puerto Rico, with 37 Terminal Doppler Weather Radar (TDWR) sites operated by the FAA and Department of Defense. Upgraded technology and integrated algorithms have substantially reduced storm-related deaths and injuries by cutting the average warning lead time for tornadoes to 10 to 15 minutes, compared to only four minutes in 1980.

In 1842, Austrian physicist Christian Doppler initially described the "Doppler effect" in regard to the compression of sound waves caused by the motion of the source relative to an observer. The change in the frequency is perceived by the rising pitch of an approaching train whistle.

A rotating Doppler radar antenna emits pulses of microwave energy that bounce off precipitation particles. The returning signal shortens/lengthens as it measures the velocity toward/away from the from the radar, which enables the radar computer to detect rotation inside a thunderstorm. Airborne dust in a thin line is indicative of storm outflow.

Volume scans typically take up to five minutes for 14 elevation angles, although shorter scans in the lower layers are employed during severe weather.

Solid objects such as hail send back a stronger signal. An algorithm determines the size of the raindrops and snowflakes. The time it takes for reflected energy to bounce back to the radar determines the distance of the target.

Switching from reflectivity to velocity mode provides a view of the wind direction and speed inside a storm. Faster winds that change direction relative to the radar are identified by the sharp shift of the returning signal, which reveals whether a target is moving toward (green) or away (red) from the radar. Cyclonic rotation is defined by peak inbound and outbound velocity divided by two. Velocity data signatures alert forecasters to a possible tornado but cannot confirm a funnel is on the ground.

Fig. 7.2 Pittsburgh's National Weather Service radar dome.

A critical phase shift leads to a Doppler-indicated tornado warning for rotation. The storm-relative velocity mode subtracts out the mean storm motion, providing the most accurate picture of the internal storm winds because the rotational velocities are generally masked in a fast-moving supercell. Storm-relative helicity (SRH) is a parameter that predicts the vertical shear of the horizontal wind that veers clockwise with height in a twisting or corkscrew motion.

A supercell typically takes the shape of a kidney bean, displaying a curling appendage on the southwestern edge in the form of a "hook echo." The "doughnut hole" indicates a vigorous updraft sweeping most of the raindrops upwards, leaving a relatively echo-free space. A tornado vortex signature (TVS) is identified by a tight couplet, where the colors are intertwined, indicating strong rotation. Bright outbound (red) and inbound (green) velocities, with pixels of yellow and blue, reflect "gate to gate" shear as the circulation tightens.

Dual-polarization technology adds another critical piece of information by integrating the horizontal and vertical beams. Dual-pol data distinguishes between precipitation types (rain, ice, hail) and irregularly shaped tumbling non-weather targets, likely lofted debris. A polarimetric tornadic debris signature (TDS) strongly suggests a tornado that is lifting debris high enough to reflect the radar beam.

Yet even the best radar technology should be corroborated by ground truth from trained weather spotters, who report directly to the NWS. This is especially important as the distance from the radar site and target is significant. The beam height (0.5-degree elevation angle) passes above a low-level circulation, which is why spotters are essential.

The presence of a TVS or a spotter report of a funnel cloud or tornado triggers a tornado warning. The NWS disseminates storm warnings over NOAA weather radio frequencies, and Wireless Emergency Alerts on compatible mobile devices. Tornado siren systems for outdoor use are spotty in Pennsylvania.

Television meteorologists rely on storm trackers and shear markers using an algorithm that pinpoints persistent rotation, a feature that comes with most radar apps. Storm warnings are passed on by radio and television reports, newsroom push alerts and wireless alerts to cell users in the threat zone.

Tornado Classification

Tornado damage can range from twisted and sheared trees to major structural collapses.

Professor T. Theodore Fujita (1920–1998), a preeminent tornado researcher at the University of Chicago, and Allen Pearson (1925–2016), director of the National Severe Storms Forecast Center (formerly the Severe Local Storms Unit) during 1965–79, devised the Fujita Scale (F-Scale) in 1971 to quantify tornado intensity.

The original scale classified tornado damage by wind speed, path length, width and expected damage. In February 2007, a newer tornado wind-damage scale was incorporated to better account for types of building construction and materials (Table 7.1). The Enhanced Fujita Scale (EF Scale) reflects a more precise measure of the level of wind damage and has been employed in the United States and Canada since 2013. The EF Scale incorporates 28 damage indicators based on the usual construction for the type of category. The Degree of Damage (DOD) rating is used to determine the estimate of wind speed, with a lower and higher bound.

Table 7.1
Fujita Tornado Scale (F-Scale)

F-Scale	Wind Estimate (MPH)	Expected Damage	Typical Damage
F0	<73	Light	Little damage
F1	73–112	Moderate	Minor damage
F2	113–157	Considerable	Roofs gone
F3	158–206	Severe	Walls collapse
F4	207–260	Devastating	Homes leveled
F5	261–318	Incredible	Total destruction

**Table 7.2
Enhanced Fujita Scale (EF Scale)**

F-Scale	Wind Estimate (MPH)	Expected Damage	Damage Indicator
EF0	65–85	Light	Tree branches
EF1	86–110	Moderate	Mobile homes
EF2	111–135	Considerable	Roofs peeled
EF3	116–165	Severe	Walls collapsed
EF4	166–200	Devastating	Homes leveled
EF5	>200	Incredible	Total destruction

Most tornadoes occur between 2:00 and 10:00 p.m., during the warmest period of the day. The average forward speed of a tornado ranges between 20 and 40 mph, though on rare occasions a tornado is nearly stationary, and a few have traveled at highway speed.

About 80 percent of confirmed tornadoes are relatively weak (EF0/EF1), with maximum winds of less than 110 mph. Typical damage is usually evident as uprooted trees, damaged outbuildings, siding and shingles peeled off and mobile homes pushed off foundations.

Strong tornadoes (EF2/EF3) pack winds in the range of 111–165 mph, easily capable of lifting a roof off a building, toppling weaker structures, rolling manufactured homes and overturning vehicles. Debris from trees and insulation is often scattered for hundreds of yards.

Violent tornadoes (EF4/EF5) have winds of 166 mph or greater, potentially causing heavy damage. Less than 5 percent of all tornadoes are classified as violent, but these storms are responsible for 75 percent of tornado deaths in the United States. Tornadoes reaching EF5 intensity (200-plus mph) toss automobiles and cinder blocks like small toys, level sturdy buildings and remove homes from foundations, derail railroad cars, tear up asphalt and otherwise cause complete destruction.

Many smaller tornadoes likely went undetected when the country was sparsely settled in the 18th and 19th centuries and were sometimes discovered belatedly, appearing as narrow tracts of twisted trees. U.S. Army Sergeant John Finley's *Report on the Characteristics of 600 Tornadoes*, published in 1884, represented the first attempt to collect a large body of data on American tornadoes, mostly based on press accounts. Finley, with the aid of 2,000 "reporters" in the field, was able to create tornado maps that identified weather patterns capable of producing a whirling storm.

Finley's goal of providing tornado alerts for the public met intense internal resistance in 1886, when the U.S. Army officially prohibited the word "tornado" from all forecasts because of a concern that such wording could cause panic. The Weather Bureau did not formally issue tornado warnings until 1952.

Snowden (1953) catalogued damaging storms in *Tornadoes of the United States*, the first comprehensive storm record since Finley's seminal work. Much of the credit for compiling a comprehensive account of early American tornadoes belongs to the preeminent weather historian David Ludlum (1970) for his multifaceted historical accounts, published in his *Early American Tornadoes, 1586–1870*.

Pennsylvania climatologist Paul Dailey Jr. prepared the first tornado risk assessment in 1970, which was updated in detail by Nese and Forbes (1998). A national record of tornadoes was exhaustively catalogued by Grazulis (1993) in his encyclopedic *Tornado Project*.

The National Centers for Climate Information (NCEI) tornado database is comprehensive since 1950, though small storms that caused minor damage were often not counted, and discontinuous long-track tornado damage was likely the work of multiple storms comprising a tornado family. NOAA's monthly and annual *Storm Data* publication categorizes damage and losses by state commencing in 1959.

Storm Damage: Tornado or Straight-Line Winds?

Storm surveys conducted by NWS personnel usually within a day of damaging storms determine whether the ground pattern is suggestive of either straight-line winds or a tornado. Radar signatures corroborate rotation in the vicinity of the damage that matches observational evidence, confirming that a tornado touched down.

Tornado damage is easily distinguishable from straight-line winds by a swirl of debris that curls back to a convergent path. Strong tornadoes leave tracts of splintered or snapped trees without leaves that were sheared, flipped cars, flattened structures, crushed mobile homes, twisted transmission towers and utility poles, sheet metal wrapped around signs and roofs peeled off, which causes walls to collapse.

Wind-blown debris strewn across a field in a starburst pattern is indicative of a straight-line wind event caused by a downburst. A microburst is capable of felling a row of trees usually rooted in moist soil, tearing the siding off of homes, hurling outbuildings and otherwise scattering loose objects over hundreds of yards.

The Stroudsburg *Jeffersonian* relayed an interesting description of a classic "wind rush" south of Blue Mountain that struck on November 6, 1880, that originally appeared in the *Easton Free Press*:

> Reports from out toward Nazareth show that the wind was very heavy in a tract line three or four miles wide, reaching from Ehrit's woods, a mile and half from Nazareth, to Seipeville. Trees were broken or uprooted, fences laid low, haystacks and cornstalk mounds scattered, and general damage to light, movable articles done. Several stables were damaged a little, and one barn belonging to a Mr. Schlaum of Seipeville was moved out into the wood and completely demolished. The woods at Seipeville, near the school house, were also badly damaged, in having many limbs from trees broken.

A typical strong downburst pattern occurred in northwestern Pennsylvania on July 16, 1997. NWS storm surveys showed that portions of Warren, Elk and McKean counties were on the receiving end of a collection of microbursts. An Elk County damage swath six miles long and three miles wide observed near Millstone and Belltown was attributed to wind speeds of 60 mph. Three separate paths pointed to microbursts in Warren County that felled trees.

On May 31, 2002, a macroburst hit Allegheny County around 7:00 p.m. that fanned out three miles in width and traveled 10 miles. Wind gusts were estimated at 105 mph at West Mifflin when violent winds struck the Kennywood Amusement Park. The pavilion above the Whip buckled in the fierce winds, killing a woman. A total of 47 injuries were reported from flying debris, and dozens of trees were toppled in the park (*Storm Data*).

Hundreds of trees were uprooted across the Allegheny Cemetery at Lawrenceville, and at the

Pittsburgh Zoo and Aquarium in nearby Highland Park. Buildings in the eastern parts of Pittsburgh suffered varying degrees of damage, especially in areas around Wilkinsburg and Homestead.

A downward surge of dense air poses a serious aviation hazard, particularly during takeoffs and landings. A strong headwind creates sudden lift over the wings, increasing the airspeed. After passing through the middle of a spreading column of rain-cooled air, a forceful tailwind reduces the airspeed and the aircraft lurches downward.

If a pilot overcompensates for the headwind by pulling back on the throttle, the unexpected loss of initial lift with the tailwind can prove disastrous. Federal Terminal Doppler Weather Radar (TDWR) units deployed by the FAA at airports and runway sensors detect crosswinds that alert tower personnel of this potential hazard.

Historic Tornado Outbreaks

In the days before tornado warnings and modern communications systems, large tornadoes caused a heavy loss of life. On May 27, 1896, a series of tornadoes killed 350 in St. Louis, Missouri. The death toll in the Southeast from April 23–25, 1908, was reportedly 324, including 143 deaths in a single tornado that struck Amite, Louisiana, and Purvis, Mississippi. Major tornado outbreaks in May 26–28, 1917, and May 7–9, 1927, took more than 200 lives.

The Great Tri-State Tornado of March 18, 1925, killed 695 persons in Missouri, Illinois and Indiana (747 total storm deaths from multiple storms). The long-tracked supercell caused death and destruction along a path of 219 miles, likely a family of tornadoes. A reanalysis suggested a single tornado caused 175 miles of continuous damage.

The second deadliest tornado ripped through Tupelo, Mississippi, on April 5, 1936 (216 deaths), and the third (202 deaths) struck Gainesville, Georgia, the next day. On April 9, 1947, the death total in Woodward, Oklahoma was 181. Large tornadoes in the spring of 1953 killed 114 people on May 11, in Waco, Texas; 116 in Flint, Michigan, on June 8; and 94 in Worcester, Massachusetts, on June 9.

During a 16-hour period on April 3–4, 1974, 148 tornadoes touched down in 13 states and southeastern Canada, resulting in 315 deaths in the United States and nine at Windsor, Ontario. A total of 330 storm-related fatalities included victims killed by straight-line winds, accidents and a fire that claimed the lives of two Ohio Air National Guardsmen during the night of April 6, while they monitored the recovery effort in Xenia.

The severity of the storms was summarized in NOAA's *Natural Disaster Survey Report 74–1, The Widespread Tornado Outbreak of April 3–4, 1974*. Seven violent storms were rated F5 tornadoes (261 mph or greater on the Fujita-Pearson scale); 23 reached F4 intensity (207–260 mph). The 1974 Super Outbreak injured more than 6,000 and caused damage in excess of $600 million (1974 dollars).

A deadly multiple-vortex tornado storm traveled 32 miles across southwestern Ohio, with winds estimated around 300 mph (F5). The tornado contained up to five subvortices that consolidated into a half-mile-wide tornado approaching Xenia. In a matter of minutes, nearly half of the city, a community of 27,000, lay in ruins.

The Xenia tornado killed 32 people and injured more than 1,300, who were treated at Greene Memorial Hospital. A total of 1,200 homes and 180 businesses were damaged or destroyed, including nine churches and seven schools. The losses totaled more than $100 million at the time.

A massive four-day severe weather event from April 25–28, 2011, spawned tornadoes in 21 states and Ontario, Canada. The Super Outbreak of April 2011 killed 324 people (348 storm-related deaths) and caused widespread damages that totaled $12 billion (2020 dollars), establishing a national record (until December 10–11, 2021).

On April 27, a swarm of 216 tornadoes claimed 316 lives (248 in Alabama) and injured more than 3,100. Fifteen tornadoes were violent (EF4+), with three reaching EF5 power (NOAA). One singularly vicious EF4 tornado carved a path of more than 80 miles from Tuscaloosa to Birmingham, causing 64 fatalities and eight indirect deaths, plus a reported 1,500 injuries.

More than 360 tornadoes touched down during April 25–28 in a zone from Texas to New York. The record number of tornadoes during the month (758) exceeded May 2003 (542). The annual tornado total in 2011 (1,691) was second only to 2004 (1,817).

On May 22, 2011, a multiple-vortex EF5 tornado nearly a mile wide cut a deadly swath through Joplin, Missouri, killing 158 (161 total fatalities). About 75 percent of the city suffered heavy damage along a path that extended for 22 miles, including zones outside of the city. More than 7,500 homes and businesses were heavily damaged or destroyed. Losses totaled $2.8 billion, which included 553 businesses and nearly 7,500 residential structures.

The Joplin tornado occurred during a multi-day tornado outbreak (180 tornadoes) between May 22 and May 27, 2011. The damage totaled $9.1 billion and took 177 lives in a week of violent weather.

Late on December 10, 2021, an exceptionally strong upper-air flow and trough of low pressure interacted with record near-80-degree warmth in the Mid-South, triggering a devastating swarm of tornadoes in waves mostly at night, fueled by historically warm high Gulf of Mexico temperatures. One quad-state supercell thunderstorm traveled from Arkansas to south-central Ohio during an 11-hour period, spawning a family of long-tracked tornadoes that struck communities with deadly precision in the dark of night.

A violent EF4 tornado on the night of December 10-11, 2021, with winds as high as 190 mph, traveled nearly 166 miles from western Tennessee into southwestern and central Kentucky during a three-hour period. The long-tracked tornado devastated half of the community of Mayfield, Kentucky, resulting in 57 fatalities and 515 injuries. The rare December outbreak spawned 71 tornadoes in nine states, killing 89 (plus six indirect deaths) and causing nearly $4 billion in damage. The greatest loss of life occurred in Kentucky (80).

Incredibly, on December 15, 2021, an unprecedented early winter derecho fueled by record warmth in the Midwest (70s) and a moist southerly flow raced northeastward across eastern Nebraska, Iowa and Minnesota. Wind gusts reached 93 mph at Lincoln, Nebraska, and 100 mph in western Kansas, kicking up dust storms. A total of 79 tornadoes were reported amidst the assault, including 15 in Minnesota, where no December tornadoes had occurred in the historical data (since 1950). A record single-day number of hurricane-force wind gusts topping 74 mph (63) contributed to five deaths and widespread damage.

Early Pennsylvania Tornadoes

The first recorded tornado in the Philadelphia area touched down in the vicinity of Paoli and Valley

Forge in the early afternoon of August 3, 1724. The funnel took a northeasterly course through Chester and Montgomery counties roughly parallel to today's Pennsylvania Turnpike (Ludlum 1970).

Roads were blocked for days by downed trees and debris. The *American Weekly Mercury* on August 13, 1724, described a pattern of uprooted trees blown over a considerable distance, fences capsized, roofs torn off homes, a collapsed mill with millstones removed and barns destroyed.

> On the thrd instant [August 14, 1724, New Style], about the hour of 12 … a most terrible and surprizing Whirl-wind …. From the *Philadelphia Country* I obtained the following: At Plymouth the whole roof was pulled off a big barn and carried out in the lot; a woman's skirt seven or eight miles in the air, and grain stacks were strewn about the field. It took up almost all the apple trees in the orchard by the root and carried them some distance. People were in danger of being carried off right in their houses.

The first reported waterspout in the region was visible over the Delaware River on June 4, 1754, in the midst of a "violent gale of wind with rain and hail," according to the *Pennsylvania Journal and Weekly Advertiser*. The June 6, 1754, account described "considerable damage in the country," after the storm broke in the Jerseys."

A destructive squall line with embedded tornadoes raced through parts of Maryland, Pennsylvania, New Jersey and New York on June 22, 1756 (Ludlum 1970). One person was killed and several seriously injured at Moyamensing, near Philadelphia, when a barn sheltering several people collapsed. Two other fatalities may have occurred at Gloucester.

The Public Library on Fifth Street in Philadelphia was one of a number of buildings damaged in the storm, according to the *Pennsylvania Gazette* of June 24, 1756. Two hundred homes were "blown down" in St. Marys County, Maryland, the same afternoon (Ludlum 1984). Tornadoes also struck Essex County, New Jersey, as well as Jamaica, Queens, and western Long Island.

Heverly (1926) recalled a major windstorm in March 1794 in Granville Township, recalling that "a tornado swept through the southwest part of Bradford county, extending into Sullivan county, and in its path of a mile in width, almost every tree was uprooted or broken."

The storm could have happened on June 19, 1794, when tornadoes struck Brookville and also Poughkeepsie, New York (Ludlum 1970), and in Connecticut (Grazulis 1993).

June 1796
Great Windfall

Henry Bradsby's account in his *History of Luzerne County Pennsylvania*, described the severe squall line east of Wilkes-Barre in June 1796:

> The first tornado known to carry havoc through the valley was in 1796. It passed over the country from west to east, unroofing barns and dwellings, and producing on the headwaters of the Lehigh what, among the old inhabitants, was called "The Great Windfall." The road leading from Wilkesbarre to Easton was completely barricaded with fallen trees, which required several months

of labor to remove. Our county appropriated $250 toward the expense.

The remnant damage piqued the interest of renowned ornithologist Alexander Wilson (1766–1813), a Scottish-born former schoolmaster, who resided in both western New Jersey and eastern Pennsylvania. Wilson undertook an arduous journey, with two companions and a notebook, from Philadelphia up to Niagara Falls and back between August and December 1804, included in his *American Ornithology* (1808–1814).

Wilson composed a 2,219-line poem entitled "The Foresters" that appeared in serial form in 1809–1810, describing a swath of wind damage he encountered on the Pocono Plateau, in an area known as the Great Pine Swamp.

In his journal published posthumously in 1818, Wilson wrote: "These tornadoes are very frequent in the different regions of the United States. The one above alluded to, had been extremely violent; and for many miles had levelled the woods in its way. We continued to see the effects of its rage for nearly 20 miles."

The storm originated near the Lehigh River at Stoddartsville in Luzerne County, near the Monroe County line, and continued in an easterly direction, following the "brow of Effort Mountain." Historian William Lesh (1945) wrote that early settlers gathered pine knots scattered for firewood for many years along the remnant storm path.

February 1824
Violent Gale at Wilkes-Barre

Tornadoes are rare in the middle of winter in Pennsylvania, which makes the description in Henry Bradsby's *History of Luzerne County, Pennsylvania* noteworthy:

> In February 1824, a most terrific hurricane passed up the Susquehanna river, prostrating fences, trees, barns and dwellings. Such was its power that it lifted the entire superstructure of the Wilkes-Barre bridge from its piers, and bore it some distance up the river, where it fell on the ice with a thundering crash.

Harvey (1909), in a voluminous Wilkes-Barre history, noted that the storm "possessed no cyclonic tendencies and did no other damage in the neighborhood." The damage could have been caused by straight-line winds, but the force required to pull a bridge from its support and carry the structure some distance suggested a tornado.

May 18, 1825
Butler Tornado

A family of violent tornadoes ripped through the northern half of Ohio on the afternoon of May 18, 1825, wiping out the community of Burlington, about 40 miles northeast of Columbus (Ludlum 1970).

As the storm system marched eastward, at least one damaging tornado formed that struck Butler, according to the *Butler Repository* (May 21, 1825). A full account of the storm appeared in the Washington *National Intelligencer* (June 1, 1825).

The story said that "houses, barns, fences, orchards, and woods, were leveled to the ground."

On Wednesday, the 18th instance, a part of this country was visited by one of the most violent and destructive tornadoes that has ever passed through it. The storm came from the southwest and passed in a northeast direction. We have not yet learned where it commenced, nor how far it continued its destructive march. It passed diagonally through this county, and its ravages are about a mile in width. It had a huge volume of smoke, arising from a tremendous fire, which, with the vivid and continued flashes of lightning, the loud peaks of thunder, the rattling of hail, and the crash of timber, with which it was accompanied, gave to it an awful and terrible appearance, that baffles description.

March 22, 1830
Allegheny County Tornado

Damaging tornadoes struck parts of Ohio on March 22, 1830. The town of Urbana was hit by a deadly twister around 2:00 p.m. that killed several children in their homes and seriously injured many other residents in the small west-central Ohio according to the *Ohio State Journal*. The storm system moved into western Pennsylvania later in the evening, spawning a powerful tornado in southern Allegheny County.

An item in the *National Intelligencer* on April 2, 1830, included the *Pittsburg Gazette* (March 26, 1830) story of a violent tornado that struck Elizabeth-town (now Elizabeth) around 7:00 p.m.:

Fourteen houses are blown down and unroofed; five barns and stables, one boathouse, one mill and one wool-carding establishment, completely crushed, with many other houses much damaged. Many families are turned out, without a roof to shelter them from the pitiless storm. Beds, bedding, and household furniture, are to be seen hanging amongst the broken timber, and strewn along the road.

Boats were tossed about and broken to pieces along the banks of the Monongahela River. The press account added, "Thanks to that Providence, who watches over and Protects us amid such calamitous visitations, no human lives are lost, though many have received slight wounds."

Another account of the evening tornado in Washington County listed two fatalities—a father and his son—in one of several homes blown down by a tornado, which was a half-mile wide, stated the Gettysburg *Republican Compiler* (April 6, 1830).

June 19, 1835
Tornado Outbreak

Hazard's Register of Pennsylvania recounted tales of a destructive windstorm that sliced through the Alleghenies and wreaked havoc all the way to the coastal plain on Friday afternoon, June 19, 1835.

The July 11, 1835, issue carried an account of a tornado judged to be a mile wide that tracked between four and five miles. The storm commenced near Clearfield, "prostrating the stoutest forest and lifting up and driving before it every thing it encountered." A house was flipped over in the storm, and a child was trapped by fallen timbers, though surviving without serious injury.

The *Susquehanna Register* told of damage in Springville, southwest of Montrose, where several homes were damaged and barns unroofed. In the previous weekly edition on July 4, 1835, the *Register* carried a dispatch describing a tornado near Williamsport: "A destructive storm passed within a few miles of our borough … doing much injury to the property that lay in its course." The largest damage occurred in a rural section along Lycoming Creek.

A report in the *Kingston Republican and Herald* mentioned powerful winds that caused "immense injury" to property in the southern portion of Wilkes-Barre "confined to a narrow strip of country." Embedded tornadoes alongside veins of straight-line wind damage accompanied a thunderstorm complex that was labeled a "hurricane."

Accounts of unroofed homes, flattened barns and fences and damaged crops were widespread. Wind damage was reported in Columbia County, where several barns collapsed. The York *Republican* stated that "some buildings were injured" and trees were felled.

The powerful thunderstorm system spawned New Jersey's most deadly tornado, which struck downtown New Brunswick about 5:00 p.m. and continued to Piscataway and Perth Amboy. The *New York Evening Star* reported that the tornado caused "dreadful havoc" in New Brunswick, "destroying and injuring nearly one hundred and fifty houses." A few miles to the east, all but two of a dozen existing buildings were destroyed in Piscataway, and five persons died in the storm (Ludlum 1983).

July 3, 1835
Razorville (Scranton) Tornado

A powerful tornado plowed through the Lackawanna Valley on July 3, 1835, slamming into the tiny community of Razorville (Providence) in one of three settlements that would later become part of Scranton, destroying virtually every building, including a Methodist church under construction (Grazulis 1993).

Harvey (1909) consulted several sources for his storm summary, noting the tornado originated "in the neighborhood of 'Fishing creek'."

It then appears to have lifted, to again descend upon the Wyoming valley near Ashley where considerable destruction of crops and buildings, including a schoolhouse, resulted. From Ashley, the whirling cloud mass moved along the base of Wilkes-Barre mountain and reappeared, according to the more reliable accounts, in the Lackawanna valley. No loss of life is recorded as a result of this tornado and damage pertained more to forests than otherwise owing to the fact that but few settlements existed in its pathway.

Harvey (1909) presented a written account from an eyewitness to the storm, who reported that "several barns and other outbuildings were torn to pieces," and there was wreckage strewn "as far north as Laurel Run." The storm was rather remarkable for its ability to move up and down over mountainous terrain.

The path of the tornado, or by whatever name it might be called, seemed to be in a direct line up the valley along its eastern side, passing back of Pittston and entering Lackawanna Valley at about the mouth of Spring Brook, touching lightly on its way further north, nor striking Hyde Park at all, but exerting its expiring force on ill-fated Razorville, now a portion of Scranton City. Hyde Park and Razorville were at the time bustling villages on the stage route between Wilkes-Barre and Carbondale. Scranton proper was only Slocum Hollow and of little consequence.

July 8, 1840
York County Tornadoes

Two tornadoes cut a swath of destruction across York County on the evening of July 8, 1840. The *York Gazette* (July 14, 1840) story stated: "We have been visited by one of the greatest whirlwinds that has ever been witnessed in Pennsylvania," with the following descriptive account:

> Yesterday [July 8, 1840], between three and four o'clock, p.m., a tremendously large and dense cloud arose in the direction about South West from the village of Newberry … Instantly, hundreds of thousands of the largest forest trees were hurled to the ground … Those at a distance of two miles could see the tops of the trees and grain in the sheath whirled along high in the air.

The storm narrowly missed Newberry, coming "within half a mile" of the town, which could explain the absence of casualties, given the intensity. Grazulis (1993) estimated that the path length was about two miles.

Remarkably, a few hours later, a second twister hit York County, this time carving a lethal course through the town of Shrewsbury, causing "great destruction of property, personal injury, and loss of life."

A letter published in the *York Gazette* stated that the storm struck around 8:30 p.m. One person was killed and several children gravely injured as fierce winds destroyed buildings and barns. The account continued: "In a few minutes the whole town was thrown into confusion and uproar. … Nearly every house in the place was submerged, and a number entirely destroyed."

June 26, 1842
Bradford County Windstorm

The late spring and early summer in 1842 brought several notable windstorms and severe thunderstorms that struck eastern Pennsylvania.

The *United States Gazette* (June 3, 1842) carried an account of a thunderstorm that lashed South Mountain and the countryside north of Lebanon on May 29, 1842: "A storm of unusual violence passed over a section of country … prostrating trees and fences, and doing other damage." Large hail up to two inches in diameter bombarded Lebanon.

Heverly (1926) described damage on June 26, 1842, as "a tornado three miles wide" that crossed southeastern Bradford County. He wrote that the storm "levelled forests, scattering fences, wrecking buildings and destroying crops." The wide path of destruction suggests a downburst but does not exclude an embedded tornado.

On July 1, 1842, an intense thunderstorm pounded Philadelphia between 7:00 and 10:00 p.m. The *Philadelphia Gazette* said the storm dumped 5.13 inches of rain, based on Pennsylvania Hospital records, flooding city streets and swamping hundreds of cellars near the Delaware River.

August 5, 1843
Cloudburst Tornadoes in Southeast

Five tornadoes, and probably more, formed near a frontal boundary stretching from Pennsylvania to Long Island on August 5, 1843.

A period of very heavy rain commenced in the morning in southeastern Pennsylvania that would culminate in deadly flooding, which took the lives of 19 people west of Philadelphia (Ludlum 1970). There was a report of up to 16 inches of rain.

A survey was conducted by Y. S. Walter (1844), with the Delaware Institute of Science:

> In the township of Bethel, not far from the line of the State of Delaware, a hurricane of great violence occurred about four or five o'clock in the afternoon. The wind blew from different points at different places in the neighborhood, as is manifest from the position of uprooted trees, &c…The wind came from the south east, and tore up a large quantity of timber (said to be about two hundred cords) all in a narrow strip, not more than two hundred yards in width. A valley of woodland, bounded by pretty high hills, had nearly all of its timber blown down, and what is very remarkable, the trees are not generally laid lengthwise of the valley, but across it, with their tops toward the north-east, while on the adjacent hills but few trees were uprooted …

The longtime records of William Whitehead, in Newark, New Jersey, showed a spectacular rainfall total of 22.48 inches in August 1843, exceeding modern records.

July 25, 1851
Wayne County Tornado

An account in the *Honesdale Democrat*, later recalled in the *Wayne Independent, Bicentennial Edition* (February 4, 1978), described a "violent whirlwind" that struck Clinton in Wayne County. The storm was "attended by a shower of hailstones of enormous size."

The damage path was reportedly 300 yards wide, lifting a Baptist meeting house off its foundation; the house was "dashed to ruins." The funnel cloud traveled toward Dyberry and Honesdale, striking several homes and knocking down trees, before lifting near Narrowsburg, New York, a distance of about 20 miles.

July 1, 1853
Berks County Tornado

A *New York Times* story on July 13, 1853, blared: "PENNSYLVANIA TORNADO" in Berks County.

> Doleful accounts of the damage by the late tornado continue to come in. In Berks County, great destruction was done in Lower Heidelberg, Spring, Cumrie, Exeter, Union, Amity, and other of the lower townships; barns, dwellings, mills, and all kinds of houses being unroofed and otherwise injured, crops beaten down, &c. At Orangeville, Columbia County, a bridge over Green Creek was lifted from its abutments, and thrown down in fragments; the gable end of the Methodist Church was burst out; whole fields of wheat were beaten to the ground, fences scattered, trees uprooted and borne off, &c.

A mile-wide swath of wind damage accompanied the storm, possibly a downburst at Sunbury, attended by large hail that shattered windowpanes in Northumberland County, and "in one case the leather top of a buggy was beaten in holes, some as large as the crown of a hat."

The account continued: "The hailstones were all large. Trees were torn up bodily, and buildings stripped of roof and easements. The loss of some of the farmers is very heavy." Wind damage was reported around New York City along the advancing squall line, possibly including a few embedded tornadoes.

April 12, 1856
Philadelphia Tornado

A probable tornado struck Philadelphia on the night of April 12, 1856, that "unroofed 150 houses in different sections of the city," according to a press dispatch, without any serious injuries reported.

From the *New York Times* (April 14, 1856):

> About 10 o'clock last evening our city was visited by a most violent gale of wind, unroofing an immense number of buildings, demolishing fences ... In the northeast section of the City, comprising the former district of Kensington, the damage was most serious.

A large Presbyterian church was "partially destroyed," and a wooden building nearby was wrecked when the church's roof went flying into the frame structure. Another church was unroofed, and a boiler house 160 feet in length was demolished at the Franklin iron works.

The *Adams Sentinel* in Gettysburg (April 21, 1856) later reported that a tornado struck York on the evening of April 12. A roof was blown in at a foundry, and trees were uprooted and fences tossed away. More significantly, the account stated that "four spans of the York furnace bridge were carried away and considerable damage done to the Columbia bridge."

May 30 and June 19, 1860
Two Violent Tornadoes

The spring of 1860 brought several major tornado outbreaks in the Midwest and two death-dealing storms in Pennsylvania.

A large tornado on May 30, 1860, tracked 20 miles through Armstrong, Clarion and Jefferson counties, killing seven persons and injuring 30 (Grazulis 1993). The "cone-shaped" funnel "resembled a whirlwind of fire and smoke" as it followed an east-northeasterly pathway from near Adams, along the border of Armstrong and Clarion counties, to three miles south of Brookville in Jefferson County.

From the account uncovered by Grazulis (1993):

> Where the funnel was narrowest its force was greatest, and it plowed up the earth to the depth of two feet, hurled large stones through the air, forcing smaller ones into trees and wood to such a depth that they could not be extricated.

Caldwell's Illustrated Historical Combination Atlas of Clarion County, Pennsylvania, printed in 1877, contained a dramatic account of the path of destruction in the western highlands:

The Great Tornado of May 30th, 1860, passed over Maysville at about half-past 11 o'clock, a.m. The storm appeared in the west, and to the observer seemed to roll over, something of the fashion of a wagon wheel, carrying with it timber, boards, shingles, parts of buildings, bed clothes—and even animals of different kinds were driven by the velocity of the wind miles away from the farms on which they belonged. Horses at work in the fields were stripped their harness. In Maysville there were two hotels, one store, gristmill, four or five dwellings. In fact the whole town was completely demolished; the burrs, four and half feet in diameter, were carried about 100 feet and deposited in the river. The bridge across the creek at Maysville was completely blown away. The saw-mill was also entirely destroyed.

A five-year-old boy and two other residents were killed instantly, and his mother died "of the effects of it about two years afterwards." Twenty buildings were destroyed around Clarion, with death and destruction extending to a mile north of New Bethlehem (Grazulis 1993). Another tornado struck southwestern New York in Cattaraugus County, killing one person near Waverly, New York.

Several weeks after a vicious storm hit northwestern Pennsylvania, another violent storm whirled across western Lancaster County on the evening of June 19, 1860, based on the following account in the *Lancaster City Express*, reprinted in the Stroudsburg *Jeffersonian* (June 28, 1860):

Last evening [June 19] between 5 and 6 o'clock, one of the most destructive hailstorms and tornadoes which has ever visited this vicinity, passed over the townships of Mount Joy, Rapho, West Hempfield, Manor, and Conestoga, doing great damage to property and crops.

Large hail pelted Mount Joy, 15 miles northeast of Lancaster, breaking windows and damaging yards and gardens. Hail covered the ground to the depth of three inches in places as the storm traveled southeast. At Mountville, there was also evidence of a tornado where crops were stripped.

A press dispatch from the western part of the county reported at Safe Harbor the "destructiveness to property and crops is incalculable." At Conestoga, the storm "completely lifted the entire body of water from its bed, so that those who were on the banks of the creek at the time could see the bottom."

Substantial storm damage occurred with the tornado over small islands in the Susquehanna River, where the funnel reportedly expanded to three-quarters of a mile. By the time the storms reached the Maryland border, the fierce winds had left behind considerable crop destruction, but no serious injuries were reported.

A little more than a week later, the *Easton Argus* described "Almost a Tornado" on the afternoon of June 29, 1860, in an entry published on July 5, 1860:

The dust flew in clouds, filling the whole atmosphere, and the violence of the wind did considerable damage. A number of trees were torn up by the root and large limbs twisted as easily as though they were mere pipe-stems. It caught a number of farmers loading and hauling in their hay. In some instances, it swept the entire load from the wagons…

A barn in Bushkill Township three miles from Easton was toppled, resulting in a serious injury; another person was hurt by a frightened horse. Many fruit and shade trees were damaged in the path of the storm. The *Easton Sentinel* called the storm a "young tornado" that was responsible for several buildings being "blown down" across the countryside.

May 11, 1865
Philadelphia Tornado

A squall line developed over eastern Pennsylvania during the late afternoon on May 11, 1865, accompanied by downbursts and tornadoes, causing extensive damage in pockets from Philadelphia to New York City. While most of the damage was probably caused by straight-line winds, there were areas of destruction associated with tornadoes.

One apparent tornado emerged from roiling clouds over Philadelphia around 6:00 p.m. on May 11, 1865, damaging or destroying 23 homes in the Fairmount Park section in the northwestern part of the city. The storm traveled northeast, wreaking havoc over several city blocks.

At the Reading Railroad Depot, the roof of a water tank, weighing several tons, was lifted and deposited onto the tracks 150 feet away. After pounding the city, the funnel crossed the Delaware River into North Camden, New Jersey, before lifting and dissipating. A 15-year-old boy was killed, and another woman may have drowned while crossing the Schuylkill River from the west bank.

The *Philadelphia Inquirer* reported "a number of miraculous escapes" from serious injury in the storm account:

The tornado seemed to confine its worst ravages to a circumscribed space, for while a number of trees were blown down, fences carried away and awnings shivered to pieces in the most populous portion of the city, the Nineteenth Ward presents a scene of terrible destruction. The worst effects of the unwelcome visitor are visible from the corner of Cumberland and Sepviva streets, and extending from there in a northeastern direction to the Delaware River.

Dispatches suggest several additional embedded tornadoes contributed to swaths of damage in sections of Newark, New Jersey, and Brooklyn, New York.

June 13, 1872
Milford Tornado

The *New York Times* (June 16, 1872) carried news of a squall line that caused extensive damage from Maine to Delaware on June 12, 1872.

A Milford resident, in the upper Delaware Valley, described the scene:

About 6 o'clock last night a terrible tornado passed over the village, leaving great destruction in its wake. In the midst of this blinding storm, and the roar of the wind, the crashing of trees, roofs of buildings, fences … The tornado lasted for about ten minutes, and when it had passed over, a scene of destruction was presented in all parts of the village.

Although the duration of the howling wind and pouring rain suggests a violent thunderstorm and downburst, there were instances of a possible tornado. The story continued: "Huge trees were twisted off and carried some distance. Valuable fruit trees were leveled to the ground." The roof of a home was torn away and carried 400 yards, injuring the occupant. Buildings were knocked down as the storm crossed the Delaware River into Sussex County, New Jersey.

July 4, 1874
Lewistown Tornado

Tragedy struck Lewistown in Mifflin County in south-central Pennsylvania on the Fourth of July 1874. A rough weather day saw numerous severe storms cause wind damage from Washington and Baltimore northward to Pennsylvania.

A disaster unfolded on the Lewistown Bridge that carried Pennsylvania Railroad trains over the Juniata River, when a tornado blew the bridge to pieces, killing three people on the bridge and injuring four others. The casualties included boys who played for local baseball teams from Mifflin and Lewistown.

Nineteen railcars were derailed on or near the bridge (Grazulis 1993). Three persons died at an iron-furnace company that was partly wrecked by the storm, among the more than 50 buildings in Lewistown that were destroyed, resulting in losses totaling in excess of $100,000. A train carrying 23 empty freight cars was derailed.

The storm was mentioned in *An Illustrated History of the Commonwealth of Pennsylvania*: "A terrific tornado swept over the town with irresistible fury." The account stated that seven people died in the storm (Egle 1876).

June–July 1877
Severe Storms Hit Southeast

A tornado that touched down 11 miles northeast of Reading near Fleetwood on June 21, 1877, traveled six miles to the northeast. The F2 tornado hit Lyons and Topton (Grazulis 1993). A storm account in the *Chester Daily Times* said, "a heavy rain storm, that, for violence and duration has seldom been equalled."

The *New York Times* summarized the storm damage: "Three large barns were completely destroyed, houses were unroofed, shedding demolished, trees uprooted, and fences blown down. The roads were very much obstructed by the debris of fences, trees, & etc."

The closing days of June 1877 were an active period of severe weather in the Midwest. Several tornadoes were reported from Illinois to Ohio on June 25, 1877, and again on June 30. The stormy pattern progressed eastward on July 1, when tornadoes were sighted in Pennsylvania, New York and New Hampshire in the late afternoon and early evening.

On July 1, 1877, a dispatch carried the story of "a spectacular and widely viewed tornado" that tracked slowly southeast across Lancaster County into Chester County (Grazulis 1993). Two fatalities occurred in Chester County, where "at least ten homes were leveled" and "a third person may have died from injuries." One victim was killed near Parkesburg and another near Ercildoun, in southwestern Chester County, where a woman died attempting to close a second-floor window.

The account in Grazulis (1993) stated that five other persons on the first floor of the home were injured. Twenty buildings were destroyed, including the Ercildoun Seminary dormitory, "where fifty students were living just a few days earlier, and had just left for summer vacation."

The tornado passed north of the community of Atglen, traveling through western and southern parts of Parkesburg, destroying four houses. The F4 storm concluded its 19-mile track of death and destruction eight miles southeast of Ercildoun. The storm had a relatively narrow path of 200 yards and injured 25 persons.

July 4, 1878
Sunday Tragedy

Another weather disaster struck on the Fourth of July, only four years after the Lewistown tornado. A tornado or downburst winds hit Sharpsburg, seven miles northeast of Pittsburgh, about 3:00 p.m. A large tree smashed into a wagon at a Sunday school picnic, killing seven children and adults. Grazulis (1993) noted, "This event may not have been a tornado. At least five other people died in a flash flood from the same storm, several miles away."

June 30, 1882
Coaltown Tornado

The springtime of 1882 was unusually chilly and damp in Pennsylvania. A southward displacement of the jet stream also led to some strong clashes of air masses.

On April 19, 1882, a tornado in Fayette County developed a mile west of Pennsville, cutting a significant 200-yard-wide damage path through the northwestern part of town, along an eight-mile track (Grazulis 1993). The F2 tornado killed one person in Mount Vernon and battered 50 buildings, with a report of $75,000 in damage.

A second deadly twister struck on June 30, 1882, about 30 miles west of Coaltown, in Lawrence County, in the northwestern corner of the state. The F3 tornado killed three residents and injured 36, damaging 15 homes and a dozen businesses along a 10-mile track (Grazulis 1993). The large path width was reportedly 400 yards.

June–July 1884
Deadly Wind and Rainstorms

The volatile spring season in 1884 brought multiple severe storm episodes in the Ohio Valley and Mid-Atlantic.

On the night of June 10, 1884, a violent windstorm with a half-mile-wide damage path slammed the Cumberland Valley and proved "very destructive to buildings, fences, and grain crops," according to the summary in the *Monthly Weather Review*. Two persons died in the Harrisburg area. There were numerous lightning strikes accompanying a probable downburst associated with a convective system.

The storm system pounded Curwensville in Clearfield County, lasting past midnight into June 11, 1884, causing Anderson Creek to overflow. Shortly after daybreak, a dam broke eight miles above town, washing away a huge quantity of lumber and destroying small bridges. At Bridgeport, "a number of houses and barns were swept away." A cloudburst at Brookville would cause the Allegheny River to rise 10 feet in less than two hours, causing dams and mills to crumble in the rush of water. A government observer caught eight inches of rain at Grampian Hills in Clearfield County.

A strong cold front brought frost in the north June 15–16, 1884, described as "very injurious to fruit and vegetables in exposed places," with less fanfare.

Low pressure churned eastward on June 24–25, 1884, triggering torrential downpours that inundated southeastern Pennsylvania, flattening crops and causing widespread flash flooding, leaving rail line and travel routes impassable.

The *Jeffersonian* (July 3, 1884) carried a Pittsburgh dispatch that recounted deadly lightning strikes in northwestern Pennsylvania and northeastern Ohio, killing several people and many livestock on June 24, 1884. The storm complex wrought "much damage to property, principally by lightning," according to the *Monthly Weather Review*. Large hail was reported at Wellsborough (now Wellsboro).

As much as 11 inches of rain near York caused dams to collapse on Codorus Creek (Shank 1988). Bridges and a part of the Schuylkill canal embankment were washed away and crops irreparably flooded. The *York Gazette* reported on the "Great Flood," described as the worst in county history. An observer in Chester County recorded 6.08 inches of rain that caused Chester Creek to overflow.

Violent weather returned in early July. A tornado outbreak on July 5, 1884, was accompanied by hail the size of "hazel nuts" at Reading. The winds "assumed the nature of a tornado" in northern Lancaster County, "leveling whole orchards, uprooting groves of shade trees and blowing down barns, tobacco houses and other outbuildings" (*Jeffersonian*, July 10, 1884). Flash flooding occurred at Chambersburg and trees were uprooted at Philadelphia, where roofs were torn off by the strong winds.

August 3, 1885
Philadelphia-Camden Tornado

A tornado outbreak in the mid-Atlantic region on the afternoon of August 3, 1885, included a deadly twister that struck Philadelphia and Camden, New Jersey. Six persons were killed and more than a hundred residents were injured, some seriously, on both sides of the Delaware River (Nese and Schwartz 2002).

The *Monthly Weather Review* described the F2 tornado that originated over South Philadelphia around 3:20 p.m., first striking Greenwich Point, where a child and railroad worker died. The funnel crossed into West Camden, New Jersey, following the contours of the river northward for two miles. Three lives were lost in Camden, and a fourth fatality occurred when the steamer *Major Reybold* was torn apart by the fierce winds.

The large tornado then crossed back into Philadelphia near Port Richmond. The width of the damage path was stunning (1,200 yards), tearing roofs off homes, splintering and tossing trees and toppling hundreds of electrical wires. The cost of the storm was estimated around $500,000, including $150,000 in the Philadelphia area. In Camden, 300 homes were damaged, with losses put at $200,000.

A vivid and chilling account of the rare deadly tornado appeared in the *Philadelphia Inquirer* the next day:

> The rush of the cyclone was estimated to be 500 feet wide. Its appearance was that of a dense black cloud revolving at a terrific rate. In the heart of it the gloom was like the darkness of midnight, and eyewitnesses describe the air as so black that they could

not see their hands before their faces. The bottom of it moved over the river like a rolling ball of smoke. The phenomenal force of the wind can only be imagined from the visible evidence of its destruction, and its power seemed to be almost supernatural. In recollection of no one in this city has a phenomenon of such character and ruinous results ever visited the neighborhood of Philadelphia, and by those who were the victims of its work it will never be forgotten. After the cyclone passed, a heavy rain storm set in, which lasted during the early part of the evening with frequent sharp flashes of lightning.

The August 3, 1885, tornado outbreak commenced with a small storm around 11:30 a.m. in Montgomery County that may have caused straight-line wind damage.

A squall line spawned a tornado in Juniata County around 2:30 p.m., less than an hour before the Philadelphia–Camden twister struck farther east.

Tornadoes were also sighted around 4:00 p.m. north of Lansdale in Montgomery County and north of Feasterville in Bucks County. The Bucks County storm was judged to be the strongest tornado (F3) of the group, traveling north-northwest for 11 miles to Solebury, damaging two homes and several barns (Grazulis 1993). Two veins of tornado damage were evident in Bucks County, which was substantial, with reports of 16 homes demolished by the high winds. A tombstone was carried 100 yards in the whirlwind.

Two tornadoes were sighted in Chester County, around 5:00 and 5:30 p.m., west of Unionville and in East Nantmeal Township. The path length of the first storm was a little less than two miles. The second storm that struck near Pocopson traveled two miles. Several barns sustained damage in both storms. Another tornado was reported in Berks County, though no time frame was given in the *Monthly Weather Review* summary.

Tornadoes touched down the same afternoon in Maryland (four) and Delaware. The F2 tornado that struck New Castle County around 4:45 p.m. damaged two houses and three barns (Grazulis 1993).

November 18, 1886
Late Autumn Tornadoes

The *Monthly Weather Review* contained a dispatch from Wilkes-Barre that described significant structural damage caused by a windstorm that had the "characteristics of a tornado," after striking the city a little past 8:00 a.m. on November 18, 1886.

Three miles north of Wilkes-Barre, in the small mining community of Parsons, two churches and several coal breakers were damaged. A high school in Miner's Mill was destroyed, along with a church under construction in Kingston. The likelihood of a tornado was confirmed by this comment: "Many substantial buildings were moved from their foundations and numerous light structures were completely destroyed." Widespread damage was done to barns, fences and orchards.

Another storm inflicted significant damage in Franklin County shortly after daybreak: "numerous buildings, school-houses, and barns were demolished and trees and fences blown down" outside of Chambersburg. Witnesses reported seeing "black whirling clouds" in the direction of the debris, almost certainly a tornado.

July 16–17, 1887
Tornadoes Hit Easton, Carlisle

During the midst of an oppressive heat wave that lasted from July 16–26, 1887, the boundary between hot, humid air over Pennsylvania and cooler air to the north became the focus of severe thunderstorms.

A fierce storm struck Easton on July 16, 1887, between 6:00 and 7:00 p.m., causing about $500,000 in property damage (Flora 1953). A small tornado blew a railroad car down an embankment and several homes were unroofed. Another tornado was sighted at Mechanicsburg that slammed into a wagon and carriage factory and blew down a steeple in front of the Methodist Church. One serious injury was reported.

The following day a squall line developed ahead of a cold front in the afternoon, spawning a tornado at Carlisle around 3:00 p.m., causing considerable damage to several schools and homes. Buildings were reportedly twisted or demolished, and roofs were blown off other buildings at Dickinson College. Widespread damage and property losses included dwellings, crops and livestock.

Matawan, New Jersey, was drenched with 5.40 inches of rain on July 22, 1887, triggering flooding that caused $10,000 in damages. Powerful storms struck again on July 23–24, dumping 3.98 inches of rain on Philadelphia, flooding streets and swamping storm sewers.

July 31, 1888
Pocono Plateau Tornado

A highly unusual high-elevation tornado touched down on the western edge of the Pocono Plateau near the Monroe/Luzerne County line and traveled about 20 miles to near Effort (Grazulis 1993).

Historian William Lesh (1945), a Stroudsburg *Record* contributor, wrote about the substantial tornado damage done to the farm buildings on the property of Hiram Hay, who resided at Houser's Mills (now Pocono Lake), which injured all nine family members. A survivor told Lesh that "All the buildings were lifted from their foundations, torn apart and strewn about."

January 9, 1889
Great Winter Tornado Outbreak

A powerful cyclone over the Great Lakes and southeastern Canada triggered a rare mid-winter outbreak of severe weather on January 9, 1889. Heavy snow accompanied by winds caused considerable damage around the Great Lakes on January 9–10, 1889, as hurricane-force wind gusts battered the Ohio Valley and Northeast.

A line of strong storms early on January 9. 1889, in western Pennsylvania turned severe, attended by downburst winds and a few probable tornadoes. Tragedy struck shortly after noon, when an unfinished seven-story building near Pittsburgh collapsed, taking down adjoining structures. Fifteen people died and 49 suffered injuries. One fatality occurred in another unfinished building that collapsed during the storm.

Twelve buildings in Pittsburgh and neighboring Allegheny City to the north were damaged by a probable downburst. Total losses reported in the *Monthly Weather Review* were estimated at $165,000.

A dispatch in the *New York Times* (January 11, 1889) described the damage:

The storm had many of the characteristics of a tornado. Its path was about two miles wide and its centre passed a little south of the city. So depressed was the conditions of the atmosphere that the wind rushed down upon the city as water would run down a declivity.

A report from East Brady in Clarion County in the *Pittsburgh Leader* stated that "a terrific hail storm passed over this place at 2 o'clock."

The action shifted to eastern Pennsylvania by late afternoon. At 4:18 p.m., a wind gust estimated near 100 mph carried away the anemometer cups at Harrisburg, where there was widespread property damage, according to the *Monthly Weather Review*. A dispatch from Sunbury reported that winds toppled two smokestacks at the Sunbury Nail Mill, which smashed through a roof, killing two workers and injuring 10 others, with four laborers reported missing. Wind damage was also reported at Williamsport, Carlisle and York along the advancing squall line.

A horrific disaster unfolded on the north side of Reading at 5:40 p.m. A tornado with a path width of 60 to 100 yards toppled the Grimshaw Silk Mill, where 275 employees were working at the time. Eighteen young women working inside the mill when it caved in were crushed by falling debris, and at least 117 more were injured, many grievously hurt.

The Reading Company paint shop nearby was blown down and caught fire, burning five laborers to death who were trapped in the store. Grazulis (1993) put the number of silk mill fatalities at 23. Subsequent accounts listed 28 deaths, including the five railroad workers who perished in the paint building.

Later, a detailed *Associated Press* (AP) account of the disaster in the *Fresno Weekly Republican* (January 11, 1889) provided a thorough picture of the calamity:

The rainfall, which continued all morning, ceased during the afternoon, and by half-past 4 'clock the sun made an effort to penetrate the clouds, when suddenly the sky changed and with appalling rapidity the storm was seen coming from the west, cutting a swath 200 feet wide and spreading destruction. In the country, farmhouses were unroofed, buildings blown down and the crops uprooted. In Reading the Mount Penn stove works, J.S. Hornberg's rolling-mill and a number of dwellings were at once wholly or partially unroofed, and a passenger car of the Reading railroad company's track was blown into splinters. In the meantime the rain was falling in torrents and it was almost as dark as night.

Hazen (1890) put the eventual death toll at 40, with damage at the time estimated at $200,000, to which the AP account gives credence:

A funnel-shaped storm-cloud struck the building directly in the center of the broadest side, and it fell to pieces as if composed of so many building-blocks. Nearly 200 human beings went down [sic] in the awful wreck.

At 6:40 p.m., as the wind shifted to the northwest at Philadelphia, a "whirlwind" hit the east side of Camden, New Jersey, stripping the roof off an icehouse, tossing it against a row of brick homes and crushing the front of another building. Wind damage was widespread over Pennsylvania, New Jersey, and New York City around sunset on January 9, 1889.

From the *Jeffersonian* (January 17, 1889):

> Prior to the storm reaching this neighborhood the day was stormy. Occasionally it rained very hard and then it could come down moderately, when towards evening it cleared away and the sun shone for a few minutes, when it clouded up again, the clouds coming from the southwest, and about 5 minutes after 6 o'clock, the gale or tornado struck this place or vicinity, accompanied by a roar similar to running cars, and began to tumble things around.

Downburst winds hit Snydersville, southwest of Stroudsburg, blowing out the windows of a home. Following a northeasterly course, a corridor of wind damage included a farm in Stroud Township and items tossed around at the fairgrounds on the southwest side of Stroudsburg. One home had "a large hole torn into it" and roof damage, and monuments at a cemetery were blown down. Damages were mostly confined to rooftops, trees, and fences.

Twenty-five workers trapped on a suspension bridge at Easton faced a harrowing situation when the guy wire snapped. The span suddenly started to "sway and shake heavily." One man was tossed eight feet into the river, but managed to reach the bank. Two more clung perilously to a railing, and the rest laid down until the storm passed, luckily escaping imminent danger.

A report filed by a correspondent for the *Monthly Weather Review* described a rare New York City tornado that struck Brooklyn at 7:40 p.m., with a path width ranging from 500 to 600 feet along a two-mile course, lifting "here and there a roof" and blowing down fences. Upon reaching Citizens' Gas Company, the storm clearly resembled a tornado, lifting a gas tank, which triggered a major explosion.

> The iron pillars not less than two feet in diameter and perhaps forty feet high, were thrown principally in a northerly direction, one or two of those nearest the tornado track being thrown in a westerly direction, thus showing clearly enough the whirl of the tornado.

The U.S. Army Signal Service officer at New York City reported that the storm in South Brooklyn destroyed several homes moving from "south to northeast; its path was well-defined, and houses were unroofed over its entire cost. The damage will probably reach $500,000."

The editor of the *Jeffersonian* commented on the abnormal winter: "The East is rapidly taking on the peculiar traits and fashions of the West. Last winter [1887–88] we beat the West on snow, while this winter we are beating it on tornadoes. It is about time to call a halt."

May 1889
Volatile Spring

The stormy weather pattern in May 1889 was almost certainly due to robust jet stream circulation directing vigorous disturbances across Pennsylvania along a sharp temperature boundary.

On May 10, 1889, an intense thunderstorm line wreaked havoc during the middle of the afternoon in eastern Pennsylvania, causing widespread wind damage. Violent wind gusts downed trees and electric lines, leaving communities in the dark for a day or more.

High winds tore down tents at the Barnum & Bailey's circus show in Williamsport, seriously injuring six performers, according to press reports. A New York man was killed by lightning in Susquehanna, and several persons were injured by high winds at Pottsville and Easton.

A violent storm struck Shamokin, in Northumberland County, on the western edge of the Anthracite Coal Region, and then hit Pottsville, where a suspected tornado tore roofs off and several buildings were "blown down." The *Monthly Weather Review* contained a notice of a tornado in Susquehanna County that destroyed a handful of buildings at Hop Bottom between 2:45 and 3:35 p.m.

A small tornado likely hit Reading, the second storm to strike the city in five months. A notice in the *Philadelphia Inquirer* stated, "Four of a handsome row of houses covering an entire block on Eleventh, from Center to Douglass Streets, had their roofs torn off" in the city.

The *Inquirer* described a "furious windstorm" that later struck Philadelphia around 5:00 p.m. and lasted about 20 minutes. The report said that "thousands of tons of dirt were whirling in the air, and made a cloud so thick as to be almost impenetrable and of a peculiar yellow appearance," possibly a funnel on the ground. Trees and signs were blown down and many homes were unroofed in the "lower section of the city."

A crowd of about 2,000 at the Philadelphia Base Ball Grounds huddled in protected areas and on the field as powerful winds blasted the ballpark. Widespread wind damage was observed in Germantown, Chestnut Hill and Olney, and several skiffs were overturned in the Delaware River. Surprisingly, no serious injuries were reported as the squall line advanced through the city from the northwest. Press accounts stated that upwards of 50 buildings suffered roof damage.

Seventy-five miles northward of Washington, D.C., a group of 25 laborers on a railroad trestle over the Potomac River were tossed into the water 60 feet below by the wind, killing several in the group. In Geneseo, New York, three workers were blown off a building under construction, and one man died in the fall. In northern New Jersey, a young girl was killed by lightning at Chadwick, and high winds caused "considerable damage" to homes and trees around Orange and Newark shortly after 5:00 p.m.

On May 19, 1889, severe storms struck again. The *Monthly Weather Review* noted a storm near Norway, in Chester County, where a lightning strike ignited an oil tank that exploded into flames. A church in Rouseville, Venango County, was burned to the ground after lightning hit.

The next day, an apparent tornado on May 20, 1889, caused damage in farm country northeast of Scranton between Clarks Summit and Clarks Green, noted in the *Pittsburgh Commercial Gazette*. The account said, "Several buildings were demolished, and every orchard in the path of the storm was wrecked, trees being torn up by the roots and carried many yards."

The storm system continued through the Wyoming Valley as a "terrific wind and rain storm, which afterwards turned into a hail storm." An embedded tornado in Wilkes-Barre destroyed several houses. Two homes were "swept from their foundations and all the inmates injured."

May 10, 1890
Deadly Northwest Tornado

A killer F3 tornado ripped through parts of Mercer and Venango counties at 5:30 p.m., and continued for 15 miles, lifting four miles southeast of Franklin on May 10, 1890. The *New York Times* reported that "everything in its path was demolished."

More than a dozen homes and 100 oil derricks were damaged by the 300-yard-wide tornado. A Venango couple milking cows died in the storm, and 17 people were injured (Grazulis 1993). The same storm system had spun off a deadly tornado on the south side of Akron, Ohio, an hour earlier, injuring eight persons.

August 19, 1890
Wilkes-Barre Tornado

A major tornado outbreak hit eastern Pennsylvania during the late afternoon of August 19, 1890, in eastern Pennsylvania. The first significant damage occurred in Berks County around 4:30 p.m., affecting barns northwest of Reading along a five-mile path.

The action shifted to the northeast around 5:00 p.m., when a tornado struck north of Shickshinny and traveled 10 miles to near Silkworth, "probably a family of small but intense tornadoes" (Grazulis 1993).

A tornado originated in Columbia County (Harvey 1909) west of Nanticoke that grew into a large vortex with a path width of 500 yards, killing "at least three people" and injuring 10 as it crossed Luzerne County, where the damage totaled $25,000 at Harveyville (Grazulis 1993).

A second EF3 tornado tracked farther south on a parallel path that struck Wilkes-Barre at 5:30 p.m. The funnel appeared as a "column of smoke" over the south side of town, possessing a path width of 300 yards between Main and Franklin Streets. The widening vortex tore through the city with deadly precision, killing 16 people and seriously injuring 15, with dozens of minor injuries. The tornado lost force after exiting the city, passing in the vicinity of Sugar Notch and Laurel Run three miles to the east, before ending on the north side of Bald Mountain.

The *Philadelphia Inquirer* carried a vivid account of the Wyoming Valley tornado:

> The first premonition of danger was a sudden darkening of the heavens which made it black as night. Then the wind increased in velocity and in a few minutes it was blowing a terrific gale. The sound of crashing thunder and falling walls, added to the wind's fury, made a noise which resembled the cannonading of a great fort by a mighty artillery. In half an hour everything was quiet and then the stillness of death fell on the stricken city.

Harvey (1909) relied on "Notes on the Tornado of August 19, 1890, in Luzerne and Columbia counties," prepared by Professor Thomas Santee, principal of Central High School, who read his paper before the Wyoming Historical and Geological Society on December 12, 1890. An appendix detailing the Wilkes-Barre damage would be supplied by Harry R. Deitrick for publication the following year.

Santee estimated "the rate of passage to have been practically a mile a minute, damage being done within a few seconds time at any given point." Homes were lifted from their foundations and in some instances hurled hundreds of feet, and brick buildings were crushed by the force of the wind.

The Pennsylvania Railroad telegraph office clock halted at 5:31 p.m., indicating the time that the building sustained major damage.

Santee estimated the damage conservatively at $240,000 in 1890 dollars, which would exceed $6 million today. At least 260 buildings were destroyed, including many homes, churches, schoolhouses and factories.

Grazulis (1993) noted several other significant tornadoes occurred in eastern Pennsylvania on August 19, 1890. A powerful (F3) tornado struck Susquehanna County around 6:00 p.m. several miles west of New Milford, tracking five miles to a point two miles east of Summerfield. Two children died and "at least two homes were levelled."

Two days later, on August 21, 1890, storms hit Fryburg, in Clarion County, and near Richland, in Lebanon County, the latter storm injuring four people, as "a barn and two homes were unroofed and torn apart" on the west side of Sheridan.

November 17, 1890
Erie Tornado

Unseasonably warm weather and a powerful storm moving through the Great Lakes on November 17, 1890, spawned a tornado four miles from Erie that started near the Catholic Cemetery. Eleven large trees were twisted off at the stump.

The *Monthly Weather Review* described a funnel that moved quickly eastward along a three-mile path, with a swath of damage a half-mile wide:

> Fence rails were blown about like straws, and shocks of corn were carried up. The instruments at the Signal Office in Erie a marked disturbance, and two special observations were taken. The clouds were low and of a gray, angry appearance, and their movement was cyclonic. ... The tornado was attended by a deep roaring sound. No lives were lost.

December 4, 1891
Rare December Tornado Outbreak

One of the rarest weather events of all in Pennsylvania is a tornado in December during a period of declining daylight and few intrusions of adequately unstable air.

Yet on December 4, 1891, an evening squall line blasted across the state, accompanied by damaging winds and a few tornadoes. The *Monthly Weather Review* reported widespread wind damage extending east to New York City, with the worst effects in the Cumberland Valley.

The storm burst upon Shippensburg, where "the clouds formed in a funnel shape" and demolished the Western Maryland and Pennsylvania and Reading railroad houses, resulting in one serious injury. A Carlisle dispatch in the *New York Times* called it "a terrific wind and rain storm, the severest for many years. ... several new buildings were partially wrecked, trees were uprooted, houses blown to the ground, and streets flooded with water." As the storm continued east to Mechanicsburg, 15 buildings were "totally wrecked" but fortunately, only minor injuries were reported.

September 7, 1893
Bradford County Tornado

A strong tornado struck Bradford County on the afternoon of September 7, 1893, destroying homes and barns near East Troy, Granville Center and Franklindale.

A broad pattern of damage initially suggestive of a downburst or squall line crossed the northern highlands of Pennsylvania. Heverly (1926) wrote that the storm evolved into a "terrific whirlwind" around 4:00 p.m., killing four people. Trees were uprooted, barns blown down and roofs torn off homes.

May 28, 1896
Destructive Southeast Tornadoes

Multiple tornadoes developed with a storm system between May 24 and 28, 1896, in a broad area from the southern Plains to the Mid-Atlantic states. The storms killed 255 people at St. Louis on May 27.

A squall line spawned an initial tornado in Adams County at 1:00 p.m. several miles east of Gettysburg, which traveled 13 miles into York County, northeast of Hanover. Four persons were injured as barns were blown down. Grazulis (1993) stated that "one home was leveled, and its furniture was carried for over for over a half mile."

Thirty minutes later, around 1:30 p.m., a tornado was spotted in York County west of Wrightsville. The storm path was east-northeast into Lancaster County, striking Wrightsville, on the west bank of the Susquehanna River, where "a school and four homes were unroofed."

The tornado destroyed a rolling mill in Columbia, 10 miles west of Lancaster, killing one person and destroying three homes. The storm reportedly pulled a column of water upward into the funnel as it crossed the Susquehanna River, creating a waterspout and "leaving the bed of the river visible nearly its entire width."

Fifteen workers were trapped in the ruins of the rolling mill in Columbia, among the 20 injuries (Grazulis 1993). Damage was also reported at Mountville and Rohrerstown. The *Philadelphia Inquirer* reported a second fatality at Columbia.

A deadly F3 tornado touched down shortly before 3:00 p.m. south of Ambler in Montgomery County. The storm continued northeast striking Jarrettown, two miles east of Ambler, before crossing Bucks County and passing over the Delaware River four miles south of Trenton, New Jersey. The *Philadelphia Inquirer* reported damage at Jarrettown included a schoolhouse, church and several buildings.

Grazulis (1993) concluded that "a family of two or three small tornadoes" caused more than $200,000 in damage and four deaths. Two persons died in Montgomery County and two in Bucks County, with 15 injuries. The damage was extensive: "At least sixteen barns were destroyed and all of the deaths may have been in barns or stables."

The parent storm skipped across southwestern New Jersey 35 miles. Another vein of tornado damage occurred near Bordentown, New Jersey, where a barn was blown down, killing one person. The storm damaged several homes and ripped trees out of the ground at Rutherford, New Jersey. Quite possibly the same supercell was responsible for a deadly tornado that struck Allentown, New Jersey, killing two men and causing $100,000 in total damage. Damage was also reported at Asbury Park.

Another tornado was sighted near Littlestown in Adams County, as a storm crossed the Maryland border into Pennsylvania (Grazulis 1993). Tornadoes were observed at Nutley, New Jersey, where damage was confined to a little less than a mile, according to the press. A funnel viewed at Perth Amboy turned into a waterspout.

Heavy downpours accompanied the violent thunderstorms. Flooding was reported at Hamburg, Reading and throughout the Lehigh Valley. Forty residences sustained water damage in Bethlehem.

To the south, high winds ripped through Washington, D.C., toppling trees on the White House grounds. A press dispatch described the damage path as "a space less than 100 yards wide [that] marked the progress of its fury." The roof of a church at Vermont Avenue and Fourteenth Street was blown off, and a young boy died after being struck by debris from a chimney that was blown down. Considerable wind damage was reported in the nation's capital, mostly fallen trees and branches.

September 6, 1898
Bradford County Tornado

A sticky, protracted late summer heat wave was brought to a halt by a series of squall lines on September 6, 1898, that developed over southern New York and northern Pennsylvania.

The first of two deadly twisters on September 6, 1898, struck western New York near Geneva, killing three at Phelps. One man was carried several hundred feet in the air over an orchard and died under a pile of debris.

The storm line barreled eastward during the late afternoon. A tornado developed in Bradford County, tearing apart several buildings and barns near Troy. As the storm continued southeast, three lives were lost in Springfield Township. One man died when the roof of a barn caved in and two Mansfield residents were killed in another barn collapse,

according to an account in the Stroudsburg *Jeffersonian*. The path of the storm was a quarter-mile wide and reportedly lasted about 15 minutes.

Strong winds raked the Northeast again on September 7, 1898, from Vermont to New Jersey, resulting in several deaths.

March 27, 1911
Philadelphia Tornado

A violent thunderstorm moved through Philadelphia around 6:00 p.m. on March 27, 1911, killing one person, injuring more than 100, and causing $100,000 in damage (Flora 1953).

The storm hit Overbrook, Germantown and West Philadelphia, before striking northeast Philadelphia, following a southwest to northeast path. The whirling winds took the form of a tornado, filling the air with debris, as the sky turned black as night, according to the press. Hundreds were injured, some seriously, by flying debris, falling trees, crashing telephone poles and a series of lightning strikes that destroyed a number of buildings.

A fatality occurred at the Tacony Station of the Pennsylvania Railroad, when the roof was blown off and carried for the distance of a city block, landing on a Pittsburgh traveler.

The *Philadelphia Inquirer* (March 28, 1911) carried a vivid storm account:

> Never before perhaps in the history of the city has a more violent storm been experienced here. Great steel signal towers stationed along the New York division of the Pennsylvania Railroad were torn from their concrete foundations and swept away like sapling trees, while telegraph poles were blown

down in rows. Frame and even brick dwellings failed to withstand the fury of the gale, which swept over the section like a tornado. Hardly a building of any kind in Tacony and the upper portion of Port Richmond escaped damage, while everything movable along the highways was carried away by the gale.

April 2, 1912
Philadelphia–Camden Tornado

A cluster or line of severe thunderstorms spawned one or more tornadoes as it crossed northeastern Delaware, southeastern Pennsylvania and southwestern New Jersey during the early evening hours on April 2, 1912.

The storm struck Philadelphia around 7:00 p.m. and crossed the Delaware River, wreaking havoc in neighboring Camden, New Jersey, where the damage was described to be worse than the August 1885 tornado.

A vivid account of the storm in the *Philadelphia Inquirer* stated: "Traveling at a terrific velocity, a miniature cyclone of seven minutes' duration encircled the city last evening, leaving in its wake a path of destruction and injury." Streets were littered with shattered glass.

The storm path was surveyed by the Weather Bureau. The path was defined as "starting in the vicinity of Delaware avenue and Market street and traveling south along Delaware avenue to South Philadelphia, round West Philadelphia to Frankford and Manayunk, where it crossed the river again and continued southward into New Jersey."

The storm initially "came up the Delaware" from the southwest, making it likely that the circuitous damage path was part of a broader downburst pattern, since no mention of a funnel cloud in Philadelphia press has been uncovered. A wind gust of 32 miles per hour (corrected) was measured on the roof of the Federal Building downtown but was estimated to have "doubled this in the sections where the damage was wrought," according to the Weather Bureau report.

The storm spawned a tornado by the time it blew through Camden, New Jersey, where "the northwest section of the city was laid in ruins." Two nurses were crushed when debris landed on a trolley car they were riding, and one later died from her injuries. Reportedly, they were caught in "the center of a maelstrom of flying roofs, falling telegraph poles and trees and odds and ends of debris."

The *Monthly Weather Review* noted a total of 200 buildings were damaged in the storm that left 100 residents homeless northeast of downtown Philadelphia and in Camden. There were several severe injuries in addition to the trolley accident.

August 21, 1914
Wilkes-Barre Tornado

History repeated itself almost to the date of the August 1890 killer tornado that struck Wilkes-Barre. On August 21, 1914, seven persons died and more than 50 were injured when a tornado struck the southwest side of the city.

The following account of the storm, culled from local reports, appeared in the Stroudsburg *Morning Press* the following day:

Shortly after 6 o'clock, while an electrical storm swept the entire valley, the tornado came out of the southwest. It struck in the Blackman patch and Blackman and New Empire Streets. A row of twenty dwelling houses, of the type occupied by miners, was first struck by the wind. They were torn from their foundations of concrete like so many card houses.

Next in line was the Laurel Silk Mill, where 14 young women were preparing to leave work. Two workers were killed in the ruins of the mill. The storm managed to skip the densely populated New Empire neighborhood before slamming into the hill section, where trees were uprooted, and a portion of the Welsh Congregational Church on Hillside Avenue lost its porch. The state police at Wyoming, across the Lackawanna River, joined local law enforcement officials to aid in the rescue work.

A retrospective in the Wilkes-Barre *Times Leader* (June 9, 2015) reviewed local accounts of the violent tornado that struck around 5:00 p.m. beneath a "blackened sky … laying waste to block after block in both the city and adjacent township." The account stated that "approximately 130 homes, businesses, churches and industrial buildings were damaged or destroyed."

The August 1914 U.S. Weather Bureau's *Climatological Data: Pennsylvania Section* report summarized damage caused by a squall line: "On the same date two men were killed by lightning at Shenandoah, a church was struck and nearly destroyed at Shamokin, five barns were destroyed near York and six buildings were struck and burned in the vicinity of Bloserville." Storm damage was also reported in Bradford County.

August 22, 1915
Hanover Tornado

The Second Galveston Hurricane slammed into the southeastern Texas coastline on the night of August 16, 1915, with 120 mph winds, dumping up to 19 inches of rain and killing 275 persons (Longshore 1998).

The remnants of the storm spun northeast through the Middle Mississippi Valley, traversing northwestern Ohio on the afternoon of August 21, 1915. Torrential downpours swamped St. Louis on August 19–20, and rounds of heavy thunderstorms pounded southern New York and Pennsylvania on August 21–22.

The cyclonic swirl spawned a twister in southeastern Pennsylvania at Hanover in York County. An account in the *New York Times* reported: "Hundreds of houses were unroofed and scores of structures and manufacturing plants were wrecked, while trees were uprooted and many poles blown over."

Twelve homes were demolished by the storm, as two funnels may have merged over the city of Hanover. Strong winds toppled streetcars. High winds and possible tornadoes occurred in southern New York in Sullivan and Ulster counties, where flash flooding resulted in $500 million in damage.

November 17, 1918
Harrisburg Tornado

A tornado swirled through the Riverside section of Harrisburg shortly before midnight on November 17, 1918, causing extensive damage but miraculously only one minor injury.

The Harrisburg *Patriot* (November 18, 1918) covered the storm's destructive force:

Riverside is in ruins. Caught in the swirl of the terrific wind an electrical storm that swept across the Susquehanna at 11:50 o'clock last night, the little suburban section to the north of the city is a wreck. More than a score of houses are [sic] almost completely wrecked. Dozens more are unroofed. One house was completely turned over. Others were swept from their foundations. Scarcely one of the picturesque little homes of the suburb but has not been damaged to the extent of hundreds of dollars.

Other reports mentioned a streetcar entangled "in the network of fallen wires and tangled debris on North Sixth street," and that a "granite stew pan was blown through a parlor window of a house on Lewis street into the midst of a family which had huddled before the fireplace during the storm." Huge trees were uprooted across different parts of the capital city, signs were blown down, and fallen wires created widespread outages.

May 18, 1918, and July 10, 1919
Two McKean County Tornadoes

Rugged McKean County in northwestern Pennsylvania, where the terrain rises above 2,000 feet, is normally safe from destructive tornadoes. Yet on May 18, 1918, a strong F2 tornado developed southeast of Aiken in the late evening and tracked seven miles that terminated west of Smethport. A barn was destroyed, along with "timber, oil drilling equipment, part of the Poor Farm, and the roof of a factory" (Grazulis 1993). Damage totaled about $150,000.

A singular occurrence of a powerful tornado crossing the Allegheny Plateau is noteworthy, but a second instance a year later was remarkable. On July 10, 1919, a funnel touched down a mile southwest of Bradford, traveling east-northeast for six miles along the south side of town. This storm in retrospect was judged to be F3 by Grazulis (1993): "Eight homes and a warehouse were destroyed, and 30 others were damaged as the tornado periodically lifted and touched down again."

May 12, 1923
First Recorded Pittsburgh Tornado

The *Pittsburgh Chronicle Telegraph*, on May 14, 1923, covered a rare funnel touching down in the metropolitan area, noting that the "first 'official tornado' in Pittsburgh goes on record."

A funnel touched down in the western suburbs of Pittsburgh at Squirrel Hill around 4:00 p.m. and traveled three-quarters of a mile, causing $20,000 in damage, according to the *Monthly Weather Review*. There was "considerable damage" done to roofs and trees.

Three days later, on May 15, 1923, a tornado struck Ellwood City in Beaver County, about 30 miles northwest of Pittsburgh, taking the life of a farmer. John Lessnet of "Camp Run" was described as "the victim of a terrific wind," reported the *Ellwood City Ledger* on May 16, in an account that was reprinted on May 16, 1973. A barn was reportedly "torn from its foundation and scattered to the winds in piece" and several buildings were knocked down.

This must have been an active weather day, because a tornado also struck Byesville, Guernsey County, in east-central Ohio which left a 50-yard-wide path that damaged about 50 buildings, many unroofed, and caused several injuries, which were reported in the Cleveland *Plain Dealer*.

June 28, 1924
Meadville Tornado

The first week of summer in 1924 took a stormy turn in parts of Pennsylvania. Damaging winds blasted areas around Erie on June 20, 1924, causing significant property damage.

Five days later, on June 25, 1924, a significant F2 tornado developed northwest of Gettysburg and traveled through the northern fringe of town, covering nine miles before ending near New Oxford (Grazulis 1993). Six barns were destroyed and several homes were unroofed, with reports of furniture being lifted from the upper stories.

A few days later, a line of violent thunderstorms formed in the sultry air over northern Ohio on the evening of June 28, 1924, spawning several deadly twisters. A devastating tornado formed near Sandusky, Ohio, around 4:35 p.m. on June 28, 1924, tearing through the lakeshore city with deadly swiftness, carving a path of death and destruction nearly a half-mile wide.

The tornado then spun out over Lake Erie around Cedar Point for about 25 miles before blasting into the city of Lorain around 5:15 p.m. Eight persons died in Sandusky and 64 were killed in Lorain; the final death toll in the Sandusky–Lorain tornado reached 85 (Grazulis 1993). A large tornado, or possibly two funnels, destroyed nearly 200 businesses and 500 homes and damaged upwards of 1,000 homes (Flora 1953).

As clusters of storms advanced southeastward into northwestern Pennsylvania, a strong tornado (F3) struck Meadville in Crawford County around 6:30 p.m. The storm track was along a line from two miles south of Geneva to a little south of Frenchtown. One person was killed in the storm, and two homes were destroyed near the beginning of the 10-mile path (Grazulis 1993). Five others died in the aftermath of downburst winds. A train derailment was attributed to the tornado in the *Monthly Weather Review*.

Extensive wind damage was also reported near Erie in the Lawrence Park area shortly after 7:00 p.m., where 37 homes were damaged by another tornado (F2) that was on the ground for about a mile (Grazulis 1993).

A terrific cluster of thunderstorms lashed the Pittsburgh area, bringing a heavy loss of life from a combination of flash flooding, collapsed buildings and electrocutions as wires were toppled by high winds. The storms claimed 11 lives. Two family members died and six others were injured in a single home that caved in at North Braddock after the foundation was loosened in a landslide.

A dispatch in the *Philadelphia Inquirer* (June 30, 1914) summarized the disaster:

> Eleven lives were lost, many persons were injured, heavy property damage was inflicted, trolley and railroad schedules were demoralized and telephone and telegraph service were crippled as the result of the terrific rain, wind, and electrical storm that swept this region late last night and early today. Almost four inches of rain fell during the few hours of the storm's duration.

April 1, 1929
Portland Tornado

A powerful tornado touched down west of Portland on an unseasonably warm spring afternoon on April 1, 1929, that crossed the Delaware River, killing three people in New Jersey (Flora 1953).

The East Stroudsburg *Morning Sun* described the storm as "Sweeping down from an almost clear sky" that appeared as a "peculiar, milky-white haze." The account continued: "Houses were torn to bits … Barns and lighter buildings were destroyed. Thousands of trees and poles were strewn about."

The Stroudsburg *Morning Press* story recounted: "Leaving a trail of damage and destruction behind, a tornado which did damage amounting to nearly $100,000 cut across Portland and Columbia, New Jersey, about 4:20 o'clock … and just as quickly as it started, just that quick it was all over, for it only lasted about five minutes."

Large hail was reported in western Monroe County as a squall line traveled toward the Pennsylvania–New Jersey border. Minor damage occurred at Wind Gap. A discrete cell strengthened near Johnsonville in Northampton County, where trees were uprooted and chicken coops unroofed.

The intensifying storm swept through Portland, before crossing the Delaware River. Several barns and homes were damaged, and outbuildings were blown down. The tornado inflicted moderate damage to the superstructure of the historic bridge spanning the Delaware River at Portland. Windows were blown out, and a section of the roof was torn off the covered bridge that opened in 1869.

The F2 funnel, with a large path width of 200 yards, traveled a little more than 10 miles across northwestern New Jersey, dissipating south of Blairstown (Grazulis 1993). The *Monthly Weather Bureau* summary stated that 30 buildings were damaged on both sides of the Delaware River. The *Climatological Data: New Jersey Section* report noted that two homes were damaged, along with several barns and smaller structures.

A farmer was killed instantly by flying timber from a collapsed barn when the storm struck three miles north of Polkville, New Jersey. Another man suffered a fractured skull while working outdoors at the same farm and was not expected to survive, according to the Stroudsburg area press accounts. A motorcyclist died in a crash during the storm, accounting for three deaths in the New Jersey side of the river (Flora 1953).

May 20 and 24, 1933
Reading Tornado, Philadelphia Storm

A line of violent thunderstorms containing strong downdrafts and embedded tornadoes raked areas of southeastern Pennsylvania on the afternoon of May 20, 1933.

A narrow tornado 15 yards wide touched down in Berks County around 5:30 p.m., injuring six persons seriously. The storm tracked east-northeast across Berks County, and it was responsible for $750,000 in storm damage at Reading, including a church. Many buildings were unroofed, and reports indicated considerable crop losses.

High winds, frequent lightning, and perhaps a few tornadoes were responsible for total damage estimated at $1 million. A marauding squall line concentrated its wrath in Adams, York, Bucks and Northampton counties, where two fatalities were reported (*Monthly Weather Review*).

On May 24, 1933, a windstorm with pockets of hail struck Philadelphia at 4:15 p.m., associated with a squall line, blasting out windows, bringing down wires and hurling debris. Six persons were killed and 14 suffered serious injuries. The storm crossed the Delaware River and pounded Camden and Moorestown, New Jersey, causing a total of $1 million in property losses.

June 13, 1939
Philadelphia Tornado

On June 14, 1939, the *Philadelphia Inquirer* described a powerful storm that struck the city that perhaps contained a small tornado:

> A severe electrical storm, freakishly combined with high, twisting, tornado-like winds, last night ripped the roofs from four houses in two sections of Philadelphia, partially deroofed a dozen other houses, and spread a wide trail of damage through the city and its suburbs.

The intense thunderstorm that hit Philadelphia around 9:30 p.m. did not cause any serious injuries. Hundreds of trees were uprooted across the city and roads were flooded, as trolleys were "marooned in several feet of water." Grazulis (1993) listed the storm as a narrow funnel (80 yards wide) that hit West Oak Lane and traveled a half-mile.

July–August 1941
Southeast Tri-State Tornadoes

Several significant tornadoes struck Pennsylvania in the late summer of 1941.

On July 18, 1941, "an apparently intense path was cut through eight miles of forest" in the Allegheny Plateau, leaving a swath of damage in Clinton and Potter counties, passing several miles east of Cross Forks (Grazulis 1993). There were no injuries despite a wide damage path of 800 yards through mostly uninhabited terrain.

Another large funnel struck on July 30, 1941, at Mechanicsburg in Bucks County, north of Philadelphia, causing a substantial path of destruction up to a quarter-mile wide. The storm destroyed several small buildings, causing some property and crop damage.

On August 25, 1941, a hit-skip tornado caught a section of southeastern Chester County, where it unroofed 15 homes near Kemblesville along a four-mile path. Damage was reported to be $75,000 and there was one injury.

The storm churned northeastward across the northwest area of Wilmington, Delaware, causing $150,000 in damages and injuring seven people. One person died at Swedesboro, New Jersey, as the storm rolled into southwestern New Jersey. The tristate tornado traveled 32 miles, though not continuously in contact with the ground, reaching a maximum path width of 300 yards.

June 23, 1944
Deadly Southwest Tornado Swarm

A series of death-dealing storms developed in the early evening of June 23, 1944, over eastern Ohio, western Pennsylvania and northern West Virginia, spawning seven significant tornadoes. A few other areas of damage could have been work of smaller tornadoes that were not classified at the time. Some of the tornadoes climbed up and down rugged Appalachian hillsides, inflicting widespread damage.

Four tornadoes were on the ground between 6:30 and 9:30 p.m. The tornado rampage on June 23, 1944, killed 45 people in southwestern Pennsylvania. The storms also took a terrible toll in West Virginia, killing 104 persons, and five more in Maryland. A reported count of 846 serious injuries, and property losses that exceeded $5 million, attested to the magnitude of the tornado swarm (Flora 1953).

A review of the chronology of the June 23, 1944, tornado outbreak by Grazulis (1993) placed the first in a series of large tornadoes around 5:30 p.m. in Armstrong County. The twister headed southeast from Rural Valley to Twin Rocks, damaging 50 homes and farms, causing three injuries.

As the storm progressed into Indiana County, another 15 homes were destroyed, and two people died. Damage extended into Cambria County, possibly a downburst. The F3 tornado had a path length of 30 miles and caused 19 injuries. Around 6:00 p.m., a smaller storm touched down in northeastern Ohio before entering Pennsylvania and injuring eight people south of Palmyra.

A supercell near Wellsburg, West Virginia, spawned a deadly family of tornadoes beginning at 6:11 p.m. (Grazulis 1993). One death occurred at Wellsburg, where 20 homes were unroofed. The violent tornado entered southwestern Pennsylvania, killing four people in Washington County and 22 in Greene County. A letter from YMCA Camp Buffalo, which took a direct hit, was found about 100 miles away.

A total of 86 homes were destroyed in Pennsylvania, where the twister reached F4 strength, possessing wind topping 206 mph. The tornado passed eight miles southeast of Uniontown, lifting briefly and then re-entering West Virginia, where four homes were destroyed, injuring 10 people. Another storm in the tornado family touched down in northwestern Maryland that killed three people and injured 25 in an area two miles north of Oakland, Maryland.

At 6:30 p.m., a violent F4 tornado, one-half mile wide, formed eight miles southeast of Pittsburgh. The deadly whirlwind traveled 50 miles, killing 17 and injuring about 200 in Allegheny County. Eighty-eight homes were destroyed and 306 were damaged south of McKeesport as the storm ripped through the small communities of Dravosburg, Port Vue, Versailles, Boston and Greenock, where most of the fatalities occurred. The path was a half-mile to one mile wide.

Homes "collapsed or were blown apart" by the ferocity of the wind as the strong tornado roared southeast parallel to the Turnpike through Westmoreland and Fayette counties, inflicting the significant damage near Donegal and northwest of Somerset.

More killer tornadoes developed in northeastern West Virginia. A powerful F4 storm struck at 6:30 p.m. northwest of Wyatt that left 100 people dead. One remarkable aspect of the storm was the 60-mile path over rugged terrain while maintaining contact with the ground.

The last major tornado (F3) of the evening of June 23, 1944, struck Tucker, West Virginia, at 10:25 p.m., taking three lives. At the end of the horrific night, the tornado outbreak claimed 154 lives and injured nearly 1,000 people in Pennsylvania, West Virginia, western Maryland and Delaware. Surveys revealed that 1,456 buildings were damaged, including 691 homes (Flora 1953).

June 7, 1947
Sharon Tornado

The 1947 tornado season was quite active in Pennsylvania, especially in the late springtime. On May 21, 1947, a tornado (F1) developed south of Centerport in Berks County around 4:45 p.m. and traveled six miles, killing a boy and injuring his brother in a garage demolished by the storm. The path was judged to be 700 yards wide (Grazulis 1993).

A tornado sighted in Berks County on May 25 between Hamburg and Lenhartsville caused only minor damage. On May 29, a tornado in Schuylkill County at Pine Grove about 4:30 p.m. took the life of a farmer and killed two horses in a barn. A storm at Carlisle damaged dozens of buildings along a 15-mile track.

However, the deadliest tornado of the season in Pennsylvania gathered over northeastern Ohio around 3:30 p.m. on June 7, 1947. Forming a few miles southeast of Warren, Ohio, the tornado hit Sharon, Pennsylvania, at 3:55 p.m. The violent F4 tornado packed winds in excess of 207 mph. The storm killed three south of Vienna, Ohio, with 40 reported injured, and damaged or destroyed 150 buildings. The losses were put at a substantial $1 million at the time. In Pennsylvania, another 325 homes would be impacted.

The deadly tornado continued along a 40-mile path through the middle of Sharon, causing more than $1 million in damages to property in Mercer County. Two people lost their lives and 70 homes were destroyed. Home and factories collapsed, as the tornado raced eastward at 75 mph. The storm carved a path of devastation and injured at least 300 people from Mercer to Grove City.

Near the end of the summer, a powerful F3 tornado blasted Washington County in southwestern Pennsylvania late at night on September 2, 1947. The storm crossed Eldersville–Burgettstown Road around 2:15 a.m., traveling east-northeast for nine miles between Eldersville and Raccoon Township.

Five houses and 40 buildings, including quite a few barns, were damaged or destroyed. Two persons died in their homes and 40 others were injured, with the worst damage centered at South Burgettstown. A roof was flung a mile from its original location, and boards landed four miles away (Grazulis 1993).

May 22, 1949
Statewide Tornadoes

A severe weather outbreak across the midsection of the country on May 20–21, 1949, marked the beginning of a week of powerful storms. On May 22, a convective system caused wind damage and spun up several tornadoes in southwestern Pennsylvania.

A downburst pattern about 20 miles long and 12 miles wide felled hundreds of trees, knocked down power lines and caused property damage from Waynesburg to Uniontown and Dunbar, as highlighted in *Climatological Data—Pennsylvania*. An embedded tornado could have caused some of the damage.

A 100-yard-wide tornado that likely formed in Beaver County tracked through Zelionpole in Butler County a little before 5:00 p.m., about 30 miles north of Pittsburgh. A Lutheran senior home and children's home on Main Street sustained damage. No injuries were reported.

A family of tornadoes struck portions of Cambria, Blair and Bedford counties on May 22 from 5:30 to 6:30 p.m., summarized in a detailed blog by Nick Wilkes on the website *Tornado Talk*.

The first in a series of tornadoes felled trees just east of Johnstown that brought down power lines along Route 219. Another tornado touched down in Carson Valley, before entering the southwestern part of Altoona, causing pockets of damage that consisted of uprooted and downed trees and minor property damage.

A trail of storm damage, which impacted several farms, extended across Somerset County, although mainly straight-line in nature. The destructive winds injured six persons in the storms; one person was electrocuted after touching a downed wire and died after the storm passed.

The *Altoona Mirror* reported a "narrow path" caused by a probable tornado at Imler in Bedford County. The path width was 500 feet and left a pattern of twisted trees, tore the roofs off several barns, and "damaged several homes."

Grazulis (1993) plotted a seven-mile F2 storm path embedded in a squall line into Berks, Lehigh and Northampton counties, beginning at 7:30 p.m. The twister formed between Bally and Clayton in Berks County. In *Climatological Data*, reported damage included "a large barn" and hosiery mill that "had walls blown out and the roof caved in." The storm moved northeast: "Many houses were damaged and partly unroofed." The funnel would pass near Palm: "a stone house was unroofed; its stone walls badly cracked." Grazulis (1993) stated the five people were injured in the home. Another tornado that formed in Lehigh County, resulting in two injuries at Hosensack.

A small tornado struck Harrisburg at 8:30 p.m. and tracked east. Trees felled by the wind blocked Front Street at River Park. A second storm around Hummelstown destroyed several homes and a few barns. Allentown reported a gust of 85 mph.

April 5, 1952
York and Lancaster County Tornadoes

An early spring tornado outbreak in eastern Pennsylvania on the afternoon of April 5, 1952, left veins of wind damage along a path across southeastern Pennsylvania, as low pressure crossed the state.

The summary in *Climatological Data—Pennsylvania* listed the first significant damage at Waynesboro, Franklin County, where two concrete block buildings were damaged, probably by a tornado. A church was unroofed and lost a wall a little farther away at Gettysburg, and several buildings were damaged, injuring one man. This storm could have also been a small tornado.

The first confirmed tornado (F2) was observed between Hanover and Spring Grove at 1:15 p.m. in York County in a discontinuous northeasterly 20-mile track from York to Wrightsville. Several factories lost roofs that were blown away in the city of York. Other structures lost walls and chimneys. Barns, homes and garages were damaged, and four persons were injured.

A crane that weighed several tons was moved 40 feet across a plant yard and a greenhouse frame was twisted off its foundation, which shattered more than a thousand panes of glass. In one home, the wind "rolled up a linoleum runner in a hallway and blew a 13-year-old girl standing in the hallway off her feet" (Flora 1953).

A second path of structural damage was traced from Wrightsville to Steelton, where a building lost a roof that was "tossed onto a bus and into stores windows, and a coal shed roof was picked up and blown through power lines which led to 3 suburbs." Three injuries were reported. The total storm damage was reportedly $2 million.

A half-hour later, a Lancaster County twister (F2) hit north of Lititz, injuring six more persons. A Lawton man was carried 100 feet in the air before passing through wires and a fallen tree. A Lancaster resident was tossed 65 feet "through the air." Several buildings were damaged, and a 420-foot tower was blown down, with damage estimated at $150,000.

Weaker storms passed through Dauphin County (2:00 p.m.) and Suedberg in southwestern Schuylkill County (2:30 p.m.), damaging three buildings and causing an airplane hangar to collapse. Farther northeast, a tornado touched down briefly around 3:30 p.m. near Mountainhome in the southern Poconos and then hit Paradise Valley near the Paradise Brook Trout Company, pulling water out of a nearby reservoir that showered a nearby field.

Straight-line wind damage was reported in southeastern Pennsylvania. In the central mountains, hailstones at Loganton in Clinton County measured 1.75 inches in diameter, large enough to break windows.

March 22, 1955
Chester County Tornado

March 1955 brought severe weather toward the end of the month with the arrival of spring. Two strong disturbances tracked across the Midwest and Great Lakes carrying destructive winds.

An intense squall line blasted through western Pennsylvania on March 22, 1955, accompanied by powerful winds. A woman was killed in Pittsburgh, and a man died near Brookville, after being struck by falling debris, according to the storm summary in *Climatological Data–Pennsylvania*. The frontal system soon marched into eastern Pennsylvania.

Tragedy struck again in the southeastern part of the state, where a significant tornado resulted in one fatality. The strong storm, listed as F2/F3 intensity, traveled nine miles from West Chester to Paoli, with a path width of up to 600 yards. A tractor-trailer driver died two miles east of West Chester after his cab was crushed by a falling tree on Route 202, according to the *Chester Times*.

An 18-month-old in High Meadows barely escaped death or injury when a chest weighing several hundred pounds blew over in a carport and landed on top of the stroller.

The press account reported damage totaling $100,000 at Coatesville and West Chester.

A 48-hour period of destructive high winds during March 26–28, 1955, followed in the wake of a strong low-pressure system that tracked just west of Pennsylvania, ushering in arctic air with blowing snow. Temperatures hovered in the 20s across the western and central counties on March 27, 1955, and winds gusted past 50 mph.

May 12–13, 1956
Southwestern Pennsylvania Storms

A line of powerful thunderstorms raked southwestern Pennsylvania and the Pittsburgh area on the night of May 12–13, 1956, inflicting an estimated $1 million in damage and causing 10 injuries, according to the *Climatological Data—Pennsylvania* "Weather Summary."

Trees and utility poles were downed by high winds, putting many thousands in the dark across a swatch of southwestern Pennsylvania. The greatest damage occurred at Aliquippa, West Mifflin, Duquesne and Windber. Torrential rains caused flash flooding and landslides as storms ripped through Beaver, Allegheny and Somerset counties.

Duquesne took the biggest hit, with damage stretching over a dozen city blocks, with a path width of two blocks. About 75 buildings had substantial damage and 200 reported some storm damage, including roofs torn off and smashed windows. Homes, two churches, a high school, post office, rectory and funeral home, along with numerous local businesses, were hit hard by the powerful winds.

The storm toll at Aliquippa was put at $500,000, including $50,000 at the Aliquippa-Hopewell Airport, which was leveled, destroying 14 airplanes and severely damaging three more. Lesser damage occurred at West Mifflin and Windber.

The "path of destruction was discontinuous for 80 miles" as documented in the state report, varying from 30 to 300 yards, caused by a "suspected tornado" (or possibly multiple tornadoes) embedded within broad downburst damage.

March 21, 1976
Eastern Tornado Outbreak

Seven confirmed tornadoes, and several other likely funnels, were reported in the early afternoon of March 21, 1976, on an unseasonably warm day, with a cold front approaching from the west and a strong jet stream disturbance overhead.

The tornado outbreak felled thousands of trees and knocked out power to about 60,000 customers. The monthly *Storm Data* report counted extensive damage affecting "10 barns, 25 other buildings and 20 house trailers" that tallied up to $1 million in damage. Flying debris caused two fatalities and 20 injuries. One fatality occurred in Chester County, where a man was struck and killed by a piece of roof that blew off his barn.

A Philadelphia youngster died from injuries after being struck by a building drainpipe, where a wind gust of 67 mph was recorded. Peak gusts reached 90 mph at Soudertown in Montgomery County, 81 mph at Lancaster, and 70 mph at Chester and York.

The first tornado touched down at Scotland in Franklin County at 11:50 a.m. Small tornadoes also struck Adams and York counties. Two tornadoes were recorded in Monroe Township, Cumberland County, around 1:00 p.m.

At 2:00 p.m., a tornado was sighted at Freemansburg in Lehigh County that traveled across Interstate 78 and continued for eight miles into Northampton County. The storm crossed the Lehigh River, where witnesses reported that the winds were "sucking a plume of water an estimated 150 feet in the air." Pieces of a barn on the opposite side of the river were blown 400 feet.

A compact tornado touched down in downtown Stroudsburg shortly before 2:00 p.m. from a funnel cloud that was first spotted over Route 611 near Stroud Mall. Monroe Silk Mills on Ann Street took a direct hit from the storm that tore apart the second floor of the building and hurled a cast-iron weaving loom 100 feet (*Storm Data*).

In addition to the silk mill, seven homes in the neighborhood sustained damage totaling $500,000 along a half-mile path, with several minor injuries sustained by residents. The supercell passed over Delaware Water Gap a short time later, knocking down a tree that seriously injured another person.

The *Wayne Independent* in Honesdale noted damage near Lake Henry, about 20 miles east of Scranton. The downburst, which very possibly was a small tornado, scooped up seven unoccupied trailers and tossed them 30 feet in the air. A garage in Waymart was lifted more than 10 feet above the concrete floor, and large trees were toppled on Park Street in Honesdale. Twin tornadoes were also confirmed in Ulster County, New York.

June 3, 1980
Natrona Heights Tornado

An outbreak of tornadoes struck Natrona Heights and the Kiski Valley, northeast of Pittsburgh, starting around 12:30 p.m. on June 3, 1980, that injured more than 140 people.

A story recounting the Pittsburgh tornadoes that appeared in the Post-Gazette (June 3, 1998) reported that "nine tornadoes ripped through four counties in Western Pennsylvania, leaving a 50-mile-long trail of damage," which started in Allegheny County. The NOAA Storm Data report counted two tornadoes. The most intense storm traveled 14 miles through Allegheny, Armstrong and Westmoreland counties.

The tornado, or more likely tornado family, initially possessed winds estimated near 200 mph (F3) and injured 42 people at a shopping center in Natrona Heights. Another funnel touched down at the Edgewood Estates Trailer Park in Apollo in Armstrong County, which sustained a direct hit. A total of 77 mobile homes were destroyed and 46 were damaged.

Seven homes were "badly damaged" by a tornado that hit Vandergraft, Westmoreland County, about 30 miles northeast of Pittsburgh. A tornado listed in the NOAA report in Indiana County demolished three mobile homes, a house, and a helicopter north of Saltsburg. The total cost of the storm was put at $6 million.

On June 7, 1980, two tornadoes occurred in Adams County around 6:00 p.m. One storm had a path width of 800 yards, tracking from Bermudian to near Wellsville, in York County, flattening barns and damaging a dozen airplanes and five hangars. A second funnel touched down near East Berlin on a 1.5-mile course, causing minor damage. A third tornado collapsed two barns and leveled small structures around Brownstown in Lancaster County shortly after 8:00 p.m.

May 1983
Western Storms

On May 2, 1983, a tornado hit near Erie around 3:45 p.m. near I-90 and Route 19, destroying seven mobile homes and damaging 50 others. A second tornado traveled along a 10-mile path through Tioga County around 6:40 p.m. that destroyed a trailer, camper, house and 10 outbuildings (Grazulis 1993). Six F3 tornadoes struck western New York, killing three persons.

On the afternoon of May 22, 1983, a storm rated F2 touched down around 2:00 p.m. near Elizabeth in eastern Allegheny County, moving over the Youghiogheny River to Westmoreland County, with a width of 200 yards. The tornado passed between Jeanette and Greensburg, and ended near Black Lick in Indiana County.

The tornado traversed a remarkable hit-skip path of 45 miles, injuring 10 persons, causing major damage to 75 homes and minor damage to 325 others. Two funnels were visible over the Greensburg Country Club. The path width reached 600 yards. On May 23, a tornado rated F1 struck three barns and two houses, crossing from Montour into Northumberland County, impacting nine properties.

August 11, 1983
Allentown Tornado

A small tornado was initially apparent five miles northwest of the Lehigh Valley International Airport on the afternoon of August 11, 1983. The storm traveled 11 miles southeast through Catasaqua, then dissipated over Freemansburg. The damage path was discontinuous but reached a path width of 150 yards in places.

A total of $3 million in losses occurred at the airport, where 20 planes and 15 hangars were damaged by the storm, according to the *Pocono Record*. Automobiles parked at the airport were damaged, and two persons were injured. Three other tornadoes were sighted in eastern Pennsylvania on August 11, 1983, north of Tobyhanna, and at South Bethlehem and Hellertown, causing minimal damage.

July 5–6, 1984
Eastern Tornadoes and Downbursts

A stormy pattern developed over Pennsylvania on July 5, 1984. A tornado touched down at High Point Lake in Somerset County and traveled 10 miles, destroying a barn and two mobile homes. This was the beginning of an active weather day across the state.

Grazulis (1993) reviewed the NOAA *Storm Data* summary about a storm outbreak on July 5–6, 1984, and wrote that "six homes were destroyed and 40 homes were severely damaged" in northern Berks County in the early evening by four tornadoes. The storms affected an area seven miles long and three miles wide. Two persons were injured by F2 tornadoes that had a damage path 300 yards wide embedded within a broader downburst extending for 2.5 miles.

The first in a series of tornadoes struck near Princeton, 10 miles northeast of Reading, at 7:30 p.m., with a discontinuous path length of seven miles. Fifteen minutes later, a second twister traveled three miles between Lyons and New Jerusalem, damaging several homes and twisting off hundreds of trees on a seven-mile track.

A third tornado developed two miles south of Topton, tracking eight miles from Berks into Lehigh County shortly before 8:00 p.m. A fourth tornado unroofed homes along a five-mile course southwest of Macungie about 8:15 p.m. as it also moved from Berks into Lehigh County.

On July 6, 1984, a Luzerne County storm (F2) was confirmed at Sweetwater, hitting around 5:15 p.m., that destroyed six homes and 10 farm buildings and damaged 55 homes and 45 cars. Three injuries occurred in a car that was tossed into a pond.

May 31, 1985
Pennsylvania's Deadliest Outbreak

An unprecedented outbreak of violent tornadoes blasted northern Ohio and Pennsylvania on May 31, 1985, leaving large swaths of devastation.

Fourteen tornadoes in Canada caused twelve fatalities. During the next eight hours, individual supercells formed south of Lake Erie, spawning unseasonable long-tracked storms across the Alleghenies. The storms took 89 lives in the United States and Canada, resulting in damages exceeding $600 million ($1.5 billion in 2020 dollars).

Eleven tornadoes touched down in Ohio, killing 12 persons and injuring 340. Three tornadoes occurred in southwestern New York, including a large F3 tornado (750 yards wide) that formed in Pennsylvania near the state line that damaged 70 homes. Six injuries were reported across southern Chautauqua County, New York.

Pennsylvania suffered the greatest destruction and loss of life—65 killed and 527 injured (NOAA *Storm Data General Summary*, December 1985). A total of 1,009 homes were destroyed, resulting in more than $376 million in total damage, which would surpass $1 billion today.

Twenty-one tornadoes touched down over Pennsylvania, including four storms that began in northeastern Ohio and crossed the state line. The outbreak brought the only F5 tornado ever recorded in Pennsylvania and seven F4 storms. A huge F4 funnel blew through the Moshannon State Forest area that spanned 1.25 miles, which grew to two miles at the end of a 69-mile trek.

At 4:00 a.m. on Friday, May 31, 1985, the National Severe Storms Forecast Center (NSSFC) in Kansas City, Missouri, highlighted a significant risk of severe thunderstorms in northeastern Ohio and northwestern Pennsylvania later in the day.

Forecasters at the Severe Local Storm Warning Center, or SELS, as it was known then, anticipated that a capping inversion of warm air several thousand feet aloft would eventually break with the arrival of an upper-air disturbance coupled with upper-level winds exceeding 100 mph.

SELS issued Tornado Watch Number 211 on Friday, May 31, 1985, at 4:25 p.m. for the period of 5:00 to 11:00 p.m. EDT that covered western Pennsylvania and northern West Virginia. By late afternoon, the thermometer reached 85°F at Erie, under a nearly cloudless sky.

A cold front crossing Illinois trailed low pressure that migrated from near Duluth, Minnesota, across the northern Great Lakes. The pleasantly warm late spring afternoon would soon turn violent in southeastern Ontario. A mid-level jet streak, with 80 mph winds, caused updrafts to rise rapidly, releasing convective energy by 3:00 p.m.

At the same time, the relatively clear skies over northeastern Ohio filled with towering clouds as thunderstorms erupted on an unseasonably warm and sticky afternoon. A stable elevated mixed layer a mile above the surface was roiled by deep-layer wind shear above a warm front lifting northward across the southern Great Lakes, causing the air to turn back toward low pressure, inducing rotating updrafts.

Before the first supercell erupted in Ohio, 14 tornadoes had already caused death and destruction in southeastern Ontario, Canada. Eight people died in a violent F4 tornado at Barrie, along with 281 injuries, in a storm that caused more than $1 million dollars in damage (2020 dollars).

The first severe thunderstorm warning was issued by the NWS in Cleveland at 4:10 p.m. for Ashtabula County, Ohio. Within an hour, a tornado vortex signature was spotted on NWS Erie radar near Monroe Center, Ohio, two miles west of the Pennsylvania border. A tornado warning was issued at 5:05 p.m., as the supercell storm entered Crawford County from Ohio. The tornado then consisted of two funnels that merged into a massive F4 storm that expanded to a width of 1,200 yards.

The wedge tornado pulverized 20 square blocks in Albion over one square mile, damaging or leveling 309 structures, killing nine people. Three also died at Cranesville.

Eighty-two people were injured during the 12-mile path inside Pennsylvania, after the storm left Ohio on its 14-mile journey.

Fig. 7.3 May 31, 1985. Two funnels merged into one violent F4 tornado approaching Albion, Erie County. The tornado slammed into Albion, then roared east for three more miles, ending northeast of Cranesville. Twelve persons died and 82 were injured. (Courtesy Erie's History and Memorabilia)

Fig. 7.4 May 31, 1985. The large tornado caused total devastation in the community of Albion, where nine persons died. (Courtesy Erie's History and Memorabilia)

The deadly tornado embarked on a 56-mile course for a little more than one hour. Widespread damage affected homes and businesses at Jamestown in Mercer County.

Fig.7.5 May 31, 1985. Homes were reduced to rubble along the path of the storm in a one-square-mile section of the Borough of Albion. (Photo by Dick Ropp)

Fig. 7.6 May 31, 1985. A total of 309 homes and businesses were destroyed or severely damaged in the path of the tornado, leaving 1,800 residents homeless. (Courtesy Erie's History and Memorabilia)

Around the time Albion was struck, a tornado formed in Crawford County, striking Linesville at 5:10 p.m., where 75 campsites and several homes were damaged along a four-mile path. One woman was crushed to death by a falling tree.

The tornado family spawned a large F4 tornado just east of Kinsman, Ohio, at 5:17 p.m., several hundred yards west of the Pennsylvania line.

The tornado proceeded east-northeast and crossed Route 322, leveling everything in its path when it plowed through Atlantic, killing five persons. Another man was killed five miles north of Sheakleyville, and two died in Cochranton. As the storm moved into Venango County, five fatalities were reported at Cooperstown, and another occurred in a mobile home park eight miles north of Franklin. Two fatalities were reported at Cherry Tree, before the tornado ended a few miles south of Tionesta.

The long-tracked tornado took a total of 16 lives and injured at least 125 persons. The damage path was up to 350 yards wide. Half of the deaths occurred in Crawford County, and eight fatalities occurred in Venango County. Massive damage from the F4 storm totaled 371 homes, resulting in $8 million in losses caused by wind speeds in excess of 207 mph.

A strong (F3) tornado formed about 5:23 p.m. two miles south of Saegertown and tracked 23 miles to a few miles east of Centerville, where two persons died. A NWS survey confirmed that there was a short-lived second tornado that hit a few miles south of the F3 storm, which formed near the end of the parent storm track.

The northernmost storm in Pennsylvania—the second F4 tornado to hit Erie County—touched down two miles southeast of Waterford at 5:25 p.m., passing north of Union City between Wattsburg and Corry, in the eastern part of the county. Four homes and barns were severely damaged on Route 8, injuring two persons, and three dozen cows perished at one of the destroyed farms. A house and wagon went airborne.

The storm continued east, north of Elgin and Corry, leveling 50 buildings and damaging many more, then crossed northwestern Warren County. The tornado tracked nearly 20 miles in Pennsylvania and eight miles in southwestern New York. Hundreds of trees were felled southeast of Clymer, New York. A total of 17 injuries were caused by the storm, but no deaths were reported.

The fourth tornado in the Ohio family of storms that entered Pennsylvania touched down northwest of Dorset, Ohio, in Ashtabula County, shortly before 5:30 p.m., and ended one mile into Pennsylvania, north of Pennline in Crawford County, where several homes and buildings were damaged. Homes were sheared away in northeastern Ohio along the 14-mile path, along with numerous downed trees and power lines. At least 15 injuries were caused by the F2 tornado.

Another powerful Crawford County tornado rated F3 developed two miles west of Centerville at 6:12 p.m. and tracked eight miles east-northeast to one mile east of Buells Corners: "Numerous houses, trailers and barns were destroyed." The PennDOT building incurred $500,000 in losses (*Storm Report*, December 1985). The path width was 427 yards.

A weaker F1 tornado, with a width of 130 yards, touched down near Buell Corners at 6:30 p.m. and traveled five miles to Grand Valley, causing minor damage and no injuries. In all, more than 1,000 homes and businesses were damaged in Crawford County, many beyond repair.

A large F4 tornado developed around 6:30 p.m. east of Oil Creek State Park, south of Pleasantville. The storm traveled 29 miles through rural Forest County, killing seven and injuring 30 people. Four deaths occurred around Tionesta, and three died when their car was blown hundreds of feet off a road. More than 700 buildings were damaged, 125 beyond repair. The storm had a path of 800 yards, passing north of Marienville in Forest County before crossing the Allegheny River and lifting in western Elk County.

The most powerful and deadliest tornado on the evening of May 31, 1985, developed around 6:35 p.m. near Ravenna, Ohio, in Portage County, about 30 miles west of Youngstown, and reached

F5 intensity over southern Trumbull County, with winds approaching 300 mph. The monster storm tore apart homes and businesses in Newton Falls and Niles. Police Captain Clayton Reakes was credited with saving many lives in Newton Falls by climbing atop the Municipal Building and ordering the sounding of the town's tornado siren (Witten 1985).

The storm tore through the heart of Newton Falls and then slammed into Niles, where eight people were killed near the business district and another died from a heart attack shortly after the storm passed. Two deaths occurred near the Ohio–Pennsylvania border in Hubbard Township.

The powerful wedge tornado entered Pennsylvania one mile west of Wheatland and tracked 14 miles across southern Mercer County south of Farrell. Six persons died in Wheatland, and a seventh fatality occurred a little farther east at Greenfield, according to the *Sharon Herald*. At least 60 people suffered injuries of varying severity. The tornado was 450 yards wide at Wheatland, where more than 50 homes were destroyed, leaving parts of the community in ruins.

The force of the wind peeled asphalt from the surface. The large funnel also struck Hermitage, destroying more than 70 homes and businesses, leaving a scene described as "a bombed out battle field" in NOAA's *Storm Data* report. The wreckage at nearby Greenfield (15 homes destroyed, 30 moderately damaged) was massive. The tornado finally lifted south of Mercer after traveling 47 miles. The violent tornado that claimed 18 lives in Ohio and Pennsylvania is the only F5 in Pennsylvania history.

A strong tornado F2 tornado formed in eastern Forest County, four miles southwest of Chafee, around 6:50 p.m. and tracked 19 miles, moving through northern Elk County near Lamont, and into McKean County, ending at Kane. The tornado was 300 yards wide yet caused surprisingly little damage due to the hit-skip path in a remote area.

A powerful F3 tornado touched down three miles west of Tidioute about 7:30 p.m., taking a 17-mile track that ended three miles south of Cherry Grove. A total of 32 buildings were heavily damaged and eight persons were injured in the rural tornado that spent the bulk of its life cycle in the Allegheny National Forest.

The widest tornado in Pennsylvania history formed around 7:35 p.m. a mile south of Penfield in Clearfield County, about five miles north of Interstate 80, which is the highest point along the interstate highway east of the Mississippi River. A dozen homes were damaged. The monster tornado was on the ground for 85 minutes and gashed Parker Dam State Park north of Clearfield.

The tornado became even larger in Moshannon State Forest, attaining a breathtaking path width of 2.2 miles along a 69-mile course. Moshannon State

Fig. 7.7 May 31, 1985. The Beaver County Times *headline, following the passage of a deadly tornado. (Courtesy* Beaver County Times*)*

Fig. 7.8 May 31, 1985. Large F3 tornado that destroyed Big Beaver Plaza, taking three lives. The tornado "roared like a freight train at full throttle—bent on a 13-mile trail of destruction." A total of 260 North Sewickley Township homes were damaged or destroyed in the path of the storm. (Courtesy Beaver County Times*)*

Forest and Sproul State Forest absorbed the heaviest blow, where an estimated 88,000 trees were destroyed. One injury was reported along the storm path that traversed sparsely populated forestland. Only a handful of tornadoes have been judged to have had a larger damage path in the United States: El Reno, Oklahoma, in 2013; Mulhall, Oklahoma, in 1999 and Hallam, Nebraska, in 2004.

The F4 tornado widened to 1.25 miles when it snapped and uprooted trees in Parker Dam State Park, where bare tree trunks are still visible to this day, and then became wider again crossing Cameron, Clinton and extreme northern Centre counties at elevations between 1,500 and 2,000 feet. The massive tornado twice forded the West Branch Susquehanna River. Ten homes were damaged or wrecked, but the huge tornado lifted seven miles north of the city of Lock Haven.

Dr. Gregory Forbes, then on the Pennsylvania State University faculty, and later the Severe Weather Expert for The Weather Channel, participated in the storm survey directed by the NWS and National Academy of Sciences. Teams fanned out along the damage paths to conduct numerous eyewitness interviews in the wake of the historic tornado outbreak.

Forbes described the damage in his blog, published on the 25th anniversary of the disaster. Focusing on the devastation in Black Moshannon State Park, he wrote: "Every tree except for 2-inch-diameter saplings was snapped or uprooted along much of the path, with the saplings permanently bent toward the ground. The trees were stacked on top of each other 10 feet high like huge piles of sticks in the game pick-up sticks."

At Parker Dam State Park in scenic Clearfield County, two scouts and three scoutmasters huddled in a small cabin at a Boy Scouts camp when the storm roared through like a giant circular saw. Forbes noted the damp, muggy weather conditions moments before the storm appeared on the horizon, appearing "as an avalanche of falling trees coming toward them."

A scout leader clung for his life in the doorway before "being levitated like a flag by the tornado winds." The towering trees initially afforded some protection, but then toppled like dominoes. Forbes added: "The last gasp of the tornado ripped off the cabin's roof. None of the occupants was seriously injured."

A tornado (F2) touched down five miles to the northwest of Emlenton in southeastern Venango County near Big Bend, at 7:54 p.m., destroying a trailer and a farm where two persons—a father and his son—were seriously injured. Five homes received some damage. Hail two inches in diameter accompanied the funnel. A weak F0 storm was noted at 7:56 p.m. in far southeastern Venango County, two miles west of Emlenton, ending in Clarion County three miles east of the Venango County line.

Shortly before 8:00 p.m., a huge F4 tornado took shape several miles south of Sheffield in Warren County. The storm barreled southeast for 29 miles to Elk State Park, arriving there 40 minutes later. Along the way, the tornado blasted through Kane and East Kane in McKean County, destroying homes and businesses and snapping off trees and power lines. Vehicles were destroyed as garages collapsed in the extreme winds.

Four persons died in their frame homes and trailers in the vicinity of Kane and at least 39 were injured. Storm damage totaled $15 million, affecting several businesses and schools, and 53 homes in Kane and East Kane, where vehicles were crushed in caved-in garages.

At 8:10 p.m., a powerful F3 tornado formed a mile south of Darlington in Beaver County, passing north of Beaver Falls, then hitting portions of Zelienople, Evans City and Saxonburg, ending a mile south of Sarver in Butler County, covering 39 miles. The deadly tornado spanned 243 yards, ripping up trees and power lines before crossing the Pennsylvania Turnpike and Marion Township.

Fig. 7.9 May 31, 1985. The aftermath of a 2.2-mile-wide F5 tornado that sliced through Moshannon State Forest. (Photo by Greg Forbes. Courtesy "The Tornado Outbreak of May 31, 1985," Paul Markowski, Pennsylvania State University Department of Meteorology and Atmospheric Science, 2015.)

Near the beginning of the path, residences were struck by the storm near Brush Creek Park. The tornado tracked several miles north of Beaver Falls, slamming into the Big Beaver Plaza shopping center north of West Mayfield. Twelve shops were destroyed and two badly damaged in the shopping center, where two persons died. Sixteen antique cars were destroyed, among more than 100 cars that sustained damage.

One man was killed by flying debris in North Sewickley Township. Hail the size of golf balls occurred at Midland and Monaca. In all, three died and 107 people were injured in Beaver County, where more than 200 homes and businesses were damaged or destroyed, adding up to $5 million in total costs at the time.

The storm continued across Interstate 79, crossing into Butler County, where two persons died at Evans City in a mobile home park. Forty residences sustained damage around and south of Callery. Four deaths occurred near and south of Saxonburg, among the six persons killed in Butler County. Along the path of the storm,

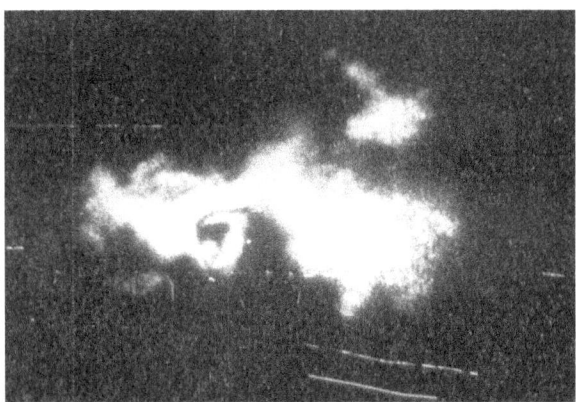

Fig. 7.10 May 31, 1985. Radar image (WSR-74C) from the top of Walker Building, Pennsylvania State University around 8:00 p.m., revealing a classic hook echo, as a massive tornado tore through Moshannon State Forest. (Courtesy Paul Markowski)

more than 120 persons were injured and damage totaled $10 million.

The supercell responsible for the mighty Allegheny National Forest tornado spawned a family of tornadoes farther east, including a strong F3 west of Bastress in southwestern Lycoming County. The storm touched down around 9:25 p.m., weakening while ascending North White Deer Ridge six miles south of Williamsport. The tornado turned right (southeast) into Union County.

The massive tornado reached F4 intensity for 1.5 miles as the storm crossed the Susquehanna River near Watsontown, before ending a 19-mile path of destruction in Northumberland County, where the path width reached 910 yards. Dozens of campers and trailers, eight homes, commercial buildings, garages and 18 vehicles were wrecked, and many more were damaged.

Thousands of trees were also uprooted, according to the NWS storm survey. Spring Lake Village Mobile Home Park was badly hit, where two fatalities occurred. The large tornado claimed a total of six lives in Union and Northumberland counties from flying debris and in overturned campers and a few mobile homes. About 60 persons were injured in the storm. Grazulis (1993) wrote that 190 homes and buildings damaged or destroyed, along with 50 vehicles, resulting in $16 million in total damage in Lycoming, Union and Northumberland counties. In Northumberland County, the cost was $10 million.

Two smaller tornadoes touched down later that evening farther west. A weaker tornado (F0) struck in Indiana County around 9:53 p.m., eight miles east of Indiana, tracking six miles while remaining mostly aloft and ending near the Cambria County line, with some light tree damage reported.

In eastern Pennsylvania, a tornado (F1) with a path width of 540 yards touched down at 10:45 p.m., tracking from Wapwallopen to near Hobbie and Freeland in Luzerne County. Some mobile homes and trailers, along with a barn and trees, were destroyed along the 11-mile trek judged to be partly the work of downburst winds.

The final tornado in the historic outbreak touched down five minutes after midnight (June 1) between Tobyhanna and Mount Pocono, resulting in relatively minor damage ($5,000), marking the end of a deadly outbreak of storms.

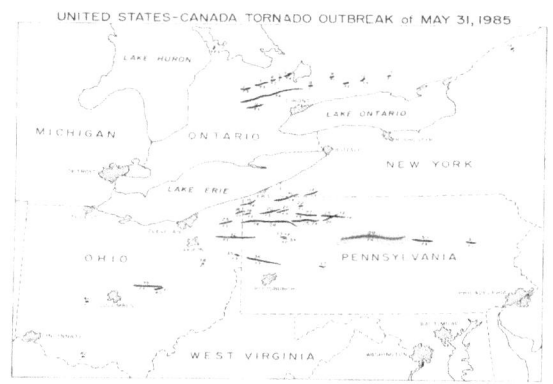

Fig. 7.11 NOAA Storm Data mapping of the tracks of 42 tornadoes recorded on May 31, 1985. (Not shown: the 43rd tornado that touched down after midnight on June 1 near Tobyhanna and a 14th tornado in southeastern Canada, later identified by Environment Canada.)

The total of 34 tornadoes in Pennsylvania during 1985 was the largest on record at the time. The State College NWS office gathered a series of interviews following the tornado outbreak that described the horrific experiences of survivors.

June 9, 1989 Philadelphia Tornado

A strong tornado, with winds estimated as high as 120 mph (F2), struck the Society Hill portion of Philadelphia around 7:15 p.m. on June 9, 1989, and traveled 0.5 mile to the Delaware River. Eyewitnesses spied one or more funnels as the storm crossed the water and passed over the Ben Franklin Bridge before touching down one more time. One man sustained serious head injuries from flying debris.

The following week, a tornado outbreak in portions of the Mid-Atlantic spawned eight tornadoes, including one in south-central Pennsylvania. A strong F2 tornado touched down on the eastern side of York on June 15, 1989, shortly before 6:00 p.m. near Wrightsville. The "intermittent 15-mile path that crossed the Susquehanna River, with a width of 100 yards" terminated in Lancaster County, two miles west of Millersville (*Storm Data*).

Sixteen homes were damaged in York County, some significantly, in addition to a church, one business and some garages and barns. Four homes were destroyed in Lancaster County and three others moderately damaged. Three injuries occurred in York County at Mountville, and four injuries were reported at Columbia in Lancaster County.

Another (F2) storm caused sporadic damage in Perry County around 6:15 p.m. on June 20, 1989, and a weaker storm (F1) touched down in Franklin County.

November 16, 1989
Late Autumn Squall Line

A powerful squall line raced across eastern Pennsylvania and New York on November 16, 1989, during the morning, accompanied by powerful straight-line winds and tornadoes ahead of a sharp cold front. November is not a time normally associated with severe weather, but on occasion a deep late-autumn storm draws tropical moisture northward into Pennsylvania in an environment with strong wind shear.

The interaction between tropical and polar air turned deadly at lunchtime in Newburgh, New York, on November 16, 1989. High winds caused a cafeteria wall at East Coldenham Elementary School to collapse. Nine students were killed and 18 others were injured. At the time, the storm was called a tornado (F1), but an analysis by Ted Fujita determined that the tragedy was caused by a downburst.

A pattern of damaging wind gusts began shortly after daybreak in eastern Pennsylvania. A downburst around 8:40 a.m. in Adams County, 280 yards wide and five miles long, destroyed a house and two mobile homes. Powerful winds that reached 76 mph at WGAL-TV in Lancaster around 9:45 a.m. felled trees and damaged roofs, including at a junior high school. The damage continued past Quarryville to north of Churchtown and may have included an embedded tornado.

Tornadoes were reported east of Centerville (F0) in Chester County and at King of Prussia (F1) in Montgomery County around 10:30 a.m. A small tornado (F0) touched down near Perkasie, Montgomery County, at 11:00 a.m. and traveled more than five miles, ending in Bucks County, damaging several homes, trees and 25 roofs.

A fourth tornado (F1) was confirmed in Bucks County at Yardley about 11:15 a.m., spawned by the supercell responsible for damage in Lancaster County, and the two Chester County tornadoes. The Bucks County tornado tracked only a mile in Pennsylvania, before crossing the Delaware River into Mercer County, New Jersey, continuing four more miles. With a sizable path of 100 yards, the storm toppled trees and damaged a school roof. A woman suffered serious injuries when her car was struck by a falling tree.

Up to 3.5 inches of rain fell in less than four hours, prompting flooding in Lackawanna and Luzerne counties, as storms passed through the Scranton and Wilkes-Barre areas. Powerful downdrafts in the Poconos shortly before 11:00 a.m. caused widespread power outages after trees landed on power lines. A trailer was pulled off its foundation at McMichael in southwestern Monroe County and turned 90 degrees, likely the work of a spin-up tornado.

Wind gusts reached 75 mph at West Chester and 61 mph at Lehigh Valley International Airport, where a funnel cloud was sighted at 10:43 a.m. High winds shattered windows in several of Philadelphia's skyscrapers and tore a portion of a roof off the overseas terminal at Philadelphia International Airport, where a gust of 69 mph was recorded.

Historically powerful F4 tornadoes so late in the season occurred in Schoharie County, New York, and near Hamden, Connecticut. Seven tornadoes were counted in southeastern New York and another seven in New Jersey. One of those storms flipped a tractor-trailer near the George Washington Bridge on the New Jersey Turnpike at Fort Lee.

Four days later, a strong cold front and line of thunderstorms blasted across the state on the evening of November 20, 1989, with damaging straight-line winds that reached 85 mph at University Park near State College, 78 mph at Lehigh Valley International Airport and also at York, 73 mph at Doylestown and 60 mph at Wilkes-Barre/Scranton Airport. Widespread wind damage was reported, mostly fallen trees, downed utility lines, a few broken plate glass windows and roof damage.

A tornado rated F2, with winds reaching 150 mph was confirmed in Huntingdon County near McAlevys Fort, with several barns and outbuildings showing damage. The path width was wide (0.7 mile) and revealed several embedded subvortices.

April 9, 1991
Early Spring Storms

On April 9, 1991, 11 tornadoes were reported across southwestern and central Pennsylvania, with seven injuries, though no loss of life. The strongest storms only reached F1 intensity. Tornadoes were sighted in Bedford, Cambria, Centre, Cumberland, Fayette and Westmoreland counties, including in the city of Johnstown.

Four family members suffered minor injuries in the town of Scottdale, 15 miles south of Greensburg, when their mobile home was lifted by a tornado (F1). Small tornadoes (F1) touched down in Johnstown and 15 miles northwest of Uniontown in Fayette County, uprooting trees, damaging a barn and causing minor damage in the evening hours. Three tornadoes were confirmed in Bedford County.

July 1992
Active Storm Season

July 1992 brought a series of stormy periods in Pennsylvania, beginning with a squall line on July 10 that caused wind damage across the southern counties. Pittsburgh had a record gust of 83 mph.

On July 15, 1992, the atmosphere turned more turbulent. Five twisters touched down over a four-hour period during the afternoon over the eastern half of the state. The first storm was sighted in Clinton County (F1), and another at Columbia, east of Berwick, across the Susquehanna River. A third tornado came down over Northampton County two miles north of Chapman, and a fourth briefly touched down in a parking lot in downtown Bethlehem. A fifth storm was reported near Philadelphia. Numerous reports of straight-line wind damage came from other sections of eastern Pennsylvania.

Two days later, another series of severe storms on July 17, 1992, brought eight tornadoes in the eastern part of the state, injuring four persons. The first funnel touched down in Snyder County, and the strongest storm (F2) struck near Unionville in Chester County, one of two F2 storms.

On the last day of the month, a funnel cloud lowered to the ground east of Levittown in Bucks County, north of Philadelphia, and was later witnessed east of Trenton, on the other side of the Delaware River. Twenty-one tornadoes touched down in Pennsylvania in July 1992, besting the previous monthly record of 11 (July 1976). The 1992 calendar year had a total of 31 tornadoes in Pennsylvania.

July 27, 1994
Mid-Atlantic Tornado Outbreak

A stationary front stretching across the Mid-Atlantic region became the focus of a severe weather event during the late afternoon on July 27, 1994, that would affect a wide area from North Carolina to Pennsylvania. Clusters of severe thunderstorms spawned five tornadoes in northern Virginia, three in Delaware and 14 in Maryland (state record).

Low pressure tracking along the boundary brought the ingredients together for a prolific tornado outbreak

The focus of severe weather shifted north into southeastern Pennsylvania after dark, at a time when the most severe storms tend to settle down with nighttime cooling. A supercell that had formed near Chesapeake Bay in northern Maryland spawned a powerful tornado that reached F3 strength at 10:55 p.m. in eastern Chester County south of Avondale.

Hunter's Run development suffered the most damage, where six homes were destroyed and 23 more were damaged. Eleven were injured in the path of the tornado, which traveled 3.5 miles and reached a width of 100 yards. Wind damage extended into New Garden Township. Farther to the east, straight-line winds trees tore down trees and wires in White Clay Preserve. Nine homes or buildings were destroyed and 49 others reported damage, with losses of $3.5 million (*Storm Data*).

As the parent storm progressed to the northeast, a tornado (F1) dipped down south of Downingtown. The supercell split into two discrete cells upon exiting Chester County. The western storm spawned another tornado at Royersford in Montgomery County, before developing into a deadly F3 tornado in Limerick Township, just east of Pottstown, shortly before midnight (11:52 p.m.). Four homes were leveled and 16 endured heavy damage in an uninhabited portion of The Hamlet subdivision, causing $5 to $5.5 million in damages along a two-mile path 100 yards wide.

Tragically, three members of family died when they were "blown out of their house" (*Storm Data*). A U.S. Savings Bond "in the child's name" traveled 50 miles northeast, and was found in Bath. A man survived after being thrown from the second story window into a neighbor's living room.

After the tornado lifted, the parent storm caused damage in northwestern Montgomery County.

June 22–24, 1996
Tornadoes, Wind and Hail

June 1996 was a particularly stormy month in the upper Ohio Valley and Mid-Atlantic region. The unseasonal strong upper-level flow provided ample deep-layer wind shear that interacted with surface boundaries that focused instability and lift.

The NOAA Storm Event Database listed more than 250 individual damage reports, including hail, high winds, flooding rains and tornadoes that caused nearly $50 million in damage. Excessive rainfalls were common (four to 12 inches) in several hours' time.

Central and southeastern Pennsylvania were raked by rounds of severe weather on June 7–8, 11–12, and 18–19, 1996. Thunderstorms on June 17–20, 1996, swamped central and western Pennsylvania. A cloudburst over State College on June 17 sent torrents of water coursing down Beaver Avenue that affected nearly 100 homes on the edge of the Pennsylvania State University campus.

Franklin County was one of several areas hit hard by flash flooding on the night of June 18, 1996. Torrential rain pounded St. Thomas and the vicinity, west of Chambersburg, where up to 12 inches fell in several hours. Homes were flooded and roads were washed out, and one woman drowned in her car on a flooded roadway in the darkness. Damage was estimated at $1 million.

Serious flooding also developed in Adams County in the late evening hours of June 18, 1996. Rainfall greater than 11 inches fell between 10:00 p.m. and 2:00 a.m., causing about $5 million in damage in the community and surrounding townships. Every major road leading into Gettysburg was closed at one point.

A total of 69 homes and business were damaged by water, along with eight bridges. The NOAA *Storm Report* added, "Two teenage girls were swept away in the waters of Stevens Run, but were rescued by firemen. The Sachs Mill Covered Bridge across Marsh Creek which predated the Civil War was washed off its piers and lay crumpled on the stream bank."

Widespread flooding accompanied training thunderstorm cells in Bucks County in the hours before midnight. A weather spotter in Langhorne recorded 11.55 inches, and a total of 18.48 inches in July 1996 was mentioned in the monthly *American Weather Observer* publication.

Serious flooding also developed in central and southeastern Bucks County, where two drowned. The losses were put at $14.5 million. Widespread flooding was reported along the Neshaminy, Mill, Brock, Queen Anne, Indian and Silver Creeks, and Deer Run, with the most substantial damage in Middletown Township along Mill Creek. Several bridges disappeared in the rising waters.

In Yardley, rising water along Brock and Silver Creek and in the canal swamped neighborhoods for the second time in the year. The NOAA summary stated: "About 200 to 300 persons were rescued. Cars were buried in water. Evacuations were ordered between the Canal and Delaware River."

Severe storms struck again in Pennsylvania on June 19–20, 1996, causing flash flooding and considerable damage at Freedom ($7.5 million) and McKeesport ($3.1 million). A late afternoon storm on the 20th at Loganton was responsible for $15 million in damage.

A warm front draped across Pennsylvania on June 22, 1996, resulting in three small tornadoes. One touched down in Bradford County in Wyalusing, tossing a 25-foot trailer, leveling a garage and also impacting Laceyville. A second tornado was reported at Duryea and West Avoca in Luzerne County shortly before 4:00 p.m.

Clusters of thunderstorms aligned with the surface warm front, dumped out hail the size of softballs and downpours of two to four inches in southern Monroe County. A hailstreak marked the path of the storm's fury between Kresgeville and Saylorsburg that continued for more than 10 miles, as large hail bombarded the countryside.

Another round of severe weather swept across portions of Ohio, Pennsylvania, West Virginia, Maryland and Virginia on June 24, 1996. A long squall line with bowing segments was responsible for concentrated pockets of wind damage. Small circulations developed along this line, resulting in a rash of tornadoes.

In western Pennsylvania, hail and strong winds wreaked havoc and caused power outages. Part of a school roof was torn off in Ambridge. Wind damage consisted of downed trees and power lines that was worst in Allegheny and Westmoreland counties. Wind gusts topped 60 mph at Penn Hills.

Flash flooding caused upwards of $150,000 in damage, after 1.55 inches of rain fell at Pittsburgh International Airport in 29 minutes, and local storm totals exceeded three inches. Roads were flooded, sending a nasty combination of mud and water into dozens of homes.

Storm damage was reported across Indiana, Washington, Jefferson, Fayette, Somerset, Blair, Bedford and Cambria counties, including flooded roads and basements. Winds gusted to 60 mph three miles south of Johnstown, and 70 mph at Greencastle, as the storms pounded Franklin and Adams counties. Trees and limbs came down in Mercersburg, Caledonia State Park, and in an area nine miles southeast of Shippensburg.

Several tornadoes were reported in Maryland and northern Virginia, where wind damage was extensive.

July 19, 1996
Tornadoes Strike Southwest

Eight tornadoes struck in the late morning and early afternoon of July 19, 1996, in southwestern Pennsylvania. There were four injuries but no fatalities, despite four F2 tornadoes.

The first tornado touched down at 10:20 a.m. in Clarion County, around six miles north of New Bethlehem, and continued in a southeasterly direction for 10 miles into Jefferson County through Langville, ending 1.5 miles west of Oliveburg. Trees were uprooted and a mobile home was rolled north of Truittsburg. The path width of 100 yards was not especially wide, but it resulted in damage to property and crops ($30,000).

Another tornado developed 1.5 miles southeast of Punxsutawney at 10:45 a.m. and crossed into Indiana County, passing through state game lands along a nine-mile path to a point northeast of Glen Campbell in Indiana County. The storm lifted the roof off a barn east of Rossiter, causing three walls to collapse. A child seeking shelter in the barn was not hurt, but a horse was sucked out of the structure and carried about 75 feet in the air, before landing in a field.

Three persons were injured, one seriously, in a mobile home that was destroyed, traveling 75 yards and ending in a grove of trees. The F2 tornado damaged or destroyed about a dozen homes and caused $200,000 in total damage. The NOAA *Storm Report* included the lifting of 30-pound cinder blocks pulled from a garage that were thrown about 300 yards along a hillside. Trees 100 to 150 feet tall were twisted in State Game Lands, as the storm briefly entered Clearfield County, which was attributed to a "strong suction vortex."

Storms struck Adams, Bedford, Blair, Cambria, Clarion, Clearfield, Franklin, Fulton, Indiana and Jefferson counties. The F2 tornado that started in Jefferson County and tracked 30 miles, ending in Clearfield County, caused $500,000 in damage.

Destructive floods accompanied slow-moving thunderstorms throughout western Pennsylvania, resulting in rainfall amounts of four to six inches. The hardest-hit region was Jefferson County, where losses tallied $100 million, primarily around Punxsutawney and Brookville. Data showed 92 homes were destroyed, including apartments and mobile homes, and 311 sustained heavy damage, plus 792 with minor damage. Seven bridges were washed out, along with countless road miles.

Extensive flooding occurred in Oil City, Franklin (French Creek), and along Sugar Creek, where streams overflowed. The NOAA *Storm Report* listed that five homes were destroyed, and 60 others had major damage. An additional 200 reported minor flood damage in Venango County. Eleven bridges were wiped out along the 100-mile corridor of damage.

Severe weather returned to Pennsylvania on July 29–31, 1996, with heavy storms causing minor damage, fostered by a fast-moving, southward-displaced jet stream. On July 30, two tornadoes were reported, in Jefferson and Franklin counties.

November 8, 1996
Late-Season Tornado Outbreak

Severe storms blasted the Ohio Valley southward to Mississippi on November 7, 1996, with hundreds of reports of damaging winds, large hail and tornadoes. Seven tornadoes touched down in Pennsylvania on November 8, setting a record for the month.

In Bedford County, one man was injured when an F1 tornado damaged several barns and a home. Another twister hit south of Old Fort and headed toward Spring Mills, causing pockets of property damage, mainly to trees and roofs.

A subsequent tornado (F1) affected Mifflin County, hitting Burnham, damaging the local YMCA roof and nearby trees. Another tornado formed along the Clinton-Wyoming County line near Pine Station and traveled to Jersey Shore. Damage was noted on the north side of Bald Eagle Mountain, mostly to trees, crops, a barn, and the roof of a home. Lycoming County was the site of a short-lived F1 tornado northwest of Linden.

The strongest tornado (F1/F2) in Pennsylvania on November 8, 1996, came down in Northampton County near Blue Mountain Ski Resort at Danielsville, where a church sustained structural damage and a car was totaled. One woman was injured in her mobile home, which was shoved off its foundation. An unoccupied mobile home was tossed 50 feet into a ditch, and the roof of another home ended up 100 feet away. Homes sustained window damage and cracked chimneys.

The same supercell spawned a tornado (F1) in southwestern Monroe County that touched down in Kunkletown and tracked 10 miles to Snydersville. A total of 15 homes were affected, five of which received major damage. An airplane hangar and planes inside were damaged, trees and utility lines were toppled, boats were overturned at a marina, and trailers were turned on their sides.

Wind damage from a combination of straight-line winds and possible touchdowns—mostly felled trees—extended from Brodheadsville and Saylorsburg northeastward to Beacon Hill in Stroud Township. There were two injuries—one serious—after a tree crash-landed on a van on Route 447 in Price Township, Monroe County. Storm damage amounted to $400,000 in a series of "brief, but numerous" touchdowns.

Flash flooding was reported in several communities on November 8, 1996. Damage in Tioga County alone totaled $2.5 million, where two dozen homes were destroyed and many more had serious water damage. Several bridges and many roads were washed out.

1998: May 31 and June 2 Historic Tornado Swarms

Two historic tornado events impacted Pennsylvania in the late spring of 1998, leaving multiple trails of destruction and establishing monthly state records for confirmed tornadoes.

Twenty-two tornadoes touched down in the state during the outbreak that commenced on May 31, 1998, and extended a few hours past midnight on June 1, 1998—a single-event record for the Keystone State. Twenty-two tornadoes touched down on May 31 Eastern Standard Time (EST), plus one on June 1 at Philadelphia. The May tornado count in *Storm Data* listed a tornado in Centre County on May 29, 1998, adding up to a state record monthly tornado count (23) that surpassed May 1985 (21).

Incredibly, on June 2, 1998, a second major outbreak spawned 22 tornadoes. The tornado total in June set a new state record (29), which included the final tornado of the May 31–June 1 outbreak, based on EST record-keeping per the NWS. There were four weak tornadoes on June 16 and two more on June 30. For the year, one tornado was reported in March 1998, four in April, one in July, and four in September, establishing a state annual record of 62, according to the State Climatologist office.

May 31, 1998

The weather pattern on May 31, 1998, featured an unusually robust jet stream that would provide plenty of energy as a series of disturbances dove southward in a northwesterly flow, creating a highly-sheared environment conducive to tornadoes. The "high risk" outlook for portions of Pennsylvania from the Storm Prediction Center (SPC) in Norman, Oklahoma, was a first-time issuance.

The storm system that would trigger the historic tornado outbreak in the Northeast was located over South Dakota on May 30, where a half-mile-wide tornado destroyed most of the town of Spencer. An unseasonably strong low-pressure system tracked across southern Quebec, Canada, placing Pennsylvania squarely in the warm sector. The arrival of an upper-level disturbance during the heat of the day unleashed powerful updrafts that encountered strong winds shifting with height. Activity developed ahead of a cold front in an environment classically suited for the formation of discrete supercells.

The first (F1) tornado in Pennsylvania on the afternoon of May 31, 1998, was spotted west of Buttonwood in Lycoming County at 5:20 p.m., causing minor property damage, traveling east across Route 15 for a half-mile. A weak F1 tornado was sighted three miles southwest of Salladasburg, tracking northeast for seven miles and causing sporadic damage. Wind damage was more widespread in Lycoming County and in Williamsport, where fallen trees blocked roads.

The atmosphere had become energized with the arrival of upper-air energy, setting the stage for the development of rapid updrafts that spawned nearly 20 tornadoes and gustnadoes in three hours from 7:00 to 10:00 p.m. in northwestern Pennsylvania.

Two tornadoes were confirmed in Elk County (F0/F1), mostly damaging trees and causing minor property damage near Ridgway and St. Marys. Bow echoes containing embedded tornadoes occurred in Cameron County, coupled with localized downbursts in excess of hurricane force, extending east across Clinton County to near Lock Haven.

At 7:00 p.m., a small (F0) tornado in eastern Luzerne County, two miles northwest of Pittston, twisted off treetops and sheared the tops of utility poles. An hour later another supercell would spawn the first of four storms that raked Pike County during a chaotic two-hour period, accompanied by large hail and damaging winds.

The first Pike County twister touched down at 8:08 p.m. three miles southwest of Blooming Grove. The storm reached F2 intensity at Promised Land and proceeded to carve a path of destruction up to 550 yards wide and two miles long in the state park. Hundreds of trees were mowed down. The damages were estimated at $200,000.

A second tornado (F1/F2) descended from the roiling skies around 8:20 p.m. near Blooming Grove, south of Interstate 84, cutting a swath of damage three miles long and 200 yards wide through wooded areas, before lifting east of Route 739 near Cranberry Ridge in the eastern part of Pike County. This storm caused $400,000 in damages.

The supercell that spawned the Promised Land tornado would soon be linked to another strong (F2) tornado near Pecks Pond. The tornado had staying power, cutting a continuous damage path 20 miles long, finally lifting a mile northeast of Dingmans Ferry near the Delaware River. Virtually every tree in the 200-yard-wide path was damaged. Routes 402 and 390 were closed for several days after the storm. The NWS field surveys indicated the tornado briefly attained F3 intensity around the vacation community of Blue Heron Lake.

An estimated 1,000 trees were toppled or sheared off on both sides of the lake, and several homes were damaged. Fortunately, the storm caused only minor injuries despite substantial damages ($1 million). Yet another Pike County storm touched down briefly near Greentown at 9:53 p.m. in the southern part of the county.

Two smaller funnels developed from a supercell over southern Lackawanna County near Old Forge and Elmhurst around 8:20 p.m. The north side of Scranton and the communities of Waverly and Taylor suffered tree and power line damage as storms with powerful straight-line winds swept through the region between 9:45 and 10:00 p.m.

A powerful tornado formed over high terrain near Mount Davis in Somerset County about 8:50 p.m. and traveled east-southeast for 15 miles into downtown Salisbury. The 50-yard-wide tornado (F2) damaged up to 15 homes and businesses in Salisbury and forced 150 residents to spend the night in a nearby shelter.

Fig. 7.12 May 31, 1998. Tornado damage on Cody Street, one block east of Route 219 in Salisbury, Somerset County. (Photo by V. W. H. Campbell, Pittsburgh Post-Gazette *Photo Archives, copyright 1998, all rights reserved)*

A 13-year-old girl died when a tree landed on her family's vehicle, and 15 others were injured. The storm continued on for another eight miles, demolishing a barn in Pocahontas, where winds approached F3 intensity, with damages nearing $4 million.

Strong thunderstorms with winds topping 60 mph and hail caused pockets of damage across the south-central part of the state, mainly to trees and cars. Later in the evening, an intense F3 vortex emerged near Evansville in northern Bucks County around 9:00 p.m., east of Lake Ontelaunee. The whirling storm traveled eight miles into Lyons, a community that took a direct hit on the north side. More than 40 homes were damaged or destroyed, and seven persons were injured. The total damage was put at $1.4 million.

A tornado touched down on June 1 (EDT) a little south of Quarryville, Lancaster County, around 12:30 a.m. and tracked seven miles to Ninepoints, where the F2 storm expanded to a width of 100 to 200 yards. Six homes received major damage, including one that lost a roof and second floor. There was only one minor injury reported.

Two more tornadoes developed on June 1 in Chester County. One storm touched down at 12:43 a.m., staying on the ground for six miles in Londenderry Township before vanishing five miles north of Avondale. Trees were toppled along a 100-yard-wide path.

A second Chester County tornado followed a three-mile course through eastern Pocopson Township, after touching down around 12:52 a.m. east of the Brandywine Creek. Hundreds of trees were damaged and the roofs of five homes were torn off before the storm ended west of the Dilworthtown Inn. Vehicles were battered by falling trees. Damage caused by the two tornadoes and straight-line gusts was near $1 million.

A squall line swept across Philadelphia County, spawning a tornado (F1/F2) that possessed a width of 200 yards, passing through Upper and Lower Moreland Townships and far northeastern Philadelphia around 1:20 a.m. near Bustleton and Lunar Park. The tornado intensified within the city limits, damaging 35 commercial buildings. Ten homes were hit by fallen trees.

The tornado had a discontinuous path of nearly six miles, causing $1.8 million in damage. Some 34,000 PECO Energy customers lost power in the storm that ended near the Bucks County line. A wind gust of 71 mph was recorded at Philadelphia International Airport.

June 2, 1998

A second widespread tornado outbreak struck the region during the evening of June 2, 1998, even as residents were still picking up the pieces from the May 31 storm rampage. Discrete cells tapped into upper energy, where winds turned with height.

Two tornado watches were issued by the Storm Prediction Center (SPC) for portions of Pennsylvania and several surrounding states. Although moisture transport was not unusually pronounced, brisk southwesterly winds in the lower levels represented deep-layer wind shear beneath powerful west winds aloft. During peak daytime heating, two lines of thunderstorms developed ahead of a strong cold front moving through western Pennsylvania, bringing rounds of torrential rain and large hail.

The first evening tornado (F1) on June 2, 1998, approached Shippingport, northwest of Pittsburgh in Beaver County, around 5:30 p.m. Four homes and a trailer sustained damage. The storm hit the edge of Raccoon Township, damaging 19 homes ($400,000). Another tornado (F0) visited the same area two hours later, knocking down upwards of 100 cherry trees and damaging a home.

A tornado (F1) formed in Allegheny County at 5:55 p.m., five miles northwest of Carnegie, and continued for a substantial distance of 32 miles, crossing the southeastern part of Pittsburgh. Hundreds of homes were ravaged, resulting in damage totaling $13 million and causing 51 injuries.

Remarkably, no deaths were reported in this storm despite tracking through southern portions of metropolitan Pittsburgh. The funnel cloud passed near Three Rivers Stadium while fans and players for the New York Mets and Pittsburgh Pirates waited out a rain delay, watching the low-hanging and twisting cloud pass nearly overhead.

Sixteen injuries occurred in Mount Washington as the F1 storm traveled along the Parkway west, ripping off roofs, which caused walls to collapse in several homes. Trees were uprooted and twisted and limbs sheared off. Widespread damage extended to Hazelwood and Rankin as the storm's path widened to 800 yards. The tornado contracted in size over the next six miles across Westmoreland County, affecting Irwin and Manor.

More than a thousand structures experienced varying degrees of damage (roofs, walls, siding, windows, doors) along the entire length of the storm path. Several railroad cars were blown over north of Baldwin. The storm dissipated over Westmoreland County at 6:28 p.m., after causing damage totaling several million dollars.

A weaker tornado briefly touched down in western Allegheny County and crossed into Westmoreland County between Greenock and Sunset Valley about two hours later, with little evident damage.

A tornado (F1) was spotted just before 7:00 p.m. in southeastern Westmoreland County, west of Donegal, that crossed the Pennsylvania Turnpike and overturned a tractor trailer, injuring the driver. The subsequent eight-mile path was intermittent through densely wooded terrain, uprooting and shearing trees, some landing on houses. One mobile home was overturned.

Fig. 7.13 June 2, 1998. A rare sight of twin funnels, traveling northeast along the Ohio River approaching Mount Washington, photographed by a KDKA-TV cameraman. (Courtesy Pittsburgh Post-Gazette *Photo Archives, copyright 1998, all rights reserved)*

In Fayette County, a tornado touched down around 7:00 p.m. four miles southeast of Markleton. The F2 storm traveled 12 miles, striking Boyntown and causing heavy damage, weakening near Pocahontas. The tornado crossed the same path taken by the tornado that struck on May 31, 1998, six miles east of Salisbury.

Ironically, a carpentry shop in Pocahontas—that was undergoing repairs by Amish farmers, after being nearly demolished—lost the new frame structure on the evening of June 2, 1998. The storm's path reached 880 yards wide in a wooded section as it traveled 12 miles in Pennsylvania, and then continued for another dozen miles across the mountainous panhandle of western Maryland.

A smaller tornado (F1) touched down four miles west of Custer City, crossing Route 219 and paralleling Route 770 along an eight-mile track. Thirty-eight homes were damaged, many struck by falling trees. Seven homes were no longer habitable. An oil museum in a barn-like structure was destroyed north of Custer City in McKean County. As the storm moved into Allegheny National Forest, the path width exceeded 1,000 feet.

In Jefferson County, a tornado (F1) with a 100-yard-wide path was sighted about two miles west of Ringgold around 6:45 p.m., pulling down trees. barns and silos, and tearing the roofs off of several homes and buildings. A pond was completely drained. This storm caused $250,000 in damage.

Between 7:30 and 9:00 p.m. four small tornadoes were confirmed in Clearfield, York, Blair and Centre counties. Northern Blair County sustained F1 damage northwest of Tyrone, mainly to trees. A weak F0 tornado was confirmed north of the Blair/Centre County border, and another was reported northwest of Irbona in Clearfield County.

The most damaging tornado in western Pennsylvania—the third to hit Somerset County in two days—touched down near Seven Springs at about 9:00 p.m. This large F3 twister (158–206 mph) was a mile wide (1,760 yards) as it roared 26 miles through the southern portion of the county to the Maryland line, east of Pocahontas, then continued for five miles to Frostburg, Maryland.

The track closely paralleled the storm that had struck two hours earlier, three quarters of a mile north, intersecting the path of the May 31 storm at a point three miles east of Salisbury. The very wide path, approaching one mile, resulted in overlapping damage in Somerset County, all within two hours on June 2, 1998.

Boynton residents had emerged from their battered properties to clean up from the first tornado around 7:30 p.m., with chainsaws and front-end loaders, only to retreat to shelter after being warned that another tornado was imminent. The second tornado broke all the windows from one house and ripped the roof off another home and a barn.

The NOAA *Storm Data* summary described how a Laurel Falls family had sought shelter in their basement behind a television set, only to watch the three-story dwelling and eight rows of foundation blocks lifted right above their heads:

> As the twister passed, they looked up to find all three stories of their house gone, along with eight rows of foundation blocks. A battery operated clock found the next morning had stopped at 9:38 pm. A neighbor told of losing electricity, then getting a phone call from his brother to warn him. He and his family took shelter in a hall closet because they had no basement. They told of hearing a buzzing noise, like a giant bee's nest. Another

neighbor found her mobile home flipped on its roof after taking shelter at her son's house. When the first tornado of the evening missed a Laurel Falls family mobile home, they proceeded to a neighbor's home. The second storm blew the trailer off its foundation.

About 40 properties sustained damage, which included "permanent and seasonal residences and farms." Many barns were destroyed in the rural section of Somerset County. Luckily, there were no injuries given the strength of the storm, though more than 100 head of cattle died in a barn that was destroyed. Many farmers reported heavy livestock losses.

Yet another strong F2 Fayette County tornado developed a little past 9:35 p.m. near Elliotsville, in the southwestern part of the county. The storm traveled five miles, before entering a sliver of Preston County, West Virginia, and a small part of western Maryland in Garrett County, covering 12 miles. Two dozen structures were damaged or destroyed, and a cow was thrown more than 100 yards. No one was injured, and damages to property and farmland were estimated at $5 million in Pennsylvania.

Discrete supercells continued to form ahead of the cold front. A tornado in Northumberland County (F1) tracked over Big Mountain, striking Sagon. A prison north of Shamokin sustained damage, along with multiple vehicles. The damage in Kulpmont totaled $1 million. Wind damage occurred in Schuylkill County, and hail piled up to sufficient depth to create slick roads.

Thunderstorms with hail and high winds raked central and northeastern Pennsylvania throughout the evening of June 2, 1998. One intensifying supercell over eastern Bradford County became tornadic in southwestern Susquehanna County around 9:05 p.m., damaging several structures and mobile homes and shearing trees as it moved up and down the ridges over a discontinuous path of a dozen miles.

At 9:45 p.m., a storm touched down east of Robinwood Mountain, near Terrytown, in Bradford County, blasting a trailer eight feet off its foundation. The two occupants inside the trailer survived because a farm vehicle, pick-up truck and storage shed prevented the trailer from rolling down a steep hill. The tornado tracked six miles through eastern Bradford County, flattening a barn and shearing trees.

The same tornado intensified dramatically as it entered northern Wyoming County east of Valentine Hill, where it was judged to be of F3 intensity as it brushed the northern part of Stevens Lake, just west of the resort community of Lake Carey. The unusually strong tornado briefly turned south and then east-southeast, moving across the lake.

Major destruction led to two deaths: a grandmother and her grandson who remained in their home were killed by falling debris. The greatest damage occurred on the land linking two sides of the lake as the path expanded to one-third of a mile, damaging 42 homes.

As the storm continued farther east, mostly F1/F2 damage occurred around East Lemon, Nicholson and Tunkhannock Township, where a home lost three of four walls and a 108-year-old barn was destroyed, along with numerous trees. In all, the twister was responsible for $2 million in property damage to 50 homes, taking two lives and causing 15 minor injuries.

The supercell traversed northern Lackawanna County, spawning a weaker F1 tornado near Factoryville that crossed Routes 6 and 11 around 10:40 p.m., mowing down trees in Lackawanna State

Forest along a three-mile path. The storm finally lifted west of Abington Road, but not before causing damage at Keystone Junior College, where windows were blown out and gutters twisted.

A tornado developed in Susquehanna County at 9:05 p.m. and traveled for 12 miles across ridgetops. Trees were twisted and snapped off. Headstones in cemetery near Route 29 at Lynn were blown over, and a vacant trailer was hit. The funnel eventually dissipated four miles east of Springville.

Five tornadoes occurred in eastern Pennsylvania, where the only two fatalities in the outbreak were reported in Wyoming County.

The NWS office at Pittsburgh confirmed a tally of nine tornadoes in the coverage area that includes most of western Pennsylvania and sections of West Virginia and the Maryland Panhandle. One of those storms was confined to Garrett County, Maryland. A tristate tornado started in Fayette County. The June 2, 1998, storms injured 77 people in nine states and caused an estimated $40 million in losses.

November 26, 1999
Honey Brook Tornado

A late-autumn tornado swirled through the small farming community of Honey Brook around 6:00 p.m., 50 miles northwest of Philadelphia. The EF1 tornado injured a dozen people along a path of 1.8 miles, destroying four homes and two barns, with another 25 homes receiving some damage. The roof of a three-story warehouse was torn off. The rare November tornado formed on a record warm day (Philadelphia reached 70°F).

The tornado left about 6,000 customers in the area without power in western Chester County.

April 28, 2002
Indiana County Tornadoes

A supercell spawned two tornadoes over southern Armstrong County that tracked eastward across Indiana County, beginning around 3:45 p.m. near Spring Church (F0), damaging trees and causing minor property damage.

A second, stronger tornado (F2) formed about six miles southwest of Indiana around 4:00 p.m., lifting south of Indiana after traveling five miles. Winds reached 135 mph as the storm reached a width of 250 yards, damaging 20 homes and many adjacent structures, including barns and garages. Eighteen vehicles were damaged and two were crushed. Only one minor injury was reported, and the damage totaled $750,000.

July 21, 2003
Tornado Destroyed Kinzua Bridge

The Kinzua Bridge in McKean County, completed in 1882, served as a railroad trestle over the Kinzua Creek Valley. The viaduct was dubbed the "Eighth Wonder of the World" at the time, a towering 301 feet that spanned 2,053 feet above the gorge.

The bridge was reinforced in 1900, after heavier locomotives raised concerns about the integrity of the ironwork, which was refurbished with steel. In 1959, the bridge was closed to rail traffic, and later purchased by the state, creating a state park.

The bridge closed in 2002 for repairs, one year before a tornado carrying 90 mph winds moved through the valley, striking the rusted middle of the structure on July 21, 2003. More than half of the spans and supporting towers came tumbling down in about 30 seconds shortly after 3:00 p.m

The thunderstorm complex developed over northeastern Ohio, before heading into western New York and northern Pennsylvania traveling at 40 mph. The storm system caused extensive damage and spawned five tornadoes. The MCS displayed a characteristic comma-shape vortex with embedded mesocyclones in the northern part of the system and long-lived rotation. The wind gusts reached 70 mph along a squall line, toppling trees and knocking down power poles.

A tornado (F1) first touched down in Kinzua Bridge State Park around 3:15 p.m. about a mile west of the famous bridge. The storm already had a history of producing damaging gusts two miles west of Kane, about 15 miles west of the state park. The tornado traveling northeast caused a damage path 600 yards wide over 3.5 miles. Straight-line wind damage was evident for another two miles, before a funnel came down two miles north of Smethport, trekking three miles. The path width was 150 yards and contained wind gusts of 90 mph. Two homes and two barns were damaged, and a third barn destroyed, though no injuries were reported.

An NWS survey team from State College found evidence of another tornado, with wind gusts as high as 160 mph (F3) in Potter County, which touched down around 4:00 p.m. The storm first struck about two miles southwest of Ellisburg, blowing down hundreds of trees. The funnel traveled northeast, destroying a barn and house, and damaging a second home. A car was lofted and tossed 600 yards away, along with a farm tractor. The path was determined to be 2.5 miles long and 200 yards wide, with F2 damage.

Another tornado (F1) touched down two miles southwest of East Salem in Juniata County, traveling three miles and causing relatively minor structural damage. An embedded tornado was confirmed in Lycoming County about 5:30 p.m., taking a five-mile path, with minor damage.

The line of thunderstorms brought high winds and flooding downpours in central and eastern Pennsylvania throughout the evening hours of July 21, 2003. On August 6, Governor Edward G. Rendell signed a Proclamation of Disaster Emergency for parts of northwestern Pennsylvania to facilitate recovery efforts.

July 14, 2004
Campbelltown Tornado

The strongest recorded tornado to strike Lebanon County touched down in the western part of the county on July 14, 2004. The National Weather Service issued a tornado warning for Lebanon County at 2:59 p.m., after detecting strong rotation southwest of Hershey.

The National Weather Service survey confirmed that an F3 tornado first touched down around 3:05 p.m. about a mile west of Campbelltown. The storm hit a housing complex near Route 322, destroying 32 homes and damaging 37 others. Downburst winds damaged another 50 homes, along with at least nine farm buildings.

Twenty-four persons were hurt, one seriously. About 30,000 customers lost power in the storm. A storm survey determined that the tornado traveled 7.5 miles before lifting two miles northwest of Cornwall. The tornado had a width of a quarter-mile, and winds were estimated between 175 and 200 mph.

September 17, 2004
Ivan's Tornadoes

The remnants of Hurricane Ivan spawned nine F1 tornadoes in south-central Pennsylvania on the evening of September 17, 2004. Funnels touched down in Fulton, Franklin, Bedford, Cumberland and Juniata counties.

The storms traveled 0.5 to 4.5 miles, damaging barns and knocking down trees. Crop damage was common, with reports of corn and tall grass flattened. No injuries were reported.

December 1, 2006
Rare December Tornado Outbreak

Tornadoes are rare in Pennsylvania during the winter because the short daylight period significantly limits heating and instability.

However, on the first day of December 2006, a line of severe thunderstorms with storm tops up to 30,000 feet advanced from Ohio into western Pennsylvania during the midday hours. An approaching cold front collided with a corridor of unseasonably warm air, where temperatures rose into the 60s and low 70s. Strong vertical wind shear associated with a powerful jet stream (125 mph) spawned the first confirmed December tornadoes in state history.

More than 65 reports of severe weather across Pennsylvania were received on December 1, 2006. The first confirmed tornado (F1) struck Greensburg, damaging a hospital and some trees. Discrete cells were moving at incredible speeds of up to 80 mph. Another tornado (F2) touched down in Luzerne County at Fairview Heights, damaging homes and knocking down trees. Farther south, storms intensified in south-central Pennsylvania, spawning a third tornado (F1) in Dauphin County. One person died at Halifax after being struck by a falling tree.

High winds of more than 60 mph knocked out power to about 65,000 Pennsylvania Power and Light (PPL) customers. Hundreds of trees were toppled in southern Monroe and Northampton counties, some ending up on top of homes and vehicles.

July 29, 2009
Cherry Valley Tornado

Longtime residents of Cherry Valley, a beautiful countryside nestled between Godfrey Ridge and Kittatinny Mountain on the edge of the Poconos, had not witnessed a tornado since June 12, 1959, when a small storm caused minor damage.

But that all changed shortly after noontime on July 29, 2009, when a large funnel swept down from the clouds five miles southwest of Stroudsburg around 1:35 p.m. The trigger was a warm front lifting north, introducing requisite low-level wind shear.

The first sign of damage occurred at the Pocono Wildlife Rehab Center near Bartonsville, where straight-line winds estimated at 70 mph were reported. Ten minutes later, a classic funnel formed two miles east of Bossardsville, southwest of Kemmertown Road, touching down in a narrow valley between Kittatinny and Godfrey Ridge.

After snapping a few rows of trees, the tornado expanded to a damage path of 100 yards, heading along Middle Road. In an instant, three barns were destroyed at the Blakeslee Farm, and a fourth was damaged. Two residents were injured, after seeking cover from the whirlwind in a garage. The roof of a nearby home was torn off, and another home nearby was damaged. Twenty-five utility poles were toppled.

Fig. 7.14 Aftermath of a tornado that hit Cherry Valley on July 29, 2009. (Photo by David Kidwell, Pocono Record)

Several homes were marred by falling trees as the EF2 tornado, packing 120 mph winds, traveled 4.6 miles, according to the NWS storm report. As the storm crossed Stroudsmoor Road traveling uphill, more trees were toppled. The funnel lifted near the intersection of Routes 191 and 611 near Foxtown Hill Road. Spotty wind damage occurred farther east in the vicinity of Delaware Water Gap. Power outages affected about 10,000 homes and businesses, and the damage totaled around $1 million.

Fig. 7.15 Home destroyed by a tornado in Cherry Valley on July 29, 2009. (Photo by Adam Richins, Pocono Record)

July 23, 2010
Northern Tornado Outbreak

The first in a series of evening tornadoes (EF1) on July 23, 2010, touched down in Susquehanna County just before 6:00 p.m. a little east of Union Dale. It traveled nearly seven miles along Route 171, taking out a trailer and numerous trees near Stillwater Lake, before ending 5.5 miles from Forest City. The storm was 200 yards wide and caused $100,000 in damage, with top wind speeds estimated at 105 mph.

Three tornadoes touched down not far from Honesdale in the northern Poconos that felled trees and causing some structural damage. PPL reported that 4,300 customers lost power in the storms, according to the *Wayne Independent*.

The first tornado confirmed in Wayne County hit north of Bethany at 6:15 p.m., leaving a path of minor damage that covered a little less than two miles. At about the same time, a significant EF2 tornado, with 115 to 120 mph winds, formed near Honesdale that created a more substantial damage path four miles east of town.

Tornadoes are uncommon in the rolling hills of northeastern Pennsylvania. The only previous F2 tornado in Wayne County traversed a four-mile path in the evening of September 18, 1991.

Four structures sustained heavy damage, beginning at a location 1.5 miles west of Beach Lake. A total of 40 structures had minor damage, and many trees were felled in an area around Woodloch Pines. The tornado tore up trees and damaged barns heading into Pike County, finally dissipating about two miles north of Greeley, or about 20 miles south of Lackawaxen in State Game Lands. The path length was a little more than 17 miles (*Storm Data*).

A spin-up (EF1) tornado formed in the vicinity of a rear flank downburst at 6:30 p.m. southeast of White Mills, between Hawley and Lackawaxen, tracked two miles in Wayne County and continued for nearly two miles into Pike County. Total losses in the EF1 storm (70 to 90 mph winds) were put at $50,000. Trees were toppled onto power lines near the Lackawaxen River.

The following day, on July 24, 2010, a tornado (EF1) struck two miles southwest of Galeton in Potter County, destroying more than 1,000 trees and knocking down about 10 utility poles along State Route 2002.

March 23, 2011
Pittsburgh Area Tornado

A damaging early spring tornado (EF2) struck 15 miles southeast of Pittsburgh on March 23, 2011, in Sewickley Township about 4:45 pm. that sliced through Hempfield Township, following Route 136 in Westmoreland County. At Rillton, houses were damaged. Roofs were torn off homes at Fort Allen and Irwin. Hempfield High School and the stadium sustained losses.

The *Pittsburgh Post-Gazette* wrote that some homes were torn off their foundations, and total property losses were later put at upwards of $4.5 million. Thirty homes were destroyed and another 60 were damaged. The newspaper reported rows of downed trees and power lines on a seven-mile path surveyed by the NWS.

The affected areas were between Route 30 and Interstate 70 in Sewickley Township. Winds reached an estimated 120 mph where the path of the storm widened to 300 yards.

May 2011
Multiple Tornadoes

The period of May 21–26, 2011, spawned deadly tornadoes in Kansas and Missouri on May 22, including the devastating Joplin, Missouri, EF5 tornado (158 fatalities), and Oklahoma, Kansas and Arkansas two days later. The 17 tornadoes recorded in Pennsylvania in May 2011 ranked third in state records for the month, following May 1998 and 1985.

On May 23, 2011, an EF2 tornado (111 to 135 mph) touched down near Kellerville in Juniata County at 5:23 p.m. that covered 4.7 miles, with a path width of 250 yards. Seven homes and 15 farms and outbuildings were damaged or destroyed, according to the NWS State College. Seven dairy cows died in one barn that was wrecked by the powerful winds. EF1 tornadoes touched down in Juniata, Union/Northumberland and Schuylkill counties, plus an EF0 storm in Carbon County.

A slow-moving low-pressure system and trailing cold front tapped plenty of instability on May 26–27, 2011, spawning 10 tornadoes across the heart of Pennsylvania. One of two Lycoming County tornadoes damaged or destroyed around 400 trees. Storms also touched down in Crawford, Franklin, Bradford, Cumberland, Perry/Dauphin, Schuylkill and Huntingdon (two) counties during a volatile spring tornado season.

North of Springbor in Crawford County, 10 buildings sustained damage. An EF1 tornado in New Franklin, Franklin County, caused pockets of damage to trees and homes. Around 8:10 p.m., an EF1 tornado touched down near Cressona, Schuykill County, and traveled 18 miles. Trees and power lines were toppled and more than 20 homes were affected, including four that had substantial damage. The storm ended in West Penn Township.

June 13, 2018
Wilkes-Barre Tornado

Low pressure tracking across Upstate New York pulled a warm front northward across the Northern Tier of Pennsylvania. The added spin of low pressure and strong low-level shear ahead of a cold front spawned damaging tornadoes in Bradford and Luzerne counties just after nightfall on June 13, 2018.

In Bradford County, about a dozen buildings, mostly homes, along with a business and church, were damaged as the storm roared through Granville, Leroy and Franklin Townships, starting at 8:16 p.m. near Troy, taking a five-mile path. Trees were toppled and power lines were downed by the F2 tornado (120 mph winds) along a path that widened to 250 yards. Fortunately, no injuries were reported.

Around 10:00 p.m., the same pattern of circulation was responsible for an EF2 tornado (130 mph winds) that touched down just beyond the Wyoming Valley Mall in Luzerne County. The twister was 200 yards wide and traveled .75 mile, lifting before reaching Interstate 81.

Storefronts in nearby strip malls endured heavy damage; roofs were torn off, concrete blocks tossed a quarter of a mile, and vehicles were overturned and crushed in the parking lot. Twenty-three buildings were damaged or destroyed. A sign from Dick's Sporting Goods was carried from Wilkes-Barre to Bear Creek Township, a distance of several miles. Six persons sustained minor injuries.

Fig. 7.16 Tornado that formed near New Stanton and moved to near Norvelt and Unity, lifting near the Westmoreland County Fairgrounds around 8:00 p.m. on June 27, 2018. (Photo by John Mark Benson)

October 2, 2018
Record Autumn Tornado Outbreak

A low-pressure area crossing New York roiled the atmosphere on October 2, 2018. An unseasonably strong southwesterly flow created deep-layer wind shear beneath westerly mid-level winds across the region, stirring the unstable environment.

A total of 16 tornadoes were later confirmed by the multiple NWS survey teams fanning out across the Keystone State, establishing a single-day record for an October event in Pennsylvania, which began in the northwest and ended in the Poconos.

The first tornado (EF2) on October 3, 2018, touched down east of Conneautville in Crawford County at 2:33 p.m., in the northwestern part of the state. The storm snapped trees and damaged a group of homes and structures, with one minor injury. The narrow funnel (50 yards) continued along a 4.1-mile path, packing winds of 115 mph. The most substantial damage occurred to a duplex and nursing facility. A second tornado (EF1) touched down 20 minutes later north of

Blooming Valley, traveling 3.5 miles and causing minor property damage.

Two tornadoes were confirmed by the NWS survey team near Brookville in Jefferson County between 5:00 and 6:00 p.m., damaging a few residential structures and trees. The storm that formed west of Brookville appeared to have crossed Interstate 80 near the North Fork Creek bridge along a four-mile path.

A second EF2 tornado was traced with the help of drone video from just north of Route 28 at Baxter to a point south of Brookville. The damage swath was judged to be 200 yards wide. Another tornado (EF1) was confirmed near Graysville in Greene County at 5:20 p.m. that caused minor damage despite a path width of 200 yards. Three weak (EF0) tornadoes were discovered in Westmoreland County with corroboration from drone and Civil Air Patrol views.

Farther east, a tornado (EF2) was sighted north of Williamsport at 5:03 p.m. south of Hartsfield, along the border between Lycoming and Tioga counties. The damage path was 300 yards wide as the storm traveled for two miles packing winds of up to 120 mph. Several hundred trees were toppled in Lycoming County. Snapped trees were evident just north of the Tioga County border.

A second EF2 Lycoming County twister, with winds of 115–120 mph, touched down at 5:15 p.m. between Liberty and Buttonwood, leaving a trail of tree damage. The storm was also captured on video.

Two weak EF0 tornadoes were confirmed in Springbrook Township, Lackawanna County, shortly after 6:00 p.m. At 6:37 p.m., a tornado (EF1) came down in Lake Harmony, Carbon County, starting at Big Boulder Ski Area, which remained on the ground for 0.9 mile. Just after 7:00 p.m., the same supercell spawned a tornado in Jackson Township, Monroe County, snapping and uprooting trees for a little more than a mile.

April 14–15, 2019
Cross-State Tornadoes

On the night of April 14–15, 2019, tornadoes touched from one end of the state to the other as low pressure moved north across the Ohio Valley, adding additional spin to the unstable atmosphere. Strong winds aloft and directional shear in the warm sector ahead of a cold front during the evening hours led to nine reported tornadoes.

The first touched down at Springboro in Erie County around 6:48 p.m., damaging several homes and leaving street signs and telephone poles "twisted or bent." The strongest tornado struck Starbrick in Warren County at 8:18 p.m., rated an EF2 with 130 mph winds. The storm traveled northeast for 16 miles, with a path width of 400 yards at its maximum. Significant damage was done to Carter Lumber, but otherwise the storm affected rural areas and caused no injuries. Two weak EF0 tornadoes were counted in Venango County.

Later in the evening, tornadoes touched down in the central mountains at Buffalo Valley in Union County and in Sullivan County. The Sullivan County storm reached EF1 strength and traveled 10 miles from World's End State Park to Dushore. An EF2 tornado at Benton in Columbia County was stronger, with peak winds of 115 mph, tracking 3.8 miles, but again with minimal damage and no injuries.

The final tornadoes occurred at 1:45 a.m. on April 15 in Lackawanna County and 1:50 a.m. in Susquehanna County. The Lackawanna County storm caused generally minor property damage on the south side of Scranton and at Dunmore. The storm, 100 yards wide, had a hit-skip path of 3.2 miles. The Susquehanna County storm was rated as EF1, which affected Harford Township, producing 90 mph winds and traveling a little more than two miles before lifting over Interstate 81. The path width was 150 yards.

The total of nine tornadoes in Pennsylvania on April 14–15, 2019, was the greatest in the state on an April day since April 9, 1991, when 11 storms were documented.

A second outbreak on April 19, 2019, spawned five tornadoes, including storms that reached EF2 strength in Mifflin and Franklin counties. EF1 tornadoes struck Fulton, Huntingdon and Juniata counties.

A narrow EF1 tornado that came down a little southwest of Knobsville, Fulton County, uprooted trees and damaged one home, traveling 1.5 miles. Farther east, an EF2 tornado (120 mph) came down along Route 30 in St. Thomas Township a little west of Chambersburg in Franklin County about 7:10 p.m., damaging C&R Produce and several homes and businesses along a 3.5-mile path.

The EF2 tornado in Mifflin County was first sighted south of Route 103 in Granville about 7:57 p.m. and moved into Lewistown, snapping and mowing down several hundred trees over a course of 4.5 miles before lifting at 8:05 p.m. Peak winds reached 115 mph, uprooting trees in the Riverside Campground that crushed several trailers and flipped one over. The Brooklyn Fire Station lost a portion of its roof.

The total of 14 tornadoes confirmed in Pennsylvania in April 2019 set a state record for the month.

May 2019
Stormy Spring

The active tornado season in 2019 continued to feed off waning El Niño warmth, which energized the subtropical jet stream atop the northern edge of a heat dome. Clusters of thunderstorms feasted off a steady supply of extremely muggy air. Disturbances cruising in a strong northwesterly mid-level flow with ample deep-layer wind shear.

Tornadoes were reported in the Commonwealth on May 19 (Lancaster and Bucks counties), May 23 (Cambria County), and May 25 (two in Indiana County). On May 28, clusters of storms raced eastward just south of a stalled front boundary. Downpours pounded the Pittsburgh area, totaling 4.75 inches on the northern edge of the city. Flooding was widespread at Zelienople, 30 miles north of Pittsburgh, resulting in several rescues of stranded motorists. Many trees and power lines came down in strong winds.

Tornadoes touched down across the state in a number of areas: Indiana, Bradford, Wyoming, Lackawanna and Lancaster counties. Among the the five tornadoes confirmed in Pennsylvania on May 28, 2019, the most intense storm that day formed in Lancaster County (EF1) at 5:52 p.m. and moved into Berks County (EF2) near Morgantown. Twenty homes were damaged in the Valley Ponds development, and about 30 percent of those were uninhabitable. A number of trees came down on cars, and roofs were ripped off some buildings that partially collapsed. Residents were quoted in the *Reading Eagle* that the tornado sounded "like an explosion" and "like a train was coming."

The stagnant pattern lingered on May 29, 2019, as another strong upper-level trough traversed the region. Six tornadoes were reported across the state, with touchdowns in Tioga, Lehigh/Bucks, Sullivan, Perry, Fulton and Dauphin counties. Very intense thunderstorms hammered eastern Pennsylvania with hail up to three inches in diameter in Northumberland County and two inches at Pleasant Valley in southern Monroe County. A total of 16 tornadoes touched down in Pennsylvania in May 2019.

The active pattern continued in June 2019, when three short-lived tornadoes touched down in western Pennsylvania on June 16. One storm moved into Mercer County from Ohio, another crossed from Butler into Armstrong County, and a third occurred in Indiana County. The storms developed along a line of severe thunderstorms in an environment with strong wind shear. The maximum winds were estimated at 105 mph around Parker, tearing the roof off of a home built in 1824.

October 31, 2019
Halloween Storms

A strong cold front trailing low pressure over the eastern Great Lakes spawned a long line of strong to severe thunderstorms on Halloween night that raced through Pennsylvania.

A powerful EF2 tornado (120 mph) descended from a squall line at Glen Mills in Thornbury Township, Delaware County, around 11:25 p.m., in the Philadelphia suburbs. The tornado had a path width of 250 yards and traveled a half-mile. The extensive straight-line wind damage was driven by unseasonably strong low-level winds. Damage was substantial at Warrington and Warminster in Bucks County, where the NWS survey estimated peak winds of 110 mph.

PECO reported 107,000 homes and businesses lost power in southeastern Pennsylvania. More than 800,000 customers were without electricity in 14 Eastern states after the storms struck. Two dozen homes were damaged by wind and falling trees, including eight that were heavily damaged by the tornado. Two injuries were reported.

The intensity of the Halloween tornado was rare for so late in the season. The 35 tornadoes in Pennsylvania in 2019 was the second largest total of confirmed storms, surpassing 1985, though well short of the record number of 62 in 1998.

August 4, 2020
Isaias Spawns Tornadoes

On August 4, 2020, a tornado spawned by Tropical Storm Isaias briefly touched down on the edge of Bensalem and Northeast Philadelphia. A stronger EF2 struck Doylestown, in Bucks County, moving rapidly northwest, with damage reported on the Doylestown Hospital Campus, where part of the roof of the Children's Village daycare was torn off. Several minor injuries were reported. The storm traveled 20 miles and reached a path width of up to 500 yards. Top winds were estimated at 115 mph.

July 29, 2021
Late July Tornado Outbreak

The 2021 tornado season brought 44 confirmed storms in Pennsylvania, taking over second place.

A thunderstorm complex that moved through eastern Ohio during the midday hours of July 29, 2021, had a long lifespan, charging east into central and eastern Pennsylvania, extending all the way to the southern New Jersey coast.

For the first time in the history of the NWS office in Mount Holly, New Jersey, a "particularly dangerous situation" tornado warning was issued for a potentially destructive and life-threatening storm. The setup included ample wind shear, tropical moisture pooling along a warm front and a well-defined convective vortex that was weakening but remained a potent force when all the ingredients were present for rotating thunderstorms.

In western Pennsylvania, the first EF1 tornado tracked from Acme to Donegal in Fayette County, covering 4.9 miles, with a path width of 100 yards. A second EF1 tornado struck Laurel Hill State Park at 2:25 p.m. Additional embedded EF0 tornadoes occurred in Crawford, Greene, Venango, Cambria and Lebanon counties. Most of the damage was limited to snapped trees and minor property losses.

Mid-level spin and energy from the disturbance caused updrafts to strengthen in southeastern Pennsylvania and New Jersey, setting the stage for a significant tornado outbreak that flattened homes and businesses, uprooted trees, and tore off pieces of roofs, siding and shingles.

A tornado touched down in Kempton, Berks County, at 4:04 p.m., uprooting around 15 trees. The EF1 tornado tracked 2.8 miles, heading east into southern Lehigh County across Weisenberg Township, possessing a path width of 90 yards.

Another tornado landed at Slatington, Lehigh County, at 4:25 p.m., rated EF1 (100 mph), which left "considerable tree damage" over a two-mile track, with "multiple uprooted hardwood trees." As the storm proceeded east, the swirling winds broke trees "on the grounds of Northern Lehigh High School" and damaged the roof of a dugout. The funnel ended at the Slatington Airport, after damaging an airplane hangar door and trees.

In Bucks County, a tornado touched down in Plumstead Township at 5:40 p.m., heading into Buckingham Township, shearing and uprooting trees on a 2.6-mile path. Greenhouse roofs and a barn were damaged near a farm market.

The supercell cycled, spawning a strong EF2 tornado around 6:00 p.m., with top winds of 120 mph at New Hope, and wide path width of 400 yards. The tornado continued northeast across Washington Crossing State Park, leaving a swath of tree damage in residential areas close to the Delaware River. The funnel crossed the river and lifted in Hopewell Township, New Jersey, where the wind convergence pattern terminated.

The most destructive tornado in the outbreak packed 140 mph winds, rated EF3, the strongest in the state since July 14, 2004. The tornado first struck Somerton in Philadelphia County shortly after 7:00 p.m., and entered Bucks County near Trevose. The storm briefly revisited Philadelphia County, before moving southeast across Bucks County. The long-tracked tornado navigated 35 miles of terrain.

The tornado intensified and expanded, with a storm width of 530 yards, carving a 3.5-mile path through an industrial complex, causing structural damage. The storm moved northeast and struck the Faulkner car dealership complex, where the damage was extensive. Five people were injured. The southwestern wall of the service garage was blown down, causing the building to collapse, as roofing material was torn off. Vehicles were also damaged nearby amidst the pattern of scattered debris.

The powerful tornado knocked down trees at a mobile home park and deposited more debris from the dealership, while damaging several mobile homes. The storm weakened over the Old

Lincoln Highway as it moved to Bensalem before lifting.

A brief tornado in Northeast Philadelphia touched down in the Bustleton neighborhood at 7:07 p.m. at Grant Avenue, causing minor EF0 damage to siding and roofs after passing Roosevelt Avenue.

The supercell continued to spin off funnels through central New Jersey to the shore region. An EF2 tornado occurred on the western edge of Barnegat Bay, becoming a waterspout, then emerging at High Bar Harbor. Two tornadoes were reported in Mercer and Burlington/Ocean counties, and an EF0 in Essex County. The July 29, 2021, event spawned more than two dozen tornadoes from the Ohio Valley to Maryland.

The tally of 14 tornadoes on July 29, 2021, was the most in the month of July since 1992 in Pennsylvania. In 2023, an August record of 12 tornadoes occurred in the state.

September 1, 2021
Ida Remnants Trigger Tornadoes

Rotating thunderstorms that developed along a warm front east of the center of low pressure spawned several tornadoes in central Maryland on September 1, 2021. An EF2 storm (125 mph) touched down at Annapolis around 2:00 p.m. and tracked more than 11 miles, damaging more than 100 homes and buildings on a destructive course that left many structures uninhabitable.

As the remnants of Ida moved north, storms with semi-discrete and embedded supercells in New Jersey and Pennsylvania spawned seven tornadoes. At 4:15 p.m., an EF1 tornado touched down at Oxford, Chester County. EF1 tornadoes between 6:00 and 6:30 p.m. hit Buckingham and Upper Makefield Townships in Bucks County.

A strong EF2 tornado at 5:35 p.m. with winds of 130 mph tracked from Fort Washington to Upper Dublin Township and Horsham Township across Montgomery County, covering an eight-mile path. Substantial damage was evident after the tornado crossed the Pennsylvania Turnpike, snapping and uprooting trees in residential areas. A field survey reported "minor to moderate roofing and siding damage to a number of homes."

One home showed EF2 damage that included partially collapsed exterior walls. The storm path extended along Route 309 to Upper Dublin High School. A woman was killed by a falling tree at a nearby residence, the only fatality in the outbreak. The tornado traveled northeast to the Temple University Ambler Campus, causing roof damage to homes and university buildings.

The Buckingham Township tornado was first spotted at 5:59 p.m. at Doylestown, tracking 4.4 miles, with a path width of 300 yards, and winds of up to 100 mph. Damage in a "heavily wooded area was mainly confined to trees."

Shortly before 7:00 p.m., an EF1 tornado with 90 mph winds hit Burlington Township, New Jersey, near Edgewater Park, associated with a supercell that had spawned a powerful EF3 tornado that struck near Mullica Hill less than an hour earlier. The storm turned north and was viewed over the Burlington-Bristol Bridge, from the Pennsylvania side of the Delaware River, lifting before reaching downtown Bristol, after traveling 2.8 miles, with a path width of 200 yards. This was the fifth confirmed tornado in Pennsylvania caused by the spin of Ida's remnant circulation.

Three tornadoes occurred in New Jersey, including the Burlington–Bristol storm that crossed into

Pennsylvania. The most powerful tornado (EF3) struck Mullica Hill at 6:10 p.m., beginning about 23 miles southeast of Philadelphia with winds reaching 150 mph, flattening or damaging nine homes and buildings and toppling large grain silos along a 12.6-mile path from Harrisonville to Deptford Township. The storm was 400 yards wide and caused two injuries. The massive wedge tornado, a rarity in the region, lofted debris an estimated 20,000 feet.

October 2021
Another Autumn Tornado Outbreak

Back-to-back severe weather events in October 2021 spawned 12 tornadoes in Pennsylvania.

In the predawn hours on October 16, a strong low-level jet interacting with high instability set the table for a complex of strong thunderstorms, enhanced by a remnant convective vortex. Nine tornadoes occurred in Ohio. About 6:00 and 6:15 a.m., two tornadoes struck western Pennsylvania in Beaver County. The more powerful tornado of the two was rated EF1 (95 mph) at Monaca, just south of the junction of the Beaver and Ohio Rivers, that continued east to near Freedom. A third tornado occurred at Clear Creek State Park in Jefferson County shortly after 8:00 a.m.

Five days later, low pressure tracking through the Great Lakes tapped into a corridor of unstable air ahead of a cold front traveling across Ohio and Pennsylvania. Two tornadoes occurred in Beaver County, including one that continued into Butler County; two in Allegheny County; another one in Butler County; and four storms touched down in Washington County. The majority of the activity occurred between 7:00 and 8:00 p.m. shortly after sundown.

The strongest storm (EF2) tracked more than 15 miles from northeast of Bethany to northwest of Houston in Washington County between 7:30 and 7:57 p.m., with a path width of 150 yards. Damage included the unroofing of a two-story home and the destruction of a farm building. Top winds approached 130 mph.

Another EF2 storm touched down in southern Hampton Township in Allegheny County that was even wider (400 yards), with winds of 105 mph (EF1). The storm grew out of the circulation that produced a touchdown in Mount Nebo. Damage was observed along a 4.8-mile track through Bryant, snapping trees and destroying a cinder-block building.

Winter Tornadoes and Windstorms

On December 6, 1884, a tornado struck Chester at 9:00 p.m. that unroofed 25 houses and blew over a mill and four church spires (*Jeffersonian*). Trees and roofs were damaged in Philadelphia. A boy was killed by flying debris at Pittsburgh.

A deep cyclonic disturbance passed over eastern Pennsylvania and northern New Jersey on January 26, 1843, bringing destructive winds and a possible mid-winter tornado at Pottsville. The storm was noted by Peirce (1847): "On January 26th Pottsville was visited by a tremendous hurricane which swept away almost everything before it."

The strongest January tornado (F2) in modern history descended on rural Crawford County on January 17, 1952, and traveled one mile. The path width was a substantial 1,200 yards, and property losses were estimated at $500,000.

A weak tornado briefly touched down south of Scranton near Moosic on January 14, 1992, spawned by a squall line that was accompanied by exceptionally strong wind gusts of 73 mph at Scranton and 74 mph at Philadelphia. On January 8, 2019, an EF1 tornado (95 mph) was noticed at New Lebanon in Mercer County, marking only the seventh confirmed January tornado in the Keystone State since 1950.

The month of February in Pennsylvania rarely brings strong thunderstorms, but there have been a few notable exceptions. A Pittsburgh dispatch in the *Philadelphia Inquirer* described "a terrible windstorm" on the morning of February 11, 1887, that caused pockets of substantial tornadic damage in eastern Ohio communities.

As the convective system advanced east, a report from Greensburg, Pennsylvania, called it "a terrible rain and wind storm" that unroofed homes and a Lutheran church, knocked down fences and blew out windows. An oil derrick was toppled, and telegraph wires came crashing down. The windstorm took a tragic turn at Tyrone, where four railroad laborers were killed by a falling tree that they had sheltered under.

High winds, thunder and lightning and torrential downpours attended the passage of a strong cold front across eastern Pennsylvania during the afternoon of February 15, 1939, spawning at least two tornadoes. The winds "unroofed homes, collapsed farm buildings, leveled wires, telephone poles and trees" north of Ephrata in northeastern Lancaster County, according to the *Philadelphia Inquirer*.

The monthly summary in the Weather Bureau's *Climatological Data* report stated that, though no funnel was witnessed, the damage was entirely consistent with an embedded tornado. There was a "funnel-shaped formation" sighted in Bath, Northampton County, that destroyed a two-and-a-half-story frame garage, and three automobiles were lost in the ensuing fire. A two-story frame barn "collapsed like a matchbox" at the Keystone Portland Cement Company, and two other buildings and a dozen homes were unroofed by the storm. Wind damage was also discovered at Belvidere and Washington in northwestern New Jersey, likely part of the supercell that passed over Bath.

On that unseasonably warm February afternoon in 1939, the temperature at Philadelphia shot up to 66°F at 11:15 a.m. ahead of a cold front that crossed the city around 4:50 p.m., which knocked the temperature back down to 34°F as rain changed to sleet and snow. A wind gust of 65 mph was recorded, with a five-minute average velocity of 50 mph just after noontime, knocking down trees and signs.

A February twister (F1) with winds estimated at 110 mph touched down in Centre County near Oak Hall on February 16, 1990, around 7:30 p.m. The funnel formed over an open field near an elementary school and traveled three miles through a rural section of the county. The damage path was 1,050 yards wide, including an assortment of twisted trees and flattened grass and minor property damage. The thunderstorm that spawned the tornado tracked within a half-mile of Pennsylvania State University.

A squall line that extended from southern New York to Virginia on the evening of February 24, 2016, associated with a deep low-pressure system over central New York, spawned two rare winter tornadoes in Pennsylvania. The storms arrived in Lancaster County around 6:30 p.m., tearing roofs apart and downing utility lines. Rotation developed around 8,000 feet near Quarryville, triggering a tornado warning. The State College NWS office issued a tornado warning with a 17-minute lead time. The storm left a quarter-mile path of damage over a distance of six miles.

The EF2 tornado (125 mph) first touched down a mile and a half north of Route 30 around 7:38 p.m. near White Horse in Lancaster County and caused five buildings to collapse. Several buildings were toppled along Route 340 as the squall line progressed to the northeast, knocking out power to 2,400 customers in Lancaster County, according to PPL. More than 50 buildings were damaged, including homes, barns and an Amish schoolhouse that was left in ruins. Fortunately, no deaths or serious injuries were reported. An EF1 tornado with 100 mph winds in Bradford County arrived near Wyalusing, close to Route 6, toppling trees and blowing over a shed.

Almost a year later to the date, a powerful squall line on February 25, 2017, struck at the conclusion of a record-breaking three-day February heat wave. A powerful EF2 tornado touched down in Pittston Township near Plains, in Luzerne County, around 2:35 p.m. The Binghamton NWS survey report estimated that peak winds reached 120 mph in Plains Township as the funnel expanded to a width of 500 yards, following a 12.8-mile northeasterly path through Pittston and Moosic in Lackawanna County.

The track passed east of Wilkes-Barre, Pittston and Scranton. The Luzerne County Emergency Management Agency (EMA) estimated that 30 homes were affected—mostly roofs lifted off—and six sustained substantial structural damage.

Shortly thereafter, an EF1 tornado developed in the vicinity of Hellam and Wrightsville in York County at 3:15 p.m. on February 25, 2017. Across the Susquehanna River, damage was evident at Columbia that was judged to be straight-line in nature. Microburst damage was extensive across northern Lancaster County in Clay Township and West Cocalico Township that involved trees falling onto homes and silos blown over. Winds were judged to range from 75 to 95 mph along a nine-mile path that was about 0.5 mile wide. Seven buildings that were parts of farms were destroyed.

On February 16, 2018, the first confirmed February tornado in Pittsburgh NWS records since 1950 made headlines—the third consecutive year a tornado was recorded in Pennsylvania in February. A relatively weak F1 storm touched down in Uniontown, Fayette County, at 6:43 p.m., with winds of 105 mph. The funnel narrowly missed Laurel Highlands High School. However, it tore off or damaged the roofs of several homes and a senior living center, brought down power lines and "snapped or uprooted" hardwood and pine trees along its entire two-mile path.

More Noteworthy Pennsylvania Tornadoes and Windstorms

The following collection of tornadoes that caused unusual or significant damage or loss of life in Pennsylvania were gleaned from press accounts summarized in Flora (1953), Grazulis (1993), the U.S. Weather Bureau's *Monthly Weather Review* and NOAA *Climatological Data* and *Storm Report* publications.

1818 July 7: A tornado crossed the Susquehanna River near Williamsport, destroying several buildings and revealing the riverbed (Grazulis 1993).

1824 July 6: A tornado struck near Pennsbury, Chester County, leaving an indelible mark on corn fields that looked as if they had been "exposed to intense heat" (Grazulis 1993).

1833 June 2: One or two tornadoes traveled northeast in the vicinity of the Susquehanna River, wiping out five homes in Britain Township, with damage also at Mount Carbon, Schuylkill County (Grazulis 1993).

1866 May 27: A small tornado lifted a covered bridge over the Lackawaxen River in Honesdale off its foundation at Church and Fourth streets, which was spun around and deposited in the river (*Wayne Independent, Bicentennial Edition*, Feb. 4, 1978).

1875 August 11: A tornado touched down near 23rd at Callowhill Streets in Philadelphia and unroofed more than 20 homes on the north side of the city. The F2 storm was about 100 yards wide and traveled for a mile, causing $50,000 in damage (Grazulis 1993).

1880 April 16: Severe storms spawned tornadoes at Corwin Center, Bradford and Oil City, called a "destructive wind-storm, demolishing derricks and unroofing buildings" in the northwestern part of the state. Two persons were seriously hurt. Damaging winds were reported around Harrisburg, caving in a covered bridge over the Susquehanna River and damaging several homes (*Monthly Weather Review*, Finley 1884).

1881 June 7: A classic tornado, "funnel shaped, small end toward the ground," passed northeast of Indiana and continued along a 15-mile track, with a path width exceeding 100 yards (*Monthly Weather Review*). The twister "leveled hundreds of acres of timber, managing to miss all homes until the end of its path," according to a local press account, after the storm formed seven miles northeast of Indiana around 5:00 p.m. One home and a dozen barns were destroyed (Grazulis 1993).

1890 July 17: A thunderstorm complex raced across southern Pennsylvania and New Jersey with damaging winds and embedded funnels.

The *Philadelphia Inquirer* carried an account from Ashland: "every building was wrecked, and the hailstones broke nearly every window pane." At Allentown, a 229-foot tall brick stack crashed onto the Barbour Thread Mill engine house. Losses included homes unroofed and flipped. A brick building under construction was blown apart at Hellertown, the debris killing at 13-year-old boy. The damage path was a mile wide.

1893 August 12: A tornado traveled a little more than a mile along the Susquehanna River at Nanticoke, Luzerne County, damaging six homes and uprooting hundreds of trees, resulting in multiple injuries. One home was lifted off its foundation and carried 50 feet, killing two occupants (*Monthly Weather Review*).

1905 April 30: A tornado was reported in the vicinity of Carbondale, with a path length of two miles and a funnel width ranging from 25 to 200 yards. Thirty homes and buildings were blown down, and a few roofs were carried 600 feet, according to William Dudley, head of the Scranton Weather Bureau office (*Monthly Weather Review*).

1909 October 21: A hit-skip funnel passed 25 miles south of Erie and caused $150,000 in damage. A "distinct funnel cloud" formed southwest of Cambridge Springs and tracked 20 miles to a location about five miles southwest of Union City. Eyewitnesses said that the storm "carried board, planks, and other missiles from 400 to 500 feet high" (*Monthly Weather Review*).

1912 June 16: Tornadoes swooped down in Greene and Washington counties in southwestern Pennsylvania, killing a young girl, who was struck by flying debris (*Monthly Weather Review*). The tornado traveled eight miles from north of Wind Ridge to Old Concord, resulting in $100,000 in damage—designated by Grazulis (1993) as a storm of F3 strength: "Every home at Old Concord was damaged."

1919 April 11: A fairly long-tracked tornado touched down in Bradford County and traveled "seven or eight miles" from Rome to Orwell, flattening barns and orchards, but did not cause any injuries.

1921 March 9: A late-winter tornado skipped along over a 12-mile path through Chester County around 4:00 p.m., passing south of Thorndale and damaging nine homes north of Romansville (Grazulis 1993).

1925 April 19: Multiple tornadoes skipped across Washington, Centre and Mifflin counties, with reports of unroofed houses. Several homes and barns were destroyed (*Climatological Data: Pennsylvania Section*). A tornado at Newell, Fayette County, had a path width of 300 yards ($25,000 damage). Possible tornadoes hit Irwin in Westmoreland County and Belleville in Mifflin County ($100,000 damage).

1928 June 14: Violent winds churned up by a severe thunderstorm on June 14, 1928, struck Newville and Bloserville, in Cumberland County. The tornado "damaged or destroyed many buildings, uprooted many trees, and beat down a considerable amount of grain and corn" (*Climatological Data: Pennsylvania Section*). Wind damage, possibly the work of embedded tornadoes, was reported in Juniata, Perry, Cumberland and York counties.

The West Chester observer recorded 13.14 inches of rain in June 1928.

1928 July 4: A severe summer storm blasted through Allegheny County. The state *Climatological Data* report stated: "At Rainbow Gardens and Olympia Park, 400 automobiles were destroyed in landslides, and 2,000 persons were marooned by floodwaters." More severe wind, caused by a downburst, occurred in Westmoreland County on July 11 and near Mifflintown on July 20 and extended for several miles.

Strong storms pummeled the state again on July 22, 1928, when hail, high wind and lightning strikes were responsible for damage "upwards of $25,000" at Carlisle.

1936 July 27: Tornadoes came down in Washington and Fayette counties that resulted in damage totaling $500,000 (Flora 1953). One storm path, with hail and high wind, covered 10 miles (*Climatological Data: Pennsylvania Section*).

1943 May 17: Two funnels were reported on the ground in Venango County, one near Pleasantville (1:00 p.m.) and another three miles northwest of Oil City at 5:48 p.m., which crossed the north side of town. Damage was estimated at $75,000 to homes and industrial buildings, with 10 injuries

(Grazulis 1993). Three persons were killed by lightning, and four additional deaths were attributed to intense local storms in the region (*Climatological Data: Pennsylvania Section*).

1945 May 17: A tornado formed 1.5 miles east of Boalsburg and tracked nine miles past Centre Hall, ending two miles southwest of Penn's Cave, with damage estimated near $100,000. The tornado occurred between 3:30 and 3:45 p.m., according to Dr. Hans Neuberger, who chaired the Pennsylvania State College (now University) meteorology section.

1963 August 3: A late summer tornado (F2) in Allegheny County moved southeast from Glassport to Carnegie, killing two persons when a three-story hotel collapsed at Glassport, and injured as many as 70 persons (Grazulis 1993). A skating rink and a number of homes were damaged.

1963 September 3: A twister injured 20 persons in a hit-skip storm pattern between Ridgeway and St. Marys in Elk County that continued for 12 miles, causing a substantial $1 million in damage. Homes were swept off their foundations or moved, garages flattened, roofs torn off and mobile homes destroyed.

1970 June 18: A tornado struck Myerstown, Lebanon County at 6:15 p.m. that killed a 13-year-old boy, injured five persons and virtually destroyed five farms along a three-mile path. The tornado was nearly 900 yards wide.

1976 July 11: July 1976 was very active, with a total of 11 tornadoes confirmed in Pennsylvania. On the afternoon of July 11, five tornadoes were sighted in the western part of the state. A powerful F2 tornado carved a path of destruction in Jefferson County starting around 3:00 p.m.

The Jefferson County storm touched down near Cook Forest State Park and tracked for 28 miles to Brookville, killing one person and injuring seven. A path width of 100 yards resulted in total damage of $250,000.

In Westmoreland County, a stronger F3 tornado struck at 4:45 p.m. Two persons were killed and 17 injured on the northern edge of Latrobe along a 12-mile path (*Storm Data*). A total of 37 houses and 12 mobile homes were destroyed, and five furniture company buildings were damaged.

1978 June 7: A small tornado rated F1 briefly visited an area in southwest Philadelphia on June 7, 1978, damaging a portion of Bartram Botanical Garden at the intersection of 54th Street and Lindbergh Boulevard, proving that tornadoes do indeed touch down in urban centers.

1981 July 26: The Indiana Sheraton–Evergreen Express Inn near Edgewood lost its roof at 1:30 p.m. on July 26, 1981, in a tornado (F2) that also damaged several homes. Two tornadoes were reported in Cambria County.

Later in the evening, a narrow, powerful F3 tornado formed near Berlinsville, in northwestern Northampton County, at the opposite end of the state, around 8:35 p.m., tracking four miles and destroying three homes, while damaging two others near Danielsville. A barn floor sailed 1,000 feet and was found intact, and a mattress was discovered a mile away lying on a telephone pole.

Downburst damage, with possibly an embedded tornado, continued for another 20 miles across northern Northampton into southern Monroe counties, as far north as Brodheadsville, continuing east to Bangor, where two mobile homes were flipped. Three injuries were reported, and the damage totaled $750,000.

Oddly, less than a week earlier, on July 21, 1981, "two distinct funnels side by side 50 feet apart" touched down in the same region at Mount Bethel in Northampton County, where 200 trees were uprooted. Another tornado was confirmed in Schuylkill County.

1984 May 23: A clustering of thunderstorms caused $150,000 in damage in Pike and Monroe counties, largely due to a pattern of downbursts. Interviews that were published a day later in the Pocono Record indicated there were some funnel cloud sightings in areas of considerable damage. A home was wrecked near Brodheadsville along Route 715, resulting in one injury, where a "black funnel" was apparently rain-wrapped. A camp on Sugar Mountain, in the southeastern part of Pike County, was struck by a very powerful windstorm that "tore a grove through a 1,600-foot stretch of woodland, ripping up trees, flipping several rowboats and tearing up a swimming dock from its foundations."

The whirling column of wind and water narrowly bypassed the Christian Herald Children's Home, where a "black funnel" was observed snapping tall trees 80 to 90 feet tall and "spewing debris for hundreds of feet." It is unclear why this storm was not classified as a tornado, despite obvious eyewitness accounts and tornadic damage.

1987 June 29–30: Bands of storms with damaging wind pounded western Pennsylvania on the night of June 29, 1987 (*Storm Data*). A tornado (F1) touched down at Paris, Washington County, around 9:15 p.m. that damaged 50 homes, 10 heavily. A weak F0 tornado had been sighted earlier in Westmoreland County near Smithton. Lightning sparked fires at Titusville, Crawford County.

High winds blasted southern Pennsylvania the following afternoon on June 30. Another short-lived tornado caused minor damage near Greensburg, Westmoreland County.

In the southeast, heavy damage from a downburst with possible embedded tornadoes occurred in northern Lancaster County between Mount Joy and Reinholds consisting of "at least 100 farm buildings, 30 silos, and 100 pieces of farm equipment." A greenhouse at Manheim was lifted over two others and smashed in a storm at 3:42 p.m. Thirty homes sustained damage as the storm passed through Elizabeth Township around 4:00 p.m. Four homes lost roofs in Lititz as the storm progressed eastward, damaging a number of barns and farm buildings.

A powerful cell raked Adams County at 4:30 p.m., claiming one victim. An anemometer at Gettysburg clocked a wind gust of 100 mph before being torn away. A woman was killed when her garage roof was ripped away and landed on her.

1991 May 6: A waterspout formed over the Susquehanna River that continued east as a tornado (F2) and slammed into a mobile home park at Bainbridge, injuring three persons.

1993 June 8: A large funnel (750 yards wide) injured two persons in Fayette County, traveling three miles around 6:45 p.m., between Everson and Wooddale, destroying three mobile homes and two bars. A smaller tornado (F1) developed west of Coatesville on June 9, 1993, near North Brandywine Junior High School around 5:20 p.m. An

AT&T employee was killed by a falling tree. The storm traveled two miles and caused $50,000 in damage.

1995 May 29: The Memorial Day tornado outbreak of May 29, 1995, spawned four tornadoes in southeastern Pennsylvania, two occurring in Lancaster County between 5:00 and 5:30 p.m. The first storm (F1) hit near Schoeneck (F1). A second tornado touched down 2.5 miles west of Millersville a half-hour later, reaching F2 strength as it crossed Little Conestoga Creek. Several homes and barns were damaged, causing minor injuries. Two F1 tornadoes were reported, one at Hosensack at 5:35 p.m. and another at Norristown at 6:00 p.m.

1998 March 9: A small F1 tornado tracked 1.5 miles, damaging two dozen Reading homes.

2002 November 10: A big tornado outbreak spawned 18 tornadoes across Ohio. Very warm, humid air surged north with low pressure in the Ohio Valley behind a warm front. After sunset, a tornado tracked from Sharpsville to New Hamburg in Mercer County, setting down at 8:54 p.m., moving through Clark with winds of 155 mph (F2).

The fast-moving storm (50 mph), 500 yards wide, killed one person and injured 19 at Clark. Fifteen homes and one business were observed to be lost, 13 houses and a business had major damage, and 29 homes had minor damage. The NOAA *Storm Report* noted that a "large number of trees were snapped or toppled." Vehicles were damaged, and one was tossed across Route 258.

A second tornado (F1) hit Crawford County at 9:30 p.m., traveling less than a mile, only causing minor damage at Cochranton.

2008 September 6: An isolated tornado associated with the passage of Tropical Storm Hanna hit East Allentown. Fifty homes, 100 trees and lines were damaged.

2011 April 27-28: Nine confirmed tornadoes touched down in northern and central Pennsylvania, spanning the night of April 27–28, 2011, as intense low pressure traversed the lower Great Lakes, with attendant low-level wind shear and moderate instability. Three to four inches of rain fell in parts of eastern Pennsylvania, resulting in flash flooding.

An EF2 touched down near the Roundtop Ski Resort in York County at 5:55 a.m. on April 28 and traveled down the slope to the parking lot. The storm traveled nearly three miles, snapping and uprooting trees and a voltage transmission tower. An EF1 tornado was confirmed in southern Huntingdon County. The last tornado in the outbreak reached EF2 strength in Lebanon County a half-hour later.

2014 June 11: An EF1 tornado formed east of Lutzville in Bedford County shortly after 11:00 p.m. and tracked nearly 11 miles to the Hopewell area, damaging two homes and hundred of trees.

2020 November 30: Low pressure advancing north through central Pennsylvania pulled up sufficient tropical air from the Gulf of Mexico in the warm sector, coupled with strong winds aloft and temperatures soaring into the upper 60s, to create a narrow corridor of instability.

A small EF0 tornado touched down in Montgomeryville at 3:47 p.m. The path was 100 yards and six to eight cars were damaged between the Texas Roadhouse and Noboru restaurants. There was structural damage to siding and equipment.

Chapter Eight

Heat Waves and Droughts

The weather has been so hot for a week past, as has not been known in the memory of man in this country, excepting the "hot summer" about seven years since.

—Pennsylvania Gazette (July 1734)

THE AVERAGE ANNUAL number of fatalities related to intense heat in the United States (183) in the past 30 years, based on NOAA data, does not tell the full story. The CDC estimates 1,220 people die from heat directly or as a contributing factor in a normal year, surpassing all other weather-related causes of death nationally.

Dangerous heat coupled with high humidity puts considerable stress on the human body without the adequate replenishment of fluids. The body tends to overheat when perspiration becomes ineffective due to limited evaporation on a muggy day, which raises the core temperature to a critically high level.

Heat illness symptoms potentially leading to a life-threatening heat stroke include the inability to sweat, rapid pulse and low blood pressure, which can cause a person to become mentally confused and result in a coma.

As the blood thickens, kidney and heart failure will commence if medical intervention is not administered quickly.

Vigorous outdoor activities on a steamy afternoon increases the risk of dehydration, heat exhaustion, or a life-threatening heat stroke. Tragically, youngsters and pets are left unattended in hot vehicles without ventilation, where the inside temperature can rise more than 30°F in less than a half-hour.

Heat waves ramp up electricity usage, creating power outages and the loss of air-conditioning. Another consequence of prolonged heat is tinder-dry vegetation that fuels wildfires spread by gusty winds. Burn scar areas in the American West are prone to flash flooding because the sparse vegetation cannot absorb runoff pouring down steep hillsides.

Since 1985, a trend in the United States is summer nighttime warming of 1.4°F— double the rate of the increase in daytime highs (0.7°F). Less nighttime cooling leads to accumulated daytime heat, which is slower to dissipate at night, especially in urban areas.

During a four-day run of temperatures in the low 90s in Philadelphia from September 7–10, 2016, the minimum on the 9th only fell to 80°F—a late-season record for such a level of warmth. This was the city's third minimum of 80°F or higher during 2016, tying 2002, 1995, 1993 and 1933.

The Philadelphia urban heat island is evident in the average number of times during the year when the temperature reaches 90°F or higher (28 days). This value doubles most locations. By comparison, the highlands in the northern and western sections of Pennsylvania experience only a few such blistering readings in an average summer.

Great American Heat Waves

The characteristic weather pattern for a heat wave is a broad high-pressure heat dome with broadly sinking air that warms and dries by compressing the lower layers. The hottest days in the Northeast and Mid-Atlantic occur in a west-northwesterly flow of relatively dry air around a subtropical ridge anchored in the Southeast.

Intense heat that brings temperatures above 100°F in the Mid-Atlantic region originates in a torrid, dry southwesterly circulation around the Sonoran heat ridge in the Desert Southwest that settles over the Four Corners region. High pressure prevails at most levels of the atmosphere, promoting high evaporation rates under nearly cloudless skies that cause soils to become progressively drier as insolation bakes the ground.

Four of the five hottest summers in 130 years of records in Pennsylvania have occurred in the present century: 2005, 2016, 2020, 2010. Also in the top 10 hottest summers are 2021, 2024 and 2002, in order of warmth, for June–August.

Farther back in time, an unprecedented string of torrid summers in the 1930s brought a record number of days with triple-digit temperatures from the Great Plains to the Eastern states. Fifty million acres of farmland were affected by the Dust Bowl drought, or "Dirty Thirties," a prolonged dry spell that did not fully conclude statistically until the fall and winter of 1940–41.

Relentless heat and scant rainfall caused crops to wither in the fields in the nation's breadbasket. Topsoil turned to dust and blew away, culminating in a series of dust storms arising over the Great Plains in the spring of 1934. Clouds of dust circulated by the westerly winds reached the East Coast, dimming the sunlight.

Generations of farming had taken a severe toll on the soil resulting from poor land management and overgrazing. During the Great Depression, thousands of families boarded up their homes and migrated westward in search of food and a new livelihood, memorably depicted in John Steinbeck's 1939 masterpiece *The Grapes of Wrath*.

Abnormally high sea surface temperatures off the Atlantic Coast and Pacific Coast contributed to unusually expansive and persistent high-pressure zones that deflected the jet stream north and diverted cooler air masses with cold fronts capable of triggering showers. The subsiding air from the mid-levels also blocked Gulf moisture from streaming northward.

Drought conditions prevailed over 65 percent of the contiguous United States in the summer of 1934. Intense heat at Fort Smith, Arkansas, brought readings of 100°F or higher on 53 days, and 45 days at Oklahoma City. From June 19 to August 21, every day a weather station registered a high temperature of 100°F or higher, with extremes reaching 118°F in South Dakota.

Two years later, in the blistering summer of 1936, temperatures reached the century mark in July and August in the Midwest and portions of the Mid-Atlantic. All-time state high temperatures were observed on July 9–10, 1936, in Pennsylvania (111°F), West Virginia (112°F), New Jersey (110°F), and Maryland (109°F). The national death toll from heat-related illnesses was estimated at 4,678, and likely was double or even triple that number.

Another withering drought enveloped the midsection of the country between 1952 and 1957, peaking in 1953–1954, when 42 percent of the nation dealt with extreme drought by the middle of 1954. The summers from 1952–1955 were notoriously hot and humid in Pennsylvania, with a record number of 100°F days

The sizzling summer of 1980 in the southern Plains brought a record 42 consecutive days with a maximum temperature at or above 100°F at Dallas, Texas, and only one day of rain. An extreme high of 113°F was reached. Washington, D.C., observed a record 67 days with 90°F heat (tied in 2010). The national death total (direct and indirect) reached more than 10,000 persons. The agricultural losses totaled $33 billion (2020 dollars).

The North American drought in 1988 impacted the central and eastern United States ranks among the worst of the century, affecting about 50 percent of the United States. The drought pattern lingered into 1989, resulting in total losses estimated to be $44 billion (2020 dollars), the costliest drought in American history. Deaths attributed to the extreme heat were between 5,000 and 10,000 (NOAA).

A blistering summer in 1995 was blamed for more than 730 heat-related fatalities in the Chicago area alone, made worse by extraordinary humidity in mid-July. Scorching heat and worsening drought recurred in July 1999 in the Midwest and East, which claimed hundreds of lives in the affected areas from Missouri to Pennsylvania.

The Southwest suffered through periods of extreme dryness beginning in 2001, and peaking in 2010–2011 and 2013–2015. Above-normal rainfall in the winter of 2015–16 helped ease the situation in Northern California, but only limited moisture reached Southern California. Heavy precipitation finally soaked the state in the wet season of 2016–17, easing the five-year statewide drought. However, heat and excessive dryness returned a few years later, exacerbating the two-decade-long megadrought and contributing to a series of devastating wildfire seasons.

In mid-July of 2021, historic drought gripped more than 94 percent of 11 western states, according to the U.S. Drought Monitor; greater than 60 percent of the region was classified as extreme or exceptional drought.

Temperatures rocketed to unprecedented levels on June 28, 2021, at Portland, Oregon (116°F), and at Seattle, Washington (108°F). The weather station at Lytton, British Columbia, hit 121°F, a record for any Canadian weather station. The hot summer of 2021 (June–August) edged out 1936 for the hottest on record in the contiguous United States by less than .01 degree, according to NOAA records since 1895; the average temperature of 74°F was 2.6°F above normal.

Historic Pennsylvania Heat Waves

John Watson's magisterial *Annals of Philadelphia and Pennsylvania, in the Olden Time* recalled the hot summers of 1727 and 1734, which caused instances of heat prostration and even a few deaths from exertion in the fields.

A spell of hot weather in July 1789 in Philadelphia caused problems with food preservation. In Watson's *Annals of Philadelphia*: "By 10 o'clock, a.m., the meats in the market putrefy, and the city mayor orders them cast into the river—merchants shut up their stores—thermometer at 96° for several days" (Watson 1868).

Early Philadelphia weather records taken at Pennsylvania Hospital do not list a single high temperature above 99°F between 1825 and 1875. However, the *United States Gazette* noted the heat on July 14, 1841: "The thermometer at noon in the shade stood at 100 degrees." Two summers thereafter, an early record of the thermometer maintained at the Philadelphia Naval Yard recorded 100°F on July 1, 1843. In the Lehigh Valley, the *Easton Express* mentioned a sizzling 102°F occurred on July 13, 1859.

1866
Mid-July Heat Wave

A hot July brought a week of highs in the 90s from July 12–18, 1866, according to the Smithsonian observer at Nazareth. High readings of 97°F and 98°F at 2:00 p.m. were obtained on July 15–16, 1866. Weather sites at Blooming Grove and Dyberry both recorded 96°F at 2:00 p.m. on July 16, 1866, in the northeastern highlands.

1868
Hottest July in the 19th Century

The mean temperature at Philadelphia in July 1868 (81.4°F) was higher than in any month in city weather records until July 1955. The 1868 heat record at Philadelphia is outstanding considering that modern urbanization has contributed to generally warmer summers in the Delaware Valley in the past century.

The reliable weather station at Dyberry registered a maximum 2:00 p.m. reading of 98°F on July 14, 1868, the warmest day in station records (1865–1903). In the west, the New Castle observer recorded 90°F or higher at 2:00 p.m. during the first five days of the month, and again from July 11–21, 1868, with extremes of 97°F and 98°F on July 14–15.

The monthly mean temperature in July 1868 noted at Pittsburgh (80.3°F) equaled the city's hottest month (July 1887) downtown. The modern readings at Pittsburgh International Airport show August 1995 (77.7°F) to have been the warmest month at the NWS site.

1876
Hot July

The centennial summer of 1876 was an especially hot one in Pennsylvania. The mercury at Philadelphia soared to 90°F or higher on 13 of the first 20 days of the month. The first heat spell arrived in Philadelphia on July 2–5, 1876, followed by a more intense blast on July 8–10, with readings of 98°F, 100°F and 95°F.

A Lehigh Valley observer at Egypt recorded 102°F at 2:00 p.m. on July 10, 1876, during a record 14-day heat spell that saw the mercury climb into the 90s by 2:00 p.m. every day from July 7–20, 1876. The Stroudsburg *Jeffersonian* noted 98°F on the 10th. In the cooler hills of Pike County, the Blooming Grove observer recorded 10 days of 90°F temperatures or higher. The Pittsburgh Weather Bureau reported six straight days from July 7–13 in the 90s.

1881
Great September Heat Wave

A period of intense heat developed toward the end of a very dry summer in 1881. Newspaper reports described wells going dry, brown pastures and endless fields of stunted corn and wheat.

Wagons traveling through the rural country stirred up great clouds of dust in a scene that must have resembled something out of the Old West. On July 10, 1881, the thermometer in downtown Pittsburgh registered a record maximum of 103°F.

A raging forest fire in Michigan on September 5–6, 1881, sent a plume of smoke that filled the sky over the Northeast during a record late season heat wave, which had developed on August 26, 1881.

The Blooming Grove observer reported 2:00 p.m. readings of 100°F and 98°F on August 30–31, 1881. The hot weather persisted through the first week of September, when the mercury reached 100°F and 99°F on September 6–7, 1881.

The Stroudsburg *Jeffersonian* described the desert-like conditions: "From one o'clock to four o'clock [September 7], was about as hot a day as Stroudsburg has any record of. The mercury ranged from 100 to 105 degrees. The air rolled along and felt as if it was heated by a furnace."

The hot, dry weather brought extreme highs of 101°F, 102°F and 100°F at Pittsburgh on September 5–7, 1881. All-time September maximum temperatures were observed on September 7 at Washington, D.C. (104°F), and Philadelphia and Boston (102°F). Temperatures soared into the 90s on September 8 in eastern Pennsylvania ahead of a strong cold front that was accompanied by heavy thunderstorms.

The months of September–November comprised the warmest autumn on record at Philadelphia (61.2°F) in the 19th century since 1825—0.1°F higher than 1900 (61.1°F).

Three years later, another outstanding autumn heat wave brought 90°F heat in southern Pennsylvania from September 4–11, 1884. A high of 100°F at Pittsburgh recorded on September 10, 1884, is the hottest reading so late in the year.

1887
Hottest July at Pittsburgh

A hot, steamy pattern wilted the East and Midwest through much of July 1887. Pittsburgh endured its hottest month on record (80.3°F), edging out July 1878 (79.5°F), reaching a maximum of 101°F on July 17, 1887.

The heat peaked during the middle portion of the month, when Philadelphia sizzled, with daily highs/lows of 100°F/73°F, 96°F/77°F, and 94°F/78°F from July 16–18, 1887. The observer reported 20 days of 90°F-plus heat in July 1887. Temperatures rose to 90°F and higher each day from July 25–31, 1887, yielding a monthly mean temperature of 80.7°F.

The Midwest also broiled on July 17, 1887, when temperatures sailed to 107°F at St. Louis and 104°F at Rockford, Illinois. Severe thunderstorms rumbled across the Northeast that evening, causing pockets of wind damage in Pennsylvania.

On July 18, 1887, readings soared to 106°F at Lynchburg, Virginia, 104°F at Raleigh, North Carolina, and 103°F at Washington, D.C. Severe thunderstorms lashed the Northeast and Mid-Atlantic on a number of days between July 17 and 26, 1887, with almost daily reports of localized flooding and wind damage.

1894
Drought Summer

Dry conditions that prevailed across the midsection of the country and parts of the East during the middle 1890s were associated with hot summers. Solar energy was not spent evaporating ground moisture, so the air above received extra warmth.

The thermometer at Philadelphia topped 90°F on June 22–25, 1894, with a maximum of 97°F on June 24. Temperatures reached the century mark at Hamburg (102°F), Aqueduct (101°F) and Lock Haven (101°F) in June 1894.

Periods of searing heat returned as the drought summer wore on, with high temperatures soaring across Pennsylvania on July 13 and 19–20, 1894. An eight-day heat wave at the close of the month from July 25–August 1, 1894, sent readings back into the 90s in southeastern Pennsylvania.

Interestingly, the state's temperature extremes in July 1894 were, coincidentally, observed at Lock Haven (102°F/31°F), a truly a singular occurrence.

1895
Record Early Heat

A second consecutive year of drought in 1895 in the East was severe enough to interrupt coal shipments heading down the Ohio River, according to an item in the *Bulletin of American Industry and Steel Association* that appeared in the Weather Bureau's *Climate and Crops: Pennsylvania Section* (April 1896).

A persistent dry pattern contributed to an intense early heat wave in the spring of 1895. The Wilkes-Barre station recorded a record eight consecutive days with highs in the 90s, from May 28–June 4, 1895. Highs reached 97°F and 96°F on the last two days of May, setting new records, as did the maximum of 99°F on June 1 (June record). The highs on June 2–4 soared to 94°F, 97°F and 95°F.

The outstanding early heat wave that commenced on May 30–31, 1895, marked a state record May maximum temperature of 102°F at Lock Haven on the 30th (equaled on May 27, 1941, at Marcus Hook). The thermometer ascended to all-time station May records of 100°F at Hollidaysburg and Carlisle, and 96°F at Williamsport.

The Stroudsburg *Jeffersonian* reported back-to-back highs of 96°F on May 30–31, 1895—the warmest ever recorded in the month of May at the climatological site that commenced data in 1910. The *Philadelphia Inquirer* reported an increase in heat-related deaths in early June.

1896
August Dog Days

A protracted heat wave in early August 1896 brought nearly two weeks of 90°F heat to much of Pennsylvania. At Philadelphia, the mercury ascended to 90°F or higher on ten consecutive days, from August 4–13, with a peak of 97°F on August 11, 1896. The highest readings were 101°F at Aqueduct on August 9 and Honesdale on August 11.

Scranton observed eight days of 90°F weather during August 5–12, 1896, with a peak reading of 95°F. The *Stroudsburg Times* reported 1896 was "the most disagreeable summer since 1876," even when compared to the hot, droughty summers of 1893 and 1894. The *Times* stated: "The daily list of deaths and prostrations is appalling in large cities, and business was almost at a standstill, while humanity gasped and prayed for relief."

1898
Hot Holiday

A four-day heat wave that commenced on July 1, 1898, culminated in extreme readings across the interior southeast of Pennsylvania, ranging as high as 107°F at Hamburg and 106°F at Aqueduct. Wilkes-Barre experienced an uncommon 100°F reading. The nation's capital roasted at 104°F on July 3, 1898, one of the hottest readings ever known in the city.

1900
Long, Hot Summer

The mean summer temperature (June-August) in 1900 was 70.9°F, the sixth warmest summer on record in the Keystone State.

The first string of 90°F-plus days arrived on July 4–7, 1900. Easton had a streak of 90°F-plus heat from July 15–22, with a maximum of 97°F on July 17. The highest readings in the state in July 1900 were 104°F at Quakertown; 103°F at Chambersburg, East Mauch Chunk, Ephrata and York; 102°F at Carlisle, Huntingdon, Lewisburg and Lock Haven; and 100°F at Wilkes-Barre.

August 1900 was hot, too, featuring an average temperature of 73.0°F, the second warmest August in the state after 2016 (73.4°F). The highest readings were 103°F at Lebanon; 102°F at Huntingdon, Lock Haven and York; and 101°F at Coatesville and Philadelphia. The warmest period lasted from August 6–12, when high readings topped 90°F in southeastern Pennsylvania.

1901
Consecutive Simmering Summers

The summer of 1901 is the tenth warmest in Pennsylvania weather history. The monthly average in July of 74.6°F (4.2°F above the long-term normal) is the second warmest month on record in the state.

A second consecutive hot summer commenced with an exceptional heat wave that lasted from June 25–July 6, 1901. Philadelphia recorded 12 consecutive days with temperatures exceeding 90°F, eclipsing the previous mark of 11 (August 16–26, 1885), standard not surpassed until August–September 1953 (13 days).

The thermometer shot up to 98°F on June 30, and climbed to 102°F, 103°F and 98°F on July 1–3, 1901. Pittsburgh observed its warmest minimum reading for any day on July 1 (82°F).

A year-end review in the *Stroudsburg Daily Times* (January 2, 1902) listed a maximum of 99°F on June 30, 1901, which exceeds all subsequent June heat waves by one degree. The *Daily Times* noted a maximum of 103°F on July 3. High-elevation thermometers registered sizzling values of 98°F at Mount Pocono and 94°F at Tobyhanna, which, although unofficial, equal the warmest days ever recorded on the Pocono Plateau.

The highest temperatures in Pennsylvania were observed on July 2, 1901, when a maximum reading of 107°F at York established an all-time local record. Other notable high temperatures included: 103°F at Coatesville, and 102°F at East Mauch Chunk, Ephrata, Huntingdon and Lock Haven. The northwestern mountains had maximum readings between 93°F and 98°F.

1911
Fourth of July Hot as a Firecracker

The Fourth of July 1911 brought the highest temperatures ever recorded in northern New England. All-time state records were noted at Nashua, New Hampshire (106°F), North Bridgton, Maine (105°F) and Westboro, Massachusetts (105°F).

Temperatures soared into the mid-90s and even higher from July 2–6, 1911, across much of

Pennsylvania. The top temperature was 105°F, first reported at Marion on July 3, while Milford hit 102°F, which was the highest on record for the station period (1903–1920).

On the Fourth of July 1911, readings reached 105°F at Brookville and Selinsgrove. The reading at Brookville was especially notable, considering the elevation of 1,400 feet. In the southeast part of the state, highs of 103°F were observed at George School, Gettysburg and Huntingdon, and 101°F at Scranton and Harrisburg. However, a value of 103°F at Pocono Lake (elevation 1,660 ft.) should be disregarded due to questionably high maximums at the site in the summer of 1911.

1918
Hottest August Day

An August heat wave in 1918 brought the highest temperatures ever recorded in Pennsylvania up to that time.

The heat peaked in western Pennsylvania on August 6, 1918, when the mercury soared to all-time records highs of 108°F at Claysville in Washington County (August state record) and 107°F at Lancaster, Coatesville and Sadsburyville (Chester County). An exceptional value of 106°F occurred at both Clearfield (elevation 1,120 ft.) and Beaver Dam on August 6. Philadelphia hit 103°F, and at Scranton the mercury ascended to 100°F.

On August 7, the eastern sections were torrid, when the temperature at Philadelphia soared to 106°F, the hottest day in the city's history, which came after a morning minimum of 82°F, setting a record daily mean (94°F). The next day's story in the Inquirer stated that "Prostrations ran into the hundreds," adding that the "furnace-breath" heat buckled the Gray's Ferry Bridge. New York City also established a record August maximum temperature (104°F).

Ten Pennsylvania weather stations reached 105°F or higher on August 6–7, 1918. Bethlehem's Lehigh University notched a high of 105°F, a mark that would not be equaled until July 3, 1966, at Allentown. Williamsport topped out at 103°F and Scranton hit 102°F. The coolest maximum was 90°F at Mount Pocono (elevation 1,720 ft.).

The heat wave at Pittsburgh lasted from August 4–8, 1918, with daily extremes of 94°F, 101°F, 103°F, 101°F and 93°F. The maximum of 103°F on August 6, 1918, established an August heat record, equaling the all-time mark (July 10, 1881)—later matched on July 16, 1988. A minimum of 81°F on August 6, 1918, established an August record, including the city's hottest daily-mean temperature (92°F).

1919
Early Hot Spell

An intense three-day heat wave in early June 1919 brought record-breaking monthly temperatures. Stroudsburg reached 95°F, 99°F and 97°F on June 2–4, 1919. The mercury at Scranton climbed to 99°F on June 3—an all-time June record.

1923
Another June Heat Wave

The heat arrived early in the late spring of 1923, sending temperatures into the lower 90s at Allentown from June 2–6. A more remarkable heat wave developed on June 17.

Allentown registered nine consecutive days of 90°F-plus heat from June 18–26, 1923. Peak warmth occurred on June 20–21 (100°F, 98°F) and June 24–25 (98°F). The thermometer hit the century mark on June 20 in Pennsylvania, with top readings of 103°F at Lancaster, 102°F at Mifflintown, and 101°F at Catawissa, Reading and Williamsport.

June 1923 ended with a cool spell and frost in the northern mountains of Pennsylvania on the 30th, with a low of 32°F at West Bingham.

1925
June Swoon

The third memorable early heat wave in seven years gripped Pennsylvania during the first week of June 1925, sending temperatures to the highest levels known for that early in the season at some eastern locations.

Almost all eastern Pennsylvania valley locations recorded seven consecutive days of 90°F-plus heat from June 1-7, 1925 (Table 8.1). New all-time June heat records were soon established at Williamsport (104°F) on June 5 and Allentown (100°F) on the 6th (tied in 1966).

Table 8.1
Maximum Temperatures (°F)
June 1–7, 1925

June	1	2	3	4	5	6	7
Williamsport	96	93	94	100	104	100	96
Harrisburg	92	94	95	98	99	99	96
Allentown	91	96	98	98	98	100	98
Philadelphia	92	98	96	98	100	100	98

1930
Blistering Summer

The summer of 1930 commenced the first in a series of torrid summers during the great Dust Bowl years that enveloped the Great Plains and Midwest. An expansive ridge of high pressure at most levels of the atmosphere became a persistent theme in the 1930s, primarily linked to warm sea surface temperatures along both coasts.

Desert-like heat periodically wafted eastward, establishing many all-time heat records. The calendar year 1930 was the driest on record in Pennsylvania, which undoubtedly contributed to the intense heat because of the lack of moisture and evaporative cooling, and limited heat-breaking thunderstorm lines.

A pattern of recurring heat developed in the middle of July 1930 and continued through early August. A sizzling stretch beginning on July 21 brought 100°F readings in southeastern Pennsylvania and an all-time maximum of 106°F at Washington, D.C.

During a noteworthy four-day heat spell, Chambersburg and Lancaster recorded consecutive 100°F-days on July 19–23, 1930, with both cities recording identical readings of 102°F, 102°F, 104°F and 100°F. Huntingdon reached 101°F on July 26, 1930, and 100°F on July 28. York endured two straight weeks of 90°F-plus temperatures (July 17–30), before readings briefly dipped into the high 80s. In northeastern Pennsylvania, the thermometer on Academy Hill in East Stroudsburg soared to 102°F on July 21, 1930, a July heat mark that lasted until 1953.

A second blast of hot, dry air in early August 1930 brought scorching temperatures. A torrid week at York lasted from August 2–10, 1930:

August 3-5 (102°F, 104°F, 103°F), and on August 8-9 (102°F, 100°F). The warmest readings came on August 4: 106°F at Lykens and Mifflintown in the Susquehanna Valley. In the southern part of the state, the dry heat was extreme: a scorching 105°F occurred at Catawissa and Chambersburg. Altoona (elevation 1,615 ft.) reported 102°F, the hottest day locally until July 22, 2011 (103°F).

Three straight days of triple-digit highs were noted at Huntingdon (102°F, 105°F, 100°F), Mauch Chunk (102°F, 103°F, 101°F), and in the river valley at Selinsgrove (102°F, 105°F, 101°F) on August 3-5, during the second driest summer in the lengthy state records (surpassed in 1966). The maximum of 108°F at Carlisle is questionable, considering a warm bias in the summer of 1930 at the station.

1931
July Heat Wave

A heat wave on June 30–July 1, 1931, sent the mercury soaring to 100°F and 102°F on consecutive days at Williamsport. The first day of July 1931 brought searing heat to the central mountains; the highest readings were 105°F at Lock Haven and Hyndman, and 103°F at Clearfield.

1933
Another June Sizzler

More scorching heat in June 1933 produced a state record maximum at Sharon, in the northwestern corner of the state, when the thermometer topped out at 107°F on June 8.

The heat peaked in the eastern part of the state on June 9, 1933, with maximum readings approaching 100°F setting numerous daily records. The mercury at Scranton and Lancaster rose to 98°F among the warmest readings. An all-time June maximum of 92°F was observed at Mount Pocono, equaled in 1952, and surpassed on June 23, 2025 (93°F).

Unusually high temperatures persisted through July 1933, in areas where soils were dry. Brookville, on the Allegheny Plateau, notched a high of 101°F on July 24, 1933.

1934
Hot, Dry Summer

The exceptionally dry weather that enveloped the midsection of the nation in the summer of 1934 brought periods of intense heat to Pennsylvania. June 1934 remains the hottest June on record in the state to this day.

In northwestern areas, the temperature soared past 90°F on the first six days of June at Corry and Franklin. A record early-season 100°F reading was observed at Franklin on June 3, 1934.

A longer heat wave commenced on June 21, 1934, when temperatures frequently mounted past 90°F over most of the state, except in the northwestern mountains. The heat peaked in the east on June 29–30, 1934, with maximum readings of 103°F at Lancaster (June 29) and 104°F at Marcus Hook.

Record June heat was also noted at Catawissa, Chambersburg, Williamsport, York and Philadelphia (102°F) on the 29th, and Harrisburg joined the century club (100°F). In the west, the peak reading at Pittsburgh was 96°F. The only Pennsylvania weather site that did not reach 90°F in June 1934 was Montrose (elevation 1,656 ft.) in the Northern Tier.

Simmering conditions continued in July 1934 during the worst of the Dust Bowl years, particularly in southern Pennsylvania. At York, the temperature topped 90°F on the first seven days of July.

High heat returned on July 14, 1934, when maximum readings climbed to 90°F or higher at York on all but two days, between July 14 and 31. Huntingdon reached 100°F on the 21st and 103°F/101°F on July 25–26, 1934. State College (elevation 1,217 ft.) had a rare instance of consecutive 100°F days on July 25–26, 1934 (100°F and 101°F). Hotter readings observed at Mifflin appear to be too warm, relative to surrounding locations, probably due to an exposure issue.

1936
Hottest Ever in Pennsylvania

Unrelenting heat gripped the Midwest during the blazing hot summer of 1936 in the heart of the Dust Bowl era. Fifteen all-time state maximum temperature records from the Midwest to the Mid-Atlantic were established in the sweltering summer of 1936, exacerbated by an ongoing severe drought.

During the second week of July 1936, July heat records were toppled. On July 9–10, 1936, an all-time state heat mark was established at Phoenixville in Chester County, when the mercury topped out at 111°F on consecutive days. An exceptional reading of 108°F was notched at Marcus Hook.

The searing heat wave originated over the High Plains in early July 1936 and arrived in Pennsylvania on July 7. During the week of July 8–15, 1936, temperatures reached unequaled levels in many communities. Pittsburgh endured eight consecutive days of 90°F-plus heat, with highs ranging from 91°F to a maximum of 102°F on July 14.

Table 8.2
Maximum temperatures (°F)
July 9–11, 1936

	July 9	July 10	July 11
Williamsport	106	104	101
Bethlehem	104	103	103
York	104	105	101
Harrisburg	103	103	100
Scranton	103	101	99
Philadelphia	103	104	98
Pittsburgh	101	101	94

The desiccating heat peaked in eastern Pennsylvania on July 9–11, 1936. Three straight days of triple-digit readings in locations below 1,200 feet (Table 8.2) in most areas was unprecedented.

Northern Pennsylvania communities reached extreme levels on July 9 at Lawrenceville (106°F), Williamsport (106°F), and Lock Haven (105°F). At Johnstown and Towanda (104°F), Scranton and Emporium (103°F), and Clarion (102°F), records were toppled. Lawrenceville and Towanda dealt with four days of 100°F heat (July 8–11, 1936).

The central mountains also experienced extreme heat on July 9, 1936, setting all-time heat records at State College (102°F) and Clearfield (103°F, tying July 1931). The hot weather eased after July 11 in eastern Pennsylvania, but the mercury soared again in the western counties. Franklin recorded seven consecutive 100-degree days (July 8–14, 1936), with maximum readings of 106°F on July 9 and 14, 1936. All-time maximum levels were observed at Corry (110°F) and Vandergrift (106°F) on July 14, 1936.

In the city of Pittsburgh, the thermometer mounted to 103°F at the Federal Building downtown and 102°F at the county airport. Southwesterly winds overcame the usual cooling effect of Lake Erie, and the city of Erie sizzled at 98°F on the afternoon of July 14, 1936, setting a longtime record heat mark that lasted until July 1988 (100°F).

Along the Atlantic Coast, New York City endured its hottest day in over a century of weather records maintained at Central Park, when the mercury reached 106°F. In New Jersey, an all-time state heat record was observed at Runyon during the three-day heat wave on July 9–11, 1936 (109°F, 110°F, 105°F). All-time state heat records were observed on July 10, 1936, at Martinsburg, West Virginia (112°F), and at Cumberland and Frederick, Maryland (109°F).

The scorching heat in July and August 1936 was even more extreme in the parched Midwest. Maximum/minimum readings at Springfield, Illinois, on July 14 (108°F/84°F) were among the hottest observed during an unprecedented stretch of 12 days at or above 100°F (July 4–15, 1936). The historic hot week accounted for a spectacular mean temperature at Springfield of 86.2°F in July 1936. The highest readings in July 1936 were 121°F in North Dakota, 112°F in Manitoba and 108°F in Ontario, Canada, reflecting the incredible extent of the massive heat dome.

Blistering heat in the Plains wafted into western Pennsylvania in periodic waves: August 1–5; August 12–15; August 20–25. Pittsburgh had five days in a row in the 90s from August 21–25.

1948
Late-August Heat Wave

Temperatures soared to the highest levels of the summer in 1948 in Pennsylvania during the last week of August. Ninety-degree readings were common across Pennsylvania over a seven-day period from August 24–30, 1948.

The highest readings were recorded in the southeastern tip of the state at Marcus Hook on August 26–29, 1948 (106°F, 105°F, 104°F, 101°F). York also recorded four consecutive 100°F days on those dates, though the maximum was only 101°F. Columbia and Shippensburg reached 102°F.

The northwestern highlands escaped the extreme heat, with readings hovering around 90°F. The Weather Bureau station at Mount Pocono (elevation 1,914 ft.) achieved an August record high of 93°F on August 26, 1948, and valley readings at Stroudsburg topped out at 102°F.

1949
Hot and Humid July

Very dry weather in June 1949 brought the lowest monthly average precipitation (1.92 inches). The dry soils provided little moisture for incoming weather systems to work with, contributing to a persistently hot, dry pattern through the summer of 1949. Selinsgrove received a scant 0.31 inch of rain in June 1949.

July 1949 was relentlessly hot in the Northeast. Temperatures exceeded 90°F across much of Pennsylvania July 2–6, 1949, with the hottest weather generally observed on the Fourth of July. Maximum readings in the southeast on July 4, 1949, reached 103°F at Columbia and 102°F at Wellsville.

Temperatures hit 90°F or higher at Philadelphia from July 18–22, 1949, followed by another round of intense heat statewide during July 25–30, 1949. Average minimum temperatures were among the highest ever recorded for any month in Pennsylvania, reflecting increasingly humid conditions that contributed to higher nighttime readings.

Hot and muggy weather continued in early August 1949, culminating in another heat wave on August 9–11, 1949. The summer of 1949 was the hottest on record in Pennsylvania since 1888 and is ranked as the fourth warmest (71.2°F).

1952
Hot, Dry Early Summer

June 1952 was the second driest June (2.32 inches) in Pennsylvania after 1949, resulting in another summer of greatly diminished crop yields. Several days in the 90s were observed in the lower elevations as early as the middle of June 1952, but that was only a prelude to record-breaking warmth during a three-day heat wave on June 25–27, 1952.

A maximum reading of 104°F was reported at Newport on June 26 and 29, 1952.

The hottest day around the state was June 26, 1952. Carlisle reported a high of 103°F and maximums of 102°F were recorded at Lewisburg, Lock Haven, Reading and Williamsport, establishing June heat records. Williamsport had consecutive 100°F days (100°F, 102°F) on June 25–26, 1952, in the hottest weather so early since June 5, 1925.

In the east, on June 26, 1952, readings topped out at 99°F at Allentown and 98°F at Scranton and Stroudsburg (June records). The downtown Pittsburgh maximum reading of 99°F also set a June record.

July 1952 was also a hot month in Pennsylvania. Stroudsburg had 17 days of 90°F or higher, with maximums in the upper 90s on July 13–15 and July 22–23, 1952.

1953
Longest Heat Wave

The summer of 1953 featured recurring high temperatures beginning at the summer solstice. Extreme heat was reported in the east on June 20–21. Stroudsburg hit a June record high temperature of 100°F on June 21, 1953, one of several Pennsylvania cities to reach the century mark.

July 1953 continued the pattern of sweltering weather, with temperatures in the 90s right off the bat. The hottest weather in the summer of 1953 arrived midmonth. During the period of July 15–21, 1953, temperatures in the east frequently topped 90°F, and some eastern locations exceeded 100°F on July 17–18, 1953.

August 1953 was not unusually hot in Pennsylvania until later in the month. A heat wave began on August 24, marking the start of a record 12 consecutive days of sizzling temperatures in the 90s or higher at Stroudsburg and New York City. The mercury at Philadelphia soared to 90°F or higher on 13 consecutive days through September 5, 1953, edging out the 1901 mark by a day (later surpassed in 1995).

All-time record September heat was observed across much of eastern Pennsylvania and northern New Jersey on September 2, 1953, with highs of 102°F at Williamsport, 101°F at Scranton and 100°F at Palmerton. The high temperature up at Mount Pocono shot up to 95°F, equaling the all-time local record (July 31, 1954).

A state September heat record was achieved at Stroudsburg on September 2, 1953, when the thermometer peaked at 105°F (Table 8.3). All-time record highs were observed in northern New Jersey, which ranged as high as 106°F at Paterson, 105°F at Little Falls and 104°F at Belvidere, Layton and Newton.

Table 8.3
Maximum temperatures (°F)
August 25–September 4, 1953

August	25	26	27	28	29	30	31
Stroudsburg	94	91	96	99	100	100	100
Philadelphia	92	93	96	98	98	100	101

September	1	2	3	4
Stroudsburg	99	105	104	93
Philadelphia	97	100	100	92

1954
Searing July Heat

Extreme drought conditions prevailed across a large portion of the eastern and central United States in the summer of 1954, contributing to extreme heat. Many town and cities in Pennsylvania received less than half of the normal July rainfall. On the whole, July 1954 was the driest seventh month in the Keystone State since July 1910.

The first hot blast in the summer of 1954 in Pennsylvania arrived at the beginning of astronomical summer (June 21–22). A hot surge on July 13–14, 1954, sent readings up to 100°F in the lower elevations of eastern and far western Pennsylvania. The highest reading on the 14th was 104°F at Sharon. In the east, maximum temperatures included 102°F at Stroudsburg and Lebanon, and 101°F at Harrisburg.

A long heat spell developed on July 26 and lasted until August 2, 1954. The hottest weather occurred on July 31, with top figures of 104°F at Phoenixville and Wellsville. Stroudsburg recorded highs above 90°F on the concluding six days of July 1954 (90°F, 96°F, 96°F, 97°F, 99°F, 102°F). Mount Pocono (elevation 1,915 ft.) set an all-time July maximum record (95°F) on July 31, 1954, equaling the September 1953 extreme.

Northern New Jersey, Newark and Plainfield sizzled at 104°F on July 31, 1954. Even those readings paled in comparison to 115°F at St. Louis, Missouri.

1955
Hottest July/August

The months of unrelenting heat in the summer of 1955 plagued much of the East. Monthly heat records were established at Allentown in both July (79.0°F) and August (75.8°F), though the August mark would fall in 1980.

Stroudsburg reported a record 25 days of 90°F-plus temperatures in July 1955, yielding a mean monthly maximum of 92.7°F. The highest reading was 102°F on July 22, 1955. Hot spells from July 3–5 and July 15–22 were worsened by the high humidity.

The average temperature at Philadelphia in July 1955 tallied 81.4°F, equaling the warmest month ever known up to that time (July 1868). The average temperature in Pennsylvania (74.9°F) was the warmest on record (later tied in July 2020).

August 1955 opened with more simmering heat. The first day of the month brought a record high of 103°F at Stroudsburg, opening a seven-day stretch with highs in the 90s. The thermometer reached the century mark again on August 5–6, 1955 (100°F, 101°F).

The heat might have lasted longer, except for the arrival of ex-hurricanes Connie (August 11–13) and Diane (August 18–19), which brought catastrophic flooding. Despite a record sopping month, Mount Pocono notched its hottest August (70.7°F).

1966
Scorching Fourth of July

The summer of 1966 occurred in a drought that had prevailed in the Northeast since 1962.

A blast of hot air arrived in Pennsylvania in late June 1966, setting the tempo for a stifling summer. On June 27, extreme heat brought a sizzling 98°F at Stroudsburg. The mercury hit 100°F at Allentown, Harrisburg and six other locations around the state, and a record-tying 101°F at New York City for June.

The high point of the summer came on the Fourth of July holiday weekend, sending Pennsylvanians flocking to neighborhood pools and mountain resorts. On July 3, 1966, the mercury reached dizzying heights in southeastern Pennsylvania. All-time hot weather marks were established at Harrisburg (107°F) and Allentown (105°F). Torrid conditions equaled the July 1936 records at West Chester (105°F) and Philadelphia (104°F). Dry soils and extreme drought played a role.

Harrisburg hit 107°F, 104°F and 106°F from July 2–4 and 100°F on July 10 and 26. In the northeast, Scranton and Stroudsburg reached 101°F on July 3. In the Poconos, a maximum of 94°F came within one degree of the record at nearby Mount Pocono in July 1954. Little rain in the south caused record low streamflow.

1980
Sweltering Drought Summer

The 1980s marked a return of the hot summer patterns that were recurrent in the 1930s, 1940s and 1950s, but generally absent in the cooler 1960s and 1970s.

Two especially hot days on July 20–21, 1980, pushed the thermometer to the century mark in southeastern Pennsylvania. The hot, dry pattern continued in August 1980, the hottest August at Philadelphia (79.9°F), until 2016 and 2022. New August heat records were logged at Allentown (78.2°F) and Scranton (75.2°F).

The mean August temperature at East Stroudsburg (74.1°F) was the warmest since 1938 (75.2°F). A heat wave brought a week of 90°F readings from August 25–September 2, 1980, that culminated with a high of 97°F on the last day of the hot spell.

The heat in the summer of 1980 was especially extreme across the southern Plains. Wichita Falls, Texas, had a maximum of 117°F in July 1980. Brutally hot conditions on July 13, 1980, saw the mercury skyrocket to 108°F at Memphis. The average temperature at Washington, D.C., established an August record (82.8°F).

1988
Extreme Summer Heat

Dry soils contributed to some unusual temperature extremes in June 1988. On June 11, 1988, the temperature dipped to 28°F at Bradford and 29°F at the Rodale Research Center in the east-central hills. A few days later, readings catapulted into the 90s in the southeast, ranging from 92°F–97°F at Philadelphia on June 13–16.

A blistering southwest flow on June 20–22, 1988, sent the thermometer soaring across the Keystone State. The maximum of 98°F at Pittsburgh tied the monthly record (June 4, 1895). A reading of 101°F was reported at Mercersburg on the 22nd. Philadelphia baked in 90°F-plus heat on June 19–23: 90°F, 94°F, 97°F, 100°F, 95°F.

A few days later, on June 25, 1988, history was made when Erie recorded its first and only 100°F reading. Pittsburgh reached 98°F, tying the June mark set three days earlier. A cool spell brought temporary relief on July 1, 1988, with frost observed at Clermont 4 NW (31°F) and Kane (32°F). In the northeast, Pleasant Mount (55°F/42°F) and Scranton (63°F/42°F) experienced an anomalously chilly July day.

The break was short-lived, because the hottest portion of the summer commenced on the Fourth of July 1988 and continued unabated through the 18th. Harrisburg and Philadelphia reached 90°F or higher on 14 of 15 days (July 4–18). Harrisburg attained 101°F on July 7–8 and July 11, peaking at 104°F on July 16. Philadelphia sweltered at 100°F on July 7–8, then soared to 102°F on July 16–17.

On July 16, Pittsburgh and Williamsport hit 103°F, marking the 13th consecutive day of 90°F heat since the Fourth of the July. (The previous record of 11 straight 90-degree days at Pittsburgh dated back to June 30–July 10, 1878.) Pittsburgh's high reading of 103°F on the 16th tied the all-time mark, last observed in July 1881 and August 1918.

Scranton had a rare triple-digit day (101°F) on July 16, 1988, and Binghamton (98°F) set a Broome County Airport (elevation 1,590 ft.) record, surpassing the previous all-time high of 96°F in September 1953. In the normally comfortable Poconos, Tobyhanna hit 94°F on July 6–7, 1988, and 93°F on July 10 and 16, last reached in July 1966. East Stroudsburg notched highs of 99°F, 99°F and 97°F on July 6–8, 1988, followed by daily-record highs of 100°F and 99°F on July 10–11, and 99°F on the 16th.

Excessive heat in the Mid-Atlantic region on July 16, 1988, brought record high temperatures of 107°F at Martinsburg, West Virginia, and 104°F at Charleston, West Virginia, Baltimore and Washington, D.C.

Heavy thunderstorms erupted during the last two weeks of July 1988, cooling things off and easing the intense drought. In August, the hot, dry circulation returned. During a 10-day stretch from August 8–17, 1988, Pittsburgh sizzled through nine days with maximums of 90°F or higher, capped off by a maximum of 100°F on the 17th.

Philadelphia recorded 49 days of 90°F-plus weather in the summer of 1988, a record that lasted all of three years. Williamsport counted a record 42 sizzling days in the 90s. In the Midwest, Chicago broiled in a record 47 days of 90°F-plus heat, including seven days when the thermometer reached the century mark.

1991
Hot and Dry

The hot, dry summer of 1991 began in the middle of May and continued with little break for the next three months. A dry spring promoted high evaporation rates that contributed to the very warm conditions in Pennsylvania in May, similar to the antecedent conditions in 1988.

Philadelphia recorded its warmest May (70.8°F) in 1991, easily besting the previous record set back in 1880 (69.0°F). The Philadelphia maximum temperature reached 90°F or higher on a record 12 days in May 1991, including a new monthly maximum of 97°F on May 30–31, 1991. On June 16, the thermometer topped out at 100°F.

The primary heat wave in 1991 arrived on July 16 and continued for more than a week. Scranton had eight consecutive 90°F-plus days, the longest such spell since 1953. East Stroudsburg hit 99°F on July 20–21, 1953, and 97°F on July 23. Williamsport had its two hottest days on July 19–20, with highs of 100°F and 101°F.

The summer months (June–August) in 1991 at Philadelphia averaged 77.9°F, eclipsing the hot summers of 1901 and 1988 (77.1°F) and 1949 (76.9°F). However, the new record would not survive the consecutive hot summers of 1993, 1994 and 1995. August 1991 marked the 11th consecutive month of above-normal temperatures at Scranton, and the heat lingered into September.

On September 16, 1991, highs soared to 95°F at Philadelphia and 98°F in Baltimore. Another scorcher the next day established a record of 53 days of 90°F-plus days at Philadelphia. Williamsport observed 40 days of 90°F-plus days in 1991, only two fewer than in 1988. New York City recorded 39 days of 90°F-plus days in the summer of 1991, two more than the previous record (1944).

1993
Hot and Muggy

After a cool summer in 1992, largely attributed to the eruption of Mount Pinatubo in the Philippines in June 1991, heat and humidity returned in earnest in the summer of 1993.

That summer featured a heat wave that lingered from July 4–14, 1993, with 11 consecutive 90°F-plus days at Philadelphia, including three days of 100°F heat on July 8–10, 1993. The maximum reading at Philadelphia was 101°F on July 10.

The mean summer temperature (June–August) in 1993 at Philadelphia was a sweltering 78.2°F, slightly warmer (0.3°F) than the previous hottest summer (1991). July 1993 (81.4°F) tied 1868 and 1955 for the warmest July, which was promptly topped in 1994.

1994
Record Heat Again

The pattern of hot summers returned even earlier in 1994. Pittsburgh endured a heat wave from June 13–20, a record that included five straight days of high temperatures at or above 95°F. Philadelphia hit 100°F on June 15 and 19.

The mercury at Newark, New Jersey, topped out at 101°F on June 15, the hottest local reading so early in the season. June 1994 was the warmest sixth month at Newark (77.8°F), beating out June 1993 by 2°F. June 1994 was also the hottest on record at Philadelphia (78.1°F), until edged out by 2010. The nation's capital stewed in 17 days of 90°F-plus, including a June record of 14 straight.

July 1994 kept up the torrid pace, becoming the warmest month ever at Philadelphia (82.1°F) and the second warmest at Newark (81.9°F), after 1993. The months of June, July and August 1994 averaged 78.4°F in Philadelphia, breaking the 1993 mark by 0.2°F. The summer of 1994 was the fifth among the previous decade to rank among the 10 warmest in Philadelphia records since 1825.

July 1995
Hot and Dry

The summer of 1995 was notoriously hot and humid in the East, following its predecessors (1993 and 1994), and the most uncomfortable since 1955 in terms of heat and humidity. The season was also the driest summer in the Northeast since 1913, generating only 64 percent of the normal rainfall.

The heat first appeared in the upper Midwest, when International Falls, Minnesota, not generally regarded as a hot spot, hit 99°F on June 17–18, 1995, on the road to seven straight record-high temperatures. On June 19, the temperature soared to 103°F at Alpena, Michigan.

The origin of a July heat wave was evident in Kansas as early as July 11 (112°F), wafting east to Toledo, Ohio (104°F), on July 14, 1995. The Midwest was gripped in a blistering summer pattern that tragically resulted in at least 750 heat-related deaths in the Chicago area alone, where the local heat index (temperature plus humidity) soared to a record 125°F. The high at Midway Airport of 106°F on July 13 set an all-time city mark.

The next day Omaha, Nebraska, reached 109°F, and Milwaukee, Wisconsin, topped out at 103°F, without the benefit of a cooling lake breeze to mitigate the heat. Burlington, Vermont, recorded rare 100°F days on June 19 and July 14, 1995.

On July 15, 1995, the misery exceeded all records at eastern Pennsylvania locations. The heat index at Philadelphia soared to an unbelievable 129°F! The dew point, which reflects the amount of moisture in the air, topped 80°F from the lower Ohio Valley to southern Pennsylvania, the equivalent of a day in the Amazon rainforest. The minimum temperature at Allentown of 79°F on July 15, 1995, and an afternoon dew point of 80°F, established all-time records (heat index of 115°F). Comparative Lehigh Valley high minimums were registered at Bethlehem/Lehigh University on July 10, 1911, and July 10, 1934 (81°F).

Williamsport and Philadelphia were sizzling at 103°F on July 15, 1995. New York City hit 102°F, and Pittsburgh, Scranton and Boston all reached 100°F. Communities such as Poughkeepsie, New York, and Danbury, Connecticut, set all-time highs of 106°F.

A lengthy spell of 90°F-plus heat in eastern Pennsylvania lasted from July 25–August 4, 1995. The mercury climbed to 90°F or higher at East Stroudsburg a record 13 times in August. Philadelphia endured 17 consecutive days with 90°F-plus temperatures in a streak that ran from July 20–August 5. A total of 21 days of 90°F-plus heat set a record at that time.

The average monthly temperature at Philadelphia in July 1995 (81.5°F) became the second warmest after 1994—a figure that was exceeded in July 2010, 2011 and 2012. August 1995 was the second warmest seventh month (79.9°F)—a mere 0.1°F below 1980.

July and August 1995 were the warmest consecutive months (80.7°F) at Philadelphia, breaking a three-month summer heat record (June–August) for the third straight year (78.6°F)—the warmest until 2010. In all, the mercury ascended to 90°F or higher on 45 days at Philadelphia in 1995.

Pittsburgh sweltered through seven straight days of 90°F-plus heat from July 28–August 3, followed by another eight in a row August 13–20, 1995, ranked as the hottest August (77.7°F). No rain fell for 25 straight days until September 7.

The thermometer ascended to 90°F or higher at Philadelphia from August 11–18, and at Williamsport from August 12–18, 1995, with readings edging into the mid- and upper 90s. The highest reading in the state was 102°F on August 14, 1995, at Emsworth Locks and Dam, downstream from Pittsburgh along the Ohio River.

Baltimore had its hottest July (81.5°F) and August (80.1°F) up to that time, bolstered by a record 25 straight days of 90°F-plus heat (July

12–August 5, 1995), as residents sweltered in the city's warmest summer (78.8°F). The scorching conditions brought scant moisture as the summer wore on.

Total rainfall in August was less than an inch at many Pennsylvania locations, including Raystown (0.34 inch) and Mercer (0.35 inch). The Strausstown observer in Berks County recorded .01 inch, establishing an August record for dryness. Pittsburgh endured 25 consecutive days without measurable rain through September 7, 1995.

New York City had its driest August (0.18 inch) in 1995, and Washington, D.C., went without measurable rain for 33 days commencing on August 7, which eclipsed the previous mark of 32 days (September 30–October 31, 1963).

July 1999
Searing Heat and Drought

The drought of 1998–99 certainly contributed to the brutal heat that developed in early July 1999 and continued into early August. High evaporation rates and parched soils allowed the lower atmosphere to heat up day after day in July and early August 1999, especially in eastern Pennsylvania.

The differences in average temperatures in July 1999 between western and eastern Pennsylvania were dramatic, reflecting local drought conditions. In the southeast, the weather station at Philadelphia's Franklin Institute recorded the highest mean temperature (84.1°F) in state weather history at the time, exceeding July 1955 (82.6°F) at Philadelphia's Drexel Institute and Point Breeze.

In Harrisburg, July 1999 (81.9°F) went down as the hottest month ever recorded in the capital city. East Stroudsburg (76.5°F) achieved its second hottest month after July 1955 (77.9°F). Allentown's misery included the driest (0.33 inch) and third warmest July (78.0°F). The only warmer months in the Lehigh Valley were July 1955 (79.0°F), July 1901 (78.4°F at Easton), and July 1949 and August 1980 (78.2°F).

In contrast, the mean temperature at Pittsburgh in July 1999 (76.0°F) merely tied for the 20th warmest July at the time.

Extreme heat developed across much of Pennsylvania in time for the Fourth of July. The first major heat wave lasted from July 3–7, 1999. The maximum reading of 104°F at the Franklin Institute in downtown Philadelphia occurred on the 5th. The weather station at Harrisburg recorded three consecutive 100°F days on July 4–6 (101°F, 102°F, 102°F). In the east on July 5, the mercury at East Stroudsburg managed to reach 100°F for the first time since July 10, 1988. Allentown reached the century mark on the 6th.

The next heat wave arrived on July 15–20 as the thermometer soared to 100°F or higher at Harrisburg on July 16–18, 1999 (100°F, 101°F, 100°F). The highest reading in the state occurred at Graterford on July 18 (106°F). Governor Tom Ridge declared a drought emergency in the last week of July 1999 for all but a dozen counties in the northwestern part of Pennsylvania.

A protracted heat wave commenced in southeastern Pennsylvania on July 23 would continue to August 3, as many locations endured 10 days of 90 F-plus heat. Philadelphia sweltered through a dozen consecutive such days during the closing days of July through August 3, only one day shy of the August-September 1953 heat. September 1999 would bring imposed water use restrictions.

In the northeast, East Stroudsburg had 11 days of 90°F-plus heat in a row commencing on

July 23, 1999, one day short of the 1953 mark. The mercury reached 90°F or higher on 22 days in July 1999, the second most after July 1955 (25 days). The mean temperature of 76.6°F at East Stroudsburg in July 1999 edged out July 1949 (76.4°F), taking second place behind July 1955 (77.9°F) for the hottest month.

By comparison, on the "air-conditioned" high Allegheny Plateau in northwestern Pennsylvania, Bradford (elevation 2,117 ft.) did not touch 90°F in July 1999 until the last day of the month.

July 2010
Blistering Heat

A sizzler of a July 2010 made more than a few folks pine for the chillier days of the snowy winter of 2009–10. In June 2010, the maximum temperature at Philadelphia reached 90°F on 15 days.

In the northeast, the thermometer at East Stroudsburg had a week of 90°F temperatures from July 3–9, 2010. The heat wave peaked on July 5–7, when Marcus Hook observed three consecutive 100°F days (100°F, 105°F, 103°F). Philadelphia hit 102°F and 103°F on July 6–7, 2010.

Also joining the century club were Allentown (101°F) and Harrisburg (100°F) on July 6. In the northeast, relatively rare century-degree days were notched by observers at Bushkill (100°F) and Saylorsburg (101°F). Even the higher elevation sites in the Poconos were uncommonly warm: Hazleton (98°F) tied the all-time mark established at nearby Freeland (elevation 1,900 ft.) in July 1955, and Mount Pocono (94°F) came within a degree of the July 1954 all-time maximum record for the month.

Philadelphia residents endured the hottest summer on record in 2010 (79.6°F), surpassing 2022 (79.3°F), 1995 (78.6°F) and 1994 (78.4°F). A record warm June (78.2°F) and hot July (81.7°F) were followed by a muggy August (79.0°F). Other historically hot summers were topped off by very warm Augusts in 2022 (81.3°F), 2016 (81.0°F) and 1980 (79.9°F). In 2010 at Washington, D.C., a hot June (80.6°F) passed 1994 by 1.2°F, giving way to the warmest July (83.1°), before 2011 (84.5°F).

Philadelphia established a new record for days of 90°F-plus heat (55), overtaking the old mark of 53 days in 1991. The hot summer in Pennsylvania in 2010 (71.1°F) ranks as the fifth warmest, after 2005 (71.5°F), 2016 (71.3°F), and the summers of 1949 and 2020 (71.2°F).

2011
Long, Hot Summer

The summer of 2011 developed a tendency toward heat and humidity on the closing days of May. The thermometer soared past 90°F over the southeastern counties on May 26, and again on May 30–31, 2011.

The most intense heat wave of the summer arrived July 17–24, 2011, with temperatures peaking on July 21–23. Shippensburg had three straight days of triple-digit readings of 103°F, 105°F and 101°F. Stevenson Dam had back-to-back highs of 102°F and 106°F on July 21–22. Huntingdon hit 102°F and 104°F, and Renovo recorded 101°F and 105°F, ranking among the highest readings ever reached in the central region.

The high of 106°F at Reading on July 22 broke the previous record of 105°F on August 7, 1918. The heat peaked at 104°F and 106°F at Lewistown on July 21–22. Philadelphia hit 103°F on the 22nd, with a sweltering heat index of 119°F.

In the lower Susquehanna Valley, Harrisburg notched 101°F and 103°F on July 21–22, 2011, and Williamsport hit 100°F and 103°F. The minimum temperature at Scranton of 80°F on July 22, 2011, was the warmest on record, eclipsing July 15, 1995 (79°F).

Some especially high readings on July 22 in the southeastern counties included 105°F at Safe Harbor Dam, and 104°F at Allentown and West Chester. Philadelphia topped the century-mark with 103°F and 101°F on July 22–23, 2011, culminating in the hottest July on record (82.4°F), sneaking past July 1994 by 0.3°F.

Scant rainfall in the western part of the state contributed to the heat as the soil became very dry. The rainfall totals at Galeton and Tionesta Dam totaled only .50 inch.

2012
Third Consecutive Hot Summer

A scorching summer in the Northeast brought stifling heat in late June and July 2012.

Philadelphia opened the month of July with nine consecutive 90-plus-degree days. The extreme heat peaked in most sections of Pennsylvania on July 7, 2012, when maximums soared to 104°F at Shippensburg, 103°F at New Castle, and 102°F at Norristown and Safe Harbor. Both Philadelphia and Lock Haven readings soared to 101°F. More heat (100°F) was felt at Philadelphia on July 18.

Excessive heat related to dry soils returned on July 17–18, 2012, in the eastern counties, with readings in the upper 90s and reaching 100°F at East Stroudsburg. Philadelphia's mean July temperature of 81.8°F was 0.6°F shy of the all-time record (2011).

2016
Hottest August

August 2016 was the hottest on record in Pennsylvania. The mean monthly temperature of 73.4°F was well above the long-term mean of 69.6°F, edging out August 1900 (73.0°F) and 1955 (72.8°F). Philadelphia's average temperature in August 2016 topped out at 81.0°F, one degree warmer than the previous warmest August in 1980.

Maximum readings were not exceptional around the state; the highest was 102°F at Norristown on August 14. The consistency of the heat from mid-month into early September coupled with high minimums in a very humid environment accounted for the historic monthly means.

2020
Scorcher of a July

The summer of 2020 brought persistent heat and humidity in July, which averaged a record 82.2°F at Harrisburg, besting 2011 (81.8°F), and 78.0°F at Scranton, surpassing 1955 (77.4°F). Pittsburgh had its hottest July (77.3°F), beating out July 2011 and 1988 (76.9°F) at the NWS airport office.

High minimum, rather than maximum, readings were largely responsible for the record monthly warmth. The high temperature in the state was 102°F at Murrysville.

The monthly temperature in Pennsylvania of 74.9°F (4.5°F above the long-term average) beat July 1955 for the hottest month in state history. Baltimore reached 90°F or higher on 25 days in a record hot July (82.6°F).

Historic Warm Winter Days

A Lebanon correspondent in *Hazard's Register of Pennsylvania* (January 16, 1830) recalled two pleasant Christmases: in 1824 and 1829 (66°F). Christmas 1964 was the warmest holiday in the 20th century, with a high of 68°F at Philadelphia, eclipsing December 25, 1889 (66°F).

The warmest location in the state on Christmas 1964 was Newell (70°F). The day after Christmas, the high of 68°F at Philadelphia tied the 1889 daily record. Harrisburg established a new Christmas mark for warmth in the state in 2020 (75°F).

A remarkable spell of early meteorological winter warmth commenced on December 2, 1982. Temperatures rose into the 70s in the western counties on December 3, as far north as Clarion (70°F). Maximum readings set numerous monthly marks: 82°F at Washington 3 N (state December record); 79°F at Confluence; 77°F at Bloserville, Uniontown and Waynesboro; 75°F at Erie and 74°F at Pittsburgh.

In the east, the warmth peaked on December 4–6, 1982, with readings in the 60s and low 70s. The warmest day on December 4 felt like spring in the region, with highs of 74°F at Holtwood, Lancaster and Neshaminy Falls. A holiday warm spell culminated in a record high of 66°F at Pittsburgh on Christmas Day 1982, and highs in the mid-60s were common in the southeastern counties on December 26.

Remarkably mild early-winter conditions prevailed again in December 1984, with very mild readings from December 12–18, and highs reaching 65°F at Erie on the 16th and 70°F at Waynesburg on the 17th. Extraordinary late warmth returned on December 28–29, with readings soaring into the 60s and 70s statewide. Pittsburgh experienced the city's latest 70-degree day (71°F) on December 28. Extreme maximums on the 29th topped out at 79°F at Mercersburg, 77°F at Reading, 75°F at Harrisburg and 72°F at Allentown and Philadelphia.

Historic December warmth returned in 1998. Balmy weather on December 4 broke or tied all-time December maximum temperature marks in the east at Philadelphia (73°F), Allentown (72°F), East Stroudsburg (71°F) and Scranton (69°F). At Harrisburg, highs of 72°F, 74°F and 73°F set new daily records.

A few days later, maximum temperatures soared on December 7 to even higher heights in places. A top reading of 78°F was observed in Adams County at the Eisenhower National Historic Site. To the east at Phoenixville, the high was 77°F. Harrisburg tied a monthly record with 75°F, as did Philadelphia (73°F) and Scranton (69 F), tying earlier days. A new record occurred at Williamsport (69°F) on the 6-7th.

The second warmest December in state history—in 2006—started on a very warm note, with a maximum temperature of 77°F at Donora on December 1. The high at Williamsport (70°F) set a monthly record, and Allentown (72°F) tied the local December mark. Harrisburg reached 74°F, one degree shy of the 1984 record.

Nine years later, a historically warm December in 2015 (42.3°F) far outdistanced 2006 (37.3°F) in Pennsylvania data. Record daily highs straddled 70°F on December 13–14, 2015, across much of the state, including an all-time monthly record high of 67°F at Mount Pocono on the 13th. During an exceptionally warm stretch, readings climbed to record heights on Christmas Eve: 73°F at Uniontown and Neshaminy Falls; 72°F at Graterford and West Chester; 71°F at Allentown and Philadelphia; 69°F at East Stroudsburg; 68°F at Scranton and Pittsburgh.

The traditional January thaw is statistically reproducible in the Northeast, centered on January 22–26, suggesting a temporary seasonal exhaustion of arctic air masses.

Two very warm winter days occurred on January 22–23, 1906, that featured record January highs of 83°F at Derry, 79°F at Irwin, 78°F at Uniontown and 75°F at Pittsburgh. Another surprisingly warm winter day on January 29, 1914, brought readings of 81°F at Irwin and 79°F at Uniontown.

A warm spell in January 1950 featured spring heat on the 25th that tied or broke records established in 1906: Johnstown 80°F; Uniontown 79°F; Pittsburgh 75°F. The warmth peaked farther east on January 26: York 78°F; Reading and Lancaster 77°F; Allentown 72°F; Mount Pocono 69°F, in a nearly snowless January.

A warm start to January 2007 brought highs of 73°F at Philadelphia and 70°F at Allentown on the 6th.

Two historic warm February days occurred in the early 1930s. On February 25, 1930, the thermometer climbed to 80°F at Quakertown and 79°F at Philadelphia. In the north, Stroudsburg hit 74°F and Mount Pocono made it to 70°F. A balmy day on February 11, 1932, saw a reading of 83°F at Hyndman, establishing a new state record for February warmth.

The state's February heat mark was nearly equaled on February 24, 1985, when the mercury soared to 82°F at Mercersburg in Franklin County. Other warm spots in the state were 79°F at Raystown Lake, 78°F at Chambersburg, 77°F at Selinsgrove and 76°F at Allentown.

February 2017 was the warmest on record in state data history (36.6°F—10.1°F above normal), culminating in record-shattering monthly highs on February 24: Uniontown 79°F; Allentown and Reading 77°F; Pittsburgh and Scranton 76°F; East Stroudsburg 75°F.

A year later, a subtropical ridge off the East Coast pumped record warmth into Pennsylvania on February 20–21, 2017. Southern Pennsylvania communities reached the 70s on the 20th, and Pittsburgh established an all-time February record with a high of 78°F three days after a light snowfall.

The next day, eastern Pennsylvania locations shattered the all-time February 2017 records by a substantial margin: Reading and Lancaster 82°F; Allentown 81°F; York 80°F; Quakertown 79°F (tied 1930); East Stroudsburg 77°F; Scranton 76°F; Mount Pocono 70°F (tied 1930).

Spring Heat

The earliest 90°F readings in Pennsylvania were measured on March 23, 1907, at Hanover (90°F) and Everett (93°F). An early heat wave on April 7, 1929, sent the mercury to 94°F at Stroudsburg.

A notable heat wave in mid-April 1896 between April 13 and 20 brought a summer-like stretch of days in the 80s and 90s, setting many all-time April heat records that still stand. On April 17, the highs reached 91°F at Stroudsburg and Philadelphia, and 90°F at Pittsburgh.

On April 18, 1896, the *Stroudsburg Times* then reported a maximum 96°F, which was "the hottest [April] day in seventy-two years [1824]," and exceeds the modern mark (2002). Extreme highs in the eastern and southern counties reached 97°F at Carlisle, and 95°F at Coatesville, Lebanon, Aqueduct and Honesdale.

Record April warmth occurred in 1915 on April 19–20 and April 24–27. A top temperature of 98°F was observed at Bloserville on the 25th, and

Harrisburg topped out at 93°F, while Punxsutawney hit 98°F on the 26th.

In March 1945, the temperature soared to summer levels in the western part of the state on March 25–26, 1945, and spread to the eastern counties a few days later. The highest reading was recorded at Marcus Hook (91°F) on the 29th.

A hot Easter weekend in 1976 tied the all-time April record. A maximum reading of 98°F on Easter Sunday, April 18, at Norristown and Port Clinton led the pace. The temperature reached the low to mid-90s in the valleys and southeastern counties on April 17–19, 1976, with maximums including 94°F at East Stroudsburg and State College.

A five-day warm spell on March 27–31, 1998, again featured some extraordinary early-season heat, tying the state record high of 91°F on March 30, 1998, at Octoraro Lake.

Historic April heat occurred on April 17, 2002, establishing an April record at East Stroudsburg (95°F). Philadelphia's Franklin Institute tied the state's April mark (98°F).

Autumn Warm Spells

An autumn blast of hot, dry air in September 1895 sent temperatures into the lower 90s for five straight days at Pittsburgh (September 18–22), and readings in the mid-90s were recorded at a number of Pennsylvania sites.

Another five-day heat wave during a dry autumn lasted from September 10–14, 1931, with extreme warmth at Stroudsburg (97°F) and Williamsport (100°F) on September 11—the warmest readings ever recorded so late in the season.

Philadelphia experienced five days of 90°F-plus heat September 22–26, 1970, with a maximum of 95°F. The average September temperature (74.1°F) was the highest since 1880 (75.4°F). The autumn of 1931 (September–November) averaged 63.6°F (5.3°F above the modern average), which eclipsed 1880 by a significant margin (2.4°F).

An early October heat spell in 1941 pushed temperatures to 100°F at Phoenixville on October 5, 1941, setting an all-time state record for the month. Other hot spots in southeastern Pennsylvania included Ardmore (98°F), Harrisburg (97°F) and Lancaster (95°F). During October 5–7, an unprecedented three-day heat wave in the 90s was observed at Harrisburg (97°F, 95°F, 97°F) and York (97°F, 94°F, 96°F), while northern and western sections cooled on the 7th.

It's worth noting that both the spring and summer of 1941 were anomalously warm, featuring record heat in mid-April, late May, and from July 23–August 1, 1941, with a peak reading of 104°F at Marcus Hook on July 28, 1941.

A slew of warm Octobers occurred between 1946 and 1956 in Pennsylvania—only two were below normal. In 1950, November opened with two days of 80°F heat, establishing all-time marks in practically every Pennsylvania community. The hottest reading was 87°F at Marcus Hook. Even the northwestern highlands did not escape the unusual fall warmth: highs of 85°F were observed at Warren and 82°F at Erie—November heat records. Mount Pocono enjoyed a balmy 77°F, a record November maximum in the northeastern mountains.

Drought Patterns

The average annual cost of damage caused by droughts in the United States is $9.5 billion (1980–2019), according to NOAA's National Centers for

Environmental Information (NCEI). A drought represents a regional moisture imbalance resulting from a persistent below-normal precipitation pattern, with a gradual reduction in soil moisture and low streamflow. The economic costs increase if the water shortage leaves fields parched, reducing yields.

Groundwater is normally recharged in the late autumn and winter, when the sun angle is lower, limiting evaporative losses. Before the ground freezes, moisture percolates into the surface and subsoil.

A drought develops in several ways. A prolonged imbalance between precipitation and moisture demand is the hallmark of an agricultural drought. Timing is critical, because the protracted dryness is exacerbated when the sun angle is high during the growing season. Heat and evaporation of moisture from the soil, and transpiration (water lost by plants through the stomata of leaves in the form of vapor), amplify a hydrological drought and lower the water table. A higher demand for water reduces the streamflow, as crops wither in the field.

Meteorological drought is fostered by a pattern subsidence beneath high pressure from the upper levels down to the surface that pushes air downward, warming and drying the lower atmosphere. A northward shift of the jet stream around extensive ridging deflects storms.

Widespread drought east of the Rockies is most common when cooler water in the eastern equatorial Pacific contributes to an expansive subtropical high-pressure belt from the Four Corners region to the southeastern states. The gradual depletion of soil moisture with a dry pattern and a limited flow of Gulf moisture leaves little fuel for low-pressure systems and local storms, initiating a self-perpetuating drought cycle.

Forecasting Regional Drought

Dry and windy conditions raise the concern for grass and wildfires during the warmer months. The combination of strong wind gusts coupled with low humidities will trigger a fire weather watch or an upgraded red flag warning by the local NWS office.

Residents are warned not to burn trash or yard debris, because embers can spread rapidly to dry vegetation (grass, leaf litter, small branches) and spark a brush fire, a message affirmed by the Pennsylvania Department of Conservation and Natural Resources.

One of the driest air masses in state history was observed on March 30, 2008. The 4:00 p.m. temperature at Scranton of 48°F and dew point (1°F) registered a relative humidity of 14 percent.

Evaporative losses from soils, lakes, rivers and vegetation are figured into various NOAA indices put out by the U.S. Drought Monitor (USDM). The Palmer Drought Severity Index (PDSI) and the Palmer Hydrological Drought Index (PHDI) quantify national drought conditions as far back as 1900.

The PSDI reflects short-term drought sooner, prior to the groundwater capacity and reservoirs reaching low levels. The PHDI weighs rainfall and ground moisture storage against evaporative losses, accounting for long-term drought based on cumulative groundwater.

Near-normal temperature and rainfall patterns register close to zero on the index. The PHDI can exceed +4 during unusually wet conditions, and -4 for extremely dry periods. Very dry conditions were observed in Pennsylvania during the following years: 1894–1895, 1908–1911, 1914–1916, 1921–1926, 1929–1932, 1953–1955, 1962–1969, 1995 and 1999–2002.

The lowest mean PHDI data in Pennsylvania records occurred in the following months: January 1896 (-4.2); November 1909 (-4.81); November 1914 (-3.36); December 1922 (-4.60); January 1931 (-7.26); February 1954 (-2.62 inches); August 1966 (-3.37); July 1999 (-3.48); February 2002 (-3.05).

More recently, hot summers in 1980, 1983, 1991, 1993, 1995, 1999, 2002 and 2012 contributed to drought that was eventually relieved by heavy rain in the late summer and autumn.

Historic Pennsylvania Droughts

Exceptionally dry periods in Pennsylvania occurred at regular intervals in the 19th century.

An item in the Towanda *Settler* dated August 31, 1822, reported on the "unparalleled drought" during a time of "extreme drought and heat." An account continued: "Week after week has passed away and not a drop of rain has reached us. The earth parched, meadows and pastures dried up, streams and springs, never before known to have failed, now dry for 40 days there has not been a cloud in the horizon."

On June 25, 1853, a *Bradford Reporter* story described a worsening "protracted drought." The account continued: "The heat is intense, the dust multitudinous, and the parched and thirsty earth fairly gasps for moisture."

Heverly (1926) recounted that the drought extended into the following growing season, when there were "weeks without rain" and "a terrible Autumn drought." On September 2, 1854, the *Bradford Reporter* commented: "The memory of the oldest inhabitant has no recollection of the like." The article described the near-complete failure of buckwheat and potatoes in an atmosphere filled with dust and smoke.

A prolonged drought in the summer and autumn of 1881 "prevailed from the first week of July till the 13th of October; a water famine ensued; the Susquehanna river the lowest in 41 years; much suffering and damage, corn and buckwheat ruined" (Heverly 1926). Creeks, wells and springs dried up and forest fires consumed pockets of woodland.

The autumn drought of 1895 disrupted the coal shipment from the Monongahela River down the Ohio River due to low water levels, which was noted in *Climate and Crops: Pennsylvania Section* (April 1896).

The lengthy precipitation records at Pittsburgh revealed very dry periods in the following years: 1839–1840, 1851, 1854, 1856, 1862, 1894–1895 and 1900. In the 20th century, pockets of severe drought occurred in Pennsylvania during the years of 1908–1917, 1949–1955 and 1962–1969.

1929–1939 Dust Bowl Drought

Dry conditions developed over Pennsylvania in the fall of 1928 and recurred in the summer of 1929. Adequate moisture in the autumn of 1929 offset moisture shortfalls temporarily, but a serious drought developed in 1930, which went down as the driest year in state history (28.85 inches).

Precipitation continued to total below normal from December 1929–March 1931. Near-average or above-average rainfall in the summer of 1931 eased the drought, until dry conditions resumed, lasting through September 1932, with only a few months achieving above-normal precipitation.

The drought was especially harsh in western Pennsylvania, where crops suffered severely. Pittsburgh had its driest year in 1930 (22.65 inches), and logged the city's fourth driest in 1932 (25.89 inches). Heavy rain in early October 1932 eased the drought situation in Pennsylvania, though moderate drought persisted before a hurricane blew in from the south in August 1933.

On May 12, 1934, the Great Dust Bowl Storm circulated darkened skies from Oklahoma to the Eastern Seaboard. The winds stripped topsoil loosened from years of overgrazing that removed anchoring vegetation in the Plains states.

1962–1966
Great Drought of the '60s

The primary reason for a substantial precipitation deficit in the Northeast in the 1960s was a persistent predominantly northwesterly flow of air aloft, coupled with the lack of landfalling hurricanes. The dry weather was concentrated during the warm season, when evaporation rates peak, depleting groundwater storage and reservoir capacity, which reached alarmingly low levels year in and year out.

The Pennsylvania Department of Agriculture placed 22 counties on drought emergency status on July 13, 1962, the first of many declarations that put water usage restrictions into effect. Farmers had to contend with corn and hay that shriveled in parched fields and a lack of replenishing fall and winter rainfall.

The driest year in Pennsylvania came in 1963, when precipitation across the state averaged below normal for 10 months. October 1963 was the driest month in state history, averaging a scant 0.24 inch. On October 19, a total of 22 towns in the state were out of water. During the entire month, 24 weather stations in the southeastern counties recorded no rain at all. Governor William Scranton sought drought aid in the face of mounting agricultural losses resulting from long-term moisture deficits.

The regional drought continued in 1964 and 1965, reaching severe to exceptional status. The weather station at Breezewood, in Bedford County, received a minimal 15.71 inches of rain in 1965, establishing an all-time state record annual minimum.

Modest precipitation the following year eased the dry conditions, but searing heat in late June and July 1966 aggravated water shortages. Heavy rainfall in September 1966 finally brought genuine relief, after years of severe and protracted drought. The winter of 1966–67 had many snowfalls and respectable moisture from coastal disturbances.

Dry spells recurred in the late 1960s but did not evolve into a severe drought. However, Philadelphia experienced a 25-day stretch from September 11–October 5, 1968, without measurable rain.

1988
Blistering Summer Worsens Drought

Very dry conditions developed in the spring of 1988 across Pennsylvania. A number of locations received less than an inch of precipitation in June. The driest spots were Honesdale (0.29 inch) and Hanover (0.34 inch).

Finally, in late July 1988, after a brutally hot three-week period, a southward shift in the jet stream brought torrential storms. Pittsburgh received only 27.09 inches of rain during all of 1988. Ample rains in 1989 and a record wet year in 1990 (52.24 inches) ended the regional drought.

1998–1999
La Niña Drought Pattern

An intense drought enveloped Pennsylvania in 1998 and lingered through the blazing hot summer of 1999. The emergence of La Niña, a cool pool of water in the tropical eastern Pacific Ocean in the early summer of 1998, coincided with a prolonged drought in the eastern United States. Precipitation in the Keystone State was below normal in five of the last six months of 1998.

A circulation pattern featuring predominant upper-level ridging at most levels of the atmosphere was responsible for sinking warm air. Droughty summers in La Niña patterns that manifested in 1988 and 1999 shared similar features—a trio of high-pressure cells from the eastern Pacific Ocean to the High Plains merging with the western arm of a Bermuda High. The polar jet stream was displaced far to the north, shunting storms into southern Canada. Prolonged dryness served up little soil moisture to evaporate, contributing to a self-perpetuating drought cycle.

A stormy January 1999 and near-normal precipitation in February and March seemed to signal a break from the dry pattern, but the respite was only temporary. Very dry conditions developed through much of the Mid-Atlantic in April and May 1999, worsening during the summer months.

Ground wells dried up in northeastern Pennsylvania, where the drought was most severe, especially in portions of Pike, Monroe and Northampton counties. During a 13-month stretch beginning in July 1998, precipitation totals were nearly 15 inches below normal in eastern Pennsylvania.

On July 20, 1999, Pennsylvania became the first state to issue mandatory water restrictions. Governor Tom Ridge declared a drought emergency in 55 of 67 counties across the Commonwealth. The only area excluded from water regulations was the northwestern corner of the state. Water restrictions were mandated in Maryland and New Jersey, much of Delaware and parts of New York and Virginia.

There were a few noteworthy storms during the height of the drought that provided local relief. A heavy thunderstorm on July 22, 1999, brought more than two inches of rain to Harrisburg. The following week, Pittsburgh was doused with 4.14 inches on July 27–28, 1999, creating a veritable oasis in a sea of drought.

Although eastern sections of Pennsylvania were still classified in the category of severe or extreme drought in early September, the remnants of Tropical Storm Dennis dropped up to seven inches of rain on western and central Pennsylvania on September 6–7, 1999, mitigating the extreme drought.

An active cold front brought additional heavy rains on September 9, 1999, but the real drought-buster came with the remnants of Hurricane Floyd on September 15–16, 1999. Floyd unloaded frequent tropical downpours on the Mid-Atlantic and Northeast, on the order of four to 12 inches, and over the drought-stricken eastern counties of Pennsylvania.

A renewed dry cycle developed in early 2001 and persisted through the middle of 2002, with peak dryness in Pennsylvania in the autumn and winter of 2001–02.

In the fall of 2024, Philadelphia did not receive measurable rain for 42 days between September 29 and November 9, 2024, a record that eclipsed October 11–November 8, 1874 (29 days).

Chapter Nine

Tropical Weather Systems

Hurricanes have played an awesome role in the American panorama. They have touched the lives of Americans great and small and, at times, changed the course of our destiny—as well as the shape of our coastline.

—Patrick Hughes, *Hurricanes Haunt Our History*

HURRICANES are the greatest storms on Earth. The environmental impacts of these fierce, sprawling tempests are manifold, posing a serious threat to communities and infrastructure in and around the path of the storm.

National Hurricane Center (NHC) data for the period of 1963–2012 indicated that 49 percent of the fatalities in the United States associated with tropical cyclones were caused by storm surge, 27 percent from flooding rainfall, and 8 percent from storm winds. Additionally, 16 percent were related to tornadoes, powerful surf, and offshore mishaps within 50 nautical miles of the coast.

NOAA researcher James Kossin reviewed track and intensity data from 1949 to 2016, determining that there was a slowdown of 16 percent in North Atlantic storms moving over land. The decrease in forward motion can potentially double the rainfall totals and result in more severe inland flooding.

Another concerning trend observed in the past several decades is the rapid intensification process in tropical cyclones, based on maximum sustained winds increasing at least 35 mph within 24 hours.

In 2020, a record-tying 10 North Atlantic tropical cyclones underwent rapid intensification, equaling the 1995 season.

The following year, Hurricane Ida gained up to 70 mph in maximum winds, before landfall along the central Louisiana coast, according to a NOAA Service Assessment. The Category 4 hurricane hit with 150 mph winds on August 29, 2021, and peak gusts up to 172 mph at Port Fourchon.

Excessive rainfalls measured in feet are not uncommon with slow-moving tropical systems. Tropical Cyclone Hiki dumped 52 inches of rain on the island of Kauai, Hawaii, on August 14–18, 1950. That total was nearly matched by Hurricane Lane on August 22–26, 2018 (52.02 inches), on the Big Island at Mountain View near Hilo.

Hurricane Harvey dumped two to five feet of rain on parts of southeastern Texas in August 2017. Tropical Storm Imelda made landfall on September 17, 2019, as a weak system and then stalled, squeezed between a subtropical high to the east and a low-pressure trough in the Midwest. The tropical rainstorm dumped 43.35 inches of rain at North Fork Taylor Bayou, causing extensive flooding again in southeastern Texas.

Hurricane Formation

Spring predictions for an active summer and fall are based on decreased upper-level wind shear, warmer-than-normal sea surface temperatures and the potential for rising air that contains plenty of moisture in a column of low pressure.

In an average year, approximately 60 disturbances originate over northern Africa and move off of the coastline near Senegal, beginning a long trek westward across the Atlantic.

A tropical cyclone first appears in area of disturbed weather, cloaked in a swirl of clouds with broad spin and condensed air expelled aloft. The enlarging wind field which begins to resemble a pinwheel generates high seas and waves 15 to 30 feet or more in the open ocean. Large swells close to the coast produce rough surf and dangerous rip currents that pose a danger to beachgoers and small vessels.

In the center of an incipient tropical system, an annular core is the formative center surrounded by a synergistic ring of piston-like thunderstorms, or eyewall, which is the heat engine. Energy comes from the evaporation of warm water. Large storms sometimes exhibit an eyewall replacement cycle, when an outer ring forms around the inner circle in a mature hurricane.

Clusters of thunderstorms form a warm core as humid air rises and condenses, releasing heat. Less frequently, a wave of energy or meandering upper-level low becomes a subtropical, or hybrid, system with tropical characteristics in the lower levels residing beneath a pool of cool air aloft.

Meteorologists keep a close eye on a cluster of growing thunderstorms possessing a circulation in a region of little vertical wind shear, which maintains the integrity of the thunderstorm towers.

The storm receives a name when the one-minute sustained wind reaches 39 mph or higher. Tropical-storm-force winds usually extend more than 150 miles from the circulation center over the ocean.

Early in the season, tropical cyclones are prone to develop on the back end of a cold front off the southeast Atlantic Coast, in the Gulf of Mexico and Caribbean. A breeding factor for tropical cyclones is derived from the spin surrounding the Central American Gyre, a feature in the summer and early fall that lingers for days to a few weeks.

The northward shift of the monsoon trough and Intertropical Convergence Zone (ITCZ) over North Africa in late summer instigates disturbances over the Sahara Desert that depart the coast between Senegal and Guinea. About 30 percent of tropical waves traversing the Atlantic Basin grow into tropical cyclones, which depends on the prevailing atmospheric and oceanic conditions in the Main Development Region (MDR).

A large-scale wave pattern (Madden–Julian Oscillation) in the tropics is responsible for broad zones of rising and sinking motions in a 30- to 60-day cycle, which plays a role in the genesis of tropical cyclones where air is rising. A mid-level trough in the atmosphere amplified by a convectively coupled Kelvin wave also is a catalyst for a tropical cyclone.

The seedling of a tropical storm departing the west coast of Africa is a defined vortex embedded in the easterly trade winds one to three miles aloft that picks up heat from the warm sea (79°F or higher), and energy to a depth of 165 feet. The African easterly jet gives rise to a low-level flow derived from temperature differences between the Sahara Desert and the relatively cooler air circulation over the Gulf of Guinea.

If there is clockwise outflow at high altitudes, a tropical system is well-ventilated. Sinking air on the fringes of the wave creates a fan-shaped cloud pattern. The towering thunderstorms are fueled by an inflow of very warm, moist air. As the pressure in the center lowers, a central dense overcast is evident on satellite imagery.

Inhibiting factors for development are cooler-than-average surface water, strong headwinds, vertical wind shear associated with the subtropical jet stream, and ingestion of mid-level dry air often containing Saharan dust. A seasonally strong West African wet monsoon between June and September washes out the Saharan Air Layer.

The path of a tropical cyclone is mostly dictated by the strength and location of the Bermuda-Azores High. A weak subtropical ridge favors a storm path that impacts the northern Leeward Islands and Bahamas, before recurving toward Bermuda or remaining out to sea. A strong high-pressure zone guides storms toward the Caribbean Islands and either the Gulf of Mexico or southeastern United States. A slowing pace over warm water, sometimes tapping into the Loop Current that flows between Cuba and the Yucatan Peninsula, results in more rapid intensification.

In autumn, tropical cyclones tend to form near the Southeast Coast or in the Gulf of Mexico and Caribbean Sea. A generally northerly track results in landfall along the northern Gulf Coast or over the Carolinas. An overland trajectory pushes wind and heavy rain northward, straddling the Appalachians or Eastern Seaboard. Torrential downpours, aided by an upslope flow, wring out large quantities of moisture and trigger flash flooding an average of twice every five years.

The winds surrounding a post-tropical system diminish rapidly due to friction, as the circulation center crosses from the smoother water to a rougher land surface. This causes winds to shift abruptly in the lowest 5,000 feet, which is conducive to mini supercells. Converging air in the landfalling tropical rainbands—the "arms" of the storm—and the spin of the circulation center spawn short-lived tornadoes. The eventual transition to a post-tropical cyclone usually occurs six to 12 hours after landfall.

A landfalling tropical cyclone is attended by a life-threatening storm surge and whipping winds that bring down trees and power lines. The right-front quadrant of a tropical system is the most destructive portion of the storm, where the forward motion augments the wind and low-level convergence. Escalating gusts cause the water to shoal and overtop natural and man-made barriers. Battering waves and storm surge water inundate coastal communities.

Storm surge is the predicted abnormal rise in water above the astronomical tide, reflecting the total water levels above normally dry ground. A massive surge of water often overtops protective flood walls. Surge coinciding with high tide raises the flooding risks, especially during a full or new moon phase. Intense downpours within a feeder band produce prolific rainfalls and flash flooding.

Water driven into the coastline by the howling wind pushes against rivers emptying into the back bays, inundating vulnerable seaside communities. Engorged streams pouring into rivers with coastal outlets encounter encroaching storm surge, which elevates the water levels. As a lumbering tropical system interacts with terrain, friction weakens the howling gusts and the storm slowly spins down. However, the risk of serious flooding and several tornadoes persists far inland.

The ongoing linkage between a remnant storm center and frontal structure starts a transformation to an extratropical system. If a cold upper-level low merges with a warm post-tropical rainstorm, considerable instability develops as cold air sinks and warm air rises, producing bands of heavy rain that tend to move over the same region for hours.

Trees with root systems loosened in the saturated soil are more easily toppled, landing on vehicles and houses, and pulling down power lines that inflict widespread outages. Hard downpours add to the seepage that presses against basement walls.

NOAA Monitors Tropical Systems

In earlier times, there was little or no warning of a hurricane threat beyond distant ship reports, with little regard for the eventual path. Around the time of the Spanish–American War, the first United States weather stations were established in the Caribbean. Willis Moore, chief of the U.S. Weather Bureau, quoted the president, "I am more afraid of a hurricane than I am the entire Spanish Navy. Get this service inaugurated at the earliest possible moment."

After a series of strong hurricanes blasted parts of the East Coast in 1954, the U.S. Weather Bureau lobbied Congress to fund a network of operational high-powered radar systems, which were installed by the next hurricane season. However, the technology was too limited in areal coverage to effectively track another rash of devastating hurricanes in the Mid-Atlantic in August 1955.

The launch of the first weather satellite in 1960 from Cape Canaveral, Florida—polar-orbiting TIROS I—paved the way for a new frontier in tropical storm tracking. The first geostationary weather satellite (GOES-1) was sent into space on October 1, 1975, taking pictures from a singular location as it circled Earth in a geosynchronous orbit above the equator.

A fleet of NOAA geostationary environmental and polar-orbiting environmental satellites currently tracks tropical systems globally. Storm intensity is routinely monitored by 13 hurricane-hunter aircraft. Ten Air Force Reserve WC-130J jet airplanes that fly out of Kessler Air Force Base in Mississippi do the heavy lifting in the outer portion of the storm and through the eyewall. U.S. Air Force Reserve Hurricane Hunters search for a low-level circulation and measure the size and strength of a tropical cyclone.

NOAA's Gulfstream IV-SP (G-IV), based in Florida, flies around and above a storm while obtaining high-altitude data at a cruising level of 45,000 feet, revealing a picture of the upper atmospheric outflow—a strong gauge of intensification. The high-flying aircraft has been used in the wintertime to view storms over the Pacific Ocean and track an atmospheric river. Additional information is obtained from two NOAA Lockheed WP-3D Orion turboprop planes that release data-gathering probes to ascertain storm structure and movement.

A typical NOAA mission departs from Lakeland, Florida, and takes eight to 10 hours to complete a crisscross or butterfly pattern. GPS dropsondes are deployed to gather a steady stream of data from top to bottom—temperature, air pressure, humidity, wind direction and speed. The equipment falls back to the ocean on a parachute. Measurements received by the aircraft through radio transmission are relayed to NOAA and integrated in hurricane forecast models.

A scatterometer aboard a low-orbiting satellite sends a beam of radiation to the sea surface to ascertain energy from the wind-driven wave motions. A microwave radiometer records the maximum sustained one-minute wind speed and investigates low-altitude circulation features. In the lower levels, interrogation by Doppler radar analyzes the vertical wind profile of a storm by utilizing horizontal scans tilted at up to 14 elevation angles.

Upper-air observations are supplemented by data from ships at sea. A bathythermograph lowered into the water calculates pressure and temperature changes, a measure of the available storm energy.

Atlantic Tropical Cyclone Season

The Atlantic hurricane season officially begins on June 1 and ends on November 30. The most recent 30-year period produced an annual average of 14 named storms. Seven tropical storms on average graduate to hurricane strength (74 mph or greater) and three become major hurricanes (111-plus mph).

A rare winter tropical storm made landfall in South Florida on the night of February 2–3, 1952, dubbed the Groundhog Day Storm. The earliest hurricane to widely impact the northern Mid-Atlantic region moved up the Eastern Seaboard on June 2–4, 1825.

The latest tropical storm to menace the East Coast struck Cape Cod on November 27, 1888, dissipating on December 2, 1888. Hurricane Otto made history on November 24, 2016, as the latest landfalling hurricane in the Atlantic Basin, striking southern Nicaragua—the most southerly storm—before re-emerging on the Pacific side of Central America.

The earliest full-fledged hurricane that made landfall in the United States moved ashore west of Cape Hatteras, North Carolina, on May 29, 1908, and then moved back over the water. The storm struck eastern Long Island and Connecticut on May 30, bringing 40 mph gusts. In recent years, early-season landfalls of tropical storms have become commonplace. Tropical Storm Ana came ashore near Myrtle Beach on May 10, 2015. Every year since, through 2021, had at least one named storm in May.

North Atlantic hurricane activity goes through cycles of 20 to 40 years, linked to warmer water. A busy period from 1871 to 1903 was followed by fewer storms from 1904 to 1925. The succeeding cycle in the mid-20th century was characterized by frequent East Coast hurricanes. Florida took a series of direct hits between 1945 and 1950.

A positive (warm) phase that lasted from 1926 to 1969 was followed by a less active period that occurred between 1970 and 1994, which had few landfalling storms. However, two major systems struck during this cycle, named Hugo (1989) and Andrew (1992). Western Atlantic tropical activity picked up in 1995, coinciding with warmer water, then declined in seasons with strong wind shear.

In August 2004, eight named storms (five hurricanes, three becoming major storms) wreaked havoc in Florida and in the Southeast. Hurricane Charley grew to Category 4 intensity (160 mph) on August 13, knocking out power to two million Florida customers, causing 24 deaths and 792 injuries. Tens of thousands of homes and businesses were damaged or destroyed from central Florida to the Carolinas.

Most storm seasons since 2004 have produced more Atlantic tropical cyclones than the long-term average. Three consecutive near- or below-normal years ended after 2015. Record marine heat waves and pockets of warm seawater in the Atlantic and Gulf of Mexico have fueled stronger hurricanes.

Hurricane Forecast Models

Weather information collected over disturbed areas in the North Atlantic, Caribbean and Gulf of Mexico is integrated into NOAA's computer forecast models.

In 2023, a new model, the Hurricane Analysis and Forecast System (HAFS), would soon refine the forecast accuracy of the storm track and rapid intensification. Higher-resolution physics and data assimilation were made possible by the latest iteration of weather and climate supercomputers.

A combination of high-resolution satellite images and dropsonde data fed into computer programs will facilitate warnings and save lives by providing better intensity predictions. In the past 30 years, the suite of tropical dynamical models decreased the margin of error ("cone of uncertainty") from 350 to 100 miles.

NHC Tropical Weather Outlook updates come out four times daily with tropical cyclone forecasts.

Historic Atlantic Hurricanes

The costliest tropical cyclones to strike the United States, based on the 2023 Consumer Price Index cost adjustment, were Katrina (2005, $195 billion), Harvey (2017, $155 billion), Ian (2022, $115 billion), Maria (2017, $111.6 billion) and Sandy (2012, $85.9 billion).

Emergency management officials are especially concerned about the combination of rising sea level and a higher storm surge, coupled with an increase in the population in harm's way in recent decades.

The Saffir-Simpson scale was initially developed in 1969 by engineer Herbert Saffir and refined in 1971, with the assistance of meteorologist Robert Simpson, former director of the National Hurricane Center. The hurricane wind scale characterizes storms by the lowest pressure, maximum sustained one-minute wind speed, potential damage and the possible height of storm surge above normal high tide.

The force of the wind increases three-fold at higher speeds; a 75-mph wind has nine times the force of a 25-mph gust. The strongest hurricanes are classified as Category 5, with sustained winds of 157 mph or higher (Table 10.1). Although a handful of Atlantic storms achieved Category 5 status over the open water, only four have made landfall in the contiguous United States. The historical NHC tropical cyclone information is based on the revised post-storm analysis hurricane database (HURDAT 2).

Table 9.1
Saffir-Simpson Hurricane Wind Scale

Category	Pressure (mb/in)	Winds (mph)	Surge (ft. above low water)	Flooding
1	980+/ 28.94+	74–95	4–5	Minimal
2	965–979/ 28.50–28.91	96–110	6–8	Moderate
3	945–964/ 27.91–28.49	111–129	9–12	Extensive
4	920–944/ 27.17–27.90	130–156	13–18	Extreme
5	919–less/ 27.16 or less	157+	18+	Catastrophic

The Texas coast has been raked by many violent hurricanes. The Indianola Hurricane that destroyed the community on August 19, 1886, possessed a very low pressure of 925 millibars (27.31 inches) at landfall, with estimated winds around 150 mph, which were likely stronger.

The Great Galveston Hurricane came ashore on the night of September 8, 1900, as a Category 4 hurricane, with winds estimated at 145 mph, and turned into the deadliest natural disaster in American history, taking an estimated 8,000 to 12,000 lives. A 15-foot wall of water easily overwhelmed the 27-mile-long barrier island.

The Great Miami Hurricane, with 150 mph winds at landfall (Category 4), devastated South Florida on September 18, 1926, taking at least 372 lives and leaving 25 percent of the Miami–Fort Lauderdale area homeless. Such a storm today in the Miami area would likely cause damage upwards of $160 billion, normalized for population and economy.

The San Felipe/Great Okeechobee Hurricane made landfall north of Miami Beach on September 16, 1928, between Jupiter and Boca Raton, causing 2,500-3,000 deaths, mostly from widespread flooding around Lake Okeechobee.

The Great Labor Day Hurricane struck the Upper Keys of Florida on the night of September 2–3, 1935, with peak winds estimated at 161 mph and gusts up to 185 mph, killing at least 408 people, including 259 World War I veterans. The storm generated a storm surge of 18 to 20 feet, though it possessed a narrow damage swath of 40 miles.

The revised NHC Atlantic Hurricane database (HURDAT2) places Hurricane Camille in 1969 as the second strongest to strike land in U.S. history. Camille was a tropical wave on August 14, 1969, west of Grand Cayman Island. In only a few days, the storm intensified to a Category 5 during the afternoon on August 16, packing estimated sustained winds of 175 mph (NOAA reanalysis). The mighty storm made landfall just before midnight at Waveland, Mississippi, dropping more than 10.60 inches of rain at Hattiesburg.

Camille generated a deadly storm surge of 24.6 feet at Pass Christian, Mississippi, surpassed only by Katrina in August 2005. The storm killed 143 people along the Gulf Coast and destroyed 5,238 homes and damaged 11,667 more.

The remnants of Camille moved across western Tennessee and Kentucky as a tropical depression on August 19, 1969, encountering a frontal system over the Blue Ridge Mountains. A massive five-hour deluge (27 inches) was unleashed on Nelson County, Virginia, with rainfall estimates as high as 31 inches on August 19–20, washing out 100 bridges in the region. Flash flooding killed at least 113 people, with 39 missing added to the death toll. The damage total was estimated at nearly $100 billion (2020 dollars).

Hurricane Ivan landed near Gulf Shores, Alabama, on September 16, 2004, packing 120 mph winds, eventually crossing the Appalachians and reaching the Mid-Atlantic Coast two days later, causing major flooding. The storm performed a loop back into the Gulf of Mexico and made a final landfall in Louisiana as a tropical depression on September 24.

During a relatively quiet Atlantic season, Hurricane Andrew left an indelible mark on southern Miami–Dade County (formerly Dade County) in South Florida early on August 24, 1992. Maximum sustained winds at landfall around Elliott Key and Homestead were reassessed at 167 mph, with a central pressure of 27.23 inches (922 millibars).

Storm surge reached 16.9 feet at Biscayne Bay, causing devastation in large portions of mainland South Florida. Total damage in the United States was estimated by the NWS at $26.5 billion (1992 dollars), which was the costliest natural disaster at the time. Hurricane Andrew made landfall again on August 26, 1992, at a Category 3 hurricane at Morgan City, Louisiana.

The historic 2005 North Atlantic hurricane season witnessed a record 28 tropical storms and hurricanes (27 named storms and one subtropical system), which featured monsters such as Katrina, Rita and Wilma.

Hurricane Katrina first made landfall in South Florida on the morning of August 25, 2005, initially without major problems. But after re-emerging in the Gulf of Mexico, the storm intensified rapidly into a Category 5 hurricane, with top sustained winds of 175 mph and a central pressure of 26.64 inches (902 millibars).

During the evening hours of August 29, 2005, Katrina churned toward the northern Gulf Coast, making landfall at Buras, Louisiana, revealing top sustained winds of 125 mph, with fierce gusts of 140 mph, according to the NWS office at Mobile-Pensacola. A second landfall occurred a little later along the Louisiana-Mississippi border.

Katrina generated a storm surge of 19 feet near Bay St. Louis, Mississippi. East of the center, record storm surge of 27.8 feet occurred at Pass Christian, Mississippi, after the water moved through several bays. The death toll from Katrina was estimated at 1,833, with more people missing and presumed lost.

New Orleans suffered catastrophic flooding that covered more than 80 percent of the city, displacing one million residents. More than 50 federal levees failed in northern and western portions of the city, submerging neighborhoods under 10 to 20 feet of water. Water from Lake Pontchartrain overwhelmed drainage canals. The historic financial losses totaled more than $125 billion at the time.

The exceptionally active 2005 Atlantic hurricane season brought two more Category 5 storms—Rita and Wilma—that also battered the northern Gulf Coast. Late on September 21, 2005, Hurricane Rita deepened to a minimum barometric pressure of 26.43 inches (895 millibars), packing winds of 180 mph, with additional energy drawn from the Loop Current. The storm weakened at landfall to a still formidable Category 3 in southwestern Louisiana early on September 24, 2005.

On October 18–19, 2005, Wilma exceeded all previous Atlantic storms in intensity, exploding during a 30-hour period from a tropical storm to a Category 5 hurricane (175 mph) in the northwestern Caribbean. The central pressure dipped to an amazing historic low of 26.05 inches (882 millibars).

Wilma was downgraded after an eyewall replacement cycle, before landfall over Mexico's Yucatan Peninsula on October 21, 2005, carrying sustained winds reduced to 85 mph. An incredible 64.30 inches of rain fell in 24 hours at Islas Mujeres in the northeastern Yucatan, the greatest 24-hour rainfall on record in the Northern Hemisphere, according to the World Meteorological Organization (WMO). Wilma re-intensified over the Gulf of Mexico on October 23 to Category 3 intensity, with landfall at Cape Romano, Florida (120 mph winds). Storm damage totaled more than $25 billion (2020 dollars).

Hurricane Felix landed along the Nicaragua–Honduras border on September 4, 2007, as a Category 5 hurricane (160 mph).

A relatively quiet season in 2009 brought nine storms, but the next three years each recorded 19 tropical cyclones. Five major hurricanes packing sustained winds surpassing 110 mph were observed in 2010, three in 2011, and two in 2012.

A corridor of dusty air hindered storm development in 2013 in the North Atlantic, with the total number of tropical storms (14) and hurricanes (2) declining. The storm count dipped to eight in 2014—the lowest since 1997—after sea surface temperatures cooled slightly. Atlantic tropical activity remained below normal in 2015 (11 named storms) in a season noteworthy for a zone of dry air over the eastern Atlantic.

The 2015–16 El Niño—one of the most intense ever—generated wind shear that further disrupted storm formation. The historic period without any major hurricanes hitting the United States held. On October 23, 2015, Hurricane Patricia achieved Western Hemisphere records, with top sustained winds of 215 mph and a lowest pressure of 25.75 inches (872 millibars) off the southwest coast of Mexico. Patricia's rapid intensification (120 mph in 24 hours) set a modern record.

The action resumed in 2016 (15 named storms). Hurricane Matthew slammed into Haiti and eastern Cuba on October 4, 2016, as a Category 4 hurricane, having briefly attained Category 5 status on the last day of September 100 miles west of the coast of Colombia, the strongest storm to form in the Atlantic Basin since 2007, when Dean and Felix roamed the waters.

Hurricane Matthew headed west and then north along the Southeast coastline. Torrential rain caused devastating flooding in eastern North Carlina on October 8–9. Strong winds left nearly three million residents from Florida to the Carolinas displaced.

After 2005, no major hurricanes made landfall in the United States through 2016; then, from 2017 to 2024, 10 major hurricanes hit the Gulf Coast.

On April 20, 2017, Arlene became just the third tropical cyclone in April in the satellite era. In 2017, there were 17 named Atlantic storms, and, notably, 10 consecutive storms reached hurricane strength for the first time since 1893. Four hurricanes made landfall in the United States for the first time since 2005.

Major hurricanes Harvey, Irma and Maria went on to cause deadly destruction during an interval of four weeks in August and September 2017 that reached from the Caribbean islands to Texas. The season featured six major hurricanes, with a total cost of more than $294 billion, including damage and cleanup efforts, which exceeded the previous hurricane season record in 2005 ($172.3 billion).

Hurricane Harvey reached Category 3 intensity (132 mph) a few hours before the eye of the storm came ashore at San Jose Beach, Texas, followed by a second landfall at Holiday Beach. The storm crawled up the coast and stalled for nearly four days, unloading a record 32.47 inches of rain at Houston's Hobby Airport from August 26–28, 2017, and 60.58 inches at Nederland, Texas, near Beaumont, setting a national record for a tropical cyclone. (The previous rainfall record for the lower 48 states was 48 inches at Medina, Texas, in 1978 during the passage of Tropical Storm Amelia.)

The storm spent two days over land before drifting back over water. A fifth and final landfall occurred in Louisiana on August 30, 2017. Total damage caused by Harvey reached $131.3 billion—the second costliest national weather disaster after Katrina. The death toll reached 89 persons—all but one caused by flooding.

Less than two weeks later, Hurricane Irma devastated a swath from Barbuda to the U.S. Virgin Islands—St. John and St. Thomas—and proceeded west to strike Cuba. Maximum sustained winds of 185 mph tied Wilma (2005) as the strongest hurricane in the open Atlantic—comparable to Gilbert (1988) and the Labor Day Hurricane (1935).

The monster Hurricane Irma killed at least 134 people in the Leeward Islands and maintained 185 mph winds for a record 37 hours, spending more than three and a half days as a Category 5 hurricane (tied with a storm that struck Cuba in 1932)—longer than all other Atlantic hurricanes except Ivan (2004).

Cuba and the Bahama Islands endured the wrath of Hurricane Irma in September 2017. As a Category 5 storm, Irma raked the islands from Antigua and Barbuda to Puerto Rico, with wind gusts of 177 mph, before hitting the Florida Keys on September 9, 2017. Hurricane Irma landed as a Category 4 storm at Cudjoe Key, packing wind gusts of 130 mph. The financial costs from wind and water were $62 billion (2023 dollars). The northerly path blasted areas in western Florida and triggered record flooding on the eastern side in Jacksonville. The loss of life in the United States was 91.

Hurricane Maria grew into a tropical storm east of the Lesser Antilles on September 16, 2017. Hurricane Maria intensified from a Category 1 to a Category 5 storm (175 mph) in little less than 18 hours. The massive hurricane roared through Dominica on September 18—the eighth consecutive hurricane and fourth major hurricane of the busy season.

Maria then charged toward the U.S. Virgin Island of St. Croix—the second landfall in two weeks. Maria came ashore at Yabucoa, Puerto Rico, accompanied by 155 mph winds, and up to 37 inches of rain that triggered devastating flooding and mudslides. The massive damage exceeded $100 billion, the costliest hurricane to strike the island of Puerto Rico.

The death toll from Maria in Puerto Rico was drastically readjusted in a late August 2018 report to an estimated 2,975, factoring in the catastrophic human toll that followed the nearly complete loss of power that lasted for more than six months. The total losses, including recovery efforts, exceeded $100 billion, behind only Katrina, Harvey and Ian.

Hurricane Jose (155 mph) narrowly missed the Leeward Islands in mid-September 2017. Hurricane Ophelia formed near the coast of Africa in early October, the farthest east that a major hurricane was tracked in the Atlantic Basin. The storm headed northeastward and struck the southern coast of Ireland, where gusts hit 119 mph. The 2017 hurricane season brought 16 billion-dollar disasters. The total damage from weather and climate disasters in 2017 surpassed 2005 ($215 billion).

Hurricane Michael roared ashore on October 10, 2018, near Mexico Beach, Florida, with fierce Category 5 winds of 160 mph lashing the Florida Panhandle—rated as the fourth most powerful landfalling storm in the United States since 1900. Storm damage totaled $30.2 billion (2023 dollars).

Above-average hurricane seasons in 2019 (18 named storms) and 2020 (30 named storms) continued the very active Atlantic storm pattern. The next season brought 21 tropical and subtropical storms in 2021, third most after 2020 and 2005.

Hurricane Dorian intensified rapidly into the strongest Atlantic storm to occur east of Florida. At landfall in the northwest Bahamas, sustained reached 185 mph at Great Abaco on September 1, 2019, tying the Labor Day Hurricane of 1935 that pounded the Florida Keys. Only Hurricane Allen in 1980 produced higher sustained winds (190 mph) in the history of recorded storms in the tropical Atlantic Basin.

By September 1, 2019, Hurricane Dorian would gain intensity (150 mph to 185 mph) in less than 24 hours. The Category 5 hurricane storm drifted over the northern Bahamas for 40 hours, which resulted in massive damage. Dorian would be a Category 2 storm at landfall along Cape Hatteras on September 4. Hurricane Dorian was the fifth Category 5 Atlantic hurricane since 2016, including Matthew (2016), Irma and Maria (2017), and then Michael (2018).

In 2020, a record 11 tropical cyclones made landfall in the continental United States, which surpassed the 1916 benchmark of nine storms. Eight named storms (five hurricanes) pounded the Gulf Coast. On August 27, Hurricane Laura blasted southwestern Louisiana with 150 mph winds. Six hurricanes hit the contiguous United States in 2020, tying 1886 and 1985.

A year later, Hurricane Ida hit Louisiana on August 29-30, 2021, strengthening rapidly in the Gulf of Mexico from a Category 1 (80 mph) to a Category 4 hurricane (150 mph) in fewer than 24 hours.

Hurricane Ida claimed 30 lives in Louisiana and two in both Mississippi and Alabama. The battering winds knocked out power to at least a million households. Torrential rain triggered inland flooding that caused severe damage.

The remnant system traveled 1,000 miles as a tropical rainstorm, unleashing heavy flooding from southeastern Pennsylvania and New Jersey to the New York City area, causing widespread damage and significant loss of life.

Storm-related deaths from tornadoes, wind and flooding surpassed 50 in the Mid-Atlantic region. Total losses reached $75 billion, the sixth costliest storm in United States history. The formation of Wanda in early November marked the 21st storm of the 2021 Atlantic season.

Table 9.2
Most Powerful Atlantic/Caribbean Hurricanes by Pressure

Rank	Name	Year	Pressure (mb/in)	Peak Intensity (mph)
1	Wilma	2005	882/26.05	185
2	Gilbert	1988	888/26.22	185
3	Labor Day	1935	892/26.35	185
4	Rita	2005	895/26.43	180
5	Milton	2024	895/26.43	180
6	Allen	1980	899/26.55	190
7	Camille	1969	900/26.58	175
8	Katrina	2005	902/26.64	175
9	Mitch	1998	905/26.73	180
10	Dean	2007	905/26.73	175

Table 9.3
Most Powerful Landfalling U.S. Mainland Hurricanes by Pressure

Rank	Name	Year	Pressure (mb/in)	Peak Intensity (mph)
1	Labor Day	1935	892/26.35	185
2	Camille	1969	900/26.58	173
3	Michael	2018	919/27.14	160
4	Katrina	2005	920/27.17	127
5	Andrew	1992	922/27.23	165

Source: NOAA/National Hurricane Center

Hurricane Names

Tropical storms were not formally named until 1950. In the first half of the 20th century, tropical systems were identified by using the Joint Army/Navy phonetic alphabet, with "Able" and "Baker" at the head of the list. In 1953, female appellations were used for the first time. The following season in 1954, Carol, Edna and Hazel became household names because of the devastation wrought by three major hurricanes that slammed into the Mid-Atlantic coastline northward to New England.

In 1978, a new system incorporated alternating between male and female names, paying heed to gender equality. Storm names were rotated every six years by the United Nation's World Meteorological Organization, unless a particularly severe or deadly storm resulted in the retirement of the designated name for historical purposes. The "I" storms represent 14 out of the 99 retired names (2024).

The terminology was refined in 2010 to account for a transition to post-tropical cyclone status. The Greek alphabet was first employed in 2005 beginning with the 22nd named storm of the historic season, and again in 2020. (The letters Q, U, X, Y and Z are not employed.) The policy changed in 2021, dropping Greek names in favor of a supplemental list.

October 18, 1703
Pennsylvania's Earliest Record Event

Accounts of tropical cyclones that hit settlements in Virginia and the Plymouth Bay Colony in New England caught the attention of the press. (Dates before 1752 were adjusted for the adoption of the Gregorian calendar by adding 11 days to an event.)

On October 18, 1703, a tropical storm caused limited damage at Philadelphia, mainly removing a roof and knocking down large trees (Ludlum 1963). A reference appeared in *Transactions of the American Philosophical Society* (1818), prepared by Nicholas Collin, under "Observations made at an Early Period, on the Climate of the country about the River Delaware," an early Swedish Colony publication.

August 23, 1724
Great Gust

A tropical storm or hurricane affected the Mid-Atlantic region in the vicinity of Chesapeake Bay on August 23, 1724, followed by a second storm a few weeks later (Ludlum 1963). A notice of the activity appeared in the *Virginia Gazette* (December 12, 1744), which was reprinted in the *Pennsylvania Gazette* (January 28, 1745).

November 2, 1743
Eclipse Hurricane

A powerful hurricane struck the Mid-Atlantic and New England coastal areas on November 2, 1743. The *Pennsylvania Gazette* reported that "a violent Gust of Wind and Rain attended with Thunder and Lightning" passed over Philadelphia.

The advancing cloud shield blocked the view of a predicted lunar eclipse at Philadelphia. The disappointment led Benjamin Franklin to study the effects of the late-season hurricane. Franklin learned that his brother in Boston witnessed the the event despite a northeast wind, meaning the storm's circulation was separate from its motion.

Franklin wrote, "The Storm did a great deal of Damage all along the Coast, for we had Accounts of it in the News Papers from Boston, Newport, New York, Maryland and Virginia." Franklin would have another opportunity to observe the effects of a hurricane that tracked up the Mid-Atlantic Coast on the evening of October 18-19, 1749. The storms confirmed to him the nature of nor'easters.

September 7–8, 1769
Chesapeake Bay Hurricane

A coastal hurricane on September 7–8, 1769, uprooted trees and sent rivers out of their banks in southeastern Pennsylvania. The *Pennsylvania Gazette* reported that the storm brought the highest tide ever recorded.

The *Pennsylvania Chronicle* described the damage: "Numbers of Mill Dams are broke, and two mills carried away. Several Vessels drove ashore in our river, and many in the Bay, we are fearful, have met the same fate. We hear of terrible devastation in all Parts of the Country, various orchards being tore up, and great Quantities of Indian Corn and Buck Wheat destroyed, and that two Men were drowned."

September 2–3, 1775
Independence Hurricane

The Independence Hurricane traveled over eastern Pennsylvania early on September 3, 1775. The wind caused considerable damage at Philadelphia, where wharves and boats were battered.

From the weather diary of Israel (Phineas) Pemberton at Philadelphia (Ludlum 1963):

Sept. 3—Stormy & showery. A violent gale from NE to SE the preceding night with heavy rain, lightning and thunder—a remarkably high tide in the Delaware this morning. Flying clouds & wind with sunshine at times.

September 23–24, 1785
Equinoctial Storm

The Equinoctial Storm that struck the Carolinas on September 23–24, 1785, caused major storm damage from Baltimore northward to Philadelphia and New York City. Remnants of the storm slammed the Mid-Atlantic on the 24th (Ludlum 1963). The hurricane arrived just past the equinox, bringing "the highest tide ever before known in Norfolk."

Summer of 1788
George Washington's Hurricane

Two hurricanes impacted the Mid-Atlantic in the summer of 1788. Both storms attracted comment by future presidents at their respective homes.

A storm that became known as "George Washington's Hurricane" took shape near Bermuda on July 19, 1788 (Ludlum 1963). The hurricane came ashore near Cape Hatteras, North Carolina, on the night of July 23, 1788, affecting Mount Vernon, Virginia, after midnight, which we know from Washington's journal.

The storm followed an inland path, pounding the Chesapeake Bay region, where it "blew a perfect hurricane, tearing down chimneys, fences, etc." Trees and crops were toppled, homes shifted off their foundations, and ships were destroyed at Portsmouth, Virginia. The storm remnants tracked north into eastern Pennsylvania and New York on July 23–24, 1788 (Schwartz 2007).

Washington remarked on the impact of the storm in his journal entry on July 24 (Jackson and Twohig 1979):

> A very high No. Et. Wind all Night, which, this morning being accompanied with Rain became a hurricane—driving the Miniature Ship Federalist from her Moorings, and sinking her—blowing down some trees in the groves & about the houses—loosening the roots, & forcing many others to yield and dismantling most, in a greater or lesser degree of their Bows, & doing other and great mischief to the grain, grass &ca. & not a little to my Mill race. In a word, it was violent and severe— more so than has happened for many years. About Noon the Wind suddenly shifted from No. Et. to So. Wt. and blew the remaining part of the day as violently from that quarter. The tide about this time rose near or quite 4 feet higher than it was ever known to do driving Boats &ca. into fields … must as it is to be apprehended have done infinite damage on the Wharves at Alexandria—Norfolk—Baltimore &ca.

A smaller but intense tropical disturbance affected eastern Pennsylvania on August 18–19, 1788, leaving a path of significant wind damage from Cape May, New Jersey, to New York City.

From David Ludlum's *Early American Hurricanes:* "It is probable that the area of high winds did not exceed 100 miles in breadth, but in about a fifty-mile-wide path the speeds must have been well in excess of 75 mph to cause such destruction, especially in the forests." On the western edge of the "compact hurricane" there were estimates of as much as seven inches of rain falling in the Philadelphia area (Ludlum 1983).

On August 20, 1788, Thomas Jefferson marked the breadth of the storm in his personal papers (Cantanzariti 2000): "To the Northwest as far as Pennsylvania, we learn that there has been an almost universal destruction of mills and forges."

August 1795
Two Hurricanes

Two hurricanes slammed into the Mid-Atlantic Coast on August 2–3 and August 13, 1795, plowing through North Carolina and Virginia (Schwartz 2007).

Elizabeth Drinker in Philadelphia had this to say about the storm in her diary, on August 4, 1795 (Crane 2010):

> This evening's paper says: "The rains for a few days have been greater and the floods higher than ever before known in Pennsylvania. The mails and public stages, which set out for different parts of the United States, were all obliged to return to the city, finding the roads impassable."

October 9, 1804
Snow Hurricane

The October Snow Hurricane in 1804 brought strong winds and heavy rain to Philadelphia on the morning of the 9th. The most interesting feature of this storm was the arrival of cold Canadian air into the circulation, bringing heavy snow to parts of New England and Upstate New York. Ludlum (1963) placed the storm track "between Philadelphia and Atlantic City" continuing northward "very close to New York City."

September 3, 1821
Norfolk–Long Island Hurricane

The Norfolk–Long Island Hurricane initially brushed the Outer Banks of North Carolina on September 3, 1821, racing northeastward, with the eye passing just west of Cape May, New Jersey, in the afternoon of September 3, 1821 (Ludlum 1983).

The powerful hurricane reached western Long Island before evening and took an inland path through central Connecticut to southeastern Maine early on September 4, having caused extensive wind damage to communities from North Carolina to Massachusetts.

The storm hit New Jersey hard as it brushed past eastern Pennsylvania, dropping up to seven inches of rain on the Philadelphia area. The *Sussex Register*, in northwestern New Jersey, described a "very severe gale of wind from the S.E., accompanied with heavy rain … prostrating fences to the ground, uprooting and twisting from their trunks the largest trees …"

A comment in Watson's *Annals of Philadelphia* recalled: "A great storm of rain and wind from the north-east destroyed many trees, blew down many chimneys, and unroofed the bridge at the Upper ferry. The Schuylkill dam rose much."

June 4, 1825
Rare June Hurricane

The earliest hurricane in the past two centuries to make the journey up the East Coast traveled from St. Augustine, Florida, to New England on June 2–5, 1825.

The storm ruined crops, flattened trees and caused extensive property damage from Washington to Philadelphia on the morning of June 4. Shipping fleets also suffered in the storm and many sailors were lost at sea.

September 11–12, 1838
Hurricane Brushes Northeast

A dry pattern in eastern Pennsylvania and New Jersey beginning on August 12, 1838, and lasting until September 11, resulted in "probably the largest fire in the first half of the nineteenth century" in Burlington and Monmouth counties in New Jersey (Ludlum 1983).

Weather records taken by Dr. John Conrad at Pennsylvania Hospital, in downtown Philadelphia, reported nary a cloud in the sky from September 3–8, 1838.

However, on the night of September 11–12 the tinder-dry conditions were relieved by the passage of an offshore hurricane that brought 6.01 inches of rain at Philadelphia (4.19 inches in eight hours). The strongest winds remained offshore, minimizing the damage.

A tropical cyclone moved through eastern North Carolina and along the East Coast on August 28-29, 1839, that merged with a cold front on the 30th, bringing a rare August snowfall in the Catskills.

October 13, 1846
Great Havana Hurricane

The Great Havana Hurricane devastated Cuba, after making landfall south of Havana on October 11, 1846, likely as a Category 5 storm.

The large and powerful hurricane blasted Key West, with another landfall farther north in northwestern Florida (Ludlum 1963). The storm headed northeast and raked the Atlantic Coast on October 12–13, 1846, after getting caught up in the jet stream, pounding the modern Interstate 95 corridor cities.

The winds at Philadelphia were deemed "the most destructive storm in 30 years." Trees blown down and damaged structures caused a backup on the Delaware River that inundated wharves, according to the Philadelphia *North American*.

The records of Pennsylvania Hospital observer John Conrad noted, "a tremendous gale from the southeast" on the early afternoon of October 13, 1846, as the storm passed off to the northeast of the city, although the rainfall was a relatively light 1.25 inches.

The storm tide was the greatest ever known in New York Harbor until Superstorm Sandy's landfall in October 2012, estimated as high as 11.2 feet above normally dry land, wiping out several hundred feet of the sea wall at Battery Park in Lower Manhattan.

July 18–19, 1850
Early-Season Hurricane

Three hurricanes struck the Mid-Atlantic region in 1850, two having a considerable impact on Pennsylvania.

A heat wave in July 1850 ended with the arrival of a tropical disturbance that struck the Carolinas and Virginia on July 18. The storm blasted the Chesapeake Bay region northward through Delaware Bay during the afternoon hours. High winds and heavy rain swept over southeastern Pennsylvania and western New Jersey and continued overnight on July 18–19, 1850.

Widespread damage occurred as streams and rivers poured out of their banks and inundated the countryside. Tropical downpours affected eastern Pennsylvania, resulting in the worst flooding on the Lehigh River since January 1841. At least 20 lives were lost after boats capsized and sank in the Schuylkill River (Ludlum 1963).

A Philadelphia dispatch in the Stroudsburg *Jeffersonian Republican* (July 25, 1850) provided the chronology:

> After a week or two of hot and scorching weather, we were favored with a refreshing shower on Monday [July 15] …. Then for nearly three days every cloud dropped down more or less rain. On Thursday [July 18] we had heavy and quickly succeeding falls of heavy rain, which soaked the ground thoroughly. In the evening the wind from the South-east, and throughout the night blew a perfect hurricane, accompanied by a very heavy fall of rain.

Agricultural losses were substantial. Corn was "blown down" and "oats and grain were leveled to the ground, and the wheat and rye ungathered in much the same condition." The high winds took down trees and branches, and the torrential rainfall flooded creeks and rivers in eastern Pennsylvania. A dam break at Mauch

Chunk (now Jim Thorpe) resulted in heavy damage to the Delaware Canal.

Press accounts from throughout eastern Pennsylvania described widespread damage to trees, fences, roofs and small buildings. Rising floodwaters swept away houses and bridges and sank small craft along the Delaware and Schuylkill Rivers. The Schuylkill River "was never known higher" at Pottsville and Reading.

The impact of the summer hurricane was detailed in the *Pennsylvania Inquirer* (March 18, 1851), called a "terrible storm ... uprooting trees, wrecking vessels, and carrying away houses and bridges in every direction. It is thought more damage has been done by this storm than by any other for very many years. The freshet appeared to be universal."

A hurricane moved up the Eastern Seaboard on August 23–25, 1850, after making landfall in the Florida Panhandle on August 22, 1850, moving back out over the Atlantic off the coast of New Jersey on August 24 (Ludlum 1963). A third East Coast hurricane blew past New York City on September 7–8, 1850.

November 2–3, 1861
Expedition Hurricane

One of the latest storms to impact the Mid-Atlantic region lashed the New York City area northward to southeastern New England in early November 1861, triggering major flooding at Newark, New Jersey, and New York City. The flooding peaked on November 3, after 20 hours of torrential rain (Ludlum 1963).

The offshore hurricane destroyed two Union vessels and damaged several others east of the Outer Banks of North Carolina on the Great Expedition to South Carolina. The objective was maintaining a blockade of Port Royal Sound. The outcome of the mission, directed by sea, resulted in the eventual capture of Port Royal Sound on Hilton Head Island from the Confederacy on November 7, 1861.

October 3–4, 1869
Saxby Gale

A British naval officer, Lieutenant Stephen M. Saxby, achieved notoriety by predicting a year in advance that a coastal storm of historic proportions would occur in early October 1869, based on an unusual alignment of the Earth, sun and moon during the time when the moon was at its closest approach to Earth (Schwartz 2007).

The Royal Navy instructor's bold projection was printed in the *London Standard* in November 1868. Newspaper columnist Frederick Allison, who wrote a weather section for the *Evening Express* in Halifax, Nova Scotia, warned in a September 1869 op-ed of possible "fierce gales" and high "equinoctial tides" in a storm on October 4–5, based on Saxby's letter.

A tropical cyclone roared up the Atlantic Coast on October 3–4, 1869, that made landfall in Nova Scotia and crossed the Bay of Fundy on October 5. The hurricane collided with an upper-level trough, accounting for torrential rains that triggered floods from Virginia to Maine (Ludlum 1983), with totals in excess of six inches.

In New Jersey and eastern Pennsylvania, the average rainfall of four to eight inches sent many rivers into high flood. In the southeastern part of Pennsylvania, the Mount Joy observer reported 8 inches, forcing rivers over their banks, leading to extensive destruction. A Smithsonian observer at

Plymouth Meeting measured six inches of rain in about 24 hours, contributing to a "disastrous freshet on the Schuylkill" that caused serious damage from Norristown to Philadelphia and took several lives.

The *Philadelphia Inquirer* called it a "freshet of unprecedented violence, so great a one, in fact, that bridges are carried away, factories, dwelling houses, ice houses, &c., are submerged, boats are swamped, and the river is swollen to three times its normal size …"

From all accounts, a light rain commenced on Saturday evening, October 2, 1869, across eastern Pennsylvania, becoming a downpour by the evening. The Dyberry observer, in the northern Poconos, logged 4.50 inches in a little more than 24 hours.

On the night of October 3–4, 1869, flooding was perceived to be imminent, according to a retrospective in the *Stroudsburg Times* (February 13, 1896). Around daybreak on October 4, water breached the cribbing of a dam in South Stroudsburg, taking out a bridge on Main Street and flooding the lower portion of town.

The story continued: "The great volume of dark, angry waters soon swept away the old grist mill, with all its machinery, swirled down on the woolen mill and Wallace's sawmill, past J. O. Saylor's building, carrying ruin with its fearful march." The *Times* report added that "the current set across Main street with its tremendous power" wiping out storefronts, before being "diverted by the construction thrown up by the Methodist church parsonage."

The damage was estimated at $40,000 (Lantz 1897). Subsequent construction of flood protection in the lower portion of Stroudsburg was "strongly cribbed and filled with stone and faced with plank, at the most exposed point. This was to be followed, in after years … with a heavy stone sloped wall."

The town council believed that the low-lying area of Stroudsburg had been made "absolutely freshet-proof." The bulked-up construction would pass the next test in February 1896, reported the *Jeffersonian*, after a winter flood on February 13.

In southern New England, 12.35 inches of rain fell at Canton, Connecticut. The Saxby Gale claimed 100 lives and pushed the water in the Bay of Fundy between Nova Scotia and New Brunswick an incredible 70.9 feet past normal low tide (Schwartz 2007).

September 17, 1876 Centennial Gale

The San Felipe Hurricane blasted Puerto Rico on September 13, 1876, taking hundreds of lives, before turning north past Florida and making landfall near Wilmington, North Carolina.

The remnants of a hurricane blew into eastern Pennsylvania on September 17–18, 1876, at the end of a hot, dry centennial summer. As the storm disintegrated over the interior, wind and heavy rain pelted eastern sections of the state, resulting in considerable property damage around Philadelphia.

The *Philadelphia Inquirer* reported, "All over the city trees were blown down, and travel by horse cars was greatly interrupted. Chimneys toppled over and the roofs of houses came down without ceremony." A Milford dispatch called it a "severe northeast rainstorm," adding that this was the first significant rain since June 1876.

At New York City, the wind was "blowing fearfully" and accompanied by "torrents of rain." The Signal Service observer reported a peak wind speed of 40 mph (corrected) at New York City at 7:15 p.m. on September 18, 1876. In western Pennsylvania, closer to the track that veered across western Maryland, Pittsburgh received 4.69 inches of rain from September 16–18.

October 4, 1877
Tropical Storm Rail Disaster

A major hurricane in the Gulf of Mexico, with winds estimated as high as 115 mph, made landfall at Panama City, Florida, on October 3, 1877.

The remnant storm center curved northeast, roughly paralleling the current path of Interstate 95, before heading out to sea south of the Delmarva Peninsula. To the left of the storm track, high winds and torrential rain pelted southeastern Pennsylvania and New Jersey northward to New York City on the late evening of October 4, 1877, as the post-tropical storm raced northeastward.

The wind gusted to 80 mph (64 mph corrected) at Philadelphia. Flooding was widespread in southeastern Pennsylvania and western New Jersey, with rainfall totals as high as 10 inches (Schwartz 2007), which undermined roads and railroad tracks.

A railroad disaster along the Pickering Valley Branch Railroad (Philadelphia and Reading Railroad) was widely covered in horrific detail by the local press and recounted in the *New York Times*.

The "Wreck of '77" occurred in a ravine, triggered by a washout one mile east of Kimberton in Chester County. A large number of Henrich Pennypacker's descendants, who had gathered in Schwenksville to mark the site of General Washington's encampment during the 100th anniversary of the Battle of Germantown, departed for home in a heavy rainstorm.

The ill-fated Pickering Valley ride turned into a disaster. An embankment over a steep ravine collapsed, just as the train approached, which the engineer did not see in time. The railroad tracks were suspended some 60 feet in the air, causing the engine and several cars to plummet. The conductor traveled to nearby Phoenixville to get assistance for the mortally wounded passengers.

A report issued by the Pennsylvania Department of Internal Affairs tallied the loss of life at five passengers and three crewmen, with 17 more passengers "more or less" injured. The total number of injuries was placed at 40 (Schwartz 2007).

October 23, 1878
Gale of '78

The remnants of a Category 2 hurricane, which made landfall at St. Augustine, Florida, on September 11, 1878, headed northward along the western slopes of the Appalachians, eventually dropping 4.48 inches of rain on Pittsburgh from September 11–13.

A little over a month later, a storm that formed in the western Caribbean on October 18, 1878, crossed southeastern Florida. A second landfall occurred north of Wilmington, North Carolina, on the night of October 22.

The storm took an inland route and caused more damage in eastern Pennsylvania than any previous tropical system since the region was settled. The circulation center passed west of Richmond, Virginia, where the barometric pressure dipped to 28.78 inches. The post-tropical storm weakened as it passed west of Washington, D.C., early on October 23, 1878, before lurching into south-central Pennsylvania.

Extensive wind damage occurred in Philadelphia and vicinity on the eastern side of the storm path trajectory. A retrospective in the September 1889 *Monthly Weather Review* reported that "substantial buildings were totally destroyed or severely damaged." About 700 buildings sustained significant damage. Twenty-two vessels (steamers and ships) were sunk and 17 were damaged. The total losses were estimated around $2 million at the time. The press reported eight deaths and 75 injuries, mostly resulting from flying debris and the collapse of small buildings.

The incessant wind and rain battered Philadelphia for six hours, according to the *New York Times* account the next day:

> The tempestuous gale of unprecedented severity swept over this city early this morning, commencing about 2:30 o'clock, and reaching its height between 6 and 7:30 a.m., subsiding with a heavy rainfall toward 9 o'clock. The velocity of the gale at 2:34 was 25 miles an hour, blowing in a westerly direction. Its fury gradually increased, and from 7:25 until 7:55 a.m. it attained a velocity of 72 miles an hour. Not in many years, if ever before, has there been so much damage done to the streets, the public squares, and along the river fronts. Many of the public school buildings are damaged, and the public squares devastated. Over 40 churches are more or less damaged by the demolition of their steeples, &c., several of which are expected to fall at any moment.

The government Signal Service observations were taken every half hour between 5:00 a.m. and noon. Powerful easterly winds ("blowing in a westerly direction") shifted to the south around daybreak as the storm approached the city. At 7:40 a.m., the barometric pressure dipped to 29.12 inches, and a peak wind of 72 mph (58 mph corrected) was recorded.

Fifty churches lost their spires, and train sheds at the Pennsylvania Railroad Depot in West Philadelphia collapsed. Homes along the Delaware River were partly submerged where the water reached its highest point at noon on September 23, 1878.

A press notice from Reading stated: "The city is strewn with broken and uprooted trees, fences, and awnings. The damage to private dwellings, and business houses is very great." A report from Harrisburg noted the onset of the storm around midnight, October 23, 1878, lasting until the forenoon. The rain "descended in torrents, flooding streets, cellars, and sewers."

Major flooding occurred at the convergence of the Schuylkill and Delaware Rivers, covering an island with muddy water. Multiple shipping interests took a heavy toll, and many boats were sunk. Two lives were lost when an oyster bank went under. Heavy wind damage extended north and west, from the Lebanon, Schuylkill and Lehigh Valleys to Scranton (Roth and Cobb 2000). Trees were uprooted and homes lost roofs and partially collapsed. Upwards of 100 buildings were unroofed and 15 homes destroyed at Chester, and two persons died at Reading.

Small tornadoes were likely spawned by the weakening tropical storm, based on dispatches from the Wyoming Valley. At Wilkes-Barre, two persons died where "a tornado inflicted great damage … Houses were unroofed, trees uprooted, windows broken, and fences and mine-drilling apparatus demolished."

In Scranton, "a whirlwind with the force of a cyclone" demolished a pudding mill. About 60 trees were uprooted in Dunmore. The storm brought stiff southeasterly winds and a "drenching torrent of water," and buildings were "destroyed all along the valley." Reports of roofs being lifted high into the air and stacks blown down illustrate the force of low-level wind shear. The Stroudsburg *Jeffersonian* reported, "In this place it blew violently, but fortunately it did not do much damage," adding that this was "one of the worst storms we have had for years."

Pittsburgh received 3.24 inches in 24 hours and a storm total of 4.48 inches from September 11–13, 1878.

September 22–23, 1882
Equinoctial Storm

A tropical storm passed a short distance off the New Jersey coast on the night of September 22–23, 1882, bringing a tremendous volume of rain and high winds to southeastern Pennsylvania.

Philadelphia received its greatest rainstorm in modern history, which began on September 21, when 1.72 inches of rain fell in advance of the storm. The rain intensified, and city records showed 4.65 inches falling during the calendar day. Heavy rain continued to fall on the 23rd, adding up to another 3.72 inches, for a total of 10.09 inches.

Tropical downpours swamped the Passaic River valley in northeastern New Jersey, dropping a state-record 17.90 inches at Paterson over a three-day period. The Weather Bureau in New York City reported 10.62 inches at Central Park, the second greatest rainfall ever recorded in the metropolitan area.

The *Philadelphia Inquirer* had this to report: "The equinoctial storm ... was the heaviest on record in this region. In some places the rainfall measured thirteen inches. The floods that were caused by the rain did great damage, but the loss was lessened by the absence of high winds." Storm losses were judged to be nearly as great as in October 1878.

August 21–22, 1888
Gulf Hurricane Curves Northeast

A powerful hurricane lashed the western Gulf Coast during the night of August 19–20, 1888, before coming ashore near the Louisiana/Texas coast. The storm quickly weakened as it moved rapidly northeast, curving across the Ohio Valley on August 21. Flooding occurred in southwestern Ohio, where high winds caused sporadic damage.

The remnant storm traveled across Pennsylvania from southwest to northeast before departing on August 22, passing a little south of the Poconos. Soaking rains reached southwestern Pennsylvania around midnight, August 21, 1888, and became heavy in the eastern part of the state by daybreak.

A total of 3.57 inches of rain fell on August 21, 1888, causing the Ohio River to rise 18 feet in 24 hours at Pittsburgh on August 22, the highest level since February 6, 1884. More than five inches of rain fell west of Pittsburgh, where the flooding was severe. High water swamped low-lying sections of the city and the "Flats" section of McKeesport.

Twenty miles west of downtown Pittsburgh, the Washington County community of Burgettstown was submerged, where the Raccoon Creek overflowed. Rainfall totals across the southwestern and central mountains were impressive: Girardville 5.65 inches; Selinsgrove 5.25 inches; Johnstown 4.49 inches.

Destructive flash flooding occurred at Washington, Uniontown, Altoona, Reading and Philadelphia. Urban flooding extended to the eastern counties and Philadelphia, causing more than $100,000 in damage. Extensive flooding in Maryland claimed 12 lives.

The worst tornado outbreak in the history of the Chesapeake Bay area occurred on August 21, 1888. A destructive tornado formed north of Washington, D.C., in the southeastern quadrant of the storm. A supercell continued east past Chesapeake Bay, spawning a tornado six miles southwest of Wilmington, Delaware, which tracked into the south side of the city. The damage path was 200 yards wide and five miles in length. One person was killed and 22 were injured. Forty buildings were damaged or destroyed, and losses were estimated at $200,000, according to the *Monthly Weather Review*.

Grazulis (1993) wrote that about two dozen funnels and waterspouts were sighted in the region. Eleven fatalities occurred in Kent County, Maryland. A dispatch in the *New York Times* noted a "cyclone" at Salem, New Jersey, another likely tornado.

August and October 1893 East Coast Hurricanes

The 1893 hurricane season was active in the late summer and early autumn, with multiple strikes along the Eastern Seaboard of the United States. A series of hurricanes sideswiped the East Coast, causing varying amounts of damage from wind and high water.

On August 22, 1893, no less than four tropical storms or hurricanes were on the weather map in the western Atlantic Basin, establishing a record for tropical storm activity in the region that would not be equaled until September 24–26, 1998.

The first in a series of tropical disturbances that was tracked in the western Atlantic in late August 1893 was spotted on August 15, and passed east of Newfoundland several days later. A new tropical system soon formed in the eastern Caribbean that took a northward course, swiping the easternmost tip of Long Island on August 20–21, 1893, bringing rain and gusty winds.

On August 19, 1893, a third tropical storm located over the western Atlantic curved northward along the Eastern Seaboard. Wind damage was widespread along coastal sections of New Jersey and New York City during the night of August 23–24, 1893. The ocean tug *Panther* sank off the coast of New Jersey, taking the lives of 16 crewmen, one of many vessels damaged in the storm.

The U.S. Weather Bureau Daily Weather Map showed the hurricane straddling New York City at 6:00 a.m. on August 24, 1893, making landfall on western Long Island, where the lowest pressure (29.23 inches) was recorded. The rain gauge at Central Park in New York City caught 3.82 inches of water between 8:00 p.m. on the 23rd and 8:00 a.m. the next morning.

A storm surge pounded Brooklyn and Queens, causing widespread damage to homes and businesses and tearing up railroad lines. Hundreds of trees were felled by high winds in Central Park, and street flooding was reported in Manhattan. High winds unroofed a number of homes and buildings, particularly in Brooklyn and Queens.

The fourth tropical system to affect the East Coast in late August 1893, first plotted on August 21, turned out to be the most destructive. A full-fledged hurricane pounded the South Carolina coast on August 27–28, 1893, where an estimated 1,000 people drowned in the storm surge and tidal flooding. The remnants of the storm traveled northward across Pennsylvania on the 28th, reaching Lake Ontario before curving northeast to the Maine–Quebec border the next day.

The 1893 hurricane season remained active deep into the autumn. A mighty hurricane plowed into Louisiana on October 1, 1893, causing horrific flooding that took an estimated 2,000 lives before exiting North Carolina on the 4th.

A second October storm formed over the Central Atlantic on October 5, navigating a slow westward track that brought the water-logged storm ashore in South Carolina on October 13, 1893. The disturbance turned northward along the eastern edge of the Appalachians, reaching the mountains of western Maryland in the evening before passing over western Pennsylvania early on the 14th.

High winds blasted areas from Pittsburgh to Philadelphia, with gusts approaching 50 mph, toppling trees and telegraph lines, unroofing homes and disrupting railroad traffic. The storm traveled all the way to Ontario. (Sixty years later, a powerful mid-October hurricane named Hazel would take a similar path from North Carolina through central Pennsylvania to southeastern Ontario, Canada.)

Yet another tropical disturbance affected Pennsylvania in October 1893. A hurricane passed the Bahamas on October 21, 1893, and made landfall over North Carolina on the 22nd. The storm remnants reached south-central Pennsylvania on October 23, 1893, accompanied by wind and rain but resulted in little damage.

October 10, 1894
Autumn Hurricane

Hurricane-force winds lashed the Mid-Atlantic Coast on October 10, 1894, after a storm made landfall in the Florida Panhandle a day earlier, with 120 mph winds and torrential rain.

The tropical system moved up the Eastern Seaboard, paralleling the modern-day I-95 corridor, passing over southwestern New Jersey and New York City. Powerful wind gusts blasted eastern Pennsylvania, New Jersey and New York City. Wilkes-Barre received 4.02 inches of rain on October 10, 1894.

September 29–30, 1896
Another Gulf Hurricane

A powerful Category 4 hurricane charged ashore along the northwestern coast of Florida near Cedar Key on September 28, 1896.

A swath of heavy rain and high winds followed a northward trek across eastern Georgia. Extensive wind damage occurred as the ex-hurricane passed a little west of Washington, D.C., on September 29, 1896, where a barometric pressure of 29.14 inches was recorded.

Trees were toppled, homes damaged and barns destroyed by wind gusts approaching hurricane force in southern Pennsylvania. The wind at Philadelphia gusted to 80 mph (64 mph corrected). A five-minute average velocity of 72 mph (58 mph corrected) was registered at Harrisburg. Damage from houses and buildings became airborne, and there were several reports of train wrecks.

The observer in York issued the following remarks on the storm in his summary: "Terrible hurricane with cyclonic conditions at midnight between the 29th and 30th. Damage in York County estimated at $200,000, probably more. Greatest storm in the history of [York] county." The mile-long span of the Pennsylvania Railroad

Bridge over the Susquehanna River at Columbia in Lancaster County was destroyed by high winds.

Local press accounts in southeastern Pennsylvania reported several hours of intense southeasterly gales on the night of September 29–30, 1896, that caused widespread major damage from Lancaster to Schuylkill County. Locations west of the storm track were on the receiving end of torrential rain that triggered serious flooding. The heaviest rainfall occurred at Hollidaysburg (4.42 inches).

In *Climate and Crops: Pennsylvania Section*, the Weather Bureau estimated the damages in the state to be about $2 million. The national death toll (114, including 30 in the Mid-Atlantic region) and total losses ($7 million) reflected the severity of the storm (Schwartz 2007).

October 8–10, 1903
Tropical Storm Floods

Two tropical systems impacted the Mid-Atlantic region in the autumn of 1903.

A minimal hurricane, called the "Freakish Vagabond Hurricane" in the *Atlantic City Daily Press*, scored a direct hit on South Jersey from the southeast on September 16, 1903—the only hurricane to make first landfall in New Jersey in the 20th century.

Strong winds lashed the coast from Virginia northward to Long Island, approaching hurricane-force. Wind damage was heaviest at Ocean City. The storm veered inland after blasting southeastern New Jersey, accompanied by heavy downpours totaling more than four inches. Eastern Pennsylvania experienced strong winds on the fringe of the storm as it moved across southern New Jersey.

The *Philadelphia Inquirer* reported: "With the first sweep of the wind, borne with driving rain at a velocity of 80 mph [64 mph corrected], telegraph poles and wires were blown down and communication with the city from almost all points was cut off." Damage was estimated in the thousands of dollars.

Only three weeks later, another tropical storm brushing the Atlantic Coast would leave a lasting impression on the residents of eastern Pennsylvania and northern New Jersey in the chilly, wet autumn of 1903. A light rainfall that began on October 7, 1903, turned into a deluge over the next 24 hours. The upper Delaware Valley received four to nine inches of rain, triggering the worst flooding ever experienced in the region at the time.

Extensive flooding occurred along the Passaic River in northern New Jersey. Paterson collected a record rainfall of 15.51 inches (11.45 inches on the 9th). Total damage in the region exceeded $7 million (Ludlum 1983).

September 14–15, 1904
Inland Tropical Storm

A hurricane made landfall near Myrtle Beach, South Carolina, on September 14, 1904. Farther north, a wind gust of 100 mph was reported at Lewes, Delaware (Schwartz 2007).

The storm chugged up the Eastern Seaboard as a big rainmaker. Intense rainfalls occurred over New Jersey and southeastern Pennsylvania. A U.S. Corps of Engineers report in 1945 (*Storm Total Rainfall in The United States*), published by the War Department (NA 1-9), listed a maximum rainfall of 10 inches at Friesburg, New Jersey. Heavy wind damage and loss of life occurred in New England as the post-tropical storm continued northward.

The Philadelphia Weather Bureau recorded 6.47 inches of rain September 13–15, 1904, and 2.05 inches of rain fell in 40 minutes in a tropical downpour on the 14th. The city's Centennial Avenue site measured 5.10 inches in the 24-hour observation period on September 14–15. A peak gust of 62 mph (51 mph corrected) was observed early on the 15th.

September 19, 1928
Lake Okeechobee Hurricane

The late summer of 1928 brought several former hurricanes northward over the interior eastern states.

The first storm made landfall as a Category 1 hurricane near Fort Pierce, Florida, on August 8, 1928. Five days later, the system passed out to sea northeast of Richmond, Virginia. Torrential rains pounded the region around Chesapeake Bay, depositing 7.31 inches at Washington, D.C., and 12.76 inches in 30 hours at Cheltenham, Maryland.

On August 14, 1928, a former Caribbean hurricane made landfall along the west coast of Florida as a tropical storm. The remnants of the storm crossed the Florida Panhandle, dissipating over southwestern Pennsylvania on August 16–17. Heavy rain totaled as much as 5.30 inches at Coatesville.

The fourth hurricane of the 1928 season struck a massive blow on the island of Guadeloupe on September 12, 1928, one week after forming over the eastern Atlantic Ocean, killing more than 1,200. The storm roared onto the coast of Puerto Rico just after dawn on September 13, packing winds in excess of 155 mph—the only Category 5 hurricane to make a direct strike on the island. The storm caused catastrophic damage and claimed 300 lives.

The historic hurricane became known as the "Second San Felipe Hurricane," named for the Saint's Day visitation. The hurricane continued on a northwesterly course past Hispaniola and the Bahamas before coming ashore over central Florida near West Palm Beach as a Category 4 hurricane on the evening of September 16, 1928. Sustained winds were reportedly 148 mph, reflected by the central pressure of 27.43 inches.

Extreme wind gusts generated a 10-foot storm surge that caused the most damage around Palm Beach. Hurricane-force winds churned up Lake Okeechobee, and southerly winds pushed the lake waters over a small dike, sending floodwaters in all directions that swamped hundreds of square miles.

Water up to 20 feet deep caused homes to float, slamming into debris and other structures, causing at least 1,836 deaths and likely more than 2,000 in total—the second worst loss of life from a storm in the United States, after the Great Galveston Hurricane of September 1900. The total death toll from the hurricane, including its passage through the Caribbean Islands, was estimated then at 3,336, but probably exceeded 4,000 lives.

The storm crossed central Florida near Orlando, before turning north. The circulation center traveled through eastern Georgia, depositing 11.44 inches of rain at Savannah on September 16–17, 1928. The remains of the hurricane reached eastern Pennsylvania early on September 19 and continued into southeastern Canada. Although the system was a shell of its former self, high winds flattened crops in eastern Pennsylvania. Rainfall totals were relatively light, and no flooding was reported.

In the United States, total damage was listed at $100 million.

August 22–23, 1933
Chesapeake and Potomac Hurricane

The North Atlantic hurricane season of 1933 was the most active in the 20th century, spawning 21 named storms. The most infamous storm of the season developed on August 17 in the western Atlantic.

The tropical disturbance intensified into a Category 3 hurricane, packing winds of up to 113 mph shortly before slamming into Nags Head, North Carolina, during the early morning hours on August 23, 1933. The central pressure dipped to 27.98 inches as the center made landfall. The eye of the storm passed 30 miles west of Norfolk, Virginia, tracking over Washington, D.C., and then north through central Pennsylvania.

Winds gusted to 88 mph at Norfolk Naval Air Station. Extensive tidal flooding extended northward to the western shore of Chesapeake Bay. Damage estimates were as high as $11 million in Virginia, where 15 people died in the storm surge as waves five to eight feet high exceeded all known previous storms at Newport News. High winds battered coastal sections of Maryland and Delaware, swamping low-lying areas.

At high tide, the Philadelphia airport was inundated in a 10-square-mile area on the southwest side of the city. A wind gust at 88 mph was recorded at Wildwood, New Jersey. Wind gusts reached 60 mph over Long Island, knocking down trees and overturning small buildings.

Widespread flooding enveloped northeastern New Jersey. Shortly before midnight on August 23, 1933, the weakening storm was over central Pennsylvania, where flooding followed a second day of heavy rainfall.

The report in *Climatological Data: Pennsylvania Section* contained a concise summary:

The damage was estimated as high as 50 percent in some sections, and seems to have averaged nearly thirty percent for the eastern half of the State. This loss was partly balanced by the improvement that moderates rains caused at most places in the western half of the State. The storm covered practically all of the tobacco growing regions, and nearly ruined the crop.

Widespread downpours resulted in excessive, record-breaking rainfall in parts of southeastern Pennsylvania. The heaviest storm total was 13.82 inches at York over a three-day period from August 22–24, 1933, with a 24-hour total of 8.48 inches (August 22–23). The monthly rainfall at York of 17.70 inches set a state record at the time.

Codorus Creek crested 15 inches higher than modern records, sending 3,000 residents fleeing for higher ground (Schwartz 2007). A total of 47 bridges were destroyed in York County, and floodwaters reached to the roof of some homes.

Flooding was widespread along the Delaware, Schuylkill and Lehigh Rivers. Rain totals reached 7.63 inches at West Chester, 6.79 inches at Allentown, 6.58 inches at Harrisburg, and 5.66 inches at Philadelphia.

The Schuylkill River at Reading crested at 19.7 feet on August 24, 1933, causing extensive flooding on the south side of the city. A conservative estimate put the damage in southeastern Pennsylvania in the neighborhood of several million dollars. Total damage to Pennsylvania highways and bridges was estimated by engineers to be about $800,000.

The *Climatological Data: Pennsylvania Section* reported: "The crop damage undoubtedly exceeded the total flood damage by a wide margin," calling the storm "the most destructive that Pennsylvania has experienced during the 45 years of the State Weather Service."

September 21, 1938
Great New England Hurricane

It would be unthinkable today what a Category 3 hurricane (111 mph or higher) would do if it struck parts of the Northeast with virtually no warning. That's what happened in September 1938.

A storm that formed off the coast of Africa on September 9, 1938, reached Category 5 intensity, and also moved at an excessive forward speed of 47 mph, before plowing into Long Island around 2:30 p.m. on September 21 as a Category 3 storm (120 mph winds at landfall).

The Great New England Hurricane, later dubbed the "Long Island Express," had a second landfall between Bridgeport and New Haven, Connecticut. The eventual path took the hurricane through far western Massachusetts and Vermont and into Quebec, Canada. A peak gust of 186 mph occurred at Blue Hill Observatory west of Boston.

Pennsylvania and New Jersey were on the far western fringe of the storm, but coastal communities caught winds of 70 mph, with gusts up to 80 mph at Battery Park in Lower Manhattan. A considerable portion of the boardwalk at Atlantic City was wrecked.

Exceptionally high tides were observed at the autumnal equinox that occurred at the full moon. Storm tides greater than 14 feet caused extensive damage across a good portion of Long Island, extending northward to coastal Connecticut, and reached 18 to 25 feet as far north as Cape Cod, Massachusetts. Providence, Rhode Island, came under as much as 13 feet of water in the downtown area, with a storm surge of 15.8 feet above normal spring tides at Narragansett Bay.

Eastern Long Island and southeastern New England suffered devastating blows from the storm. Hurricane-force winds and storm surge contributed to deadly flooding throughout the region that took 682 lives. At the time, losses totaled at least $306 million, including nearly 19,000 homes destroyed and 26,000 vehicles damaged beyond repair.

The infusion of tropical moisture contributed to some hefty rain totals in eastern Pennsylvania on the closing days of summer. Doylestown recorded 6.52 inches on September 20–21 and 7.40 inches during September 19–23, 1938. Other hefty six-day totals were tallied at Allentown (6.09 inches) and Stroudsburg (6.30 inches), where the observer logged 5.00 inches in 24 hours on September 20–21.

September 13–14, 1944
Great Atlantic Hurricane

Six years after the Great New England Hurricane swept up the coast, another monster hurricane took a slightly more seaward course over eastern Long Island.

The Great Atlantic Hurricane, packing 140 mph winds, struck the Eastern Seaboard on September 13–14, 1944, accompanied by an assault of heavy rain and very high winds. New Jersey state climatologist A. E. White wrote, in *Climatological Data: New Jersey Section*, that the hurricane was the most destructive to hit the Garden State since the Civil War. The wind damage was confined mostly to the coastal plain.

The western fringe of the storm brought wind gusts at Philadelphia of 60 mph. The storm caused minor flooding in parts of eastern Pennsylvania, where as much as four inches of rain fell on saturated ground.

At Atlantic City, New Jersey, a peak gust was clocked at 82 mph as the storm passed about 50 miles offshore in the late evening on September 14, 1944. Sustained winds at New York City reached 81 mph, with a gust at 99 mph recorded as the storm crossed eastern Long Island. Rahway, New Jersey, received 11.40 inches of rain.

The Great Atlantic Hurricane made landfall near Port Judith, Rhode Island, with a central barometric pressure of 28.34 inches. The death toll of 390 included 248 sailors who died when a Navy destroyer, the U.S.S. *Warrington*, foundered about 450 miles east of Vero Beach, Florida. Total damage exceeded $100 million. There were 46 deaths reported on the mainland.

A number of hurricanes took aim on the Mid-Atlantic Coast between 1945 and 1953 in a particularly active cycle, though none had any significant impact on Pennsylvania. Tropical storms passed just east of Philadelphia in September 1945 and August 1949.

On August 31, 1952, Hurricane Able buffeted South Carolina, then headed northeast parallel to the Interstate 81 corridor. A 63-mph wind gust occurred at Washington, D.C. At 1:30 p.m. on September 1, 1952, the remnant storm center was plotted southeast of Reading. High winds (40–50 mph) caused minor damage in eastern Pennsylvania.

Four to seven inches of rain fell in an area from Franklin County northward to Union County and eastward to Pike County from August 31–September 2, 1952.

October 15, 1954
Hurricane Hazel Blasts Pennsylvania

The 1954 hurricane season brought a cluster of storms that lashed the East Coast—Carol, Edna and Hazel. The total storm damage at the time in the United States reached $780 million.

Hurricane Carol whirled past New Jersey on the morning of August 31, 1954, before charging across eastern Long Island. Wind gusts reached 135 mph at Block Island, Rhode Island. Coastal Connecticut was hit with 110 mph gusts. Storm damage totaled $462 million. In the region, 65 people died and more than 1,000 were injured. Losses included 4,000 homes and 3,500 cars.

On September 10, 1954, Hurricane Edna nipped at the Outer Banks of North Carolina, later coming ashore over Martha's Vineyard, Massachusetts, bringing peak wind gusts of 120 mph, and 100 mph at Block Island.

The third major hurricane to hit the East Coast in 1954 arrived rather late—in the middle of October. Hurricane Hazel developed into a dangerous Category 4 hurricane over the southeastern Caribbean Sea on October 5, 1954, blasting Hispaniola on October 12. Ninety-eight people died in western Haiti, where winds gusted to 100 mph.

After wreaking havoc in the islands, Hazel took aim on the Eastern Seaboard, striking the North Carolina border shortly after daybreak on October 15, 1954, with sustained winds of 130 mph and a central barometric pressure of 27.70 inches—the only Category 4 storm to ever strike the North Carolina coast.

Hazel's accelerating forward speed of 45 mph augmented the force of the winds that buffeted the Mid-Atlantic shoreline farther north, with an unofficial gust of 130 mph at Hampton, Virginia. Energy derived from the temperature difference translated into exceptionally strong winds far inland, even as the center weakened over eastern Virginia.

Swift upper-level steering currents associated with low pressure aloft and a cold front dropping south from the Great Lakes caused Hazel to accelerate along a frontal boundary (Halverson 2015). Hazel crossed into south-central Pennsylvania around 9:00 p.m. on October 15, 1954, knocking down one billion bushels of fruit in eastern and central Pennsylvania. Within an hour, Hazel was centered north of Coudersport, where the storm joined up with the cold front.

In *Climatological Data—Pennsylvania*, the extent of the damage was described in detail:

> Wind damage was apparently most general and severe in the eastern half of Pennsylvania. Peak gusts shaded from 94 miles per hour at Philadelphia to 86 miles per hour at Reading to 80 miles per hour at Harrisburg to 58 miles per hour at Philipsburg. Major damage was spotty, but more or less general. Rail traffic was delayed or halted by downed trees, wires and poles. Many from highways were blocked by fallen trees and downed wires. Ships and small craft were torn loose their moorings in the Delaware River.

Schwartz (2007) quoted observer James Weber in Lancaster County: "The worst hurricane ever to hit Lancaster County. The center moved up the Susquehanna River Valley around 7 PM Fri, Oct. 15. Between 6 and 8 PM sustained winds of 60 to 70 mph were recorded." Weber noted a gust of 85 mph and a barometer reading of 28.98 inches. Peak gusts farther north reached 82 mph as far as the Lehigh Valley at Allentown.

The U.S. Weather Bureau office in Philadelphia reported 11 deaths, and two people were missing in the aftermath of the storm. The total amount of losses at Philadelphia was listed as $1.88 million, and up to $2 million in the east-central portion of Pennsylvania, according to newspaper accounts. An additional $1 million in damages occurred in surrounding sections of eastern Pennsylvania.

The rainfall at Pittsburgh on October 15, 1954, totaled 3.56 inches. At Uniontown, a total of 4.60 inches of rain sent Coal Lick Run and Redstone Creek out of their banks. Up to seven inches of rain fell in Somerset County, and five persons died in the floods that followed (Schwartz 2007).

At the Pittsburgh Weather Bureau, the local meteorologist-in-charge, Jacob T. B. Beard, said 13 people perished in the rising waters, 29 others sustained injuries, and 2,929 houses were either damaged or destroyed. Major flooding occurred in Allegheny County along "Turtle Creek, where high waters invaded stores, homes, churches." The East Pittsburgh Westinghouse Plant flooded.

The Monongahela and Youghiogheny Rivers went into flood, and "hundreds of persons were reported left homeless due to the high water" in the aftermath. A reservoir north of Pittsburgh at Derry overflowed, cutting off the water supply. The confluence of the Allegheny, Monongahela and Ohio Rivers revealed where waters crested at 7.4 feet above flood stage on October 16, 1954. Low-lying sections of downtown Pittsburgh experienced flooding, filling basements and covering the Baltimore and Ohio Railroad Yard with 10 feet of water.

In West Newton, five feet of water covered streets as the Youghiogheny River reached a level two feet above the March 1936 flood level, judged to be the worst flooding ever known in the borough. Many communities along the river were under water. Newspaper reports estimated flood damage in western Pennsylvania to be upwards of $15 million. The Ohio River exceeded flood stage in every community from Pittsburgh to the West Virginia Panhandle.

Hazel's death toll in the United States (95) and damage ($281 million) was among the greatest in Mid-Atlantic history. The remnants of Hazel continued northwestward with unusual ferocity all the way into Ontario, Canada, early on October 16, 1954.

Torrential downpours and flooding in Ontario were deemed the worst weather disaster in Canadian history. The storm brought a record low barometric pressure of 28.96 inches, and thunderstorm winds gusted to 100 mph at Toronto. Flash flooding took the lives of 81 people in Canada, with losses estimated around $135 million.

August 1955
Hurricanes Connie and Diane

The passage of two hurricanes in rapid succession did not give the waterlogged ground enough time to fully absorb the copious rainfall, when remnants of Hurricane Diane arrived less than a week after Connie's deluge, which resulted in historic flooding and loss of life in eastern Pennsylvania.

The staggering loss of life and destruction that occurred northward to southeastern New England led to calls for improved radar tracking systems.

Hurricane Connie slammed into North Carolina during the midday hours on August 12, 1955, with a wind gust of 83 mph. The remnants of Connie turned north along the Appalachian foothills, reaching south-central Pennsylvania on the late morning of August 13. By the end of the day, what was left of Connie was centered near Erie at 7:30 p.m.

The most forceful winds and heaviest rainfall affected southeastern Pennsylvania, on the southeastern flank of the circulation center. A peak wind gust of 67 mph was recorded at Philadelphia, knocking down trees and power lines. A large fair tent was blown over in York County. Extensive damage occurred to fruit trees, as the storm chugged northwest into Erie County.

Ex-hurricane Connie became a prolific rainmaker, sowing the seeds for the terrible floods that would be wrought by the remains of Diane the following week. The rainfall from Connie exceeded nine inches in parts of Monroe, Montgomery and Chester counties, as the storm combined with low pressure that stretched the rain event from August 10–14, 1955.

Along the coast, New York City's LaGuardia Airport was swamped with an estimated 12.20 inches of rain that caused heavy flooding in New York City and Long Island. A number of creeks in eastern Pennsylvania overflowed, but the main rivers stayed within their banks, because water levels had been low in a prolonged drought.

Even as Connie disintegrated over Ontario, Canada, forecasters were concerned about Diane, which eventually made landfall on the morning of August 17, 1955, near Wilmington, North Carolina, with sustained winds of 85 mph and gusts in excess of 100 mph at coastal locations.

The general consensus was the storm would soon turn northeast and head out to sea to the south of Pennsylvania. Instead, the remnants of Diane took a more westerly path, veering across Virginia and West Virginia. Hurricane forecasters handed off the storm to local forecasters after the final 11:00 p.m. bulletin from the Weather Bureau in Washington, D.C.

The surprise came when the remnants of Diane lurched into southern Pennsylvania shortly before midnight on August 18, 1955, interacting with an area of low pressure. During a period of 30 hours, eastern Pennsylvania was pelted with torrential downpours that could not be absorbed by sodden soil. The runoff raced down waterlogged hills and swelled mountain tributaries, sending creeks past previous record flood stages.

More than 10 inches of rain fell near the junction of Schuylkill, Carbon and Luzerne counties. Scranton experienced the city's worst flooding in modern history. The Lackawanna River overflowed in southern Wayne, Lackawanna and Luzerne counties. In the Poconos, six to 13 inches fell in 24 hours.

Fast-moving torrents of water overtopped the riverbanks, culminating in a wall of water 30 feet high tearing through the lower reaches of Brodhead Creek in southeastern Monroe County. An area expanding from southern Wayne and Pike counties and across Monroe County was hardest hit, where at least 89 deaths directly attributed to floodwaters on the night of August 17–18, 1955, were counted—76 in Monroe County.

The rampaging Brodhead Creek reached a historic crest of 27 feet early on August 18 at Minisink Hills—17 feet above flood stage—with the passage of the surging waters that inundated the lower portions of the twin boroughs of Stroudsburg and East Stroudsburg.

Extensive flooding occurred along the length of the Lehigh River and the Schuylkill River, all the way to Philadelphia. The swollen Delaware surged to record levels, observed farther south at Pennsylvania–New Jersey stream gauges in Easton and Riegelsville, New Jersey.

The death toll from flooding in eastern Pennsylvania was listed as 90 in *Climatological Data—Pennsylvania*, with 10 people missing and likely drowned. In addition, five casualties were indirect deaths listed by various press sources during the recovery effort. An estimated total of 150 bridges (rail and road) and 30 dams were destroyed, as widespread losses were tallied up to $70 million in the months after the calamitous storms.

Extensive flooding extended into northwestern New Jersey, where hundreds of residents were forced to evacuate. The remnants of Diane passed over the length of Long Island, swamping southeastern New England, causing disastrous flooding.

In the Berkshires of western Massachusetts, Westfield received 18.15 inches of rain in 24 hours. The storm total of 18.76 inches set a state monthly rainfall record (26.85 inches). Hartford, Connecticut, recorded 12.50 inches in the storm.

Flooding in southern New England killed 91 persons. Diane claimed more than 200 lives and displaced 35,000 families in the Northeast and Mid-Atlantic region, causing $832 million in damage, revised upwards in later reports to nearly $1 billion.

June 29, 1957
Hurricane Audrey

Hurricane Audrey struck the coast of southwestern Louisiana on June 27, 1957, with 125 mph winds, before turning northeast and lumbering through northern Mississippi, Tennessee and Kentucky as a post-tropical storm, with heavy wind and rain.

Audrey's rapid intensification in the warm waters of the Gulf made it a particularly deadly storm, taking at least 416 lives in the United States, mostly in Cameron Parish, Louisiana.

The remnant storm hit areas west of the Appalachians hard, traveling along the upper Ohio River valley and across far western Pennsylvania and southwestern New York. The storm spun off 22 tornadoes, a record at that time for a remnant system, including one that damaged a few buildings at Elkins, West Virginia (Schwartz 2007). The post-tropical low became enmeshed with a wave on a cold front, strengthening as the low moved across northwestern Pennsylvania.

Lines of strong storms strafed Pennsylvania, blowing down a drive-in movie screen at Beaver Falls, and adding wind energy to a squall line that brought a wind gust of 75 mph at Allentown. A wind gust reaching nearly 100 mph was reported at Jamestown, New York, as the circulation center tracked north over Lake Ontario into Canada.

September 12, 1960
Hurricane Donna

Hurricane Donna pounded the Florida Keys on September 10, 1960, with 180 mph winds, bringing a storm surge of 13 feet. Donna continued north along the east coast of Florida before turning up the Eastern Seaboard.

Wind gusts reached 121 mph at Charleston, South Carolina, 93 mph at LaGuardia Airport in New York City and 138 mph at Blue Hill Observatory, Massachusetts, on September 12, 1960, attesting to the ferocity of the storm, which maintained its integrity up the coast. The anemometer at Philadelphia clocked a wind gust of 59 mph, and the rainfall totaled 5.62 inches at Marcus Hook. Precipitation figures in New Jersey ranged as high as 8.50 inches at Seabrook.

Hurricane Donna packed 100 mph winds at landfall on September 12 on Long Island, with a very wide eye more than 90 miles in diameter. Storm damage was estimated at $300 million. The center of the storm passed east of Bangor, Maine. Fifty lives were lost in the United States.

August 27–28, 1971
Tropical Storm Doria

Hurricane Doria was downgraded to a tropical storm by the time the storm made landfall near Kennedy International Airport on the evening of August 27, 1971. The wind gusted to 80 mph along the coast as hard rains pounded New Jersey and eastern Pennsylvania all day long and into the night on August 27–28. Streets and subways filled with water and became flooded around New York City early on the 28th from the storm that traveled along the coast. Damage was heaviest in New Jersey ($138 million), primarily due to flooding of the central and northeastern sections of the Garden State, which extended into southeastern Pennsylvania.

The greatest 24-hour rainfall was 11.42 inches at Neshaminy Falls in Montgomery County. Tropical Storm Doria triggered Philadelphia's second heaviest August rainstorm (5.68 inches) which nearly equaled the August 3, 1898, intensity record (5.98 inches). Farther north, 4.19 inches of rain fell at Stroudsburg.

June 20–25, 1972
Tropical Storm Agnes

An early-season hurricane born in the warm Gulf waters on June 16, 1972, east of the Yucatan Peninsula will be forever remembered for the catastrophic flooding and loss of life (122) from Virginia to New York. Hurricane Agnes caused $4 billion in damage—a record for a tropical system in the United States up to that time.

Agnes struck Cuba before reaching the Gulf Coast. The storm crashed ashore near Valparaiso, Florida, early on June 19, 1972. Agnes weakened over Georgia and the Carolinas but gained a second wind as the circulation center emerged over the Virginia Capes and interacted with several features approaching from the west.

Upper-level low pressure captured what remained of Agnes, leading to a very slow movement parallel to the Atlantic Coast. Orographic lift contributed to intense cloudbursts over the Northern Appalachians from southern New York to northern Virginia along a stalled frontal boundary. The northward progression of the storm was blocked by an area of high pressure off the coast of New England, setting the stage for catastrophic flooding in the Middle Atlantic and Northeastern states.

The remnants of Agnes, now a weak post-tropical low, veered northwest and performed a loop from near New York City across southern New York and the mountains of northeastern Pennsylvania. Drenching rains exceeded one foot in the Susquehanna Valley. A rain gauge monitored by the United States Department of Agriculture (USDA) in far western Schuylkill County recorded a spectacular 14.50 inches in 24 hours on June 21–22, 1972. NOAA *Storm Data* estimated that the total probably exceeded 19 inches.

Much of the region between the central mountains and the eastern highlands received six to nine inches of rain, heaviest in an area a little north of Williamsport south to Harrisburg. Raging waters poured into Wilkes-Barre, which was devastated by historic flooding, as the Susquehanna crested at 40.91 feet on Saturday, June 24, 1972, reaching nearly 19 feet above flood stage—a record until September 2011 (42.66 feet).

The state capital of Harrisburg was engulfed by floodwaters that swamped about half of the city, where the Susquehanna River crested at an all-time record stage of 32.57 feet—3.34 feet higher than the previous record flood stage attained in March 1936. The streamflow at peak discharge was measured at one million cubic feet per second. During all of June 1972, 18.55 inches of rain fell at Harrisburg—an all-time record for any month.

In Pennsylvania, a total of 68,000 homes and more than 3,000 businesses were damaged or destroyed by water, and fire in some instances, and 220,000 residents were left homeless. Storm-related deaths were higher in Pennsylvania (50) than any other state, followed by New York (25). Total damage in Pennsylvania was put at $2.1 billion, which would equal about $14 billion today.

The Flood Plain Management Act enacted in 1978 focused on state floodplain regulations.

September 23–26, 1975
Hurricane Eloise

Hurricane Eloise lashed Puerto Rico on September 17, 1975, before moving on to blast Haiti and the Dominican Republic. Taking a westward track, the storm slammed Mexico's Yucatan Peninsula and then veered northwest into the Gulf of Mexico.

Just before daybreak on September 23, 1975, Eloise made landfall near Fort Walton Beach, Florida, as a Category 3 hurricane accompanied by 130 mph winds. The remnants of Eloise promptly turned north, traveling along the western flank of the Appalachians.

Although the storm was downgraded to an extratropical system later on September 24, 1975, potent low pressure crossed the mountains of West Virginia two days later and triggered copious downpours from the nation's capital to central Pennsylvania. On September 25–26, 1975, nearly 20,000 Pennsylvanians were evacuated from their homes.

The moisture-laden tropical storm unleashed seven to 14 inches of rain between September 23 and 26 from Virginia to Pennsylvania, causing widespread major flooding for the second time in four years.

The highest rainfall totals in eastern Pennsylvania included 7.75 inches at Gettysburg and 6.34 inches at Philadelphia, but areas to the west received up to a foot of rain in the Susquehanna and Juniata River basins. Seven persons were killed by flooding from Eloise in the Susquehanna region, and the total damage in the state was estimated in excess of $150 million.

September 5–6, 1979
Hurricane David

Hurricane David formed west of Africa on August 28, 1979, then traveled west on the wings of the easterly trade winds through the islands of Dominica, Guadeloupe, Martinique and the Bahamas. Sustained winds reached 150 mph as David pushed on to the northwest, into the Dominican Republic on August 31. By then, the staggering toll of 2,052 persons in the Caribbean attested to the storm's deadly ferocity.

A weakening David made landfall over southeastern Florida on September 3, 1979, before moving back over the ocean. A rare second landfall occurred along the South Carolina coast the next day. On September 5–6, 1979, the storm turned inland and headed north through Virginia and Maryland into central Pennsylvania, passing near Harrisburg, accompanied by tropical downpours and strong winds.

The storm rainfall in northeastern Pennsylvania added up to 4.22 inches at Stroudsburg and 5.24 inches at Tobyhanna. The heaviest rainfall was 6.32 inches at South Mountain, near the Maryland border. Flash flooding and power outages were widespread in the east.

The remnants of David spawned a tornado in Oley Township that destroyed several buildings, and another twister killed one person and injured four in a mobile home park at Avondale in New Garden Township. Four tornadoes touched down in Pennsylvania, and tornadoes were also spotted in New Jersey and Delaware. A total of 34 tornadoes spawned by David in the Middle Atlantic region established a record for a tropical system at this latitude.

One week later, Hurricane Frederic plowed into the west side of Mobile Bay, Alabama, shortly after daybreak on September 12, 1979, accompanied by wind gusts as high as 145 mph. The remains of Frederic journeyed north, pelting eastern Ohio and western Pennsylvania with more than five inches of rain. The storm passed over northwestern Pennsylvania on September 14 and continued into Canada, causing $2.3 billion in damage.

During the course of the 1979 season, a record four tropical systems crossed Pennsylvania, including the remains of Hurricane Bob and Tropical Storm Claudette in July.

September 26–27, 1985
Hurricane Gloria

Hurricane Gloria raced up the Eastern Seaboard, with sustained winds of 130 mph on September 26, 1985, grazing the Outer Banks of North Carolina.

Gloria sped northward, threatening the New York City area with the prospect of being struck by a major hurricane. The storm made landfall over Jones Inlet in central Long Island during the forenoon on September 27, 1985, though fortunately at low tide, sparing the metropolitan area a more destructive scenario. Sustained winds diminished to 90 mph, sufficiently strong to uproot thousands of trees and cause considerable property damage, leaving 1.5 million without power.

Eastern Pennsylvania was on the left side of the storm track and subjected to tropical downpours. Several cities established new record 24-hour rainfall totals at the time: Allentown 7.85 inches; Stroudsburg 6.59 inches; Scranton 6.52 inches. In the western Poconos, 9.11 inches fell at Long Pond and 7.83 inches at Freeland.

September 21–22, 1989
Hurricane Hugo's Inland Trajectory

Hurricane Hugo developed near the Cape Verde Islands on September 9, 1989, reaching Category 5 intensity along its trek across the Atlantic.

After passing over Puerto Rico as a Category 3 hurricane, the storm weakened over the Atlantic, only to strengthen again to a Category 4 storm, crashing into Charleston Harbor on September 21, 1989.

The storm took an unusual inland trajectory, racing northwest and then turning north on the opposite side of the Appalachians. Six hours after landfall, the tropical storm reached Charlotte, North Carolina, and then curved north through Virginia to Parkersburg, West Virginia, eventually weakening near Erie on September 22, 1989. Western Pennsylvania caught gusty winds and received several inches of rain.

September 5–6, 1996
Hurricane Fran Goes Inland

Hurricane Fran lashed the North Carolina coast near Cape Fear on the evening of September 5, 1996, as a Category 3 storm, with 115 mph winds. Fran took a route in the same fashion as Hugo.

The remnants of Fran moved northwest into the Appalachians as a prolific rainstorm, cutting through western Pennsylvania and hugging the Ohio border. The post-tropical system transited eastern Lake Erie and Ontario, Canada.

Torrential rain broke out across southwestern Pennsylvania. Downpours in the southern tier of of the state dropped 9.80 inches at Newell, 5.80 inches at Biglerville, and 4.38 inches at Gettysburg.

Flash flooding impacted more than a dozen counties in the region, as the Juniata River went out of its banks, resulting in two deaths.

One young woman died in her flooded vehicle in Newville, and a resident drowned outside her home in Perry County. The storm's residual circulation spawned a tornado north of York Springs in Adams County.

An upper-level low pressure system absorbed some of the remnant moisture from Fran on September 7–9, 1996, triggering more downpours. Two persons died in Abington Township in southern Montgomery County, where up to 10 inches of rain fell in three hours on September 8, resulting in serious flooding along Sandy Creek.

The NWS storm summary listed eight homes destroyed and about 550 homes and apartments damaged. Seventy-five businesses were also damaged or destroyed. Dozens of homes sustained losses elsewhere in the county, and the total damage was $20 million.

September 15–17, 1999
Hurricane Floyd

A drought left Pennsylvania soils parched by the end of August 1999, but two tropical systems brought an end to the dry spell in September.

On September 4, 1999, a weakening Hurricane Dennis would make landfall along the North Carolina coast and was later swept northwest by high-level winds across western Pennsylvania. The remains of Dennis collided with an eastward-moving cold front, producing heavy rain in central Pennsylvania. Williamsport was deluged with 6.29 inches on September 7, the second heaviest single-day rainfall in city history. (The previous daily-rainfall record of 8.66 inches occurred on June 22, 1972.)

As Dennis fell apart over eastern Pennsylvania, Hurricane Floyd strengthened in the eastern Atlantic. Floyd appeared on the weather map on September 2, 1999, in the vicinity of the Cape Verde Islands, later becoming a Category 5 hurricane with sustained winds of 155 mph as the storm passed over the Bahamas on September 13–14.

The eye widened to a diameter of 50 miles, and the storm extended for a distance of 700 miles.

Floyd was deemed the largest Atlantic storm of the century, four times the aerial size of Hurricane Andrew in 1992. The largest peacetime evacuation in United States history occurred along the East Coast, when Floyd turned northwest, though the effects ultimately bypassed Florida and Georgia.

Floyd eventually crashed into the Carolinas with more force than Hurricane Fran in 1996. Floyd struck Cape Fear, North Carolina, at 3:20 a.m. on September 16, 1999, with sustained winds at 110 mph. The weakening hurricane crossed over eastern North Carolina, re-emerging at Virginia Beach. Wilmington, North Carolina, was swamped in a record 19.06 inches of rain (13.38 inches in 24 hours) from September 14–17. A state record total of 24.06 inches fell at Southport 5 N.

Floyd accelerated to the northeast at 30 mph, caught up in the steering currents around an upper-air low. The center passed east of Atlantic City in the early afternoon of September 16, 1999, making landfall over western Long Island in the evening. The track took the circulation center 25 miles east of Hartford, Connecticut, just before midnight and then near Worcester, Massachusetts, and Portland, Maine.

Intense bands of rain pelted the coastline from the Carolinas to Maine. Ten days prior, the remains of Hurricane Dennis dumped 6 to 16 inches of rain in eastern North Carolina, leaving the soil sodden. Floyd's 12- to 20-inch deluge spawned widespread flooding that killed 52 people, causing $6 billion in losses.

Eastern Pennsylvania and New Jersey lay in the northwestern semicircle of the storm path, which normally receives the heaviest rainfall. A cool, dry Canadian air mass wedged against the Appalachians, denoted on weather maps by a stalled cold front that was overrun by tropical moisture.

The heaviest rainfall amounts in the Northeast were in excess of one foot in 24 hours. Some of the greatest totals were 14.13 inches at Little Falls, New Jersey, and 13.70 inches at Brewster, New York. The biggest rainfall in Pennsylvania was 12.13 inches at Marcus Hook, and in Maryland, Chestertown had 12.59 inches. The urban corridor from Washington, D.C., to New York City received a half-foot of rain.

Wilmington, Delaware, logged a record 24-hour rainfall of 8.79 inches on September 15–16, 1999. The storm total at the Philadelphia International Airport of 6.98 inches was the heaviest in modern records, and a 24-hour measurement of 6.70 inches, ending at 8:00 p.m. on the 16th, exceeded the previous record of 5.98 inches (August 3, 1898).

More than 10 inches of rain fell in Delaware County, including 10.66 inches at Newton Square and 10.52 inches at Radnor. Northwest of Philadelphia, rainfall amounts were as high as 10.07 inches at Doylestown (Bucks County), 10.04 inches at Valley Forge Park (Chester County), and 9.25 inches at King of Prussia (Montgomery County).

Farther north, six to nine inches pelted the Lehigh Valley and southern Poconos. In Lehigh County, a total of 9.05 inches fell at Coopersburg. The total of 7.61 inches at Lehigh Valley International Airport was comparable to the amount during the passage of Hurricane Gloria on September 26–27, 1985 (7.85 inches).

The East Stroudsburg observer measured 7.12 inches of rain on September 15–16, 1999, including 6.56 inches in 24 hours (.03 inch shy of the September 1985 record). In the higher elevations, 6.87 inches filled the gauge at Tobyhanna. Totals were generally in the three- to six-inch range in the Susquehanna and Lackawanna Valleys.

Due to a recent drought, low streamflow provided room to contain the heavy runoff, though flash flooding was widespread. Eight Pennsylvanians died in storm-related accidents and flooding, as Floyd moved up the coast on September 16–17, 1999. More than 2,000 homes and businesses were destroyed. Operations along the Southeast Pennsylvania Transportation Authority rail service were brought to a halt by heavy flooding. More than 410,000 Pennsylvania customers were without power during the storm.

Conditions were much worse in eastern New Jersey, where the U.S. Geological Survey (USGS) reported that the flooding along the Raritan River Basin was the worst in more than 200 years. At Manville, the river reached a height of 27.5 feet, surpassing the high-water mark of 23.8 feet, a level attained during the passage of Tropical Storm Doria (August 28, 1971). The situation at Bound Brook was disastrous, where 800 families were trapped in the upper stories of their homes by swirling floodwaters. A water level of 42.13 feet was 20 feet above flood stage, breaking the 1971 record of 37.5 feet.

The total death toll in the United States from wind, flooding and accidents caused by Floyd would reach 84, the largest loss of life during a tropical cyclone in the country since Hurricane Agnes in 1972 killed 122 people.

June 16–17, 2001
Soggy Remnants of Allison

Tropical Storm Allison formed in the western Gulf of Mexico on June 5, 2001, and proceeded to swamp portions of southeast Texas and southwestern Louisiana with one to three feet of rain.

A historic total of 37 inches fell in the Houston metropolitan area. The storm caused 41 deaths and $5.5 billion in damage.

The remains of Allison turned up the Eastern Seaboard on the weekend of June 16–17, 2001. A cold front approaching from the west intercepted the remnant circulation near the Atlantic Coast, producing copious rains from the Carolinas to Rhode Island. Rainfall rates in excess of three inches per hour were observed in southeastern Pennsylvania, triggering heavy flooding in low-lying areas.

Six residents of the Village Green Apartments in Upper Moreland Township, Montgomery County, died in a natural gas explosion, caused by floodwaters that broke a natural gas line, which sparked a fire. One motorist drowned in Wissahickon Creek. Excessive rains swamped Bucks and Montgomery counties. Top totals included 10.17 inches at Chalfont and 9.30 inches at Doylestown, both in 24 hours.

September 18–19, 2003
Hurricane Isabel

On September 18, 2003, Hurricane Isabel landed in eastern North Carolina as a Category 2 storm a few days after having attained Category 5 strength for 48 hours.

Isabel turned northwest into Virginia, moving through the Shenandoah Valley, after leaving a trail of massive tree damage that knocked out power to more than 700,000. Gusts reached 83 mph at Norfolk, Virginia. The soil was saturated from the passage of Tropical Storm Henri a few days earlier that flooded the southeastern part of the state (6.02 inches fell at Pottstown).

The eventual track of the post-tropical cyclone paralleled Interstate 79 from Pittsburgh to Erie, downing trees and power lines. PPL Electric Utilities reported that 495,721 customers lost power in the state. The wind picked up during the evening of September 18, 2003, and howled through the night, subsiding the following morning. Peak gusts averaged near 60 mph and reached 63 mph in Lancaster County, according to observer James Weber (Schwartz 2007). The south-central mountains tallied two to four inches of rain.

Hurricane Isabel was responsible for $3.6 billion in total damage and knocked out electricity to six million homes and businesses. The death toll from the storm was 51.

September 2004
Hurricanes Frances and Ivan

September 2004 brought a tropical storm and four hurricanes to Florida that caused dozens of fatalities and more than $40 billion in total damage.

Hurricane Frances hit Florida's east coast on September 5, 2004, before turning north-northeast, tracking west of the Appalachians across eastern Ohio and western Pennsylvania. Widespread flooding resulted in insurance claims in 60 of the state's 67 counties.

The remnants of Frances crossed the eastern tip of Lake Erie on September 8. Heavy rain that pelted the state totaled four to eight inches and brought widespread flooding in parts of western and central Pennsylvania. New Castle picked up 8.07 inches and Altoona received 5.81 inches.

Hurricane Ivan came ashore along the coast of Mississippi west of Gulf Shores, Alabama, on September 16, 2004, toting 120 mph winds. Ivan quickly became extratropical as the storm lumbered along the crest of the Appalachians. The remnants would turn southeast on September 17-18, joining a frontal system lying across the Delmarva Peninsula. Ivan circled back to the Gulf of Mexico after passing over southern Florida several days later, returning as a tropical depression on September 24 in southwestern Louisiana.

The Commonwealth lay on the soggy northwestern side of the storm, receiving four to 10 inches of rain atop ground still saturated from Frances, which resulted in widespread flooding from the Pittsburgh area to the Susquehanna Valley. The Susquehanna River crested 7.4 feet above flood stage at Harrisburg on September 19, 2004.

The storm spawned 120 tornadoes along the path from September 15–18, a record for a tropical cyclone. Nine small tornadoes were counted in south-central Pennsylvania, with touchdowns in Cumberland, Franklin and Bedford counties. A total of 19 Pennsylvania counties were declared a disaster area in the aftermath of Ivan, on top of the damage caused by the passage of Frances.

Hurricane Ivan—the second powerful storm to impact Pennsylvania in nine days—claimed 12 lives. Total damage nationally exceeded $28.1 billion (2020 dollars). The Pennsylvania Emergency Management Agency reported a peak of 250,000 customers lost power in the storm.

September 30–October 1, 2010
Remains Of Nicole Bring Downpours

An early autumn tropical rainstorm tracking north caused flooding in eastern Pennsylvania. Tropical Storm Nicole merged with a north-south frontal boundary near the coast, becoming the pathway for several waves of low pressure along the Mid-Atlantic Coast. Heavy rain developed on September 30, 2010, and continued October 1.

Rainfall totals averaged five to 10 inches in eastern Pennsylvania and three to six inches in the central part of the state, sending streams and creeks out of their banks. Evacuations were required along parts of the Schuylkill River as water covered roads in the Manayunk section of Philadelphia. Moderate flooding occurred along the Schuylkill and Lehigh Rivers.

The heaviest event totals were: Kennett Square 9.77 inches; Graterford 9.55 inches; Pennsburg 9.36 inches; Chadds Ford 9.04 inches in Chester, Montgomery and Delaware counties. Philadelphia International Airport recorded 5.41 inches of rain. In the northeast, 10.45 inches of rain fell at Blakeslee and 7.26 inches at Mount Pocono. Heavy flooding in the Perkiomen Valley in Montgomery County took the life of a woman who drowned in floodwaters from a swollen creek, and two men were rescued eight hours after being swept into the East Branch Perkiomen Creek. The King of Prussia Mall and a number of homes and businesses in the area reported flood damage.

August 27–28, 2011
Tropical Storm Irene

Irene officially became a Category 1 hurricane on August 22, 2011, while crossing Puerto Rico. As the storm churned northward, high waves, pounding surf and torrential rain caused widespread damage from North Carolina to New Jersey.

Hurricane Irene briefly made landfall on North Carolina's Outer Banks on August 27, 2011, before moving out to sea. The downgraded system made a second landfall as a tropical storm over southeastern New Jersey, passing over New York City on the 29th. The strongest winds remained well offshore, but heavy rainfall from eastern Pennsylvania northward to Vermont triggered deadly flash flooding. The devastating combination of wind and water knocked out power to 7.4 million customers.

Creeks and rivers overflowed their banks across eastern Pennsylvania on August 29, where four to eight inches of rain fell on saturated ground. (Hard downpours on August 13–15 totaled 5.95 inches at Neshaminy Falls and 5.80 inches at Philadelphia.)

The greatest 24-hour rainfall on August 27–28 from Irene occurred at Phoenixville (8.00 inches) and Springtown (7.73 inches). The August rainfall totaled a record 19.34 inches at Philadelphia and 19.31 inches at Neshaminy Falls for any month.

Powerful winds uprooted trees and knocked out power to more than 750,000 customers at peak. Pennsylvania Utility Commission reported that 1.3 million Pennsylvania electricity customers were affected during the 12 days it took to fully restore power.

Six deaths were attributed to storm-related incidents in Pennsylvania, among the 41 total fatalities caused by Irene. Damage statewide totaled about $425 million in losses covered by relief agencies and private insurers and reached $16.6 billion in the eastern United States. Power outages in Pennsylvania were most severe in the Allentown area, where PPL Electric Utilities reported 428,503 outages out of a total of 1.4 million customers, costing $32 million in repairs and overtime.

The remnants of Irene triggered catastrophic flooding in Vermont, the worst since 1927. Eleven inches of rain fell in 24 hours, wiping out more than 200 bridges and 500 highway miles and leaving thousands homeless.

October 29–30, 2012
Superstorm Sandy

Hurricane Sandy was a tropical wave on October 22, 2012, off the northern coast of Africa. The storm reached hurricane strength on the morning of October 24 while crossing the Atlantic, before plowing into Jamaica and western Cuba.

The circulation of Hurricane Sandy passed east of North Carolina on October 28, 2012. The massive storm encountered blocking high pressure over the Canadian Maritimes, which caused an abrupt left turn, putting the huge storm on a collision course with the populous Northeast.

The breadth of the hurricane extended over an incredible distance of 943 miles as it lurched northwestward. Maximum sustained winds reached 115 mph, when the central pressure off the New Jersey coast plummeted to 27.76 inches (940 millibars)—the second lowest pressure observed north of Cape Hatteras, North Carolina. In early October 1975, Hurricane Gladys had a minimum pressure of 27.73 inches (939 millibars).

Feasting over the warm Gulf Stream waters, Sandy developed a new eye shortly before landfall near Atlantic City, New Jersey. The central pressure of 27.92 inches (945.5 millibars) at 7:30 p.m. near Brigantine, New Jersey, was the second lowest for a hurricane coming ashore north of Cape Hatteras, North Carolina, only exceeded by the Great New England Hurricane of 1938 reading of 27.79 inches (941 millibars).

The powerful winds and long fetch churned up seas 30 feet high, heightened by the difference between high pressure over Labrador, Canada, and the circulation center. Hurricane-force winds extended for 175 miles and tropical-storm-force winds reached a radius of 485 miles.

The NHC determined that Sandy had lost sufficient tropical character to be classified as a hybrid storm at landfall, despite no practical difference in the powerful winds and dangerous storm surge. Public confusion ensued when hurricane warnings were abruptly discontinued. The policy was re-examined the following year to ensure that future warnings would be left in place under such circumstances.

An all-time record low barometric pressure was observed at Atlantic City (28.01 inches) just before 8:00 p.m, as the massive storm slowly transitioned to a post-tropical system, lumbering across southern New Jersey and through eastern Pennsylvania. Peak wind gusts along the coast reached 96 mph at Eaton Neck, Long Island, 91 mph at Islip, and 90 mph at Tompkinsville, New Jersey.

Catastrophic storm surge, coastal flooding, and a devastating fire in Queens, New York, all caused severe damage. Floodwaters cascaded into many of New York City's low-lying sections, swamping LaGuardia and John F. Kennedy International Airports. Historic damage was done to the New Jersey Transit System ($400 million). A record-breaking storm surge measured above the average low tide occurred at Kings Point on the western tip of Long Island Sound (14.38 feet).

The combination of astronomical high tide due to a full moon and a storm surge of historic proportions brought a singularly high total water level, or storm tide, at the Battery in Lower Manhattan of 13.88 feet—including a storm surge of 9.23 feet. The storm tide reached 9.15 feet above mean high-tide level and 14.1 feet above mean lowest low water (MLLW)—the average height of the lowest tide recorded daily.

The devastating rush of water exceeded the level recorded during the passage of Hurricane Donna in September 1960 by four feet. Historically, the Norfolk–Long Island Hurricane in September 1821 produced a high-water mark of 11.2 feet.

Power was lost to nearly all of Manhattan below 39th Street. Around 9:00 p.m. the subway tunnels under the East River were breached by four feet of seawater, according to the Metropolitan Transit Authority. Sea walls from Staten Island to Rockaway and Lower Manhattan were overrun by the historic surge. Shoreline beaches where sediment had been removed became most susceptible to erosion and flooding.

The death toll around New York City (43) was heavy, and the damage totaled $19 billion. Strong southeasterly winds caused the waters to pile up in Delaware Bay, reaching a record height of 10.6 feet at Philadelphia, 5.4 feet above astronomical high tide, which exceeded the previous maximum of 10.5 feet, observed on November 25, 1950, and April 7, 2011.

Sandy chugged through southern Pennsylvania, with the center passing 10 miles southwest of Philadelphia around midnight, accompanied by hurricane-force wind gusts and torrential rain. Wildwood, New Jersey, received 11.91 inches of rain. A sharp thermal contrast from a merger with a cold front strengthened the already enormous storm, which trundled westward along a path that paralleled the Pennsylvania Turnpike. Trees were toppled and snapped across eastern Pennsylvania.

The South Allentown anemometer clocked a wind gust of 81 mph, and another registered 76 mph at Bensalem. Weather Service sites showed top gusts of 73 mph at Pocono Mountains Municipal Airport, 70 mph at Lehigh Valley International Airport, and 68 mph at Philadelphia International Airport.

Fig. 9.1 Satellite image of Hurricane Sandy approaching the New Jersey coast on October 29, 2012. (NOAA)

Fig. 9.2 National Guard Alpha Troop 2nd 104 assisted the Marshalls Creek Volunteer Fire Department on Mt. Nebo Road. (Courtesy Pocono Record*)*

Fig. 9.3 Pine tree on Gap View Road in Minisink Hills. Thousands of trees whipped down during Sandy's onslaught, toppling power lines. (Photo by David Kidwell, Pocono Record*)*

Adding to the eerie atmosphere, transformers exploded, and the night light took on an eerie bluish glow. Trees were stretched like rubber bands, breaking and crashing onto homes, garages, cars and power lines. The forceful tropical rainstorm soon met up with an interior cold-core system over the Appalachians, which provided additional energy and delayed the expected weakening effects of an overland journey.

In western Pennsylvania, 3.90 inches of rain fell at Greensburg. More than 30 people in eastern Westmoreland County had to be rescued from rising floodwaters that surrounded automobiles. Trees and power lines were knocked down in Altoona, where the high winds caused four buildings to collapse. The remnants of Sandy eventually turned north toward Lake Erie and crossed over Ontario, Canada.

On the cold side of the storm track, deep snows piled up as high as 24 to 34 inches over the mountains in eastern Tennessee and Blue Ridge of Virginia, extending west across eastern West Virginia and western Maryland. Large falls of one to two feet blanketed the Laurel Highlands in southwestern Pennsylvania.

Historic low-pressure records were established across southern Pennsylvania. The pressure dipped to 28.12 inches at Philadelphia (eclipsing 28.43 inches set on March 13, 1993); 28.46 inches at Harrisburg (28.62 inches on January 1, 1913); near-record values of 28.49 inches at Allentown; 28.56 inches at East Stroudsburg; and 28.69 inches at Scranton.

The full damage wrought by Superstorm Sandy ($85.9 billion) is ranked fifth, after Katrina (2005), Harvey and Maria (2017), and Ian (2022). In the NHC Tropical Cyclone Report, greater than 8.5 million residences and businesses would eventually lose power either during or directly after the storm, and 650,000 were damaged or destroyed.

Sandy was blamed for 159 deaths in the United States (direct and indirect) according to NOAA (233 in total, including the Caribbean Islands). The loss of life in Pennsylvania was 14, according to the state Emergency Management Agency. The storm caused $16 million in total damage, and an estimated 1.5 million Pennsylvania electric customers lost power during the storm (1.93 million in the region). The infrastructure took a huge hit; more than 200 bridges and roads were closed for repairs.

August 4, 2020
Hurricane Isaias

Tropical Storm Isaias formed on July 30, 2020, after making the trek across the Atlantic Ocean, the earliest ninth named storm, surpassing Irene (August 7, 2005).

Heavy rain and high winds lashed Puerto Rico and the Dominican Republic. Tropical Storm Isaias made an initial landfall in the Bahamas on August 1, and then took a track that paralleled the east coast of Florida and Georgia.

Isaias strengthened to a Category 1 hurricane packing 85 mph winds, before slamming into the coast of North Carolina near Oak Island Beach around 11:00 p.m. on August 3, 2020. A wind gust of 99 mph was recorded at Federal Point, North Carolina. The potent storm accelerated north-northeast parallel to the Interstate 95 corridor, bringing flooding rain and hurricane-force wind gusts on the eastern side of the track, and spawning more than 20 tornadoes.

Isaias interacted with a frontal boundary draped over the Mid-Atlantic region, which enhanced the low-level convergence, instability and wind shear conducive to tornadoes. A fast-moving segment of air acted as a further lifting mechanism on the eastern side of an upper-level trough of low pressure over the Midwest, directing the storm north near Chesapeake Bay and along the Delaware River Valley.

Flash flood and tornado watches were issued for eastern Pennsylvania and surrounding states. A corridor of torrential rain averaging four to eight inches flooded areas from northeastern Maryland through the northwestern suburbs of Philadelphia, Lehigh Valley and southern Poconos. Rainfall from August 3-4 tallied 8.85 inches at Harleysville, 4.90 inches at Stroudsburg and 4.16 inches at Philadelphia. The Little Lehigh River at Allentown and Perkiomen Creek at Graterford reached record flood levels.

Serious flooding caught motorists by surprise and necessitated high-water rescues. Winds gusted to 52 mph near Allentown and 50 mph at Mount Pocono. Along the coast, top gusts reached 70 mph at John F. Kennedy International Airport, 75 mph at Cape May, New Jersey, and 78 mph at Farmingdale, Long Island.

Two tornadoes were reported in southeastern Pennsylvania as Isaias marched northward. The strongest was an EF2 tornado that touched down northeast of Philadelphia and tracked 20 miles to Doylestown. The funnel that developed a little north of the Philadelphia Mills Mall took a rare northwesterly course, with peak winds as high as 115 mph and a path width of 200 yards. Trees were snapped and cars overturned by the twister. Six automobiles were tossed at the Doylestown Hospital complex. Children's Village daycare suffered a direct hit.

A smaller EF0 tornado advanced for 2.8 miles across parts of Worcester Township, Montgomery

County, also with a path width of 200 yards. Two EF1 tornadoes were reported near the New Jersey coast, in Strathmere and Barnegat Township, Cape May County. Another storm tracked 29.2 miles north through central Delaware—a state record.

During the height of the storm, upwards of three million utility customers lost power from North Carolina to New England—more than half of those in New Jersey—as the weakening tropical storm raced northeast at up to 40 mph across Philadelphia and northwestern New Jersey, continuing north near the Hudson Valley into western New England and Quebec, Canada.

 Excessive rainfall totals of four to eight inches accompanied the passage of Tropical Storm Isaias, as the system transitioned to a post-tropical cyclone, causing creeks and rivers to overflow and stranding motorists on flooded roads and highways.

Power outages in Pennsylvania totaled 986,000, the most since Sandy (October 2012) knocked out electricity to 1.93 million customers in the state. Total damage was assessed at $4.72 billion.

October 28–29, 2020
Tropical Storm Zeta

Tropical Storm Zeta formed in the western Caribbean on October 24, 2020, making landfall in Mexico's Yucatan Peninsula as a Category 1 hurricane. After briefly weakening over land, Zeta strengthened to a Category 3 hurricane (115 mph) over the Gulf of Mexico before making landfall in southeastern Louisiana in the early evening on October 28.

 Zeta, the 27th named storm of the busy 2020 hurricane season, was the fifth tropical cyclone to breach the Louisiana coastline. The remnant tropical low merged with an upper-level trough as it sped past the Mid-Atlantic Coast south of Atlantic City on October 29. Heavy rain and strong winds affected Pennsylvania as the storm became extratropical, with the added rarity of sufficient cold air on the northern fringe to turn rain to snow on October 30 in interior New England, where several inches accumulated in Massachusetts.

A similar event occurred with Sandy on October 30–31, 2012, as the center of circulation traversed western Pennsylvania on the fringes of Wilma October 24–25, 2005. Moisture from a tropical cyclone moving up the Eastern Seaboard on October 21–22, 1944, after a South Carolina landfall ended with a light rain/snow mix in the Poconos.

Tropical Storm Eta formed on October 31, 2020. The late-season tropical cyclone deepened explosively in the western Caribbean four days later, reaching peak sustained winds of 150 mph and a low central pressure of 27.23 inches (922 millibars), weakening prior to landfall in Nicaragua. Hurricane Eta ranked as the third strongest November hurricane on record in the Atlantic Basin, after the Great Cuba Hurricane (1932) and Hurricane Iota.

After pounding Central America with torrential rain that triggered deadly flooding and mudslides, the remnant tropical low gathered energy over the warm waters of the western Caribbean, striking Cuba as a tropical storm on November 8, later making landfall the next night in the Florida Keys. A fourth landfall occurred on November 12 near Cedar Key, Florida. Eta was the 12th separate tropical cyclone to make a landfall in the continental United States, breaking the 1916 record of nine.

A week later, Hurricane Iota grew into the sixth major storm in 2020, and set a new record when it became the 30th named storm, surpassing 2005. The powerful hurricane gained 70 mph in 24 hours (November 15–16) in the western Caribbean, peaking at 160 mph, the 10th Atlantic storm to intensify rapidly in 2020 (tying the 1995 record). Iota had a minimum pressure of 27.08 inches (917 millibars).

Hurricane Iota struck Nicaragua on November 16 as a Category 4 storm (155 mph), notably the strongest November storm since 1932. (The Great Cuba Hurricane attained winds of 175 mph, eventually hammering Cuba on November 9, 1932, as a Category 4 hurricane.)

Storms in 2020 that intensified rapidly in 24 hours (maximum sustained winds increased 35 mph or more) were: Hanna, Laura, Sally, Teddy, Gamma, Delta, Epsilon, Zeta, Eta, Iota.

This trend of rapidly strengthening hurricanes has been attributed to warming ocean waters at lower depths, which provides energy for growing tropical cyclones. Warmer water increases the risk of heavier rainfall rates by 10 to 15 percent. A study showed a tendency for tropical systems to weaken more slowly in the first 24 hours after landfall and only lose half of their strength, which compares to 75 percent 50 years ago; prolonged heavy rainfall increases the threat of flooding and loss of life.

The right combination of warmer-than-average Atlantic sea surface temperatures linked to the warm phase of the Atlantic Multidecadal Oscillation, light vertical wind shear associated with La Niña, including winds off of Africa, and a prominent West African monsoon, boosted the historic 2020 hurricane season.

August 21–22, 2021
Tropical Storm Henri

A tropical wave briefly grew into Hurricane Henri off the East Coast on August 16, 2021, with 75 mph winds, weakening to a tropical storm at landfall near Waverly, Rhode Island, on August 21. The highest wind gust was 70 mph at Port Judith, Rhode Island. The storm downed trees and power lines, with root systems softened by the recent passage of the remnants of Tropical Storm Fred on August 18–19. More than 140,000 homes lost power from New Jersey to eastern New England.

An upper-level low over the Appalachians pulled the broad corridor of tropical rains farther west, causing flooding rains in portions of southern New England, New York City, northern New Jersey and eastern Pennsylvania on August 21–22.

Pocono rainfall totals on August 20–22, 2021, topped out at 7.16 inches in Bossardsville, 6.67 inches just west of Blakeslee, 6.24 inches at Saylorsburg and 5.82 inches at Stroudsburg. Water backed up on West Main Street in Stroudsburg beneath a bridge over Pocono Creek, swamping apartment complexes and partially submerging parked vehicles. Portions of Interstate 80 above Tannersville and Route 611 south of the Delaware Water Gap were closed for a time due to high water. Bands of heavy rain pivoted southwestward, with totals of 5.82 inches at Portland in Northampton County and 5.81 inches at North Wales in Montgomery County.

New York City was swamped by a massive rainfall of 8.19 inches that triggered flash flooding and inundated parts of the subway system. Nine inches fell at South Slope in Brooklyn, and 9.85 inches were recorded in a Brooklyn Heights gauge. In northern New Jersey, 9.44 inches swamped the town of Cranbury.

Chapter Ten

Climate Trends

Climate is what on an average we may expect, weather is what we actually get.
—Andrew John Herbertson, *Outlines of Physiography*

UNLIKE THE SAYING regarding politics, all weather is *not* local. The infinitely complex coupling between the land, ocean and atmosphere creates dynamic feedback loops that drive climate trends.

In the late 1700s, a raucous academic debate in early America developed regarding a perceived change in the climate due to human influences, primarily the clearing of forests for cultivation and crop production. The modern discourse on climate change is centered around greenhouse gas levels rising at a rapid rate from the burning of fossil fuels, which is linked to a warmer Earth that poses human and ecological consequences.

Shifting weather and climate patterns impact the delicate balance in nature. The risk of intense heat waves, expanding drought, and elsewhere in more intense storms with heavier rainfall and flooding, reflects the complex interaction of the warmer atmosphere and global seas.

Land and ocean temperatures attained record highs in 2023, likely the warmest conditions in at least 125,000 years (NOAA). The average rate of warmth each decade (0.32°F) doubled after 1980, when compared to the previous 100 years. In the contiguous United States, surface warmth achieved a record high in 2012, closely followed by 2016. Summer nights warmed 1.4°F, twice the rate of daily maximum temperatures since data started in 1895.

Global Temperatures Rising

Earth's average temperature has risen more than 2°F above preindustrial levels, when the amount of atmospheric carbon dioxide was only 280 parts per million (ppm). In 2023, the average concentration peaked at 424 ppm at the Mauna Loa Observatory in Hawaii, greater than at any time since 4.3 million years ago, during the Pliocene climatic optimum.

The five hottest years in Earth's history were observed between 2016 and 2024, according to independent NOAA and NASA analyses, based on land-based observations and data obtained from ships and buoys.

The pattern has been remarkable: the five warmest years all occurred between 2005 and 2024. NOAA heat records were often topped in successive years: 2024, 2023, 2016, 2020, 2019. The next ten warmest since 1880: 2017, 2015, 2022, 2021, 2018, 2014, 2010, 2013, 2005, 2009.

Recently, June 2023 was the hottest on record globally, which was followed by the hottest month worldwide in July. Across the continental United States, sizzling heat over the southern part of the country established monthly records for extreme warmth from Phoenix, Arizona, to El Paso, Texas. Expansive heat domes developed in the American Southwest, southern Europe and portions of Asia that brought blazing heat, shattering long-term records.

Further evidence of the warming pattern comes from satellites that carry microwave radiometers to measure thermal-infrared radiation emitted from Earth's surface, which equates to the temperature. Satellite measurements obtained five miles up show an overall warming of 0.8°F since 1979, a little less than the NOAA surface-based data (1°F) for the same period.

A much longer timescale reveals that Earth is warmer than at any other time in at least 1,000 years. We know this based on the reconstruction of paleoclimates using proxy records. Ice cores are drilled and extracted from remote glaciers in far-away places and analyzed to determine the level of the oxygen isotopes and gases contained in the ice as far back as 800,000 years.

Ancient pollen grains and spores, deposited by wind and water in old lake bed sediments, reveal shifting rainfall patterns by the types of vegetation. The width of tree rings in centuries-old forests and fossil pollen contained in sediments offer evidence of multi-year megadroughts.

Trees and vegetation release billions of tons of water vapor every day that evaporate under a strong sun, causing air to rise and condense into rainclouds. Gaseous atmospheric compounds and a combination of soil bacteria, fungi, leaf particles, microbes and pollen transported by the wind act as a surface for the condensation of water vapor molecules, creating trillions of clouds droplets.

Pollen levels are rising in an atmosphere that contains more carbon dioxide, which stimulates plant growth during a longer growing season in many parts of the world.

A hotter planet is strongly linked to a greater number of extreme weather events. A pattern of rising temperatures in urban areas contributes to a higher heat output from air-conditioners, which further warms the environment. Wildfire seasons are lasting longer in autumn, exacerbated by heat and drought, emitting toxic smoke and fumes.

Warmer air has a greater capacity to absorb evaporated moisture, which is released in large, low-moving storms, reflected by the significant increase in heavier rainstorms and potential flash flooding. The human and economic impacts from floods are substantial. In other areas, extreme heat creates desiccating conditions conducive to crop losses, water shortages and wildfires.

The United States experienced 22 individual billion-dollar weather and climates disasters in 2020 (NCEI), from freshwater flooding, tropical cyclones, severe storms, droughts and wildfires. In September 2023, another record (28) was set for inflation-adjusted losses exceeding $1 billion.

The previous record for billion-dollar disasters in the United States was 16 in 2011 and 2017, the latter resulting in total damage of more than $270 billion. August 2017 brought Hurricane Harvey, a slow-moving storm that stalled over southeastern Texas while dumping more than 60 inches of rain. One report suggested that Harvey's prolific downpours were juiced up 15 percent by higher surface water temperatures in the Gulf of Mexico.

During the mid-Pliocene Warm Period around 3 million years ago, levels of atmospheric carbon dioxide levels were between 380 and 420 ppm, a possible analog to how Earth's climate system will respond to comparable forcing. At the time, global sea level was about 65 feet higher, and the mean temperature averaged 7°F warmer than today.

No single weather event or seasonal anomaly can be directly attributed to climate change without historical context, though clusters of extreme events are significant. Numerical climate models consistently point to the role of human activities as a driver behind increasing weather anomalies. Precipitation patterns were compared to periods when the average regional temperatures were a a few degrees lower than today.

Global warming has been amplified by many positive feedback loops affecting climate change in a dynamic ocean-atmosphere system. Melting glaciers in polar and mountainous areas facilitate a greater absorption of sunlight at high latitudes, and on mountaintops, with shrinking snow and ice. Thinning sea ice allows more solar energy to be absorbed in dark, open water that further warms the lower atmosphere.

Noteworthy is that the Arctic has been warming at up to four times the rate of lower latitudes. Wildland fires have increased the melting of the permafrost beneath lakes and in the soil, releasing carbon dioxide and methane into the atmosphere from the extensive peatlands.

Greenhouse Gases and Warming

Heat-absorbing chemical compounds in the atmosphere are transparent to sunlight and opaque to outgoing or terrestrial radiation.

The extraction and burning of fossil fuels (coal, oil, natural gas), industrial pollution, transportation, agricultural practices, melting permafrost and deforestation contribute to rising greenhouse gas levels that have accelerated a planetary warming. About 80 percent of global energy is linked to fossil-fuel emissions, primarily coal-fired electricity plants, gasoline and industrial operations. China is the world's largest emitter of carbon, followed by the United States and India. In the United States, 19 percent of electricity is generated from burning coal.

More than 50 billion metric tons of greenhouse gases are emitted into the atmosphere each year, the highest levels of carbon dioxide, methane and nitrous oxide in at least 800,000 years. The rate of carbon dioxide increase is 10 times greater than at any time in the past 50,000 years. About 36 million metric tons of carbon dioxide are emitted annually, triple the levels from human activities in 1965.

The Amazon rainforest, a big carbon sink, is now a net carbon emitter. Wildfires are responsible for about 290 million metric tons of carbon dioxide each year. Peatlands comprise 3 percent of the land surface but hold 30 percent of Earth's carbon. Land degradation from forest fires coupled with a rapidly-warming Arctic are causing the permafrost to melt. As frozen soil thaws, buried methane and carbon from decomposed plants and animals are converted into greenhouse gases by microbes.

The oceans absorbed 93 percent of the excess heat released by fossil-fuel emissions and other human activities between 1971 and 2015—about 20 times greater than the atmospheric uptake. The ocean water removed 31 percent of the carbon dioxide emissions between 2010 and 2019; the Southern Ocean accounted for 40 percent.

However, not without an environmental cost. Hydrogen ions are increasing acidification of the ocean, which is detrimental to the exoskeletons of shellfish and corals and harms the habitat of many marine ecosystems. A 2019 report found that since 2008, coral reefs around the world had diminished about 14 percent. Climate scientists have expressed concern that there is a finite capacity to dissolve and retain carbon dioxide in the deep seas, which is a function of water temperature, wind patterns and ocean currents.

A considerable quantity of carbon is taken up by land ecosystems and removed through rock weathering, which releases minerals that come together to form calcium carbonate (limestone). About 30 percent of global carbon emissions are absorbed by forests and soil, so deforestation and expansive wildfires are a growing concern.

Plant growth depends on photosynthesis. The food-making process involves sunlight, water and carbon dioxide to make sugar (glucose). Trapped light energy is required for the chemical reaction that nourishes the leaves and adds oxygen to the atmosphere. Plant growth accepts carbon dioxide through leaf pores that is stored in the branches.

The exchange of carbon between soils, plants, animals within the ocean and atmosphere system is governed by various feedbacks. The loss of more than one billion acres of tropical rain forests in the past 40 years has released massive amounts of carbon stored, which is oxidized to form carbon dioxide. Tropical forests are subjected to clear-cutting and burning for logging and agriculture to expand pastureland. An aggregate total of 7.5 billion metric tons of carbon dioxide was released in 2017 from deforestation, biomass and peat forest drainage.

Vast petroleum deposits have traditionally offered cheap energy. The use of fuel reserves for electricity derived from compressed animal and plant remnants (peat) are a major source of greenhouse gas emissions. Around 60 percent of carbon emissions linger for decades to as long as 150 years, which continue to warm the lower atmosphere (radiative forcing). The land and ocean soak up the remainder of carbon dioxide.

Bubbles of air lodged in ice cores extracted from mountain glaciers show that the amount of carbon dioxide in the atmosphere during the Last Glacial Maximum (LGM), about 20,000 years ago, dipped to 190 parts per million (ppm). The warm interglacial period that followed saw a rise in carbon dioxide levels to 280 ppm, which held steady under 300 ppm before the Industrial Revolution (1750–1850).

Natural fluctuations in atmospheric carbon occur with cyclical variations in Earth's orbit, mountain building and volcanic activity. Crustal uplift exposed more rocks to weathering that absorbed carbon dioxide, especially those rocks rich in silicate. Eventually, cooler conditions decreased the rate of the chemical reactions, leaving more carbon dioxide in the air, which contributed to a renewed warming trend.

Volcanic eruptions and outgassing caused spikes in carbon dioxide concentration and higher global temperatures. In a never-ending cycle, rising global temperatures instigated more rapid chemical reactions and weathering of silica in crustal rocks, which removed carbon dioxide and initiated another cooling trend.

Coal-fired power plants, factories and motor vehicle exhaust release waste gases from incomplete combustion, spewing carbon, nitrogen oxide, ozone and methane emissions in addition to fine particulates (PM2.5). Hydrofluorocarbons (HFCs) used in refrigerants, fire extinguishers and air conditioners supplanted chlorofluorocarbons (CFCs) in the 1990s to reduce damage to the ozone layer—but those chemicals turned out to be 100 times more potent as a warming factor.

Carbon and methane emissions from agricultural activities are carried by the wind. Methane is the second most significant greenhouse gas in the atmosphere, comprising 20 percent of global emissions. Methane has a warming capability 87 times greater than carbon dioxide in the first 20 years, but in a much shorter atmospheric residence time. The concentration of methane has more than doubled in the past two centuries. Nitrous oxide is 300 times more potent than carbon dioxide as a warming factor.

Livestock emissions from bacteria fermented in the guts of farm animals (cow flatulence) account for more than a third of the methane emissions related to human activities. Another example is paddy rice cultivation, when flooded fields release bacteria that emit methane. Additional methane sources include wastewater treatment plants, decomposing organic matter in landfills that release certain bacteria, and petroleum and natural gas emissions seeping out of the ground from leaky pipes. Thawing permafrost is a significant source of atmospheric methane.

Natural sources of nitrous oxide are rain forest soil, permafrost and the ocean. The gas eventually rises to the stratosphere, forming nitrogen oxides that deplete the ozone layer. The use of nitrogen-based synthetic fertilizer releases nitrous oxide in production activities. Farming equipment adds to fossil fuel emissions that affect climate change in a warming atmosphere.

In the lower atmosphere, ozone molecules are responsible for one-quarter to one-third of the warming in the past 200 years. Tropospheric ozone is a source of the hydroxide that is the main oxidant in the atmosphere and reacts with hundreds of species.

Warming Seas

Sea surface temperatures are warming faster than at any time in recent centuries, particularly since the 1980s in the Atlantic Ocean and Pacific Ocean, linked to a global rise in air and land temperatures.

Warm air has a much greater capacity to store moisture, and coupled with higher evaporation rates loads the column of air with water vapor, which is conducive to heavier rainfalls. Thermal expansion of warmer water and higher pressure contribute to a rise in global sea level, averaging more than an inch per decade, which raises the volume of water in response to a decrease in the total density, despite unchanging mass.

The world's oceans have heated up 1.5°F in the past century. Upper ocean heat content achieved a record high in 2023, tracked by NOAA. February 2024 (69.1°F) surpassed August 2023 for warmth between 60 degrees latitude (Copernicus Climate Change Service). The Atlantic and Gulf of Mexico warmed .34°F per decade since 1970.

More than 90 percent of extra heat produced by human activities is stored in the ocean. One of the impacts is that for the period of 1925–2016, marine heat waves increased 34 percent and lasted 17 percent longer, with warming now observed to greater depths between 300 and 1,000 feet below the surface.

The interplay between the atmosphere and ocean plays a major role in weather and climate. The wind drives ocean currents and affects sea surface temperature patterns that impact atmospheric stability. Cooler-than-normal water is conducive to sinking motion and high pressure (dry weather) and warm water fosters rising air (convective rain). Warmer water transfers more heat and moisture to the atmosphere, promoting clusters of thunderstorms that push air up and outwards, amplifying waves in the jet stream.

The redistribution of energy and saltwater by ocean currents with differences in density is the central driver in the ocean-atmosphere system. The large-scale circulation propels warm water to higher latitudes, and the return flow increases the equatorial trade winds that push the surface water westward. A long-distance correlation, or teleconnection, linking a warmer North Atlantic to air pressure in the tropical Pacific affects the global weather patterns. Air sinking over cooler tropical waters enhances the pressure gradient, creating robust easterly trade winds that impact sea surface temperatures, rainfall and overlying climatic features.

Precipitation Trends

Five of the state's 10 wettest years have happened in the present century, despite records back to 1895. Many climatological sites have experienced a rise in average annual precipitation of three to five inches since the mid-20th century.

Heavy rain events have increased substantially in the eastern half of the country between 1958 and 2016, according to the government's National Climate Assessment. The greatest increase in the top 1 percent of precipitation events was observed in the Northeast (55 percent) and Midwest (42 percent). In the 12-state region from Maine to West Virginia, the most dramatic precipitation uptick has occurred since 1996.

A United Nation's Intergovernmental Panel on Climate Change (IPCC) report in 2007 concluded that water vapor had increased about 5 percent in the past century, mostly since 1970 (4 percent). The decade of 2001–2010 was the wettest on record, based on records dating back to the 1860s, maintained by the World Meteorological Organization (WMO).

An understanding of the precipitation, cloud and radiation physics is crucial to any climate discussion (Groisman et al. 2004). Warmer air equates to a higher capacity for the air to absorb moisture through evaporative processes. A 1°F increase in air temperature adds about 4 percent to the moisture content of the atmosphere, which is then released in greater quantities during heavy rain events. The higher water vapor content also traps planetary heat, with an increase in buoyant air parcels that build rainclouds.

Thicker low clouds reflect and filter sunshine, cooling daytime temperatures while absorbing outgoing terrestrial heat after sunset, making the nighttime warmer. High clouds allow the most incoming sunlight to pass through to the surface while absorbing outgoing terrestrial radiation, with a slight net warming effect. Tiny particulates emitted from smokestacks and vehicle tailpipes also play a role in the water cycle by providing condensation nuclei.

Climate models have traditionally struggled with water vapor and cloud cover feedbacks that profoundly impact Earth's heat balance. Satellite data confirmed there has been greater absorption of heat by the ocean surface in the middle latitudes as cloud cover decreased. The mean cloud cover pattern shifted to higher latitudes, and cloud bases are rising. More subsidence on the southern margin increases the risk of steadfast drought conditions.

Cyclical precipitation tendencies in the United States have been well-documented. A wetter pattern east of the Rockies in the late 19th century ended a decades-long drought in the Plains states. However, by the early 1930s an expanding drought consumed the much of the Plains states, known as the Dust Bowl, when minimal rainfall and poor soil management in prior years devastated the agricultural midsection of the country.

A wetter cycle returned in the 1940s, but another severe mid-continental drought developed between 1952 and 1957. The precipitation deficits eased by the late 1950s. However, the general circulation by the early 1960s featured a persistent ridge in western Canada and a predominant northwesterly flow that brought colder winters in the Midwest and Northeast and consistent well-below-normal precipitation.

The pattern flipped in the late 1960s and 1970s, bringing wetter conditions to the eastern two-thirds of the nation. A warmer, drier pattern in the late 1980s culminated in a severe drought in the blistering summer of 1988. A much wetter southwesterly flow east of the Rockies dominated the 1990s, as temperatures warmed significantly compared to the mid-20th century.

Recent decades have featured mostly above-normal temperatures and moisture through the Eastern states. At the same time, the West dealt with persistent dryness. In 2021, 58 percent of the West endured severe or extreme drought, before consecutive wet winters replenished water levels.

Melting Ice and Global Sea Level

The Arctic region has been warming more than twice as fast as the rest of the world. Arctic sea ice and land ice are melting sooner. The June snow cover above the Arctic Circle in Eurasia in 2022 dipped to the third lowest level on record since data commenced in 1967. The loss of glacial mass has contributed to a steady rise in global average sea level, which is 3.6 inches higher than the 1993 mean, when satellite measurements commenced.

Following the Ice Age, or Last Glacial Maximum, the global sea level rebounded 400 feet and the Atlantic coastal plain retreated about 100 miles. Although stable for about 3,000 years, satellite data show that the sea level has increased 8 to 10 inches since 1880, and 3 inches in the past 25 years, with locally higher rises.

Ice melting in northern Canada, Russia, and over Greenland is a consequence of higher air and ocean temperatures. Satellite data revealed that the Greenland Ice Sheet shed 1,140 billion tons of ice since 1985, a rate seven times faster than three decades ago. Thawing permafrost releases stored carbon and methane into the atmosphere and furthers global warming.

Glaciers acquire mass when snow compacts into a granular ice, or firn. Both fresh snow and ice reflect about 80 percent of incoming sunlight. An ongoing trend of earlier melt in the spring has been accelerating the warming trend because rain and melting snows diminish the brightness (reflectivity) of the ice sheet and increase the absorption of solar radiation.

In July 2012, orbiting satellite measurements detected a surface ice melt that exceeded 97 percent during a five-day period, the greatest thaw since 1889. The loss of ice amounted to 400 billion tons, nearly four times the melt rate in 2003. Rivers and lakes formed on the surface along the ice and rock boundary leading to the ocean. The northeast and southeast portions of the ice sheet have increased, while the southwest has been losing more ice. There was a plateau of shrinkage in 2013 and 2014, but that soon changed. During the spring of 2019, a record early loss of ice was in response to temperatures rising to 40°F above normal.

Earth is now shedding about 1.2 trillion tons of ice each year. On August 14, 2021, rain came down in central Greenland at Summit Station (elevation 10,551 ft.) for several hours. The spectacle was a first in the record of the National Science Foundation Station located 500 miles above the Arctic Circle, and only the third time in the past decade that the temperature rose above freezing. A wavy jet stream that plunged south over northeastern Canada and looped back across Greenland pulled unseasonably mild and very moist southwesterly winds over the ice sheet.

Some climate scientists have theorized that a rapid Arctic warming has reduced the variation in temperature between the Arctic region and lower latitudes, which could be the driving force behind increasing episodes of an amplified jet stream. A sharper, or wavier, north-south (meridional) flow tends to bring more extreme weather conditions.

The Greenland ice sheet spans 665,000 square miles and extends up to two miles thick. The rapid loss of ice since the 1970s has raised world sea level approximately a half-inch. A pattern of milder winters and summers, which is governed by sustained high pressure, yielded more hours of sunlight, accelerating the melting, sending rivers of land ice gliding down the mountains into the sea.

The ice sheet has been losing more ice than has been added in the cooler months. The loss of ice comes primarily from iceberg calving, or splitting off, into the ocean. Additional losses occur during surface melt in the summer and from contact with the ground below the ice sheet.

The Greenland climate record is complex and highly sensitive to ocean currents and atmospheric circulation patterns that drive air and water temperatures. Ice cores containing dust and sea salts revealed a natural long-term 1,500-year oscillation, called a Dansgaard–Oeschger event, which runs opposite in phase with the average temperature in Antarctica, delayed by about 200 years.

The dynamics of Greenland's ice sheet—second in size only to Antarctica—have been the subject of intensive studies. One reason is that if the entirety of the Greenland ice were to calve or melt, global sea level would rise 23.6 feet. The ice sheet mass has declined since 2002, based on data from the Gravity Recovery and Climate Experiment, or GRACE, satellite records (NASA). The rapid loss is estimated at 234 billion tons of ice per year.

Collapsing ice shelves—floating ice platforms created by glacial runoff—provide a shield as meltwater seeks new outlets to the sea. Large land glaciers with steep slopes discharge ice through channels (fjords). Ohio State University scientists placed more than 50 GPS rovers (sensors) along the Greenland coast to monitor surface compression caused by the ice flow into the Atlantic, which provides an indirect measure of glacial thickness.

The Arctic has been losing sea ice at a dramatic rate since satellite measurements commenced in 1979. Sea ice reflects 50 to 70 percent of incoming sunlight, compared to about 6 percent reflected by open water at high latitudes. Dark-blue meltwater and older snow absorb more heat, reducing the thickness of sea ice and thinning the edges of continental glaciers, which are more vulnerable to the loss of protective snow cover. Some of the most substantial losses have occurred in the Bering Sea near the western Alaskan coast, which climate scientists have linked to unusual warmth in Siberia.

The National Snow and Ice Data Center (NSIDC) in Boulder, Colorado, relies on satellite mapping to track the minimum volume of Arctic sea ice in September, which has fallen by one-third since 1988. In 2012, the lowest sea ice extent dipped to 1.31 million square miles, leaving open water over 40 percent of the central Arctic Ocean. Abnormally warm fall and winter weather conditions are the primary reason, and sea ice is disappearing at an annual average rate of 20,800 square miles.

On September 7, 2016, the extent of sea ice shrank to 1.6 million square miles, which tied for the second lowest coverage (September 17, 2007). In November 2016, the mean temperature in the region was a staggering 23°F above normal. The annual temperature was nearly 11°F above the 100-year average. The exceptional Arctic warmth led to a record warm winter.

In early February 2017, at sites to the north of 80° N, the temperature soared to a stunning 29°F above normal, abetted by surges of mild maritime air accompanying North Atlantic storms. NSDIC satellite data revealed a record lowest maximum ice extent (5.57 million square miles) on March 7, 2017.

NASA satellite data computed that Earth's glaciers lost 279 billion tons of ice—nearly 67 trillion gallons of water—between 2012 and 2017. Melting ice and the thermal expansion of the ocean pushed sea levels upward at an annual rate of about .08 inch. If that trend continues, global sea levels could rise one to three feet by the end of the century, resulting in even more coastal erosion and tidal flooding, equating to billions of dollars in economic losses.

Alaska's glaciers have lost 83 billion tons of ice a year since 1959, and at an accelerated rate compared to the years prior to 1993 (57 billion tons). The prevalence of southerly winds reaching the Arctic are the primary reason. The temperature in Alaska has risen 2.4°F annually (3.3°F in the past half-century). Winter warming has been the most dramatic (5.4°F).

In the winter of 2015–16, an intense cyclone entered the region on December 30, 2015, and hovered for days, drawing enough heat and moisture into the region to cause sea ice to shrink dramatically. Minimum sea ice thickness has dipped 65 percent as warm air penetrated farther north. The Northern Hemisphere snowpack decreased by about a million square miles in the springtime since 1967. Alpine glaciers are receding at an alarming speed since the pace of winter warming tripled in the past 40 years at high latitudes.

In late February 2023, Antarctica recorded its lowest summer sea ice extent since records began in 1979 (691,000 square miles). Scientists are unclear if this is a climate change signal or primarily natural variability. The intrusion of warmer water has been implicated in weakening ice shelves and calving of more than a dozen large icebergs between 2000 and 2020 on the Antarctic Peninsula.

In the previous decade, the NSIDC reported that the areal extent of sea ice that surrounded Antarctica (7.78 square miles) on September 20, 2014. Prior to 2020, Antarctica gained 7,000 square miles of sea ice (0.1 percent) since 1980.

The Antarctic Peninsula has warmed at a rate five times faster than the global average, while other portions of the continent have cooled.

The large-scale circulation around Antarctica traps cold surface water, limiting the upwelling of relatively warmer water that melts ice shelves from below. However, a stronger flow linked to rising global temperatures could push colder water away from the continental shelf, resulting in greater upwelling and the outgassing of carbon dioxide.

Detailed satellite analysis revealed that the ice sheet had melted at nearly triple the previous rate since 2017, losing more than 240 billion tons of ice annually, which was hastened by the demise of protective ice shelves that held back glaciers from flowing into the Southern Ocean. The total loss of ice since 1992 was estimated at three trillion tons—40 percent occurred from 2012 to 2017. At this rate, melting could add another six inches to the global sea level by 2100. The complete loss of the West Antarctic ice sheet could raise seas up to 17 feet.

In East Antarctica, the story has been different. The region includes a portion of the South Pole, which experienced an increase in ice thickness attributed to heavier snowfall.

The Ross Ice Shelf—the largest in Antarctica—that floats above colder water grew at a record rate between 2011 and 2014. Yet the unexpected discovery in 2015 of a river flowing underneath Totten Glacier sounded alarm bells among scientists.

During the austral summer and fall (December–May), a downslope westerly wind warmed the Antarctic Peninsula and drew warmer seawater beneath the glacier. A complete loss of the Totten Glacier could spur a rise in global sea level by more than six feet, based on the catchment size.

Health Impacts of a Warming Earth

On June 20, 2020, the temperature reached 100°F above the Arctic Circle for the first time in recorded history at Verkhoyansk, Russia. An estimated 800 to 1,200 people died in several unprecedented heat episodes that scorched the Pacific Northwest in June and July 2020.

In other parts of the world, the warming trend has resulted in more frequent heavy precipitation events and flash flooding. Torrential downpours have caused catastrophic flooding, causing considerable loss of life, and overwhelmed antiquated sewage systems in urban areas.

Intense heat is slower to dissipate overnight, especially in urban areas, raising health concerns. Soil degradation and expanding drought increase societal stressors caused by water and food shortages.

Warmer and wetter conditions have contributed to the spread of vector-borne illnesses carried by mosquitoes and ticks that benefit from milder winters. Mosquitoes carry more than 30 different viruses. Ticks are responsible for a rapidly increasing incidence of Lyme disease and many other illnesses after becoming infected by bacteria, viruses or parasites.

Prolonged drought exacerbated by more frequent intense heat waves results in crop losses and water shortages. Both Lake Powell and Lake Mead, in the upper and lower Colorado River Basin, reached all-time record low levels in 2021.

The desiccating and gusty Santa Ana circulation crosses the Sierra Nevada, and is funneled through canyons to the coast, warming by compression and drying the sparse vegetation. Leaves and chaparral become tinder-dry under a scorching summer sun. Strong winds created by high pressure over the Great Basin send embers beyond the ridge lines, spreading wildfires to the urban-wildland interface, putting more homes at risk. Burn scars are susceptible to mudslides and debris flow when the rainy season ensues the middle of autumn in Northern California.

Wildfires in the American West have torched double the number of acres compared to 30 years ago. In 2018, more than 8,500 wildfires in California burned 1.9 million acres, killing 100 people, including 86 in the Camp Fire around Paradise, the deadliest wildfire in the state's history. Total annual losses were estimated to be at least $16 billion.

Two years later, the August Complex Fire that ignited during a series of thunderstorms in Northern California on August 16–17, 2020, devastated communities and ecosystems in the Pacific Coast Range. A record 1,032,648 acres were scorched. During the blistering hot summer and early fall weather in 2021, lightning sparked wildfires that killed 2,261 to 3,637 giant sequoia trees. Heavy losses occurred in Sequoia National Park and the Sequoia National Forest area.

Ecological Impacts of a Warming Earth

Rising temperatures have an indelible impact on global ecosystems. Biodiversity is threatened by the loss of habitat due to the consequences of prolonged heat and drought. Wildlife must adapt to changes in vegetation patterns, resulting in an increase in animal migration.

Insects that devour crops are increasing in number, adding to the socioeconomic consequences of global warming. Denuded oak trees and the death of pines are tracked by foresters. The hibernation and migratory patterns of animals have been altered by warming conditions. Some species of birds and butterflies have shifted their range in search of food or have to make different choices. Spring warmth leads to earlier blooms, and active pollinators are more susceptible to harm in a late freeze.

Mountain birds forage for food at higher elevations in response to changing temperature and precipitation patterns. A shift in the marine ecosystem resulted in the loss of millions of Pacific cod in the Gulf of Alaska and thousands of California sea lion pups between 2013 and 2016, as a pool of anomalously warm water formed off the Pacific Coast from Alaska to California.

Chemical cycles in response to higher ocean temperatures cause coral bleaching, which disrupts the food chain. Increasingly acidic oceans harm kelp forests and marine species that depend on phytoplankton for food, depleting the fisheries and access to protein for millions of people around the world.

The growing season has lengthened by more than a week in the United States, and in parts of the Upper Midwest by 24 days (1995–2011) and 11 to 20 days elsewhere. This has been beneficial to fruits and vegetables in some areas, except where precipitation extremes have hampered yields.

Since 1895, the last freeze in autumn occurred a week later for the period of 2007–2016 compared to 1971–1980. The warming trend has delayed the fall color change in the northern portion of the country. Monarch butterflies normally reach Mexico around the first of November. In late October 2017, the butterflies were observed in Ontario, Canada, and New Jersey, when they usually have migrated to Texas. Whether born later, hanging around longer in warmer autumn weather, or getting stuck due to gusty winds, the natural cycle is disrupted.

Increasing concentrations of carbon dioxide could cause a decline in nutrients in food crops sold as futures on commodity markets. Anomalously wet patterns in the late autumn and early spring reduce crop yields, potentially leading to food shortages, with economic consequences affecting tens of millions of people. Reduced yields drive up production costs—fertilizer, seed and fuel. Cattle and hog farmers face a loss of feed.

The recent Great Lakes Restoration Initiative launched in 2010 is funded by the U.S. Environmental Protection Agency (EPA). Federal assistance is geared to improving water quality and testing, waste-treatment upgrades, best management practices and fertilizer application certification.

Rapidly warming waters are contributing to thicker algae mats. In the hot, stormy late spring and summer of 2011, a massive algal bloom covered a substantial portion of Lake Erie extending for more than 100 miles from Toledo to Cleveland in Ohio stretching to near the northwestern Pennsylvania shoreline.

Scientists at NOAA's Great Lakes Environmental Research Laboratory in Ann Arbor, Michigan, linked the larger blooms of blue-green algae (cyanobacteria) to warmer, wetter conditions and subsequent heavy runoff. Phosphorus and nitrogen loading, transported through waterways, can turn freshwater streams and lakes into a breeding ground for toxic algae. Inhaled poisonous droplets can cause respiratory and liver illnesses that are potentially fatal.

Nitrates enter the soil and groundwater from the application of farm fertilizers and manure, and often leach down into the drinking water. High exposures makes it more difficult for red blood cells to exchange oxygen, which is why nitrate levels are monitored by the EPA.

Agricultural runoff, leaky septic tanks, storm-sewer runoff, wastewater overflows and lawn fertilizers collectively increase the amount of nitrates and phosphorus, which is flushed into watersheds during rainstorms. The excess nutrients nourish slimy, greenish algal blooms that form a scum that thrives in warm, shallow water on hot, sunny days.

Although not all algal blooms are dangerous, microcystin toxin detected in tap water automatically triggers warnings, commencing at 1 part per billion (ppb). Microcystin is a class of toxins produced by blue-green algae that peaks in August and September. When algal blooms die, oxygen is removed from the water, leaving the aquatic environment hypoxic, which creates "dead zones" in parts of the Great Lakes and a large one off the Louisiana coast that extends nearly 9,000 square miles.

Blue-green algal blooms have been observed in lakes from the Adirondacks to Utah in recent years, and the number of blooms has increased. In August 2014, nearly 500,000 residents in the area surrounding Toledo, Ohio, were advised not to drink tap water due to a major increase in the algae population over the water-intake crib. The western Lake Erie basin had a record expanse of blue-green algae in the summer of 2015.

Marine ecosystems have been influenced by rising sea surface temperatures. Some species of warm water fish have migrated north into cooler waters to survive. The consequences of a decline in marine life in certain areas extend up the food chain to seabirds, as sources of sustenance have been displaced.

Fisheries and fish farming (aquaculture) have suffered from the increase in the number of ocean heat waves—defined as much-above normal readings for at least five consecutive days. Some commercial fish and shellfish species have declined by as much as 35 percent. Lobsters normally found in the Mid-Atlantic coastal waters near New Jersey have been congregating off the Maine coast in significant numbers in the past decade due to rising sea surface temperatures.

Rising ocean surface temperatures have also caused acidification in the oceans, and the loss of kelp forests and sea grass. In a recent warm period, nearly 20 percent of the global coral reefs died. The reefs are made from coral polyps and single-celled algae, capturing sunlight and carbon dioxide to create sugars used for food by tiny animals.

The uptake of dissolved carbon dioxide reduces the pH of seawater, diminishing the ability of corals to develop the calcium carbonate skeletons that comprise a large portion of the reef foundation. As the waters warm, the chemistry is altered and algae are expelled, causing coral bleaching and die-offs. Rising sea surface temperatures also thin the shelves of some animals and oysters.

In 2013, cold water failed to move south due to persistent high pressure over the North Atlantic, which diminished the mixing of cold air and contributed to the loss of corals in the Caribbean the following year.

Record marine warmth in 2016 would affect nearly 660 miles of Australia's 1,380-mile Great Barrier Reef, which has been around for 25 million years. Coral bleaching killed about half of the shallow-water corals that are an essential part of the ecosystem. Previously, unusual tropical Pacific warmth in 1997–98 killed 16 percent of the global corals.

Wildlife biologists have recorded changes in roaming patterns related to warmer conditions and habitat changes. Invasive species and fungal diseases could further alter the forest ecosystem. Millions of trees have died in California from a combination of drought (2011–2015), a blistering summer (2017), and ravenous bark beetles.

In western North America, warmer winters are expanding the insect ranges. Mountain pine beetles have gained a foothold, ruining huge tracts of whitebark pine trees. Animals that depend on pine nuts, including bears, experienced a loss of food sources. Spruce bark beetles killed millions of white spruce trees in Alaska as the weather warmed.

The snowpack in the Sierra Nevada reached a 500-year minimum during the winter of 2014–15 during the fourth consecutive year of a severe drought, based on tree ring analysis of the blue oaks surrounding the Central Valley. The Sierra watershed is essential to California's supply of drinking water and agricultural needs. Excessive moisture in 2022–23 alleviated the shortages.

Very warm and dry conditions have plagued the Southwest, resulting in wildland fires, deforestation and the loss of river habitat and hydroelectric power. Insect blight and clear-cutting release carbon dioxide back into the atmosphere, raising the temperature. The boundary of some tree species is shifting northward. Plant diseases caused by pest infestations and the proliferation of invasive species have impacted ecosystems and displaced the habitats of animals.

Donora Smog Disaster in October 1948

Donora, Pennsylvania, a vibrant community in the Monongahela Valley, 25 miles southwest of Pittsburgh, had experienced prolonged smoggy periods for decades in the early 20th century.

The steel and zinc mills, including U.S. Steel's Donora Zinc Works and American Steel and Wire plant, periodically filled the air with sulfur emissions and other toxic gases, which would become trapped below a layer of unusually warm air just above the valley floor. The towering 150-foot-high smokestacks pumped gases and fine particles into the air that hung over the valley in a stagnant pattern with little wind for days on end.

On October 26, 1948, high pressure building southward from Ontario becalmed the region and brought a pattern of early morning fog. The industrial valley, surrounded by hills rising 400 feet above the valley floor, became entrapped in a poisonous smog. High levels of sulfuric acid, nitrogen dioxide and carbon monoxide built up over four days, as residents increasingly complained of breathing difficulties.

The atmospheric inversion—warm air overlying cool air—trapped the toxic yellow smoke. A blanket of fog on the mornings of Halloween weekend further rendered sunlight ineffective as a mixing agent by limiting solar energy, inhibiting vertical motions. The deteriorating air quality sickened thousands of residents.

By October 30, 1948, more than 20 residents had succumbed after a week of being subjected to the toxic stew. An estimated 6,000 persons would become ill in the valley with various forms of respiratory distress during and after the disaster. Rain finally arrived on October 31, 1948, clearing the air the next day and dissipating the toxic pall. Months, and probably years, later, more residents perished from the effects of respiratory diseases and lung damage acquired during the Donora Smog Disaster.

The scope of the disaster opened the door for scientists to study the relationship between air pollution and health.

Aerosol Impacts on Climate and Human Health

The burning of fuel at very high temperatures contributes to poor air-quality days and morbidity. The presence of fine particulate matter (aerosols) that reacts with hydrocarbons forms nitric acid or acid rain. Hydrocarbons are by-products of unspent fuel from incomplete combustion.

Tailpipe emissions from diesel engines, though yielding less carbon dioxide as compared to most gasoline-powered vehicles, release harmful levels of nitrogen dioxide that are measured in micrograms per cubic meter. Methane and oxides of nitrogen emitted from refineries, power plants, and gasoline-powered vehicles react with sunlight and raise ground-level ozone levels.

Smog is the poisonous pall of smoke mixed with fog hanging over cities for days in stagnant weather patterns that aggravates disease. Harmful volatile organic compounds (VOCs) are emitted from industrial sources that produce cement, asphalt, biofuels, paints, pesticides, solvents, hair spray, vinyl floors, hydrogen and sulfuric acid.

Anthropogenic sources of nitrous oxide are fossil-fuel emissions, industrial processes involving nitric acid, clearing agricultural land through burning, and fertilizer applications.

Black carbon, ash and soot from smokestacks, chimneys and wood-burning stoves have historically befouled the air. Burning coal for electricity produces sulfur dioxide, mercury and nitrogen oxides. Sulfate aerosols produced by the oxidation of sulfur dioxide from coal-burning power plants are heavily concentrated in urban areas.

Fossil fuel emissions from coal-burning power plants, smokestacks and vehicles emit particulate matter referred to as PM2.5 (2.5 micrometers or smaller). These tiny particles cross cell membranes from the lungs to the heart and other organs by traveling through the bloodstream. Air pollution from gases and particulates increases the risk of heart and lung disease, cancer, stroke, heart attack and worsening asthma caused by inflammation in the respiratory tract.

The Clean Air Act, passed by Congress in 1970, improved the air quality by limiting pollution released from industrial and energy sources, chemical factories and transportation. The network of air sensors monitor local air quality. Levels rising above AQI standards trigger statewide air alerts.

Hotter, drier weather has increased the number and intensity of wildfires. Plumes of smoke release toxic fumes carrying carbon monoxide and metals. On July 6-7, 2002, the sky darkened Pennsylvania, New York and New Jersey, and the visibility dipped to a half-mile in an acrid haze from up to 170 fires in northern Quebec ignited by thunderstorms four days earlier, fanned by strong winds. The billowing smoke reached an altitude of 12,000 feet.

More than 150 wildfires in Quebec, Canada, led to particle pollution warnings on June 7-8, 2023, across the Midwest to the Southeast. Smoke plumes were carried by a northerly flow, which cloaked skylines in an orange haze that blotted out the sun and raised the AQI to record levels (300 to 400-plus).

Certain nitrogen oxide compounds form during reactions with sunlight such as alkyl nitrates. Trees and vegetation release hydrocarbons that produce gases. Nitrous oxide rises from the soils and oceans.

Particle pollution gives rise to fiery sunrises and sunsets with the right particle size, distribution and chemical reactions. Blue light is scattered out, and longer wavelengths create a reddish-orange glow.

Trees give off oxygen, water vapor and aerosols that scatter sunlight, and react with particulates to create a bluish haze over Appalachian forests. The scent of pine trees comes from terpenes. Another compound (isoprene) reacts with nitrogen oxides to form ozone and fine particulate matter.

Aerosols emitted from smokestacks and motor vehicles scatter, reflect and absorb solar radiation in a complex interaction that has been linked to shifts in rainfall patterns in parts of the world. The climatic impact of particle pollution depends on the size and distribution of the aerosols. Organic compounds contained in some aerosols increase the probability of cloud droplet formation by reducing surface tension. Smoke, sea salt, dust and volcanic dust in the atmosphere scatter light, making clouds composed of liquid and solid droplets highly reflective.

Solar Variability and Weather

The sun fuses hydrogen fuel that is converted into helium in its core. The energy output, or luminosity, is essentially a constant, which varies by 0.1-0.2 percent.

The total amount of solar radiation striking a surface perpendicular to the top of Earth's atmosphere is given as 1,367 watts per square meter (World Meteorological Organization).

The sun is a ball of gas that generates a turbulent flow of plasma, which takes a convoluted path to the outer shell (photosphere) because the sun rotates faster at its equator. The resultant shear torques the magnetic field, causing loops to become entangled, snap and erupt at the surface from the differential rotation in large waves.

The temperature at the surface of the sun is close to 10,000°F. Beyond the photosphere is the red chromosphere, where the temperature climbs to 14,000°F at the top. This layer joins the corona that expands the solar atmosphere for millions of miles, where the temperature eventually reaches a sizzling 3.5 million °F—300 times hotter than the solar surface.

Galileo was the first to observe curious dark blemishes on the solar surface in 1610 through his telescope. The dark spots that appear singly or in clusters are associated with twisted loops of the strong magnetic fields that emerge from electrified gases. The spots are more than 1,000°F cooler than the solar surface because upward-moving energy is trapped below.

Solar flares erupt in the vicinity of sunspot groups, releasing pulses of electromagnetic energy. Filaments wrap around the surface when a hot stream of gas (prominence) rises above the limb of the photosphere. As the corona cools, a loop of plasma and hot gases is threaded across gaps in the twisted magnetic field lines. An unfathomable amount of energy, on the order of millions of hydrogen bombs, is released in large eruptions with up to a dozen solar flares in a day.

High-energy charged particles (protons, electrons, ions) and X-rays are whisked away by the solar wind and move through interplanetary space at speeds of 1 million mph, slamming into Earth's protective geomagnetic field. High-speed collisions excite gas molecules in Earth's upper atmosphere, producing the spectacular northern and southern lights.

Coronal mass ejections (CMEs) reach Earth's atmosphere in 24 to 48 hours, signaling a geomagnetic storm. Major solar eruptions are capable of wreaking havoc with power grids and satellite navigation and cause high-frequency radio blackouts. The impact of a CME depends on the orientation of Earth's magnetic field lines that normally deflect most of the incoming material as the magnetosphere is compressed. If Earth is facing in the direction of the solar wind, an impressive display of the northern lights (aurora borealis in the Northern Hemisphere) dances and arcs in shimmering green and dazzling red curtains.

Large magnetic storms cause electrons directed by Earth's magnetic field to broaden out in an oval shape, which increases the chance of seeing the northern lights farther south in the mid-latitudes if skies are clear. Charged particles that strike oxygen atoms 60 miles above Earth's surface give off pale yellow-green auroras, with a reddish veil at much higher altitudes. Blue and purplish red colors reflect solar plasma interacting with nitrogen atoms, mostly below 60 miles, which is a rarer occurrence.

Sunspot numbers vary from a maximum to a minimum, peaking every eight to 13 years. The 22-year double sunspot cycle constitutes the period for a reversal of the magnetic poles back to the original configuration. Although the variation in total irradiance, or energy received from the sun, during a sunspot cycle varies only 0.1–0.2 percent, there are considerable differences depending on the wavelength in an active solar period. Ultraviolet radiation reaching Earth's atmosphere is significantly greater, which is capable of influencing the stratospheric circulation.

A finding by astronomer John A. Eddy (1931–2009) in the 1970s linked the absence of sunspots during the 1645–1717 grand (Maunder) minimum to the coldest portion of the Little Ice Age (1570–1730). The Maunder minimum was preceded by a 5 percent decrease in the solar spin rate near the equator. Another diminution in sunspots from 1790–1825 (Dalton minimum) coincided with a chilly period in the Northern Hemisphere.

Solar luminosity reached a minimum in 2009, with 260 days showing no sunspots—the lowest value of electromagnetic activity since satellite measurements commenced in 1978. A secondary sunspot minimum was observed on May 10, 2010. Sunspot numbers peaked in December 2014 at 101 during Cycle 24, far below the historical average (160–240) observed during the solar maximum of 2014, before falling again. In 2018, many days featured a spotless sun. These longer cycles are thought to precede a pronounced minimum in solar activity, possibly related to a more meridional solar convective flux.

A proposed solar-weather connection has focused on the relationship to the speed of the easterly trade winds, which are somewhat stronger during an active solar phase. There is also evidence of an increase in precipitation in the middle latitudes of the Northern Hemisphere during high sunspot periods, mostly related to above-average pressure centers and a loopy jet stream associated with larger, slow-moving storms and greater temperature extremes.

The run-up to a solar minimum coincides with weaker circumpolar winds, favoring a high-amplitude flow, or blocking pattern, conducive to polar outbreaks reaching the eastern states. High sunspot activity has been correlated with the formation of upward-moving tropospheric waves and a stronger polar vortex, which limits the southward displacement of cold air and mild winters in the Northeast. Large waves enhance the temperature gradient between the troposphere and stratosphere and strengthen the jet stream.

Another avenue of research linked maximum sunspot activity to a global La Niña-like response, a cyclical climate system reflecting cooler waters that drift westward in the eastern tropical Pacific, and higher pressure over western Canada. This configuration favors wetter and cooler conditions in the Pacific Northwest, as Pacific storms undercut a western ridge and move across the Upper Midwest, with drier and mild winter conditions in the East.

Energy associated with a burst of ultraviolet (UV) energy during heightened solar activity can destroy parts of the ozone layer, warming the stratosphere as ozone molecules are split. A purported link exists between active sunspot periods and a decrease in global cloud cover that would favor generally warmer and drier conditions in the mid-latitudes.

A possible cause-and-effect relationship has been posited between solar activity and the solar wind deflecting galactic cosmic rays after a CME. In a sunspot minimum, the chance for a CME decreases and cosmic rays impact Earth to a greater degree. The cosmic-ray flux is thought to seed low-level clouds through the ionization of air molecules, producing more aerosols that serve as condensation nuclei, resulting in increased cloudiness and possible precipitation.

Volcanism and Climate

Volcanoes are the tumultuous manifestation of Earth's relentless spasms. Fissures spew red-hot lava that scorches everything in its path, filling the air with hydrochloric acid and fine volcanic particles composed of rock and glass. Sulfur dioxide and carbon dioxide react with oxygen and water vapor in the presence of sunlight to create a layer of volcanic smog.

Large eruptions expel huge volumes of ash, gases and pollution. Thousands of tons of fine particles float for months in the circumpolar winds. Volcanic aerosols scatter sunlight, which temporarily cools the planet after the particles reach the stable lower stratosphere.

Sulfur compounds emitted in large quantities by upheavals such as Santa Maria (October 1902), Mount Agung (March 1963) and El Chichón (1982) lowered the average global temperature by a few degrees. Sulfate particles mixed with dust and ash form sulfuric acid droplets by interacting with scant moisture. Mount Pinatubo blew its stack in the Philippines on June 12–15, 1991, lofting about

17 million tons of sulfur dioxide up to an altitude of 12 miles.

An even greater eruption that commenced on April 5, 1815, in the Dutch East Indies (Indonesia) culminated in the vast explosion of Mount Tambora, which rocked the mountain to its core six days later. Historians believe the death toll approached 100,000 people, most consumed by fire, molten rock, debris and ash. Crops and most vegetation were obliterated. Scientists estimate that 12 cubic miles of material were ejected to an altitude of at least 25 miles, circling the globe many times and cooling Earth by as much as several degrees.

An exceptionally chilly decade followed multiple volcanic eruptions in AD 536 and 541, in the Pacific Northwest and also Central America. Geologists have charted the eruption of 40 volcanoes worldwide since 2000 from Alaska to Indonesia and Eritrea. An increase in volcanic activity and aerosols (sulfates) could partly explain the short-term pause in rising global temperatures that occurred in the middle of the last century.

One of the aesthetic effects of a large eruption is marvelous red sunsets created by the interaction with tiny water droplets. A sulfuric acid haze scatters mostly orange-red light, presenting the image of a fiery sun rising or sinking near the horizon.

Major tropical volcanoes have been linked to El Niño patterns by a diminution of the West African monsoon between 9° N and 20° N, which weakens the tropical easterlies and results in a westerly flow near the equator in the North Pacific.

Ozone Hole

A protective layer of ozone in the stratosphere absorbs harmful solar ultraviolet (UV) radiation that is not screened by atmospheric gases. Direct exposure to UV radiation can cause skin cancer, cataracts and crop damage.

The NASA-NOAA Suomi NPP satellite and NASA's Aura, coupled with balloon-borne measurements, continue to monitor the seasonal ozone hole over Antarctica. In early October 2015, the Ozone Monitoring Instrument (OMI) recorded an ozone hole of 10.9 million square miles—fourth largest since 1979—at a minimum concentration of 101 Dobson Units.

The single-day record size of 11.5 million square miles occurred on September 9, 2000—larger than the area of North America (9.5 million square miles). Two years later, NASA and NOAA data showed that the ozone hole over Antarctica shrank to its smallest extent since 1988, linked to greater mixing and fewer ice crystals.

In the winter of 2020, a very strong and stable polar vortex formed, possibly linked to record warm North Pacific waters producing fewer large waves in the Arctic atmosphere. In the summer of 2021, a record cold regime at the South Pole contributed to an unusually large hole due to the strengthening of the stratospheric polar vortex and an increase in stratospheric clouds that reacted with ozone-depleting chemicals such as chlorine and bromine.

Satellite measurements that commenced in 1979 first revealed a decrease in the concentration of ozone below 220 Dobson Units, which constituted a "hole" in the ozone shield. In October 1985, a 30 percent loss of ozone was observed six to 10 miles above Antarctica. Ozone depletion has also been observed in the Arctic and mid-latitudes, but not to the same extent. In 1992, the opening expanded to the size of North America.

Refrigerant gases found in air conditioners and various home appliances, aerosol can propellants, fumigants, constituents in cleaning and industrial solvents all comprise chlorofluorocarbons (CFCs), known to be ozone-eaters. The destruction of ozone begins with chemical reactions on the surface of ice crystals in the lower stratosphere that strip CFCs of chlorine and bromine, breaking apart ozone in the coldest months, when polar stratospheric clouds are most abundant over Antarctica. Nitrous oxide released from fertilizers also destroys ozone.

A global effort to phase out human-produced halogens culminated in the signing of the Montreal Protocol in 1987, an agreement that banned CFCs internationally that went into effect in 1989. The results have been quite successful. The ozone concentration above Antarctica rose 4 percent from 2000 to 2013 at an altitude of 30 miles. Since 2000, the ozone hole has decreased 16 percent, shrinking more than 1.5 million square miles.

Pacific Climate Systems

In the early 1920s, Sir Gilbert Walker (1868–1958), a British physicist and statistician, discovered a cyclical pattern of atmospheric pressure patterns in the equatorial Pacific Ocean. Walker analyzed pressure fluctuations responsible for temperature and precipitation variability extending beyond the western Pacific Basin, now referred to as the Walker circulation.

Since the 1600s fishermen off the coastal waters of South America noticed a pattern of abnormally warm waters that peaked around Christmastime. Warmer water suppresses nutrient rich cold water from reaching the surface, resulting in a decrease of phytoplankton that fish feed on, affecting those species that consume the fish. Sailors observed the seasonal warm water cycle that hovered near Peru and off the coast of Americas, dubbed El Niño ("the child").

El Niño is a cyclical warming of the surface waters in the central and eastern equatorial Pacific. As the generally steady tropical easterlies weaken, warm water sloshes eastward from the western portion of the Pacific Basin. The warm phase of the climate pattern known as El Niño–Southern Oscillation (ENSO) has far-ranging impacts that affect the health of marine ecosystems, fisheries, agriculture and public health.

Atmosphere-ocean feedbacks are more evident where warm waters add heat and water vapor to the overlying air, which enhances upward motion and creates waves in the jet stream. Low pressure in the western tropical Pacific moves east, accompanied by clusters of thunderstorms that bring heavy rain to coastal South America in the autumn and winter. The larger-scale circulation is enhanced by an energized subtropical jet stream across the southern United States that transports moisture, fueling coast-to-coast storms and a southward shift of the polar jet stream.

The intensity of El Niño is quantified by the Southern Oscillation Index (SOI), which is based on standardized mean sea level pressure (MSLP) between Darwin, Australia, and Tahiti.

El Niño corresponds to a negative SOI value. A reversal, or positive SOI phase, indicates strengthening easterly trade winds that cause the upwelling of cooler waters, which flow westward. The sister of El Niño is called La Niña, comprising an ENSO cycle that varies from three to seven years between warm and cool phases.

El Niño brings drought in the Western Pacific Rim, affecting portions of Australia, South Asia and South Africa. The west coast of South America and the southern United States receive enhanced rainfall from an active storm track fueled by a strong subtropical jet stream. Another well-known El Niño long-distance climate "teleconnection" is a decrease in Atlantic hurricanes, because wind shear tears the tops off thunderstorm clusters and inhibits organization.

An unrelated abnormality off the West Coast from 2013–2016 occurred when an extensive pool of warm water up to 11°F above normal expanded from Baja California to the Bering Sea, nicknamed "the Blob" by a Washington state climatologist. A resilient downstream ridge of high pressure aloft in the northeastern Pacific caused intensified drought conditions in California. The anomaly that first appeared in October 2012 lingered until 2015. The average sea surface temperature in 2015–16 rose to a record-tying 3.6°F above normal, equaling the historic El Niño of 1997–98.

In the summer of 2015, a formidable El Niño took shape that brought a parade of Pacific storms inland. El Niño energy combined with an atmospheric river to boost the northern Sierra snowpack to 87 percent of the historic mean in March 2016. An active southern storm track is the hallmark of a strong El Niño. Low pressure in the Gulf of Alaska funnels warmth into Alaska and northwestern Canada and deflects polar outbreaks that are blocked by a persistent westerly or zonal flow east of the Rockies.

The strongest El Niño of the 20th century occurred in 1997–98. An endless string of severe storms lashed Southern California and the southern United States, killing 189 persons and causing $4 billion in damage nationwide from a vicious combination of torrential rain and powerful windstorms, with outbreaks of severe storms in the Southern states.

The northern tier of the states basked in exceptionally mild winter conditions. Teleconnections caused sinking air over the Asia–Pacific region, with severe drought in Indonesia, Malaysia and the Philippines, resulting in heavy crop losses. On the opposite end of the North Pacific, powerful storms battered coastal South America from Ecuador to northern Chile, triggering deadly flooding and mudslides.

A variant of El Niño, with two Walker cells labeled El Niño Modoki, is centered farther west, with an active subtropical jet stream and a ridge of high pressure in western Canada that carries a cold northwesterly flow of air into the eastern United States. Phasing of the polar and subtropical jet stream results in above-normal snowfall in the Mid-Atlantic. This outcome is conditioned on the formation of a blocking high-pressure ridge in the Arctic that forces cold air southward into the northeastern United States. A number of major East Coast snowfalls occurred in moderate El Niño winters: 1957–58, 1969–70, 2002–03 and 2009–10.

The sister circulation, La Niña, reflects strengthening easterly trade winds that push warm water westward. In the eastern tropical Pacific, the upwelling of cooler, nutrient-filled water provides a boon to fisheries. Teleconnections include a strong East Asia monsoon bringing heavy rain and cool weather. Wetter conditions affect East China, Indonesia, the Philippines, northern Australia and southern Africa, with drying in the Horn of Africa and Central Asia.

The interaction between cool La Niña waters and the overlying atmosphere displaces the polar jet stream northward, as a staunch ridge of high pressure forms east of the Gulf of Alaska. Moderate to strong La Niña cycles feature strong low pressure over the northeastern Pacific that taps atmospheric rivers, triggering lengthy bouts of stormy weather across the Pacific Northwest and northern Rockies.

Colder-than-average weather prevails in a northwesterly circulation from the northern Plains to the Great Lakes. The jet stream curls northeastward over the eastern Great Lakes, generally sparing the Northeast and Mid-Atlantic from deeper blasts of arctic air. Heavier-than-normal precipitation occurs in the Ohio Valley and eastern Midwest on the boundary between cold and mild air, with drier conditions east of the Appalachians.

Infrequent phasing of the polar and subtropical jet streams limits the number of Atlantic coastal storms with a prominent subtropical high-pressure zone near the Southeast Coast. Philadelphia NWS records since 1950 show no winter storm depositing 10 inches or more of snow in La Niña, although the great snowfall of January 31–February 2, 2021, brought massive totals ranging from one to three feet north of the Philadelphia area. A mild La Niña winter can always shift gears with the development of a North Atlantic block and southward expansion of the polar vortex, resulting in several weeks of harsh winter weather in January and February.

A lingering effect of La Niña is a tendency for hot, dry summers with the expansion of a subtropical ridge in the Southeast. A higher risk of North Atlantic hurricanes arises due to less vertical wind shear. Cooler waters in the eastern tropical Pacific promote sinking air, with the opposite portion of a broad circulation cell promoting rising air over the western Atlantic Basin.

A long-term governing climatological pattern above 20° N in the North Pacific, discovered in the late 1990s, was designated as the Pacific Decadal Oscillation (PDO). The PDO varies over a period of 20–30 years featuring warm and cool sea surface temperatures cycles.

In a positive (warm) PDO phase, cooler water resides in the central North Pacific. The temperature gradient that favors low pressure in the Gulf of Alaska is displaced southeast, pumping warm air into western North America. Western ridging leads to a northwesterly flow east of the Rockies. A positive PDO occurred from 1925–1946, 1977–1998, 2003–2007 and 2014–2019.

A negative (cool) PDO features below-normal sea surface temperatures in the eastern part of the North Pacific as far north as the Gulf of Alaska and more frequent La Niña seasons. Cool PDO phases were observed from 1890–1924, 1947–1976, and for shorter periods in 1998–2002 and 2008–2013.

The Eastern Pacific Oscillation (EPO) quantifies the distribution of pressure anomalies and sea surface temperature patterns. A negative EPO is a warm water anomaly off the western coast of North America. High pressure directs cold Canadian air into areas east of the Rockies. A positive EPO is characterized by a persistent trough south of the Gulf of Alaska, with weak ridging in the eastern North Pacific promoting a low-amplitude westerly flow of mild Pacific air across the Lower 48 states.

The Madden–Julian Oscillation (MJO) reflects phases in a 30- to 60-day eastward-moving cycle of thunderstorm activity that affects the global distribution of heat and enhances the subtropical jet stream. Tropical forcing amplifies planetary waves, particularly in winters when the polar vortex is more unstable. The added energy impinges on the upper-level flow and intensifies an El Niño rainfall pattern. Rising air also enhances tropical storm development, as moisture-laden air is drawn northward by the subtropical jet stream.

Pulses of energy in the MJO interacting with warm waters in the Indian Ocean and across the western tropical Pacific build clusters of towering clouds and storms drifting eastward. Eight phases lasting about five days are associated with areas of upward and downward motions, resulting in greater or diminished convection, respectively. A phase of rising motion linked to atmospheric rivers enhances rainfall and mountain snowfall in the western United States.

Atlantic Climate Systems

The thermohaline circulation (THC) reflects how sea water heat fluxes are regulated by differences in temperature and density, influenced by wind and tidal currents that drive ocean currents.

The Atlantic Meridional Overturning Circulation (AMOC) describes the global ocean conveyor belt. Warm water transported by the Gulf Stream northward from the equatorial region past the Caribbean Sea parallel to the east coast of North America cools in the North Atlantic, eventually sinking south of Greenland to depths of 3,000 to 16,000 feet. The deep-water current travels southward past the edge of Africa and into the Indian Ocean and Southern Ocean before rising again in the South Atlantic.

The introduction of less salty, lighter freshwater from the Labrador Sea as glaciers and sea ice melt more rapidly has slowed the conveyor belt due to the change in density of the upper layers of the ocean. Studies have shown that the AMOC has weakened more than 15 percent since the 1920s. The long-term effect of a reduction in the poleward movement of tropical heat would be a colder pattern in the Canadian Maritimes, northeastern United States and northwestern Europe.

Proxy data obtained from deep ocean sediment cores containing calcium carbonate shells of foraminifera—tiny organisms in marine environments—revealed a natural rhythm in the conveyor belt of 1,500 years, which is thought to be regulated by long-term solar and oceanic influences.

The Atlantic Multidecadal Oscillation (AMO) is a pattern of sea surface temperatures that follows a timescale of 50–80 years. A cool (negative) time period is correlated with fewer Atlantic hurricanes. Teleconnections include drier summers in faraway places, such as Great Britain, Ireland, Sahel region, west-central India, and the southern Sahara Desert. Variability in the North Atlantic Ocean sea surface temperature anomalies affects ocean heat content (OHC) that fuels tropical cyclone development.

Cool AMO phases prevailed from 1903–1925 and 1971–1994. Warm (positive) phases occurred during the periods of 1926–1969 and 1995–2000. More recently, there has been considerable variability with short-term fluctuations every few winters. A tendency toward a warm phase resumed after 2015. A relationship appears to exist between the Pacific and Atlantic when certain phases of the PDO and AMO overlap. The coincidence of a warm PDO and warm AMO corresponded with persistent dryness in the central and eastern United States and somewhat colder winters.

The Arctic Oscillation (AO) is a measure of the intensity of the polar vortex, the semi-permanent whirlpool of frigid air in the middle and upper levels of the atmosphere centered above the North Pole. A warmer (positive) AO is associated with a colder-than-normal stratosphere and a stronger, more restrictive circumpolar circulation that hems in brutally cold arctic air. A colder (negative) AO reflects anomalous high pressure over Greenland and the Arctic, forming a blocking pattern. The logjam causes the polar jet stream to buckle, opening the floodgates for intrusions of frigid outbreaks in the eastern United States, especially if the polar vortex is disturbed.

The North Atlantic Oscillation (NAO) index quantifies the pressure difference between the semi-permanent Icelandic Low near Greenland and subtropical Azores–Bermuda High. A warm (positive) NAO phase favors a persistent westerly (or zonal) flow, created by strong subpolar low pressure and above-normal pressure in the subtropical Atlantic. The jet stream and storm track are displaced northward, confining frigid air masses to mainly northern and central Canada, resulting in generally milder winters in the Northeast.

A negative NAO, which is associated with warming of the polar stratosphere and a weaker polar vortex, links higher pressure in the Arctic and lower-than-average pressure in the central Atlantic. High-latitude blocking funnels a harsh chill into the eastern United States and promotes a favorable bend in the jet stream near the Eastern Seaboard that promotes nor'easters bearing heavy snow and wind. There is evidence that a deep snow cover in portions of northeastern Europe and Asia in late autumn contributes to an increasingly negative AO/NAO pattern.

A very persistent negative NAO climate pattern between the mid-1950s and late 1970s resulted in clusters of cold, snowy winters in the Northeast. In the late 1980s, a consistently positive NAO index brought a series of mild winters. The NAO index has flipped to highly negative values for short intervals that brought severe winters and deep snows in the Northeast and Mid-Atlantic in 1995–96, 2013–14 and 2014–15. The more recent consecutive brutally cold winters were likely affected by a pool of anomalously warm water in the northeastern Pacific and subsequent ridging in the jet stream in the West, which strengthened the northwesterly flow aloft east of the Rockies.

Sudden Stratospheric Warming and Quasi-Biennial Oscillation

Normally, the polar vortex is pent up over the North Pole region most of the winter season. The occasional disruption of the circumpolar flow reaching the stratosphere disturbs and sometimes splits the polar vortex, sending bitterly cold air plunging southward into the central and eastern United States.

If the jet stream encounters vast weather systems, such as expansive high pressure over Siberia and northeastern Asia, wave energy propagates upward from the troposphere into the stratosphere and stretches the polar vortex. Over time a piece of the cold pool descends through the atmosphere and raises the surface pressure over the Arctic. During the course of several weeks the lobe of brutally cold air is directed into the northern United States and Eurasia.

A sudden stratospheric warming phenomenon begins in response to the upward transfer of wave energy, which slows the westerly winds near the stratospheric equator. The restoration of equilibrium creates a downward transfer of momentum, with implications for mid-level dynamics in the troposphere below. Wave energy moves downward at the rate of about 0.6 mile per month through the upper troposphere.

Research by Judah Cohen, a climate scientist with Atmospheric and Environment Research, has focused on a decrease in the overall temperature gradient in the lower atmosphere between a warming Arctic and the equatorial zone. The smaller temperature gradient translates into weaker circumpolar winds, occasionally destabilizing the polar vortex. Cohen's theory also incorporates Eurasian snow cover, which has a tendency to increase as the frigid air surrounding the Arctic becomes warmer and contains more moisture. Open water in the Arctic Ocean during the short summer further serves as a moisture source.

On December 31, 2018, the polar vortex situated 20 miles above the surface split into three pieces in response to a sudden stratospheric warming of 125°F, a very dynamic event moving down from the stratosphere to the troposphere (lower atmosphere). One lobe of frigid air traveled south across eastern Canada into the Upper Midwest and Great Lakes, eventually bringing bitterly cold temperatures of -20°F to -40°F and dangerous wind chill values of -40°F and -60°F on January 31, 2019. The deep dip of the polar jet and expanding frigid outbreak claimed more than 25 lives and injured hundreds in eight states, a result of frostbite, carbon monoxide poisoning and broken bones in falls and heart attacks.

A 28-month cycle that reflects a reversal of the winds in the equatorial stratosphere, which alternates between easterlies and westerlies, is linked to upward-moving gravity waves high in the tropical troposphere, influencing the polar westerlies. The long-recognized Quasi-Biennial Oscillation (QBO) affects the storm track across the North Pacific, and seasonal tropical cyclone activity in the North Atlantic.

The easterly (negative) phase is associated with a weaker jet stream that shifts poleward above the North Pacific, then dips south over the North Atlantic, sending cold air masses into the eastern United States and Northern and Central Europe. In a westerly (positive) phase, the polar jet stream becomes faster and is positioned farther south over the North Pacific, but higher up over the North Atlantic, favoring mild and wet winters in the East.

The alternating easterly and westerly phases of the QBO have different effects on the MJO and atmospheric rivers. Stronger zonal winds migrate downward in the atmosphere and create large-scale equatorial waves that impact the larger-scale circulation between the tropics and mid-latitudes.

Climate Modeling and Mitigation

Climate models simulate the weather from a set of initialized variables—temperature, pressure, wind, moisture—that require supercomputers to unravel the many threads running through climate change. Complex interactions between the atmosphere, ocean and land surface are difficult to process in the context of human activities. Researchers calculate the probable frequency of extreme weather events as possible outcomes without human influences to gauge the impact of anthropogenic climate change.

Policies designed to reduce the levels of greenhouse gases emitted from coal-burning power plants, smokestacks, cement production and transportation rely on the implementation of cost-effective adaptive strategies. Ongoing efforts to reduce global carbon emissions have been hindered by disagreements over the economic impacts of proposed regulatory measures and the viability of switching to cleaner renewable energy sources.

The 26th annual climate summit (COP26) in Glasgow, Scotland, in 2021 offered nations an opportunity to review pledges to reduce carbon emissions that were made during the 2015 Paris Agreement. In 2015, the Intergovernmental Panel on Climate Change (IPCC) urged upgrades in infrastructure and housing, and stricter regulation of industrial, agricultural and fossil-fuel emissions, which included 195 signatories. The goal is to limit the rise in the global average temperature to below 3.6°F by reducing carbon emissions to net zero by 2050.

Yale and George Mason University found that 69 percent of Americans are concerned about global warming, with an understanding that the issue is no longer something to be concerned about in the distant future. Researchers believe that people relate to climate change through the lens of extreme events they experience or view on television or the internet, which makes the message more visceral, rather than an abstract future concern.

Those who question the threat of climate change point to the fact that China emits more than twice the share of greenhouse gases as the United States, followed by India, to manage the need for electricity. In the United States, carbon emissions have been reduced from a combination of regulation and a shift to natural gas.

Nuclear power accounts for 20 percent of the nation's energy. Other renewable technologies, or clean (non-carbon) energy sources—solar, wind and hydroelectric power—are proving to be economically beneficial to local communities, creating new jobs and commerce opportunities. Wind and solar power currently provide 8 percent of our electricity. Carbon sequestration technology removes carbon from the air, which is turned into liquid and buried deep underground. Other uses extend to the fields of construction and medicine.

The ocean, land and cryosphere (snow, ice) are very sensitive to even subtle changes in the heat and moisture fluxes. As the atmosphere warms, more moisture evaporates and rises in the form of water vapor to create clouds and precipitation. In areas where the air is undergoing subsidence or sinking motions, evaporative losses of soil moisture and a vapor pressure deficit lead to expanding drought.

Earth's climate system is better understood through models that incorporate the atmosphere, land and sea measurements of temperature, moisture and wind. The trickiest aspect is modeling cloud cover. Climate models unambiguously link rising carbon emissions to higher temperatures, heavier rainfall, melting glaciers, higher sea levels and coastal flooding.

The dynamic atmosphere-ocean system is subject to natural (internal) forcing and the effects of rising greenhouse gas emissions. Global models look at the present set of variables and known climate feedbacks to simulate future circulation patterns. The response to model-driven climate data is focused on mitigating the impacts of more frequent heavy downpours and higher storm surge in coastal communities.

The construction of larger storm drains, fixing weakened and aging sewer systems to manage excess storm water runoff, and zoning law changes that prohibit construction in flood-prone areas are viable ways cities look to address climate change. Coastal communities have built larger sea walls and increased environmental buffers to protect the shoreline.

Land cover modification also affects climate. Intensive agricultural practices such as irrigation raise the humidity. In the Midwest, transpiration enhances thunderstorms that trigger substantial rainfalls. Globally, approximately 18 million acres of forest disappear each year, based on data from the Food and Agriculture Organization, an agency of the United Nations. Deforestation has reduced the extent of tropical rainforests, delaying the start of the monsoonal rains.

Urban sprawl diminishes cloud cover by raising the height of the boundary layer and cloud base. Brown particles emitted from industrial sources are transported to high latitudes by the winds aloft, coating snow and ice and reducing their reflectiveness, which increases the absorption of solar energy. A NASA study indicated that wind-blown dust has contributed to a faster rate of snowmelt that enters the Colorado River each spring.

Coal-fired power plants have been on the decline, mostly due to more stringent emissions controls and the availability of natural gas. Energy derived from the burning of coal has dropped from more than 50 percent in 2008 to 20 percent in 2022. Burning natural gas to generate electricity produces about half of the carbon dioxide as burning coal, with the added benefit of less particulate pollution.

Natural gas, which is extracted from deep rock formations, is a cleaner-burning fossil fuel. Since 2014, power generation from natural gas for the production of electricity surpassed the use of coal in Pennsylvania. Utility companies have switched to the process of hydraulic fracturing to retrieve oil and natural gas deposits buried in shale rock (Utica, Marcellus) from thousands of feet below the surface. Fracking involves drilling vertically and horizontally into shale, injecting water and chemicals to free up trapped natural gas.

There are some deleterious by-products in the retrieval of natural gas that include the release of methane, a deterioration in water quality, forest degradation and pollution from increased traffic in rural areas. The disposal of wastewater through injection wells is a concern because it becomes radioactive when interacting with shale. However, fracking is viewed by some as a bridge to low-carbon technologies.

Geoengineering solutions to lower the amount of carbon dioxide in the atmosphere include brightening the ocean surface with plankton. Making the ocean less acidic and adding nutrients to stimulate photosynthesis are among the proposals to increase the uptake of carbon dioxide in the ocean and improve the quality of marine life (animals and plants).

Planting more trees is an economical way to reduce carbon in the air. Trees take in carbon dioxide during photosynthesis to make food, releasing oxygen and water vapor in the process while filtering pollutants such as sulfur dioxide and nitrogen dioxide through the leaves. Adding minerals to increase alkalinity helps dissolve more carbon in the ocean. Marine plants and trees soak up carbon dioxide, which is stored in the wood.

Other proposals to reduce sunlight and cool the planet involve complicated schemes that inject reflective sulfate aerosols into the stratosphere to filter incoming sunlight, seeding clouds and space mirrors. In urban areas, painting roofs white reflects more sunlight and marginally reduces the heat island effect.

Renewable sources of energy are derived from the bioproducts of agricultural and forest waste. Organic matter, or biomass, is utilized as fuel at power stations to generate electricity, which is collected from yard construction, waste-treatment plants, and industrial residue. Sustainable agricultural practices such as no-till farming produce healthier soil and sequester carbon in the soil, resulting in higher yields. The use of buffer strips, two-stage ditches and cover crops reduces runoff and further contributes to a cleaner environment.

Carbon capture uses large fans and filters to draw in air and filter out carbon dioxide molecules before the emissions leave a smokestack. Carbon dioxide is heated, mixed with water and pumped deep underground utilizing geothermal energy, eventually cooling and turning into rock. Some companies have experimented with chemicals that scrub minute amounts of carbon dioxide to sequester the gas. Biomass-based power plants are another means to sequester carbon.

Energy conservation measures that promote clean air and tackle climate change include such renewable energy sources as wind, solar, water and geothermal. Proposed economic solutions to limit pollution would put a cap on emissions and create allowances for companies to buy and sell, as the market dictates.

Electric vehicles powered by batteries do not have internal combustion engines that oxidize carbon and hydrogen atoms and emit heat energy, and convert carbon into carbon dioxide. Fuel cell vehicles depend on hydrogen produced by natural gas, which is a far less carbon-intensive fuel compared to coal and oil (petroleum) products, and reduce waste emissions from manufacturing sources.

Planting winter crops to sequester carbon is another way to produce renewable natural gas. Processing crops involves crushed oilseeds that produce oils used for biofuels.

Low-carbon technologies are becoming less expensive and a more efficient transition from fossil fuels, to meet the supply of global energy through renewable sources.

Inflation Reduction Act

In August 2022, the Inflation Reduction Act (IRA) provided $369 billion to boost projects that use renewable energy sources through spending tax credits and grants, with additional incentives for electric vehicles, over 10 years.

Climate policies are designed to reduce our dependence on fossil fuels. Building the market for clean energy through tax credits is expected to stem rising global temperatures. Incentives to promote the installation of solar panels, energy-efficient technology and wind farms are thought to offer market opportunities and improve both air quality and human health.

Appendices and Climate Tables

WEATHER EXTREMES in the Keystone State are derived from NOAA climate data archived by the National Centers for Environmental Information (NCEI). Observer records have been published in a number of journals, going back to the *Monthly Weather Review* (U.S. Weather Bureau), and now by NOAA as *Climatological Data—Pennsylvania*.

State records are compiled and reviewed by the Pennsylvania Climate Office, located at the Pennsylvania State University. Modern values are based on the 1991-2020 dataset. The Northeast Regional Climate Center at Cornell University is affiliated with the NOAA/NCEI Regional Climate Center (RCC) program.

Appendix One
Monthly Weather Extremes

JANUARY

Warmest	85	Freeport	22/1906
Coldest	-34	Smethport	5/1904
Wettest	12.81	Laporte	1996
Driest	0.06	Everett	1981

FEBRUARY

Warmest	83	Hyndman	11/1932
Coldest	-39	Lawrenceville	10/1899
Wettest	11.16	Milheim	1981
Driest	Trace	Huntingdon	1968
		Hyndman	

MARCH

Warmest	92	Everett	23/1907
Coldest	-31	West Bingham	18/1916
Wettest	11.60	Galeton	1936
Driest	0.08	Mifflintown	1910

APRIL

Warmest	98	Bloserville	25/1915
		Punxsutawney	26/1915
		Norristown	19/1976
		Port Clinton	19/1976
Coldest	-5	Brookville	1/1923
		Corry	
		Saegertown	
Wettest	14.92	Ephrata	1874
Driest	0.19	Austinburg 2 W	1950

MAY

Warmest	102	Lock Haven	30/1895
		Marcus Hook	27/1941
Coldest	10	Clermont 4 NW	10/1966
Wettest	15.02	Quakertown	1894
Driest	0.06	Bethlehem	1964

JUNE

Warmest	107	Sharon	8/1933
Coldest	20	Somerset	10/1913
Wettest	20.45	York 3 SSW	1972
Driest	0.02	Gettysburg	1966

JULY

Warmest	111	Phoenixville	9-10/1936
Coldest	28	Clermont 4 NW	9/1963
Wettest	19.81	Park Place	1947
Driest	Trace	Myerstown	1955
		Middletown	

AUGUST

Warmest	108	Claysville	6/1918
Coldest	23	Clermont 4 NW	29/1982
Wettest	23.66	Mount Pocono	1955
Driest	0.01	Strausstown	1995

Appendices and Climate Tables

SEPTEMBER
Warmest	105	Stroudsburg	2/1953
Coldest	17	Hawley 1 S	28/1947
Wettest	25.41	Bear Gap	2011
Driest	0.00	Loyalhanna Lake	1985
		Honesdale	

OCTOBER
Warmest	100	Phoenixville	5/1941
Coldest	7	Coudersport	21/1952
Wettest	16.88	Stroudsburg	2005
Driest	0.00	McConnellsburg	1892
		Bethlehem	1924
		Center Hall	
		Hamburg	
		Montrose	
		Renovo	
		Unity Reservoir	
		Beavertown	1963
		Berne	
		Breezewood	
		Chadds Ford	
		Huntingdon	
		Lebanon 4 WNW	
		Virginville	

NOVEMBER
Warmest	88	Claysville 3 W	3/1961
Coldest	-15	Somerset	29/1930
Wettest	13.93	Chalk Hill 3 ENE	1985
Driest	0.12	Centre Hall	1917

DECEMBER
Warmest	82	Washington 3 N	4/1982
Coldest	-29	Clermont 4 NW	25/1980
Wettest	13.39	Joliett	1996
Driest	0.05	Carlisle	1877

Appendix Two
Annual Weather Extremes

TEMPERATURE
Highest Temperature	111	Phoenixville	July 9-10, 1936
Lowest Temperature	-42	Smethport	January 5, 1904
Coldest Month	4.3	Bradford FAA	January 1977
Warmest Month	84.1	Philadelphia (Franklin Institute)	July 1999

RAINFALL
Greatest 24-Hour Rainfall	14.50	Schuylkill Co.	June 21-22, 1972
Largest Monthly Rainfall	25.41	Bear Gap	September 2011
Greatest Annual Precipitation	89.01	Bear Gap	2011
Least Annual Precipitation	15.71	Breezewood	1965

SNOWFALL
Greatest 24-Hour Snowfall	41.0	Lakeville 1 SSE	February 15–16, 1958
Greatest Snowstorm	50.0	Morgantown	March 19–21, 1958
Greatest Monthly Snowfall	117.8	Laurel Summit	February 2010

Appendix Three
Seasonal Averages and Extremes

	Avg. Temp.	Warmest	Coldest	Precip.	Wettest	Driest
Winter	29.1	35.2/1931-32	19.5/1917-18	9.39	13.95/1979	4.49/1980
Spring	47.9	53.0/2012	43.0/1926	11.40	19.55/2011	5.68/1926
Summer	69.2	71.5/2005	64.9/1927	12.86	19.11/2018	6.63/1966
Autumn	51.6	55.5/1931	45.8/1917	11.44	19.00/2011	4.20/1908

Appendix Four
Monthly Averages and Extremes

MONTH	TEMPERATURE			PRECIPITATION		
	Average	Warmest	Coldest	Average	Wettest	Driest
January	26.6	37.5/1932	12.7/1977	3.23	6.51/1978	0.72/1981
February	28.9	36.7/2017	15.2/1934	2.60	5.40/2018	0.62/1968
March	36.8	47.7/2012	24.5/1960	3.53	7.15/1936	0.62/1910
April	48.4	53.6/2017	40.9/1943	3.82	8.06/2011	1.21/1946
May	58.6	64.4/1944	50.7/1967	4.05	7.88/1989	1.55/1964
June	66.9	71.6/1943	61.7/1958	4.43	11.22/1972	1.27/1966
July	71.2	74.9/1955	66.8/2009	4.41	7.52/2018	1.90/1909
August	69.5	73.4/2016	63.2/1927	4.02	8.51/1955	1.37/1930
September	62.8	67.1/1961	56.8/1918	4.30	10.10/2011	0.77/1943
October	51.4	57.8/2007	43.5/1925	3.86	6.91/1976	0.22/1963
November	40.7	46.5/1931	33.2/1976	3.29	8.23/1985	0.61/1917
December	31.7	42.3/2015	17.4/1989	3.50	6.23/1990	0.65/1955
Annual	49.5	52.1/2024	45.2/1917	45.03	63.97/2019	28.85/1930

Appendix Five
Climate Extremes for Selected Cities

PHILADELPHIA

Temperature

Highest:	106°F	August 7, 1918
Lowest:	-11°F	February 9, 1934
Warmest month:	82.4°F	July 2011
Coldest month:	20.0°F	January 1977

Precipitation

Greatest in 24 hours:	8.26	July 28-29, 2013
Least monthly:	Trace	October 2024
Greatest annual:	64.33	2011
Least annual:	29.31	1922

Snowfall

Greatest in 24 hours:	27.6	January 7, 1996
Greatest in a single storm:	30.7	January 7-8, 1996
Greatest in one month:	51.5	February 2010
Greatest in a single season:	78.7	2009-10
Least in a single season:	Trace	1972-73
Greatest depth:	28	January 9, 1996

ALLENTOWN

Temperature

Highest:	105°F	July 3, 1966
Lowest:	-15°F	January 21, 1994
Warmest month:	79.0°F	July 1955
Coldest month:	16.4°F	February 1934

Precipitation

Greatest in 24 hours:	7.85	September 26-27, 1985
Greatest monthly:	13.47	August 2011
Least monthly:	0.00	October 1924
Greatest annual:	71.72	1952
Least annual:	28.76	1941

Snowfall

Greatest in 24 hours:	31.9	January 22-23, 2016
Greatest in a single storm:	31.9	January 22-23, 2016
Greatest in a single month:	43.2	January 1925
Greatest in a single season:	75.2	1993-94
Least in a single season:	5.0	1931-32
Greatest depth:	28	February 12, 1983

SCRANTON

Temperature

Highest:	103°F	July 9, 1936
Lowest:	-21°F	January 21, 1994
Warmest month:	77.4°F	July 1955
Coldest month:	15.0°F	January 1977

Precipitation
Greatest in 24 hours:	6.52	September 26-27, 1985
Greatest monthly:	11.76	August 1955
Least monthly:	0.03	October 1963
Greatest annual:	61.08	2018
Least annual:	26.12	1930

Snowfall
Greatest in 24 hours:	22.4	March 13-14, 2017
Greatest in a single storm:	22.4	March 13-14, 2017
Greatest in a single month:	42.3	January 1994
Greatest in a single season:	98.3	1995-96
Least in a single season:	7.3	1988-89
Greatest depth:	29	January 13, 1996

HARRISBURG

Temperature
Highest:	107°F	July 3, 1966
Lowest:	-22°F	January 21, 1994
Warmest month:	81.9°F	July 1999
Coldest month:	19.1°F	January 1918

Precipitation
Greatest in 24 hours:	12.55	June 21-22, 1972
Greatest monthly:	18.55	June 1972
Least monthly:	0.02	October 1924
Greatest annual:	73.73	2011
Least annual:	26.02	1895

Snowfall
Greatest in 24 hours:	30.2	January 22-23, 2016
Greatest in a single storm:	30.2	January 22-23, 2016
Greatest in a single month:	42.1	February 2010
Greatest in a single season:	81.3	1960-61
Least in a single season:	8.8	1937-38
Greatest depth:	32	January 13, 1996

WILLIAMSPORT

Temperature

Highest:	106°F	July 9, 1936
Lowest:	-20°F	January 21, 1994
Warmest month:	79.6°F	July 1901
Coldest month:	14.9°F	January 1977

Precipitation

Greatest in 24 hours:	8.66	June 22, 1972
Greatest monthly:	16.80	June 1972
Least monthly:	0.16	September 1943
Greatest annual:	70.26	2011
Least annual:	25.98	1895

Snowfall

Greatest in 24 hours:	23.1	January 12-13, 1964
Greatest in a single storm:	24.7	December 16-17, 2020
Greatest in a single month:	40.1	January 1987
Greatest in a single season:	85.9	1995-96
Least in a single season:	7.0	1988-89
Greatest depth:	26	January 11, 1996

PITTSBURGH

Temperature

Highest:	103°F	July 16, 1988
Lowest:	-22°F	January 19, 1994
Warmest month:	80.3°F	July 1887
Coldest month:	11.4°F	January 1977

Precipitation

Greatest in 24 hours:	5.95	September 17, 2004
Greatest monthly:	11.05	November 1985
Least monthly:	0.06	October 1874
Greatest annual:	57.83	2018
Least annual:	22.65	1930

Snowfall
Greatest in 24 hours:	23.6	March 13, 1993
Greatest in a single storm:	27.4	November 24-26, 1950
Greatest in a single month:	48.7	February 2010
Greatest in a single season:	82.0	1950-51
Least in a single season:	8.8	1918-19
Greatest depth:	26	January 22, 1978

ERIE

Temperature
Highest:	100°F	June 25, 1988
Lowest:	-18°F	Feb. 16, 2015; Jan. 19, 1994
Warmest month:	77.6°F	July 1921
Coldest month:	12.5°F	January 1977

Precipitation
Greatest in 24 hours:	10.42	July 22-23, 1947
Greatest monthly:	13.27	July 1947
Least monthly:	0.02	October 1924
Greatest annual:	61.70	1977
Least annual:	23.84	1934

Snowfall
Greatest in 24 hours:	32.6	December 25-26, 2017
Greatest in a single storm:	48.9	December 24-27, 2017
Greatest in a single month:	93.8	December 2017
Greatest in a single season:	166.3	2017-18
Least in a single season:	19.6	1932-33
Greatest depth:	39	December 21, 1989

Heaviest Snowstorms

ALLENTOWN
Jan. 22-23, 2016	31.9
Jan. 31-Feb. 2, 2021	27.3
Jan. 7-8, 1996	25.6
Feb. 11-12, 1983	25.2
Mar. 19-21, 1958	20.3

PHILADELPHIA
Jan. 7-8, 1996	30.7
Feb. 5-6, 2010	28.5
Dec. 19-20, 2009	23.2
Jan. 22-23, 2016	22.4
Feb. 11-12, 1983	21.3

SCRANTON
Mar. 13-14, 2017	22.5
Jan. 12-13, 1964	21.1
Jan. 7-8, 1996	21.0
Nov. 24-25, 1971	20.5
Jan. 18-20, 1936	20.0

HARRISBURG
Jan. 22-23, 2016	30.2
Feb. 11-12, 1983	25.0
Jan. 7-8, 1996	24.2
Jan. 15-16, 1945	21.0
Feb. 18-20, 1964	20.8

PITTSBURGH
Nov. 24-26, 1950	27.4
Dec. 16-18, 1890	25.9
Mar. 12-14, 1993	25.3
Feb. 5-6, 2010	21.1
Jan. 8-9, 1884	18.0

Snowiest Winters

PHILADELPHIA
(Average: 23.1)

65.5	1995-96
55.4	1898-99
54.9	1977-78
49.1	1960-61
44.3	1966-67

ALLENTOWN
(Average: 33.1)

75.2	1993-94
71.8	1995-96
67.2	1966-67
65.1	1960-61
63.6	1957-58

SCRANTON
(Average: 44.3)

98.3	1995-96
90.4	1993-94
88.6	1904-05
82.8	1915-16
76.8	1969-70

HARRISBURG
(Average: 29.9)

81.3	1960-61
77.6	1995-96
75.9	1993-94
74.7	1963-64
70.6	1977-78

WILLIAMSPORT
(Average: 35.8)

85.9	1995-96
83.6	1977-78
82.6	1969-70
81.2	1993-94
80.2	1960-61

PITTSBURGH
(Average: 44.1)

82.0	1950-51
78.5	1913-14
77.4	2009-10
76.8	1993-94
76.0	1960-61

ERIE
(Average: 104.3)

166.3	2017-18
149.1	2000-01
145.8	2008-09
143.0	2002-03
142.8	1977-78

STATE COLLEGE
(Average: 43.8)

109.3	1993-94
99.0	1995-96
98.2	1977-78
92.5	1992-93
91.9	1960-61

Pennsylvania Weather Stations

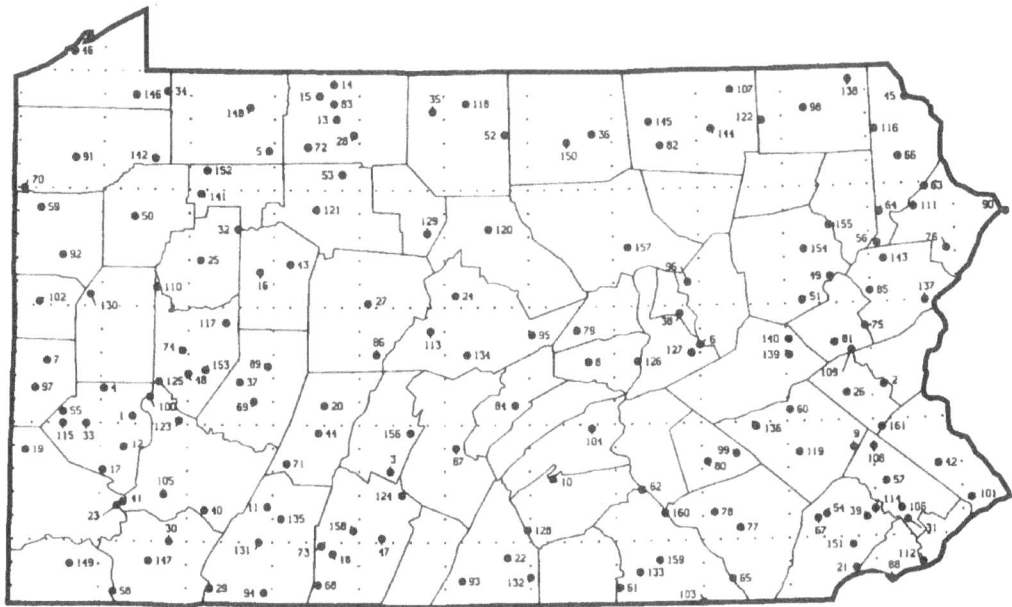

Source: National Climatic Data Center.

No.	Station Name	Latitude Deg/Min	Longitude Deg/Min	Elev	No.	Station Name	Latitude Deg/Min	Longitude Deg/Min	Elev
100	Acmetonia Lock 3	4032N	07949W	748	28	Altoona FAA AP	4018N	07819W	1476
2	Allentown WSO AP	4039N	07526W	388	29	Salina	4140N	07959W	1230
3	Altoona FAA AP	4018N	07819W	1476	30	Connellsville 2 SSW	4140N	07902W	1360
4	Bakerstown 3 WNW	4140N	07959W	1230	31	Slippery Rock	4032N	07949W	748
5	Barnes	4140N	07902W	1360	32	Allentown WSO AP	4039N	07526W	388
6	Acmetonia Lock 3	4032N	07949W	748	33	Coraopolis Neville Island	4018N	07819W	1476
7	Allentown WSO AP	4039N	07526W	388	34	Chadds Ford	4140N	07959W	1230
8	Altoona FAA AP	4018N	07819W	1476	35	Barnes	4140N	07902W	1360
9	Bakerstown 3 WNW	4140N	07959W	1230	36	Acmetonia Lock 3	4032N	07949W	748
10	Barnes	4140N	07902W	1360	37	Allentown WSO AP	4039N	07526W	388
11	Acmetonia Lock 3	4032N	07949W	748	38	Altoona FAA AP	4018N	07819W	1476
12	Allentown WSO AP	4039N	07526W	388	39	Devault 1 W	4140N	07959W	1230
13	Altoona FAA AP	4018N	07819W	1476	40	Barnes	4140N	07902W	1360
14	Slippery Rock	4140N	07959W	1230	41	Acmetonia Lock 3	4032N	07949W	748
15	Barnes	4140N	07902W	1360	42	Doylestown WSO AP	4039N	07526W	388
16	Brookville Sewage Plt	4032N	07949W	748	43	Altoona FAA AP	4018N	07819W	1476
17	Allentown WSO AP	4039N	07526W	388	44	Bakerstown 3 WNW	4140N	07959W	1230
18	Altoona FAA AP	4018N	07819W	1476	45	Barnes	4140N	07902W	1360
19	Bakerstown 3 WNW	4140N	07959W	1230	46	Erie WSO AP	4032N	07949W	748
20	Barnes	4140N	07902W	1360	47	Allentown WSO AP	4039N	07526W	388
21	Chadds Ford	4032N	07949W	748	48	Altoona FAA AP	4018N	07819W	1476
22	Allentown WSO AP	4039N	07526W	388	49	Bakerstown 3 WNW	4140N	07959W	1230
23	Altoona FAA AP	4018N	07819W	1476	50	Franklin	4140N	07902W	1360
24	Bakerstown 3 WNW	4140N	07959W	1230	51	Acmetonia Lock 3	4032N	07949W	748
25	Barnes	4140N	07902W	1360	52	Allentown WSO AP	4039N	07526W	388
26	Claussville	4032N	07949W	748	53	Altoona FAA AP	4018N	07819W	1476
27	Allentown WSO AP	4039N	07526W	388	54	Bakerstown 3 WNW	4140N	07959W	1230

Source: National Climatic Data Center

No.	Station Name	Latitude Deg/Min	Longitude Deg/Min	Elev	No.	Station Name	Latitude Deg/Min	Longitude Deg/Min	Elev
55	Glenwillard Dash Dam	4140N	07902W	1360	109	Bakerstown 3 WNW	4140N	07959W	1230
56	Acmetonia Lock 3	4032N	07949W	748	110	Barnes	4140N	07902W	1360
57	Allentown WSO AP	4039N	07526W	388	111	Chadds Ford	4032N	07949W	748
58	Slippery Rock	4018N	07819W	1476	112	Philadelphia WSCMO AP	4039N	07526W	388
59	Bakerstown 3 WNW	4140N	07959W	1230	113	Altoona FAA AP	4018N	07819W	1476
60	Barnes	4140N	07902W	1360	114	Bakerstown 3 WNW	4140N	07959W	1230
61	Acmetonia Lock 3	4032N	07949W	748	115	Barnes	4140N	07902W	1360
62	Allentown WSO AP	4039N	07526W	388	116	Acmetonia Lock 3	4032N	07949W	748
63	Salina	4018N	07819W	1476	117	Allentown WSO AP	4039N	07526W	388
64	Bakerstown 3 WNW	4140N	07959W	1230	118	Altoona FAA AP	4018N	07819W	1476
65	Barnes	4140N	07902W	1360	119	Reading	4140N	07959W	1230
66	Brookville Sewage Plt	4032N	07949W	748	120	Barnes	4140N	07902W	1360
67	Allentown WSO AP	4039N	07526W	388	121	Acmetonia Lock 3	4032N	07949W	748
68	Altoona FAA AP	4018N	07819W	1476	122	Allentown WSO AP	4039N	07526W	388
69	Bakerstown 3 WNW	4140N	07959W	1230	123	Salina	4018N	07819W	1476
70	Barnes	4140N	07902W	1360	124	Bakerstown 3 WNW	4140N	07959W	1230
71	Acmetonia Lock 3	4032N	07949W	748	125	Barnes	4140N	07902W	1360
72	Allentown WSO AP	4039N	07526W	388	126	Brookville Sewage Plt	4032N	07949W	748
73	Kegg	4018N	07819W	1476	127	Allentown WSO AP	4039N	07526W	388
74	Bakerstown 3 WNW	4140N	07959W	1230	128	Altoona FAA AP	4018N	07819W	1476
75	Barnes	4140N	07902W	1360	129	Bakerstown 3 WNW	4140N	07959W	1230
76	Acmetonia Lock 3	4032N	07949W	748	130	Slippery Rock	4140N	07902W	1360
77	Allentown WSO AP	4039N	07526W	388	131	Acmetonia Lock 3	4032N	07949W	748
78	Altoona FAA AP	4018N	07819W	1476	132	Allentown WSO AP	4039N	07526W	388
79	Laurelston St Village	4140N	07959W	1230	133	Altoona FAA AP	4018N	07819W	1476
80	Connellsville 2 SSW	4140N	07902W	1360	134	Bakerstown 3 WNW	4140N	07959W	1230
81	Acmetonia Lock 3	4032N	07949W	748	135	Barnes	4140N	07902W	1360
82	Allentown WSO AP	4039N	07526W	388	136	Acmetonia Lock 3	4032N	07949W	748
83	Coraopolis Neville Island	4018N	07819W	1476	137	Allentown WSO AP	4039N	07526W	388
84	Bakerstown 3 WNW	4140N	07959W	1230	138	Altoona FAA AP	4018N	07819W	1476
85	Long Pond	4140N	07902W	1360	139	Bakerstown 3 WNW	4140N	07959W	1230
86	Acmetonia Lock 3	4032N	07949W	748	140	Connellsville 2 SSW	4140N	07902W	1360
87	Mapleton Depot	4039N	07526W	388	141	Acmetonia Lock 3	4032N	07949W	748
88	Altoona FAA AP	4018N	07819W	1476	142	Allentown WSO AP	4039N	07526W	388
89	Bakerstown 3 WNW	4140N	07959W	1230	143	Titusville Waterworks	4018N	07819W	1476
90	Barnes	4140N	07902W	1360	144	Bakerstown 3 WNW	4140N	07959W	1230
91	Chadds Ford	4032N	07949W	748	145	Barnes	4140N	07902W	1360
92	Allentown WSO AP	4039N	07526W	388	146	Acmetonia Lock 3	4032N	07949W	748
93	Salina	4018N	07819W	1476	147	Allentown WSO AP	4039N	07526W	388
94	Bakerstown 3 WNW	4140N	07959W	1230	148	Altoona FAA AP	4018N	07819W	1476
95	Barnes	4140N	07902W	1360	149	Bakerstown 3 WNW	4140N	07959W	1230
96	Acmetonia Lock 3	4032N	07949W	748	150	Chadds Ford	4140N	07902W	1360
97	Allentown WSO AP	4039N	07526W	388	151	West Chester 1W	4032N	07949W	748
98	Altoona FAA AP	4018N	07819W	1476	152	Allentown WSO AP	4039N	07526W	388
99	Bakerstown 3 WNW	4140N	07959W	1230	153	Altoona FAA AP	4018N	07819W	1476
100	Connellsville 2 SSW	4140N	07902W	1360	154	Bakerstown 3 WNW	4140N	07959W	1230
101	Acmetonia Lock 3	4032N	07949W	748	155	Wilkes Barre-Scranton WSO AP	4140N	07902W	1360
102	Allentown WSO AP	4039N	07526W	388					
103	Coraopolis Neville Island	4018N	07819W	1476	156	Acmetonia Lock 3	4032N	07949W	748
104	Slippery Rock	4140N	07959W	1230	157	Allentown WSO AP	4039N	07526W	388
105	Barnes	4140N	07902W	1360	158	Altoona FAA AP	4018N	07819W	1476
106	Acmetonia Lock 3	4032N	07949W	748	159	Bakerstown 3 WNW	4140N	07959W	1230
107	Allentown WSO AP	4039N	07526W	388	160	Connellsville 2 SSW	4140N	07902W	1360
108	Altoona FAA AP	4018N	07819W	1476	161	Zionsville 3 SE	4032N	07949W	748

Source: National Climatic Data Center

Bibliography and Source Materials

Abbe, Cleveland. 1906. "Benjamin Franklin as Meteorologist." *Proceedings of The American Philosophical Society* 45 (183): 117-128.

Ambrose, Kevin, Dan Henry, and Andy Weiss. 2002. *Washington Weather: The Weather Source book for the D.C. Area*. Fairfax, VA: Historical Enterprises.

"A Shower of Shell-Fish." 1870. *Scientific American* 22: 386.

Bailey, J. F., and J. L. Patterson. 1975. "Hurricane Agnes Rainfall and Floods, June-July 1972." USGS Professional Paper 924. Washington, D.C.: U.S. Government Printing Office.

Barnes, John. H. and W. D. Sevon. 2014. *The Geological Story of Pennsylvania*. Harrisburg: Pennsylvania Geological Survey (4th series).

Barron, James. 2010. "Rough Morning? It could have been worse." *New York Times* (Apr. 29).

_____. 2014. "Volunteering to Document the Weather for 84 Years." *New York Times*, A16-17 (Aug. 6).

Bausman, Joseph H. 1904. *History of Beaver County, Pennsylvania: Centennial Celebration*. New York: Knickerbocker Press.

Betts, Edwin. M., ed. 1944. *Thomas Jefferson's Garden Book, 1766-1824*. Philadelphia: American Philosophical Society.

Blodget, Lorin. 1857. *Climatology of the United States and of the Temperate Latitudes of the Northern American Continent*. Philadelphia: J. B. Lippincott and Company.

Bradsby, Henry C., ed. 1893. *History of Luzerne County, Pennsylvania*. Chicago: S. B. Nelson & Company, Publishers.

Brickner, Roger. 2006. *New York City Weather: A Chronicle of Events, Volume One: 1609-1929*.

Bristow, Arch. 1932. *Old Time Tales of Warren County*. Meadville: Press of Tribune Publishing Company.

Brodhead, Daniel. 1855. "Letter from Colonel Brodhead to George Washington (Feb. 11, 1780)." *Pennsylvania Archives* 12: 206.

Brooks, Harold and Charles Doswell. 2001. "Normalized damage from major tornadoes in the United States: 1890-1999." *Weather Forecasting* 16: 168-176.

Brown, R. H. 1940. "The First Century of Meteorological Data in America." *Monthly Weather Review* 68 (5): 130-133.

Caldwell, J.A. 1877. *Caldwell's Illustrated Historical Combination Atlas of Clarion County, Pennsylvania*. Condit: J. A. Caldwell.

Catanzariti, John, ed. 2000. *The Papers of Thomas Jefferson*. Princeton: Princeton University Press.

Clayton, H. H. 1944. *World Weather Records Washington, D.C.: Smithsonian Miscellaneous Collection* 79: 891.

Cohen, I. Bernard. 1990. *Benjamin Franklin's Science*. Cambridge: Harvard University Press.

Crane, Elizabeth F. 2010. *The Diary of Elizabeth Drinker*. Philadelphia: University of Pennsylvania Press.

Creigh, Alfred. 1870. *History of Washington County, Pennsylvania*. Washington: A. Creigh.

Crowther, Hugh G. 1995. "Tornadoes: Philadelphia Story." *Weatherwise* 48 (1): 50.

Dale, Frank. 1996. *Delaware Diary*. New Brunswick: Rutgers University Press.

Davis, Emerson. 1851. *The Half Century*. Boston: Tappan & Whittemore.

Delaware County Institute of Science. 1910. "The flood of 1843 (report of a special committee [reprinted])." Chester: Delaware County Institute of Science Proceedings 6 (1): 1-46; 6 (2): 54-86.

Dolan, Edward F. 1988. *The Old Farmer's Almanac Book of Weather Lore*. Dublin, NH: Yankee Books.

Eckhart, Thomas D. 1992, 1996, 1997. *The History of Carbon County* (3 vols.). Allentown: Thomas D. Eckhart and the Carbon History Project.

Egle, William H. 1876. *An Illustrated History of the Commonwealth of Pennsylvania: Civil, Political and Military, From Its Earliest Settlement to the Present Time*. Harrisburg: Dewitt C. Goodrich & Co.

Eisenlohr, William. 1952. "Floods of July 18, 1942, in North-Central Pennsylvania." USGS Water Supply Paper 1134-B: 67-72.

Fischer, David H. 2004. *Washington's Crossing*. London: Oxford University Press.

Fitch, John. "Accounts 1784-91." *John Fitch Papers II-35-D-4*. Library of Congress.

Fleming James R. 1990. *Meteorology in America, 1800-1870*. Baltimore: Johns Hopkins University Press.

Flora, Snowden D. 1953. *Tornadoes of the United States*. Norman: University of Oklahoma Press.

Ford, Paul L. 1902. *The Journals of Hugh Gaine*. New York: Dodd, Mead & Company.

Fountain, Henry A. 2009. "A Scientist's Autopsy of the Johnstown Flood." *New York Times*, D3 (Oct. 27).

Fulks, J. R. 1954. "The Early November Snowstorm of 1953." *Weatherwise* 7 (1): 12-16.

Garriott, Edward B. 1906. *Cold Waves and Frosts in the United States, Bulletin P.* Washington, D.C.: U.S. Department of Agriculture.

Gelber, Benjamin D. 1998. *Pocono Weather: A Weather History of Eastern Pennsylvania, the Poconos, and Northwestern New Jersey.* Stroudsburg: Uriel Publishing.

_____. 2002. *The Pennsylvania Weather Book.* New Brunswick, N.J.: Rutgers University Press.

_____. 2009. "Ben Franklin on Global Warming." *New York Times* Op-Ed (Nov. 17).

Gnidovec, Dale. 2015. "Microorganism might have triggered extinction." *Columbus Dispatch*, G3 (Feb. 8).

Gordon, Thomas. 1833. *A Gazetteer of the State of Pennsylvania*. Philadelphia: T. Belknap.

Grazulis, Thomas P. 1993. *Significant Tornadoes 1680-1991.* St. Johnsbury, VT: The Tornado Project of Environmental Films.

Groisman, P. Y., R. W. Knight, T. R. Karl, D. R. Easterling, B. Sun, and J. H. Lawrimore. 2004. "Contemporary Changes of the Hydrological Cycle over the Contiguous United States: Trends Derived from In Situ Observations." *Journal of Hydrometeorology* 5: 64-85.

Haines, Benjamin F. 1902. *Centennial and Illustrated Wayne County*. Honesdale: B. F. Haines.

Hall, Jonathon P. 1858. "Register of the thermometer for 36 years, from 1821 to 1856." Boston: *American Academy of Arts and Sciences, Memories* 6 (2): 233.

Halverson, Jeffrey B. 2010. "Second Wind: The Deadly and Destructive Inland Phase of East Coast Hurricanes." *Weatherwise* 68 (2): 21-27.

Harvey, Oscar J. 1909. *A History of Wilkes-Barre: Luzerne County, Pennsylvania* (3 vols.). Wilkes-Barre: Raedor Press.

Havens, James M. 1958. An Annotated Bibliography of Meteorological Observations in the United States, 1715-1818. "Key to Meteorological Records Documentation." No. 5.11. Washington, D.C.: U.S. Weather Bureau.

Hazard, Samuel. 1828. "Effect of climate on navigation, [Delaware River, 1681-1828]." *Register of Pennsylvania* 2: 23-26 (July); 379-386 (Dec.).

Hazen, H. A. "The Most Destructive Tornadoes since 1872." *Science* 16 (390): 43-45 (July 25).

Henry, A. J. 1907. "The Cold Spring of 1907." *Monthly Weather Review* 35 (5): 223-225.

Heverly, Clement F. 1926. *History and Geography of Bradford County, Pennsylvania, 1615-1924.* Towanda: Bradford County Historical Society.

Hildreth, Samuel. 1826. *American Journal of Science* 11 (2): 232.

Hiltzheimer, Jacob. 1892. *Extracts from the Diary of Jacob Hiltzheimer, of Philadelphia, 1765-1798*. Philadelphia: The Historical Society of Pennsylvania: University of Pennsylvania Press.

Hoffman, Luther S. 1938. *The Unwritten History of Smithfield Township, Monroe County*. East Stroudsburg: Artcraft Press.

Hogeland, William. 2010. *Declaration: The Nine Tumultuous Weeks When America Became Independent, May 1-July 4, 1776*. New York: Simon & Schuster.

Hough, Franklin B. 1872. *Meteorological Observations, Made Under the Instructions of the Regents of the University at Sundry Stations in the State of New York* (2nd series). Albany, NY: Weed, Parsons and Company, Printers.

Hughes, Patrick. 1987. "Hurricanes haunt our history." *Weatherwise* 40 (3): 134-140. (Revised version of an article that appeared in *American Weather Stories* [Government Printing Office, 1976]).

Jackson, Donald and Dorothy Twohig, eds. 1979. *The Diaries of George Washington* (6 vols.) Charlottesville: The University of Virginia Press.

Jennings, Arthur 1950. "World's greatest observed point rainfalls." *Monthly Weather Review* 78 (1): 4-5.

Jefferson, Thomas. *Garden Book*, 1766-1824. Boston: Massachusetts Historical Society.

_____. *Notes on the State of Virginia*. 1787. London: Stockdale.

Johnson, Kirk. 2000. "Not complaining about the weather, doing something about it." *New York Times*, A23 (June 30).

Kalm, Peter (Pehr). 1972. *Travels into North America*. Translated by John Reinhold Forster. Barre, MA: Imprint Society.

Karl, Thomas R., Laura K. Metcalf, M. K. Nicodemus and Robert Quayle. 1983. *Statewide average climatic history-Pennsylvania, 1888-1982*, Series 6-1.

Keen, Richard A. 1992. *SkyWatch East: A Weather Guide*. Golden, CO: Fulcrum Publishing.

Kershner, Isabel. 2013. "Pollen Study Points to Drought as Culprit in Bronze Age Mystery." *New York Times*, A11 (Oct. 23).

Kocin, Paul, Alan D. Weiss, and Joseph J. Wagner. 1988. "The Great Arctic Outbreak and East Coast Blizzard of February 1899." *Weather and Forecasting* 3 (12): 305-318.

Kocin, Paul J. and Louis W. Uccellini. 1990. *Snowstorms Along the Northeastern Coast of the United States: 1955 to 1985*. Boston: American Meteorological Society.

Kury, Franklin L. 2011. *Clean Politics, Clean Streams: A Legislative Autobiography and Reflections*. Lehigh University Press (co-published with the Rowman and Littlefield Publishing Group, Inc.): Lanham, MD.

Lantz, Jackson. 1897. *Picturesque Monroe County, Pennsylvania*. Stroudsburg: Morris Evans.

Labaree, Leonard, ed. 1961. *The Papers of Benjamin Franklin*. New Haven. CT: Yale University Press, 3: 463.

Laskin, David. 1996. *Braving the Elements: A Stormy History of American Weather*. New York: Doubleday.

Lawson, Thomas. 1840. *Meteorological Registers for the years 1826-1830 (1822-1825 Appended)*. Philadelphia: Haswell, Barrington and Haswell.

Leech, Thomas. 2005. "The Collapse of the Kinzua Viaduct." *American Scientist* 93 (4): 348-353.

Lepper, Bradley T. 2014. "Ancient cultures affected by climate change, too." *Columbus* (OH) *Dispatch*, E5 (June 15).

Lesh, William S. 1945. "Landmarks of Monroe County." *The Record* (Aug. 23).

Levering, Joseph M. 1903. *A History of Bethlehem, 1741-1892*. Bethlehem: Times Publishing.

Lewis, Joseph. 1941, 1943. "Diary of Joseph Lewis." *Proceedings of the New Jersey Historical Society* 59 (3): 159-161; 61 (1): 49, 51.

Linn, John B. 1877. *Annals of Buffalo Valley, Pennsylvania, 1755-1855*. Harrisburg: Lane S. Hart.

Longshore, David. 1998. *Encyclopedia of Hurricanes, Typhoons, and Cyclones*. New York: Facts on File, Inc.

Lorditch, Emily. 2007. "The Flood City: How Johnstown, Pennsylvania, is keeping the forces of nature at bay." *Weatherwise* 60 (6): 42-49.

Lorant, Stefan. 1964. *Pittsburgh: The Story of an American City*. Garden City, NY: Doubleday & Company.

Ludlum, David M. 1960. "The Weather at Gettysburg." *Weatherwise* 13 (3): 101-105.

_____. 1960. "Big Snow of 1836." *Weatherwise* 13 (6): 248-252.

_____. 1961. "New York City Weather Highlights." *Weatherwise* 14 (2): 66.

_____. 1963. *Early American Hurricanes: 1492-1870*. Boston: American Meteorological Society.

_____. 1966. *Early American Winters: 1604-1820*. Boston: American Meteorological Society.

_____. 1968. *Early American Winters II: 1821-1870*. Boston: American Meteorological Society.

_____. 1970. *Early American Tornadoes: 1586-1870*. Boston: American Meteorological Society.

_____. 1971. *Outstanding Weather Events*. Princeton: Weatherwise, Inc.

_____. 1982. *The American Weather Book*. Boston: Houghton Mifflin Company.

_____. 1983. *The New Jersey Weather Book*. New Brunswick, NJ: Rutgers University Press.

_____. 1984. *The Weather Factor*. Boston: Houghton Mifflin Company.

_____. 1985. *The Vermont Weather Book*. Montpelier, VT: Vermont Historical Society.

_____. 1988. "New York City icebound: The Hard Winter of 1780." *Weatherwise* 41 (6): 334-336.

_____. 1989. "The Johnstown Flood." *Weatherwise* 42 (3): 88-92.

Markowski, Paul. 2015. *The Tornado Outbreak of May 31, 1985: Looking Back at One of Pennsylvania's Deadliest Weather Events*. State College: The Pennsylvania State University. (https://www.weather.gov/ctp/TornadoOutbreak_may311985).

Martin, Jere. 1997. *Pennsylvania Almanac*. Mechanicsburg: Stackpole Publishing.

Mason, Charles. 1899. "Charles Mason's Daily Journal" (National Archives), quoted in *Maryland Weather Service*. Baltimore: The Johns Hopkins Press 1: 343-344.

McCarthy, Charles. 1972. "200 years of high water along the Susquehanna River." *Wyoming Valley Observer* (July 2, 1972).

McCullough, David G. 1968. *The Johnstown Flood: The Incredible Story Behind One of the Most Devastating Disasters America Has Ever Known*. New York: Simon and Schuster.

McMaster, John. B. 1933. "The Johnstown Flood." *The Pennsylvania Magazine of History and Biography* 57 (3): 220 (July); 57 (4): 316 (Oct.).

McKnight, W. J. 1905. *A Pioneer Outline History of Northwestern Pennsylvania*. Philadelphia: J. B. Lippincott Company.

Mellick, Andrew Jr. 1889. *The Story of an Old Farm, or Life in New Jersey in the Eighteenth Century*. Somerville, NJ: *Unionist-Gazette*, 514.

Miller, Dixon R. 1958. "Spring in the Poconos." *Weatherwise* 11 (3): 132-133.

Miller, James E. 1946. "Cyclogenesis in the Atlantic Coastal Region of the United States." *Journal of Meteorology* 3 (2): 33-44.

Millikan, Frank R. 1997. "Joseph Henry: Father of Weather Service." The Joseph Henry Papers Project. Smithsonian Institution: Washington, D.C.

Monmonier, Mark. 1999. *Air Apparent*. Chicago: University of Chicago Press.

Muhlenberg, Henry Melchior. 1958. *The Journals of Henry Melchior Muhlenberg*, trans. (3 vols.). Philadelphia: The Muhlenberg Press.

Musgrove, William. 1831. Remarks on the January 1831 snowfall, from the "Meteorological Register." *Register of Pennsylvania*, 104 (Feb. 12).

Myers, Albert Cook, ed. 1912. *Narratives of early Pennsylvania, West New Jersey and Delaware, 1610-1707*. New York: C. Scribner's Sons.

_____. 1937. *William Penn: His Own Account of the Lenni Lenape or Delaware Indians, 1683*. Moylan: Albert Cook Myers.

Nese, John and Glenn Schwartz. 2005. *The Philadelphia Area Weather Book*. Philadelphia: Temple University Press.

National Climatic Data Center. 2013. "Billion-dollar weather and climate disasters." National Oceanic and Atmospheric Administration.

Newton, Larissa. 2017. "Throwback Thursday: The Start of Snow Removal in Pennsylvania." *PennDOT Way* blog (Oct. 19).

Nutting, Wallace. 1924. *Pennsylvania Beautiful (Eastern)*. New York: Bonanza Books.

Oplinger Carl. S. and Robert Halma. 1988. *The Poconos: An Illustrated Natural History*. New Brunswick: Rutgers University Press.

Owenby, James. R. and D. S. Enzell. 1992. *Monthly Station Normals of Temperature, Precipitation, and Heating and Cooling Degree Days, 1961-90-Pennsylvania*. Climatography of the United States, No. 81. Asheville: U.S. Department of Commerce, National Oceanic and Atmospheric Administration, National Climatic Data Center.

Peirce, Charles. 1847. *A Meteorological Account of the Weather in Philadelphia, January 1, 1790, to January 1, 1847*. Philadelphia: Lindsay and Blakistan.

Pemberton, Phineas. n.d. *Meteorological Observations in Philadelphia*. Philadelphia: American Philosophical Society Library.

Pershing, Benjamin H. 1924. "Winthrop Sargent." *Ohio Archeological and Historical Quarterly* 33 (3): 237-281.

Pierce, David. 2015. "Flood of '55: Bridges destroyed all along the Delaware River." *Pocono Record*, 1-2 (Aug. 19).

Potter, Sean. 2011. "Retrospect: July 4, 1776: The Declaration of Independence." *Weatherwise* 64 (4): 10-12.

Proud, Robert. 1797. *The History of Pennsylvania, in North America*. Philadelphia: Zachariah Poulson.

Rittenhouse, David. 1780. "Height of Fahrenheit's thermometer at Philadelphia, 1780." Philadelphia: Pennsylvania Historical Society.

Rosenberg, Gary D. 2009. *The Revolution in Geology from the Renaissance to the Enlightenment*. Boulder: The Geological Society of America.

Roth, David M. and Hugh D. Cobb III. 2000. "Re-Analysis of the Gale of '78—Storm 9 of the 1878 Hurricane Season." http: www.hpc.ncep.noaa.gov/research/roth/gale of 78.htm.

Rupp, I. Daniel. 1845. *History of Northampton, Lehigh Monroe, Carbon, & Schuylkill Counties*. Lancaster: G. Hills.

Schafer, Jim and Mike Sagna. 1992. *The Allegheny River*. University Park: Penn State University Press.

Schmidlin, Thomas. 1996. *Thunder in the Heartland*. Kent, OH: Kent State University Press.

Schultze, David. 1953. *The Journals and Papers of David Schultze, Volume II: 1761-1797*, ed. Andrew S. Berky. Pennsburg: The Schwenkfelder Library.

Schwartz, Rick. 2007. *Hurricane and the Middle Atlantic States.* Springfield, VA: Blue Diamond Books.

Scofield, David. 2013. "40 Years of Archaeology at the Meadowcroft Rockshelter." *Western Pennsylvania History* 96 (3): 5-7.

Serfass, Donald R. 2013. "An island in the middle of Tamaqua." *Times Herald* (Apr. 13).

Sevon, W. D., Gary M. Fleeger and Vincent C. Shepps. 1999. *Pennsylvania and the Ice Age.* Harrisburg: Pennsylvania Geological Survey (4th series).

Shafer, Mary. 2010. *Devastation on the Delaware* (2nd edition). Ferndale: Word Forge Press.

Shank, William H. 1988. *Great Floods of Pennsylvania: A Two-Century History* (2nd edition). York: American Canal and Transportation Center.

Shein, Karsten A., D. P. Todey, F. A. Akyuz, J. R. Angel, T. M. Kearns and J. L. Zdrojewski. 2013. "Revisiting the Statewide Climate Extremes for the United States: Evaluating Existing Extremes, Archived Data, and New Observations." *Bulletin of the American Meteorological Society* 94 (3): 393-402.

Shukaitis, Nancy M. 2007. *Lasting Legacies of the Lower Minisink.* East Stroudsburg: Nancy M. Shukaitis.

Sloane, Eric. 1963. *Folklore of American Weather.* New York: Duell, Sloan and Pearce.

Stevens, William K. 2000. "Persistent and severe, drought strikes again." *New York Times* (Apr. 25).

The Pennsylvania Magazine of History and Biography. 1891. "Pennsylvania Weather Records, 1644-1835." Philadelphia: University of Pennsylvania Press 15 (1): 109-121.

U.S. Department of Commerce. 1963. *History of Weather Bureau Wind Measurements.* Appendix I-C: 47. Washington, D.C.: Department of Commerce.

U.S. Weather Bureau. 1934. *Climatic Summary of the United States, Sections 87-89: Pennsylvania.* Washington, D.C.: U.S. Government Printing Office.

_____. 1942. *Daily and hourly precipitation supplement storm of July 17-18, 1942.* U.S. Department of Commerce (40 pp.).

Van Doren, Carl. 1938. *Benjamin Franklin.* New York: Viking Press.

Walter, Y. S. 1844. *Report of a committee of the Delaware Institute of Science on the Great Rain Storm and Flood of August 1843.* Chester: Delaware Institute of Science.

Walters, Mrs. Horace G. 1965. *Stodgell Stokes, Ledger-C.* Monroe County Historical Society. (Additional sources cited: *Jeffersonian* (Apr. 16, 1874; July 15, 1875; Dec. 16, 1886); and Stroudsburg Census, 1850-60.

Watson, Benjamin F. 1993. *Acts of God: The Old Farmer's Almanac Unpredictable Guide to Weather and Natural Disasters.* New York: Random House.

Watson, John F. 1833. *Historic Tales of Olden Time, Concerning the Early Settlement and Progress of Philadelphia and Pennsylvania*: E. Littell and T. Holden.

_____. 1868. *Annals of Philadelphia and Pennsylvania, in the Olden Time.* 2:347-69. Philadelphia: J. B. Lippincott and Company.

Webster, Noah. 1810. *Memoirs of the Connecticut Academy of Arts and Sciences.* Vol. 1, part 1. New Haven: Steele & Co.

_____. 1835. "Notices of extraordinary seasons of cold." *American Journal of Science* 28 (1): 186.

Weiss, Webster C. 1863. *Incidents of the freshet on the Lehigh River.* Philadelphia: Crissy and Markley.

Wilson, Alexander. 1818. *The Foresters: A Poem of the Pedestrian Journey to the Falls of the Niagara, in the Autumn of 1804.* Newtown: S. Siegfried & J. Wilson.

Witten, Donald E. 1985. "May 31, 1985: A Deadly Tornado Outbreak." *Weatherwise* 38 (4): 193-199.

Wood, Gillen D. 2014. *Tambora: The Eruption That Changed the World.* Princeton, NJ: Princeton University Press.

Zook, John. 1905. *Historical and pictorial Lititz.* Lititz: Express Printing Company.

Newspapers used as sources for historical events, listed by city:

Albany, NY: *Cultivator and County Gentleman*
Allentown: *Daily Chronicle; Morning Call*
Altoona: *Altoona Mirror*
Annapolis, MD: *Maryland Gazette*
Baltimore, MD: *National Register; Niles' (Weekly) Register*
Belvidere, NJ: *Belvidere Apollo; Warren Journal*
Bethlehem: *Bethlehem Daily Times*
Bloomsburg: *Bloomsburg Register*
Boston, MA: *Boston Gazette; Daily Evening Transcript*
Bradford: *Bradford Era*
Brookville: *Brookville Repository*
Butler: *Butler Repository*
Carlisle: *Carlisle Volunteer*
Centre Hall: *Centre Reporter*
Chester: *Chester (Daily) Times*
Clearfield: *Clearfield Republican*
Cleveland, OH: *Cleveland Herald; Plain Dealer*
Columbia: *Columbia Spy*
Columbus, OH: *Ohio State Journal*
Concord, NH: *New Hampshire Sentinel*
Doylestown: *Doylestown Intelligencer*
Easton: *Easton Argus, Easton Centinel (Sentinel); Easton Democrat and Argus; Easton Express-Times*
East Stroudsburg: *East Stroudsburg Press; Morning Press; Morning Sun*
Elizabethtown, NJ: *New Jersey Journal*
Erie: *Erie Daily Times; Erie Observer*
Franklin: *News-Herald*
Germantown: *Germantown Zeitung*
Gettysburg: *Adams Sentinel; Gettysburg Sentinel; Republican Compiler*
Hanover: *Hanover Gazette*
Harrisburg: *Daily Patriot Union, Harrisburg Evening Gazette; Harrisburg Evening Transcript; Hazard's Register of Pennsylvania; Patriot*
Hartford, CT: *Connecticut Courant*
Hemstead, NY: *New York Chronicle Advertiser*

Hazleton: *Hazleton Standard-Speaker*
Honesdale: *Wayne Independent; Wayne County Herald*
Huntingdon: *Huntingdon Gazette*
Indiana: *Indiana Democrat*
Kingston: *Kingston Republican and Herald*
Kittanning: *Kittanning Gazette*
Lancaster: *Lancaster City Express; Intelligencer Journal; Lancaster Journal*
Lewistown: *Lewistown Eagle*
Mauch Chunk (Jim Thorpe): *Mauch Chunk Courier*
Milford: *Pike County Press; Milford Herald*
Montrose: *Independent Volunteer, Independent Republican; Susquehanna Register*
Mount Vernon, OH: *Mount Vernon Democratic Banner*
Newark, NJ: *Newark Daily Advertiser; Sentinel of Freedom*
Newport, RI: *Newport Mercury*
New York City: *Evening Star; New York Commercial Advertiser; New York (Evening) Post; New York Herald; New York Packet; New York Times*
Oil City: *Derrick*
Philadelphia: *American Mercury; Democratic Press; Evening Bulletin; Evening Telegraph; Evening Traveller; Franklin Gazette; North American; Pennsylvania Advertiser and Weekly Journal; Pennsylvania Chronicle; Pennsylvania Gazette; Pennsylvania Inquirer; Philadelphia Gazette; Philadelphia Inquirer; Public Ledger (and Daily Transcript); United States Gazette; Vincent's Register*
Pottsville: *Miners' Journal*
Reading: *Reading Eagle; Gazette and Democrat*
Rochester, NY: *Genessee Farmer*
Salem, OH: *Salem Daily News*

Savannah, GA: *Savannah Republican*
Scranton: *Scranton Times; Scranton Times-Tribune*
Sharon: *Herald*
Smethport: *Pennon and Settler*
State College: *Centre Daily Times*
Stroudsburg: *Daily Record; Jeffersonian (Republican); Monroe Press; Morning Record: Pocono Record; Stroudsburg (Daily) Times; The Record*
Sussex, NJ: *Sussex Register*
Tamaqua: *Times Herald*
Titusville: *Titusville Herald*
Toledo, OH: *Blade*
Towanda: *Bradford Reporter; Settler*
Washington, DC: *National Intelligencer; National Journal; Washington Post*
Wellsboro: *Tioga County Agitator*
West Chester: *Republican*
Wilkes-Barre: *Times Leader*
Williamsburg, VA: *Williamsburg Gazette*
Williamsport: *(Daily) Gazette and Bulletin; Williamsport Sun*
Winchester, VA: *Winchester Gazette*
Wyoming: *Wyoming Valley Observer*
York: *York Gazette*

Index

A

Abbe Jr., Cleveland 25, 31
Abrams, Elliot 15
Acadian Mountains 9
Adams County 124, 270, 308, 321, 330, 333, 359, 382, 423
Adams, John 30
Adams, John Quincy 37, 144
Adovasio, Jim 15
Aiken 312
air pollution 448
Albany 100, 136, 139, 144, 161, 168, 176
Alberta clipper 54–55, 68, 105, 205–206
Albion 70, 323–324
Albrightsville 133, 139
Aleutian Low 49
Allegheny City 220–221, 302
Allegheny County 243, 266, 271, 286, 291, 316, 321, 339, 353, 357–358, 417; County Airport 108, 121, 264, 276
Allegheny Front 5
Allegheny Mountains 6, 10–11, 55, 58, 128, 131
Allegheny National Forest 5, 326, 329, 340
Alleghenian orogeny 10
Allegheny Plateau 21, 27, 47, 83, 113, 119, 122, 147, 161, 166, 171, 210, 262, 271, 280, 312, 315, 370, 380
Allentown 48, 59, 93–95, 105, 107–108, 110–113, 115, 117–118, 120, 122–123, 125–128, 131–134, 136–138, 141–143, 152, 154, 160–161, 167, 180, 182, 184, 194–195, 197–198, 201–202, 205, 219, 228, 230–231, 247–248, 258, 260, 264, 271, 275, 318, 322, 360, 368–369, 373–375, 378–383, 414–415, 417, 420, 423, 428–431
Allentown (NJ) 308
Aliquippa xvii, 319–320; Aliquippa-Hopewell Airport 320
Altoona 42, 75, 96, 100, 104, 106–107, 122–123, 130, 152, 156, 168, 193, 201, 205, 225, 229, 257, 318, 370, 409, 426, 430
Ambler 308, 352
Ambridge 334
American Philosophical Society xvii, 28–30, 33, 36, 78, 400
American Revolution (War of Independence) 78
Amity Township 137
Analomink 191, 245
Anderson Creek 299
anemometer 151, 303, 359, 420, 429
Antarctica 10–11; ice sheet 442–443, 452–453
aphelion 14
Apollo 321
Appalachian Mountains 3, 8, 84

Aqueduct 217, 366, 383
atmospheric river 49, 209, 392, 454
Arctic 11, 13-14, 16, 18, 50, 52-56, 59, 65, 69, 79, 88, 92-96, 98, 101-102, 108, 112-113, 116, 119, 121, 123, 129-130, 135, 138, 154, 168, 173, 175, 177-178, 180, 183, 185-191, 193-200, 202-203, 205-206, 319, 383, 437, 441-444, 452-458
Ardmore 384
Armstrong County 316, 321, 342, 350
Ashland 229, 356
asthenosphere 7
Atlantic Coast 18–20, 49, 51, 54, 68, 92, 94, 105, 110, 119, 128–129, 131–133, 200, 209, 232, 276, 362, 372, 390, 395, 401–402, 404, 411–412, 416, 421, 426–427, 432
Atglen 299
Austin 233–234, 241
Avalonia 9
Avoca 120, 334
Avondale 214, 332, 338, 422
Azores-Bermuda (North Atlantic) High 49, 391, 457

B

Baker Run 123
Bald Eagle Valley 262
Baltica 9
Baltimore 87, 90, 92, 95–96, 101–102, 108, 110–111, 117, 124, 127, 129–131, 133–134, 142, 144, 151–152, 161, 166, 169, 176, 185, 190, 201, 203, 298, 376–378, 381, 401–402, 417
Bangor 138, 176, 278, 359, 420
barometer (barometric pressure) 28, 37, 59, 90, 105, 119, 155, 221, 417
Barton, Benjamin Smith 35
Barton, Clara 226
Bartonsville 118, 228, 344
Bartram, Benjamin Swift 35
Bartram, John 29–30; Botanical Garden 358
Bartram, William 29
Bath 35, 332, 354
Battery (Lower Manhattan) 145, 340, 404, 415, 429
Bear Creek 131, 347
Bear Lake 69
Beaver County 270, 312, 317, 326–328, 339, 353
Bedford 87, 145, 159, 165, 195, 317–318, 344, 375, 383, 427
Bedford County 38, 75, 318, 331, 334–335, 360, 387
Bellefonte 3, 74, 100–101, 165, 194, 257
Belltown 286
Belvidere 91, 93, 107, 141, 149, 159, 165, 180, 259, 354, 373
Bensalem 350, 352, 429
Bering Land Bridge 15
Berks County 66, 133, 164, 208, 279, 294, 301, 306, 314, 317–318, 322, 349, 351, 379
Bermuda High 58, 258, 262, 388, 457
Berwick 93, 157, 180, 197, 332
Bethlehem 30, 85, 99, 105, 107, 140–141, 145, 148, 150–153, 156, 160, 162–163, 170, 189, 191–192, 198, 215–216, 228, 230–231, 236, 247, 309, 322, 332, 368, 371, 378
Big Boulder Ski Area 348
Biglerville 128, 423
Big Meadows 124, 161
Big Mountain 341
Binghamton 95, 106, 127, 136, 138, 143, 155, 158, 168, 189, 201, 260, 355, 376
Birch Creek 261
Birdsboro 133
Blakeslee 44, 111, 129, 133, 155, 157, 167, 190, 259, 344, 427, 433
Blodget, Lorin 38–39
Blooming Grove 74, 93–97, 99, 139, 150–151, 156, 159, 165, 183, 185, 188, 228, 337, 364–365
Bloomsburg 143, 148, 221, 251, 257, 260, 278
Bloserville 113, 193, 311, 357, 382–383

Blossburg 138
Blue Knob 71, 75, 100, 155–156
Blue Mountain 5, 7, 51, 93, 151, 154, 163, 295; Ski Area 335
Boalsburg 358
bombogenesis 68, 131
Boston 29–31, 68, 90, 92, 102, 110, 112, 117, 120, 122, 126, 140, 161, 164, 176, 178, 183, 194, 200, 316, 365, 378, 401, 415
Boyertown 208
Boynton 340
Braam, Jacob 27
Bradford 100, 116, 120, 123, 138, 149, 157–158, 167, 171, 184, 193, 198–201, 203–205, 217, 241, 256, 312, 356, 375, 380, 386
Bradford County 84, 93 95, 129, 138, 155–156, 158, 163, 165–166, 168. 170, 184, 218, 230, 256, 263, 289, 293, 307–309, 311, 333, 341, 346–347, 349, 355, 357
Brandywine Creek 266, 338
Breezewood 387
Bridgeport 231, 299, 415
Brodhead Creek 55, 244–247. 249, 419
Brodheadsville 105, 190, 208, 336, 359
Bronze Age 15
Brookville 97, 101, 103, 194, 196, 289, 295, 299, 319, 335, 348, 358, 368, 370, 464, 485
Brown, Charles B. 35
Brownsville 187, 254
Buchannon Valley 270
Buck Hill Falls 110
Bucks County 27, 39, 132–134, 137, 180, 213, 243, 248, 255–256, 264, 266, 271, 278, 301, 308, 315, 330, 332–333, 338, 350–352, 425
Buckstown 128
Burgettstown 317, 409
Burnham 335
Bushkill 77, 132–133, 135, 141, 190–191, 193, 205, 271, 297, 359, 380; Township 297

Bushkill Creek 232
Bustleton 338, 352
Butler 264, 271, 290, 317, 328, 350, 353
Butler County 317, 328, 353
Buttonwood 95, 336, 348

C

Callery 328
Cambria County 170, 193, 195, 253, 262, 266, 316, 329, 349, 358
Camden 25, 68, 297, 300–301, 303, 310, 315
Camelback Mountain 127, 259
Cameron County 97, 337
Campbelltown 343
Canadensis 103, 140, 205, 245–246, 264
Carbon County 7, 12, 72, 92, 133, 162, 216, 218, 228–231, 236, 240, 263, 279, 346, 348
Carboniferous Period 9
Carlisle 88, 101, 107, 146, 152, 166, 184, 278, 302–303, 307, 317, 357, 366–367, 370, 373, 383
Carnegie xvii, 222, 257, 339, 358
Casey, Robert 119, 121
Cashtown 270
Cassandra 170
Catasaqua 322
Catawissa 144, 165, 369–370
Cathcart, Robert 30
Catskill Delta 9
Centerport 317
Centerville 227, 325, 330
Centre County 3, 74, 262, 336, 340, 354
Centre Valley 74
Chambersburg 90, 98, 116, 121, 231, 300–301, 333, 349, 367, 369–370, 383
Chandlers Valley 206
Charles II 1, 26
Cheat River 254
Cherry Valley 191, 193–194, 344–345

Chesapeake Bay 20, 47, 79, 118, 137, 210, 226, 249, 332, 400–401, 404, 410, 413–414, 431
Chester 319–320, 353, 408
Chester County 40, 90, 93–94, 111, 132–134, 139, 146, 151, 164, 183–184, 243, 255, 263, 266, 271, 278, 289, 298, 300–301, 305, 315, 319–320, 330, 332, 338, 342, 352, 355, 357, 368, 371, 407, 418, 425, 427
Churchtown 330
Clarence 155, 206
Clarion 197, 205–206, 225, 241, 295–296, 303, 307, 327, 334–335, 382
Clarion County 295, 303, 307, 327, 334
Clausville 161
Claysville 368
Clayton 318, 326
Clearfield 19, 92, 124, 225–226, 291, 299, 326–327, 335, 340, 368, 370–371
Clearfield County 124, 299, 326–327, 335, 340
Clermont 43, 157, 167–168, 171, 197, 199–200, 376
climate change 10, 15–16, 32–35, 42, 62, 435, 437, 439–440, 443, 458–459, 461
Climatic Optimum 11, 15, 435
Coalport 240
Coaltown 299
Cochranton 324, 360
Codorus Creek 236, 251, 300
Coffin, James. H. 38, 93
Coffin, Seldin J. 38, 144
Colden, Cadwallader 28–29
Colebrook 157
Collegeville 101
Columbia 87, 98, 139, 217, 259, 295, 308, 330–331, 355, 372, 412
Columbia (NJ) 247, 314
Columbia County 263, 278, 292, 294, 306, 348
Columbus (OH) xviii, 128, 169, 195, 204, 290
Commonwealth 3, 5, 113, 121, 280, 298, 349, 388, 427

Comte de Buffon, Georges-Louis Leclerc 35
computer models 61–62, 124, 208; convection-allowing 273; hurricane forecast 392, 394
Conemaugh River 222–223, 225–226, 252–253
Conneautville 70, 347
Conshohocken Creek 218
Cook Forest State Park 6, 358
Coombe, Thomas 29
Corbett, Tom 72
Coriolis force 38, 46, 274
Corry 69, 75, 103, 116, 143–144, 147, 157, 161, 168, 194, 237, 325, 370–371
Coudersport 19, 141, 143, 155, 157, 161–162, 190–191, 194, 196, 206, 234, 241, 417
Covington 138
Coxe, John Redman 34
Cranesville 323–324
Crawford County 6, 70, 103, 157, 164, 179, 227, 271, 279, 313, 323–325, 346–347, 353, 359–360
Cretaceous Period (145-66 million years ago) 10
Croll, James 13
Cumberland County 134, 193, 320, 357
Cumberland Valley 5, 299, 307
cumulus 48, 267

D

Damascus 135, 231
Dansgaard, Willi 17, 442
Danville 84, 101, 217, 251, 261
Darlington 328
Darlington, William 35
Dauphin County 263, 266, 319, 344
Day, Theodore 40, 94–96, 99, 150, 159, 218
Delaware Bay 4, 25, 47, 124, 210, 404, 429
Delaware County 41, 43, 149, 213–214, 264, 350, 425
Delaware River 4, 7, 25, 27, 55, 76–79, 81, 99, 140, 147, 174–175, 180, 196, 210, 212–213, 216–217, 220,

228, 230–232, 239, 243–244, 246–248, 255–257, 259–260, 289, 293, 297–298, 300, 305, 308, 310, 314–315, 330, 332–333, 337, 351–352, 404, 408, 417, 419, 431

Delaware Valley 19, 22, 25–26, 30, 32, 76, 101, 105, 113, 116, 123, 137, 147, 169, 177, 179, 193, 205, 210, 236, 264–265, 297, 364, 412

Delaware Water Gap 7, 57, 132, 139, 216, 220, 255, 257, 259–260, 320, 345, 433

Delmarva Peninsula 54, 119, 137, 407, 427

derecho 275–276, 288

Derry 121, 161, 417

Devil's Hole Creek 244–245

Devonian Period (419-359 million years ago) 9

Dilldown Frost Pocket 44

Dilworthtown Inn 338

Dingmans Ferry 133, 187, 337

Dinwiddie, Robert 27–28

dolomite 10

Donegal 119, 316, 339, 351

Donora 382, 447–448

Doppler radar 277, 280, 282–283, 287, 393

Douglass, William 29

Dover (NJ) 104, 131, 144; DE 160

Downingtown 86, 266, 332

Doylestown 93, 331, 350, 352, 415, 425–426, 431

Drifton 140, 151, 193

drought, patterns of 384–386

DuBois 118, 120, 196–197, 200–201, 205–206

Dunlo 132

Duquesne 319–320

Dunmore 247, 349, 409

Dushore 143, 159–160, 165, 348

Dust Bowl 18, 362, 369, 371, 386–387, 441

Dyberry 40, 74, 93–97, 99–101, 103–104, 139–140, 145–146, 150–151, 156, 159, 165, 182–189, 218, 230, 232, 247, 294, 364, 406; Creek 240, 249

E

Earth 7–8, 10–11, 13–16, 45–49, 53, 55, 60–62, 85, 219, 222, 267, 279, 295, 386, 389, 392, 405, 435–438, 440, 442–445, 449–452, 459

East Benton 129, 138

East Coast 18, 91, 106, 112, 115, 131, 133, 137, 140, 168, 183, 190, 221, 362, 383, 392–393, 403, 405, 410, 416, 424, 433, 454

Eastern Seaboard 18, 20, 50, 56, 58, 68–69, 91, 102, 118, 120, 122, 130, 132, 146, 155, 185, 187, 204, 209, 217, 229, 235, 249, 254, 280, 387, 391, 393, 405, 410–412, 415–416, 420, 423, 426, 432, 457

East Nantmeal Township 131, 133, 301

Easton xviii, 38, 87–90, 92–94, 102–103, 139–141, 144, 148–150, 156, 158–159, 164–165, 176, 179–181, 183, 188–189, 194, 198, 212–214, 216–217, 228–232, 239, 241, 244, 247–248, 257, 278, 286, 289, 296–297, 302, 304–305, 364, 367, 379, 419

East Stroudsburg xviii, 45, 59, 88–89, 92, 95–97, 99, 115, 117–118, 120–125, 127–129, 131–135, 137, 141–143, 145–148, 150, 152–155, 157, 161–162, 166–169, 181, 191, 193–194, 200–202, 204–206, 245–246, 249, 258–259, 264, 314, 369, 375–384, 419, 425, 430

Ebensburg 75, 131, 192, 195

Echo Valley 191

Eckhart, Thomas 216, 218, 226, 229–231, 236, 241

Edinboro 70, 75, 137, 156, 158

Eemian interglacial 12

Effort 129, 228, 290, 302

Egypt 15, 73, 94–95, 97, 144, 184, 186–187, 364

Elizabeth 291, 320; Township 359

El Niño-Southern Oscillation 129, 349, 397, 452–455

Eliot, Jared 31

Elk County 154, 197, 286, 325–326, 337, 358

Ellwood City 312

Emlenton 327

Emporium 190, 192, 241

Endless Mountains 5
Eocene Epoch 11
Ephrata 146, 150, 182, 258, 354, 367
equator 8–10, 13, 46, 48, 50, 55, 60, 385, 392, 405, 449–450, 452–453, 456–458
Equinunk 258
Ercildoun 298–299
Erie xvii–xviii, 6, 15, 22, 48, 59, 67, 69, 85, 88, 95, 103, 105–106, 109, 116–117, 120, 123, 127, 131, 137, 144, 160, 162, 167, 185, 191, 194, 197, 199, 201, 204–206, 220–221, 235, 242, 248, 275, 307, 313, 321, 323, 356, 372, 376, 382, 384, 423, 426
Erie County xvii–xviii, 21, 70-71, 75, 114, 129, 138, 235, 243, 324-325, 418
Espy, James Pollard 36–38, 86
Etna 257
Evans, Lewis 26
Evans City 328
Eynon 271

F

Fayette County x, 121, 125, 175, 254, 299, 316, 331, 334, 340–342, 351, 355, 357, 359
Feasterville 301
Ferrel, William 38
firn 441
Fleetwood 298
Fleming, James R. 33, 35–36
floods 49, 104, 126, 207–211, 213–214, 217, 220–222, 226, 228–229, 235–237, 240, 243, 246, 248–250, 252–256, 262, 402, 409, 412, 418, 435
foraminifera 456
Forbes State Forest 6
Forest County 325–326
Fort LeBoeuf 28
Four Corners 52–53, 362, 385
Fort Washington 352
Francis E. Walter Dam 155, 171, 204

Frackville 165
Franklin 27–28, 190, 217, 223, 271, 296, 306, 325, 335, 370–371
Franklin County 121, 139, 143, 301, 318, 320, 330, 333–335, 344, 346, 349, 383, 416, 427
Franklin, Benjamin xvii, 30–32, 35, 76, 81, 400
Franklin Institute 36–37, 39–42, 86–87, 125, 379, 384
Freas, Clymer 67
Freedom 333, 353
Freeland 107, 112, 142, 145, 152–153, 157, 160–161, 166, 193–194, 196, 200, 202, 242, 329, 380, 423
French Creek 28, 217, 335
Frenchtown 313
freshet (*See also* floods) 86, 210–218, 231–232, 247, 405–406
Fryburg 307
Fujita Scale (Enhanced Fujita Scale) 284–285

G

Gainesboro 134
Galeton 113, 167, 346, 381
George School 106, 368
Germantown 29–30, 77, 79, 164, 305, 309, 407
Gettysburg 4, 39, 86–87, 90, 92, 160, 270, 291, 295, 308, 313, 318, 333, 359, 368, 422–423
Gettysburg College 39
Girard 71
Girardville 409
Gist, Christopher 27–28
Glengary 134
global warming 10–11, 441, 445, 459
Gobbler's Knob 66–67
Godfrey Ridge 72, 127, 134, 155, 344
Gondwana 10
Gordon 106, 145, 264
Goschenhoppen (Goshenhoppen) 163
Gouldsboro 23, 74, 109–118, 120, 123, 125, 127, 131, 141–142, 144–148, 152 154, 157, 160–162,

166–167, 171, 194–196, 205, 236–237, 266
Grady, Julia 197
Graterford 125, 266, 379, 382, 427, 431
graupel 269, 277
Graysville 348
Great Galveston Hurricane 395, 413
Great Labor Day Hurricane 395
Great Lakes 3, 12, 20, 26, 41, 47, 52–53, 57–59, 69–70, 98, 105–106, 128–129, 167–168, 173, 193, 202, 206, 222, 242, 258, 262, 274, 276, 280–281, 302, 307, 319, 323, 350, 353, 360, 417, 446, 455, 458
Great Miami Hurricane 395
Great Okeechobee Hurricane 395
Green, Jesse C. 43
Greencastle 121, 134, 334
Greene County 121, 254, 264, 316, 348, 351, 357
Greenland 16, 69, 77, 173, 441-442, 456-457; ice sheet 442
Greensburg 156, 252, 321, 331, 344, 354, 359, 430
Greentown 190, 247, 337
Gulf Coast 101, 110, 114, 121, 198, 200, 265, 391, 395–399, 409, 421
Gulf Stream 30, 54, 69, 428, 456
Guthsville 219, 275

H
hail 43, 47, 87, 151, 166, 269, 272–273, 275, 277–280, 283–284, 289, 291, 293–296, 300, 303, 305, 314, 319, 327–328, 333–335, 337–339, 341, 350, 356–357
Hamburg 229, 309, 317, 366
Hamlinton 102, 229
Hanover 116, 130–131, 148, 170, 308, 311, 318, 383, 387
Harborcreek 235
Hare, Robert 37–38, 202
Harrisburg (Harrisburg/Middletown) xvii, 20, 43–44, 48, 59, 74, 87, 92, 95, 97, 101, 112–116, 118, 120, 122–123, 125–128, 131–132, 134, 137, 140, 142–144, 146, 148, 150, 155–156, 159–160, 165, 167, 169, 182–183, 185, 189, 192, 195, 199, 201, 205–206, 210, 217, 221, 225–226, 228, 237, 240, 249–250, 252–253, 255–257, 261, 263, 299, 303, 311, 318, 356, 368–371, 374–376, 379–382, 384, 388, 408, 411, 414, 417, 421–422, 427, 430
Harveys Lake 138, 156, 222
Harveyville 306
Hawley 74, 103, 110–111, 116, 120, 125, 128, 134–135, 138, 144, 157, 160, 167, 197, 199, 205–206, 239–240, 242, 247, 278, 346
Hazard, Samuel 77, 81, 83, 86–88, 90, 139, 143, 148, 158, 163–164, 169, 174–177, 213, 291, 382
Hazleton 95, 116, 132, 151–152, 159, 161, 164–168, 189, 244, 380
Heidelberg 257, 294
Hellertown 322, 356
Hempfield 270, 296, 346
Hendrickson, Cornelius 25
Hendrickson, Richard G. 43
Henry, Joseph 37–38, 40–41
Hermitage 326
Hershey 257, 260, 263, 266, 343
Heverly, Clement 84, 86, 93, 95, 156, 158, 163–165, 170, 211–212, 214, 217, 289, 293, 308, 386
Hickory Run State Park 12
Hidden Valley 43
Hiltzheimer, Jacob 29, 163
Hobbie 329
Hoffman, Luther S. 88–89, 92, 95–97, 99, 150, 164, 169, 181, 193
Hogeland, William 32
Hollidaysburg 158, 366, 412
Hollisterville 120, 122–124, 126, 128, 145, 161
Holocene Epoch 15
Holtwood 111, 255, 259, 382
Homestead 287; FL 395

Honesdale 40, 96, 110, 122, 124, 126, 135, 141, 144, 146, 149–151, 167–168, 180, 189, 230–232, 240–241, 243, 247, 294, 320–321, 345, 356, 366, 383, 387
Honey Brook 133, 342
Hop Bottom 305
Hudson Valley 70, 432
Huntingdon 84, 101, 107, 146, 175, 226, 240, 367–371, 380
Huntingdon County 226, 331, 346, 349, 360
hurricanes, classification of 394;
    named storms:
    Able 400, 416
    Agnes xi, 249, 251–253, 255–256, 259, 261, 421, 425, 477
    Allen 398–399
    Allison 425–426
    Amelia 397
    Ana 393
    Andrew 393, 395–396, 399, 424
    Arlene 397
    Audrey 419
    Bob 423
    Camille 395, 399
    Carol 400, 416
    Claudette 423
    Connie 243, 374, 418
    David 422
    Dean 397, 399
    Delta 433
    Dennis 388, 424
    Diane 243, 248, 374, 418–419
    Donna 420, 429
    Doria 420, 425
    Edna 400, 416
    Eloise 252, 421–422
    Eta 432–433
    Felix 397
    Fred 265, 433
    Floyd 262, 266, 388, 424–425
    Fran 256, 423–424
    Frances 426–427
    Frederic 422
    Gamma 433
    Gilbert 397, 399
    Gladys 428
    Gloria 423, 425
    Gordon 264
    Hanna 360, 433
    Harvey 389, 394, 397–398, 436
    Hazel 400, 411, 416–418
    Henri 265, 426, 433
    Hiki 389
    Hugo 393, 423
    Ian 394, 430
    Ida 265, 352, 389, 399
    Iota 432–433
    Irene xii, 260–261, 427–428, 431
    Irma 63, 397–399, 430
    Isabel 426
    Isaias 264, 350, 432–432
    Ivan xi, 257, 344, 395, 398, 426–427
    Jeanne 257
    Jose 398
    Juan 253
    Katrina 394–399, 430
    Lane 389
    Laura 399, 433
    Lee 260–261, 263
    Maria 394, 397–399, 430
    Matthew 397, 399
    Michael 398–399
    Milton 399
    Mitch 399
    Mitch 399
    Ophelia 398
    Otto 393

Patricia 397
Rita 396, 399
Sandy 59, 62–63, 162, 365, 394, 404, 428–432
Tammy 258
Teddy 433
Wilma 396–397, 399, 432
Zeta 432–433
Hyndman 370, 383

I
Iapetus Ocean 9
Illinoian glaciation 11
Imler 318
Indiana 149, 156, 159, 165, 197, 215, 329, 342, 356, 358
Indiana County 158, 276, 316, 321, 329, 334-335, 342, 349-350
insolation 12–14, 46, 55, 57, 173, 272, 362
Intertropical Convergence Zone (ITCZ) 14, 46, 390
Irwin 339, 346, 357, 383
Islip 429

J
Jacobs, Michael 39, 92
Jarrettown 308
Jefferson County 97, 295, 334–335, 340, 348, 353, 358
Jefferson, Thomas 32–33, 35–36, 78–79, 82, 402
jet stream 12, 18, 47–54, 56–57, 69, 73, 117, 122, 127–128, 133, 138, 173, 198, 242, 272–273, 281, 299, 304, 320, 335–336, 344, 349, 362, 385, 387–388, 391, 404, 440, 442, 451, 453–458
Jim Thorpe 96, 215–216, 228–229, 231, 236, 240, 405
John F. Kennedy (JFK) International Airport 114, 134, 420, 429, 431
Johnstown 106, 122, 160, 168, 222–227, 233, 238–239, 252–253, 278, 318, 331, 334, 383, 409
Jordan Creek 228
Juan de Fuca Plate 8

Juniata County 256, 301, 343, 346
Juniata River 210, 225–226, 240, 253, 298, 422–423

K
Kalm, Pehr 33, 76
Kane 19, 23, 113, 120, 123, 153, 157, 171, 205, 326, 328, 343, 376; Kane, East 328
Kellersville 228
Kemblesville 315
Kennett Square 102, 141, 427
Keystone State 2–3, 6, 19–21, 45, 48, 74, 114–115, 124, 137, 143, 164, 169, 195, 198, 242, 256, 261, 323, 336, 347, 354, 367, 374–375, 388
King of Prussia 330, 425, 427
Kingston 86, 221, 250–251, 271, 292, 301
Kinzua Bridge 342–343
Kittanning 176
Kittatinny Mountain 5, 344
Kiskiminetas River 233
Krakatoa 18, 84
Kresgeville 118, 243, 279, 334
Kunkletown 72, 134, 138, 208, 335

L
La Anna 142
Labyrinth Garden 29
Laceyville 334
Lackawanna County 129, 143, 337, 341, 348, 355
Lackawaxen River 231, 239–241, 247, 249, 346, 356
Lafayette College 38, 93, 103, 144, 180–181, 188, 194
LaGuardia Airport 108, 110, 113, 118, 120, 126, 418, 420
Lake Carey 341
Lake Conemaugh 222
lake-effect snow 69-70, 197, 200, 203
Lake Erie 6, 15, 21–22, 28, 69–70, 75, 85, 106, 109, 116–117, 131, 235, 313, 322, 372, 426, 430, 446
Lake Ontelaunee 338

Lakeville 110, 113, 119, 167
Lake Wallenpaupack 110, 113, 268
Lake Wynonah 134
Laki volcano 32
Lancaster 80, 87, 90, 101, 110, 116, 130, 142, 152, 160, 174, 176, 179, 191–192, 198, 229, 231, 260, 264, 296, 298, 319–320, 368–370, 382–384
Lancaster County 69, 79, 118, 139, 157, 163, 165, 168, 179–180, 182, 218, 243, 258–260, 296, 298, 300, 308, 318–319, 321, 330, 338, 349, 354–355, 359–360, 405, 412, 417, 426
Landisville 130, 157, 160
landmass 8
Langhorne 333
La Niña 388, 433, 451, 453–455
Lansdale 86, 137, 301
Lansdowne 208
Laporte 126, 143, 155, 168
Laskin, David 34
Last Glacial Maximum 12, 438, 441
Late Illinoian 11
Late Wisconsinan 12
Latrobe 119–120, 358
Laurasia 10
Laurel Highlands 6, 19, 21–22, 75, 103, 118–119, 128, 155, 162, 356, 431
Laurentia 9
Laurentide Ice Sheet 12
Lawrenceburg 102, 167
Lawrenceville 19, 166, 189–190, 194, 253, 286, 371
Lawton 319
Leader, George M. 6, 110, 269, 303, 311, 327
Lebanon 5, 157–158, 176, 192, 203, 261, 263–264, 266, 293, 307, 343, 351, 354, 358, 360, 367, 374, 382–383, 408
Lebanon County 157, 263, 266, 307, 343, 358, 360
Lebanon Valley 5
LeBar, John Clyde 43, 207, 232

Lee, Jung-Eun 14
Legaux, Peter 30, 174
Lehigh Canal 241
Lehigh County 219, 264, 318, 320, 322, 351, 425
Lehigh River 7, 210–216, 229–231, 247, 266, 290, 320, 404, 419, 431
Lehigh Gap 7, 216
Lehighton 120, 122, 124, 216, 230–231
Lehigh University 236, 368, 378
Lehigh Valley 5, 72-73, 89, 93-95, 97, 99, 107, 110-111, 122, 125, 135-136, 138-139, 143-144, 150, 154, 180-181, 184, 186-187, 191, 198, 215, 218-219, 230-231, 236, 258, 262, 265, 309, 322, 364, 378-379, 408, 417, 425, 429, 431; International Airport 191, 198, 322, 331, 425, 429
Leonard Harrison State Park 155
Lepper, Bradley 15
Levittown 332
Lewisburg 183–184, 190, 225, 250, 253, 256, 367, 373
Lewis, Joseph 26, 81–82
Lewis Run 113
Lewistown 87, 139, 226, 257, 298–299, 349, 380
lightning 31, 34, 39, 60–61, 70, 96, 115, 118, 133, 154, 163, 212, 220, 267–272, 274, 278, 291, 299–301, 305, 309, 311, 314, 354, 357–359, 390–392, 400–401, 445
Limerick Township 332
limestone (See also calcium carbonate) 9, 438, 447, 456
lithosphere 7
Lititz 163, 168, 319, 359
Little Conemaugh River 222–223, 252–253
Little Ice Age 16–17, 156, 168, 450
Loch Loman 200
Lock Haven 21, 100–101, 122, 140, 142, 147, 151–152, 225–226, 327, 337, 366–367, 370, 373, 381
Logan Airport 126, 161
Logan, James 27, 76

Loganton 319, 333
Logstown 28
Long Island 38, 43, 54, 68, 90, 117, 120, 131, 135–137, 142, 289, 294, 393, 403, 410, 412, 414–416, 418–420, 423–424, 429, 431
Long Pond 44, 112–113, 120, 122–123, 125, 144, 161–162, 166, 168, 200–201, 271, 423
Lorain 313
Lords Valley 135, 168
Lorenz, Edward 61
Lottsville 70
Loyalhanna Lake 465
Loyalsock 255, 261–262
Ludlum, David M. xvii, 1, 29–31, 36–39, 49, 77–80, 82–83, 87–91, 98, 109, 149, 159, 163–165, 174–178, 182, 210, 213–214, 218, 222, 232, 237, 247–248, 285, 289–290, 292, 294, 400–405, 412
Lunar Park 338
Luzerne County 140, 151, 153, 155–156, 164, 168, 171, 204, 211, 215, 222, 266, 289–290, 302, 306, 322, 329, 334, 337, 344, 347, 355–356
Lycoming County 70, 95, 261, 278, 280, 329, 335–337, 343, 346, 348
Lykens 370
Lyons 298, 322, 338

M

macroburst 275
magma 8
Mahanoy Creek 229
Mahantango Creek 249
Makin, Thomas 27
Manatawny Creek 208
Manheim 69, 263, 359
Mansfield 124, 131, 309
Maplewood 278
Marcus Hook 19, 125, 366, 370–372, 380, 384, 420
Marienville 325

Marion 368
Markleton 6, 340
Marshall, Christopher 32–33
Mason-Dixon Line 130, 161, 276
Mauch Chunk 96, 148, 215–216, 228–229, 231, 236, 240, 367, 370, 404
McConkey's Ferry 78
McKean County 157, 190, 241, 256, 312, 326, 328, 340, 342
McKeesport 316, 333, 409
McMichael Creek 39, 54, 218, 246
Meadowcroft Rockshelter 14
Meadville 88, 103, 109, 129, 166–167, 175, 279, 313
Mechanicsburg 302, 307, 315
Medieval Warm Period 16
Mercer 197, 326, 379
Mercer County 159, 317, 324, 326, 350, 354, 360
Mercer County (NJ) 330, 352
Mercersburg 334, 375, 382–383
Mercyhurst College 15
mesocyclone 281–282, 343
mesoscale convective system (MCS) 273-274, 343
mesoscale convective vortex (MCV) 274
Mesozoic Era 10
microburst 275–276, 286, 355
Mid-Atlantic Ridge 8
Mifflin County 116, 121, 298, 335, 349, 357
Mifflintown 192, 357, 370
Milankovitch, Milutin 13
Milanville 115, 231
Milford 87, 102, 104–105, 107, 133, 138–139, 141, 152, 160, 166, 187, 190–191, 220, 232, 264, 297, 307, 368, 406
Mill Creek 235, 333
Miller, Dixon R. 43–44, 111, 157, 200
Miller, James E. 68, 138
Millstone 286
Millvale 257

Milton 97, 156, 176, 226, 250, 253
Miocene Epoch 11
Mississippi River 25, 49, 227, 326
Mississippian Period 9
Monongahela River 27, 82, 144, 210, 254, 291, 386
Monroe County xviii, 1, 39, 44, 72, 76, 99, 109, 111, 134, 157, 194, 207, 213, 217–218, 236, 243, 246, 248–249, 255, 266, 279, 290, 314, 331, 334–336, 348, 350, 419
Monroeton 75, 166, 230
Montgomery County 30, 134, 137, 208, 218, 255, 264–266, 301, 308, 320, 330, 332, 352, 420, 424–427, 431, 433
Monthly Weather Review 25, 31–32, 42, 99, 105, 140, 151, 156, 169–170, 181, 212, 221, 233, 235, 299–305, 307, 310, 312–314, 355–357, 408, 410
Montrose 84, 89, 106, 113, 115–116, 123, 126–129, 142, 144–145, 147, 149, 151, 153–155, 157–159, 161, 166–168, 190, 194, 196, 199–200, 291, 370
Montville 137
Moosic Mountains 5
Morgantown 111, 133, 161, 349
Morris, James 66
Morris Plains 134
Morrisville 30, 84, 165, 170, 180
Moshannon State Forest 323, 326, 328
Moss Grove 179
Mountain Top 168
Mount Ararat 187
Mount Davis 6, 337
Mount Joy 139, 150, 165, 179–180, 218, 260, 263, 296, 359, 405
Mount Mitchell 134
Mount Penn 5, 303
Mount Pinatubo 18, 85, 377, 451
Mount Pocono 59, 110, 115, 118, 122–125, 127, 129–132, 135–136, 138, 141–145, 147–148, 152–154, 157, 159–162, 166–168, 171, 192–196, 205, 239, 242–244, 248, 259, 261, 329, 367–368, 370, 372–375, 380, 382–384, 427, 431
Mount Union 240
Mount Vernon (OH) 299; (VA) 163, 401
Mountville 296, 308, 330
Mount Washington 339
Mountain West 52
Moyamensing 289
Mountville 296, 308, 330
Muhlenberg, Henry M. 29, 77–79
Muncy 250, 253
Murrysville 381

N

National Centers of Environmental Information (NCEI) 42-43, 45, 71, 286, 385, 436
National Hurricane Center (NHC) 389, 394, 399
National Weather Service (NWS) xvii, 41–42, 47, 61, 283, 343
Native American 5, 26, 28, 34, 52, 76
Natrona Heights 321
Nazareth 29, 138, 150, 159, 161, 179, 181, 278, 286, 364
Neola 115–116
Nese, Jon 29, 37, 158, 285, 300
Neshaminy Creek 243
Neshaminy Falls 243, 382, 420, 428
Nesquehoning 236
Newark 4, 92, 94, 108, 112–113, 122, 126, 132, 147, 154, 163, 176, 178–181, 194, 265, 294, 297, 305, 374, 377, 405
Newberry 230, 293
New Bethlehem 296, 334
New Brighton 179
New Castle 26, 96, 109, 153, 184–186, 301, 364, 381, 426
New Columbus 236
Newell 357, 382
Newfoundland 247, 268, 410

Newfoundland (Canada) 3, 16
New Hamburg 360
Newport 226, 240, 373; RI 79, 145, 401
Newton 88, 139, 158, 176, 181, 192, 326, 373, 417, 425
Newton Falls 326
Newton Square 425
New Tripoli 153
Newville 357, 423
New York City 62, 76, 79-80, 82-83, 88-92, 94-102, 104, 107-108, 110-113, 117, 120, 122, 126-127, 131-132, 134, 140-141, 143-149, 151-154, 159-160, 162-166, 168, 174-178, 180-181, 183, 185-187, 189, 193-194, 196, 201, 221, 233, 249, 252, 265, 295, 297, 304, 307, 368, 372-373, 375, 377, 378-379, 399, 401-403, 405-407, 409-411, 416, 418, 420-421, 423, 425, 428-429, 433; Brooklyn 87, 94, 100, 149, 165, 297, 304, 349, 410, 433; Central Park 94, 96, 108, 117, 126, 131, 134, 165, 168, 174, 180, 189, 193, 265, 372, 409-410; Manhattan 80, 90, 126, 174, 181, 404, 410, 415, 429; Queens 100, 129, 180, 410, 429; Queens, Jamaica 179, 298; Staten Island 80, 134, 429
New York-New Jersey Highlands 4
Ninepoints 338
Nittany Valley 84, 226
nor'easter 54, 63, 69, 78, 91, 93, 112, 118, 129, 132, 134–137, 142, 144, 147, 151–154, 158, 160–162, 197
Norris, Isaac 76, 158
Norristown 30, 91, 93, 101, 150, 165, 178–182, 218, 231, 251, 266, 360, 381, 384, 406
North American Craton (Plate) 8
Northampton County 134, 156, 277, 314, 320, 332, 335, 354, 358–359, 433
Northern California 52, 56, 363, 444–445
Northern Hemisphere 12–14, 38, 46, 53, 58, 173, 205, 396, 443, 450–451
northern lights (aurora borealis) 450
Northern Tier 58, 93, 131, 137–138, 155, 347, 370

North Pacific High 53
North Potomac 134
Northumberland 175
Northumberland County 156, 278–279, 295, 305, 321, 329, 341, 346, 350
North Whitehall 73, 93, 139, 149, 159, 165, 180–182
Norway 305

O

Oakland 132, 135, 316
Octoraro Lake 384
Ogletown 118
Ohio River 6, 9, 15, 27–28, 103, 123, 220–221, 233, 252, 339, 366, 378, 386, 409, 417–418, 420
Ohio State University xvii, 442
Oil City 217, 227, 335, 356–357
Oil Creek 6, 227, 263, 325
Oligocene Epoch 11
Oliveburg 334
Old Forge 200, 247, 337
Oswayo 206

P

Pacific Coast 362, 445
Pacific Plate 8
Paleocene-Eocene thermal maximum (PETM) 11
Palmer Drought Severity Index (PDSI) 385
Palmer Hydrological Drought Index (PHDI) 385
Palmerton 196, 205, 373
Pangaea 10
Paoli 288, 319
Paradise Creek 244–245
Paris 359; France 35, 458
Park Place 166, 242
Parker Dam State Park 326–327
Parkersburg 423
Parkersville 151
Parkesburg 298–299

Park Place 166, 242
Patton Bog 15
Pecks Pond 244, 337
Peirce, Charles 30, 84, 90–91, 143, 149, 158, 169, 175, 178, 353
Pemberton, Israel (Phineas) 27–29, 78, 210, 401
Pen Argyl 104
PennDOT 122, 130, 257, 266, 325
Penn Haven 216, 239
Penn, William 1–2, 26–27, 30, 76, 173
Pennsylvania, climate of 17–23; glaciation 12–16; Pennsylvania Hospital 17–18, 88–89, 176–177, 199, 293, 363, 403–404; physiographic provinces: Appalachian Plateaus 5
Atlantic Coastal Plain 4
Central Lowlands 6
Deep Valleys 5
Eastern Lake 6
Gettysburg–Newark Lowland 4
Great Valley 5
New England 4
Piedmont 4–5, 9, 20–21, 70
Pittsburgh Low Plateau 6
Reading Prong 5
Ridge and Valley 5
Waynesburg Hills 6
Pennsylvanian Period 9
Pennsylvania State University xvii, 43–44, 84, 107, 116, 124, 191, 195, 198, 327–328, 333, 354
perihelion 14
Perkasie 125, 133–134, 278, 330
Permian Period 10
Permo-Triassic extinction (Great Dying) 10
Perry County 271, 330, 423
Philadelphia 17-20, 23, 25-41, 43, 48, 59, 69, 72, 75-79, 81-84, 86-105, 107, 110-113, 115-119, 121, 123-128, 130-165, 167-171, 174-191, 193-195, 197-203, 205, 208, 210-212, 214, 217-219, 228-229, 233, 243, 245, 247, 251-252, 255-256, 258, 261-266, 271, 275, 277-279, 288-290, 293-295, 297, 300-306, 308-310, 313-315, 320, 329-332, 336, 338, 342, 350, 352-354, 356, 358, 361, 363-384, 387-388, 400-404, 406-414, 416-420, 422, 425, 427-432, 455; Manayunk 264, 266, 310, 336, 388, 427; settlement 15, 25, 210; Society Hill 329; South 300, 310; West 309-310, 408
Philadelphia County 338, 351
Philippines 18, 377, 451, 454
Philipsburg 123, 153–154, 162, 257, 417
Phillipsburg 232, 244, 247
Phoenixville 19, 107, 160, 171, 197, 266, 371, 374, 382, 384, 407, 428
Pike County 20, 94, 96, 99, 104, 107, 133, 141–142, 169, 200, 228, 232, 264, 337, 345–346, 359, 364, 416
Pimple Hill 44, 110–111, 127, 161, 167, 200–201
Pine Creek Gorge 5
Pine Grove 176, 317
Pineapple Express 49
pitched pine 4
Pittsburgh xvii–xviii, 6, 8–9, 15, 22, 27–28, 48, 59, 67, 74–75, 87, 97–98, 100, 103, 107–109, 112, 114, 117, 119, 121–123, 125, 128, 130–131, 140, 142, 144–145, 151, 154, 157, 160–162, 164–169, 171, 176–177, 179, 184–185, 187, 190, 195, 198–206, 208, 210, 212, 219–222, 225–226, 231, 233, 237–240, 243, 252, 254–257, 260, 264–266, 270, 276, 283, 286–287, 299–300, 302–303, 305, 309, 312–313, 316–317, 319, 321, 334, 338–339, 342, 346, 349, 354–355, 358, 364–365, 367–368, 370–373, 375–379, 381–384, 386–388, 407, 409, 411, 417–418, 426–427, 447
Pleasant Mount 19, 23, 114, 117, 120, 123, 126, 129, 142–143, 147, 166–167, 196, 200, 206, 376
Pleistocene Epoch (Ice Age) 11, 12
Pliocene Epoch 11
Pocahontas 338, 340

Pocono Creek 216, 228, 433
Pocono Lake 105, 141, 160, 170, 190–194, 232, 302, 368
Pocono Mountains Municipal Airport 127, 155, 429
Poconos 2, 5, 19–20, 54–55, 89, 94–97, 101, 106, 110–111, 117–118, 120, 122–123, 125, 131–132, 134–136, 138–140, 145, 147, 155, 157–162, 164, 166–167, 186–187, 190, 198, 201, 205, 218, 228, 230, 232, 239, 243 244, 246, 259, 264, 319, 331, 344–345, 376, 380, 406, 409, 419, 423, 425, 431–432
Pocono Plateau 5, 20, 22–23, 72, 92, 105, 111–113, 115, 122, 125, 127, 129, 141, 149, 151, 157, 160, 164–165, 168, 170, 232, 290, 302, 367, 375
Pocopson Township 338
Polar Express 195, 197
polar vortex 55, 57, 73, 85, 132, 173, 202, 205, 451–452, 455–458
Port Allegany 241–242
Port Clinton 215, 263, 384
Portland 57, 144, 146, 153, 232, 247, 259, 314, 433; steamer 140
Potter County 157, 206, 234, 343, 346
Pottstown 228, 231, 251, 263, 332, 426
Pottsville 65, 87–88, 90–92, 107, 134, 143, 148–150, 158, 163–166, 169, 176–179, 215, 230, 237, 263, 305, 353, 405
pre-Illinoian glaciation 11
Promised Land 337
Prompton 240, 247
proto-North America 8–10
Punxsutawney 66–67, 122, 334–335, 384
Punxsutawney Phil (Groundhog Day) 66–67

Q

Quakers 26
Quakertown 115, 140–141, 146, 152, 160, 189, 197–198, 367, 383
Quarryville 330, 338, 354
quartzite 9

R

Raccoon Township 317, 339; Creek 409
Radnor 425
Rainforest 378, 437
Ransom 135, 143
Razorville 292–293
Reading 5, 20, 66, 95, 98, 101, 105, 125–126, 128–129, 134, 144–145, 152, 154, 156, 165, 198 199–201, 208 215, 218, 229, 231, 236, 242, 250–251, 259, 282, 297–298, 300, 303, 305 307, 309, 314, 322, 349, 360, 369, 373, 380, 382–383,. 408–409, 414, 416–417
Reading Prong 4
Rector 122
Redfield, William 37–38
Reichle, Charles Gotthold 29
Reinholds 359
Rendell, Edward G. 259, 343
Renovo 107, 122, 138, 147, 165, 225–226, 380
Rew 197
Riceville 227
Richboro 137
Richfield 256
Ridge, Tom 125, 379, 388
Ridgway 194, 197, 206, 337
Ringgold 340
Risdon, Elisha 85
Rittenhouse, David 29, 35, 174–175
Rochester 117, 164–165, 169, 204
Rockport 180
Rocky Mountains 49–50, 124, 186
Rodinia 9
Rohrerstown 308
Rouseville 305
Rush, Benjamin 34, 174

S

Saegertown 103, 151, 157, 159–160, 188, 193, 325
Sadsburyville 368

Saffir-Simpson Hurricane Wind Scale 394
Salem Corners 101, 153
Salisbury 337–338, 340
Salladasburg 337
sandstone 3, 9–11
Sandusky 313
Sargent, Winthrop 82–83
Saxonburg 328
Saylorsburg 127, 133, 138, 155, 162, 196, 264, 279, 334, 336, 380, 433
Schultze, David 29, 82, 163
Schuylkill County 19, 65, 91, 97, 134, 142, 145, 161, 165–166, 176, 215, 230, 242, 249, 258, 263, 266, 317, 319, 341, 356, 359, 412, 421
Schuylkill River 29, 175, 177, 208, 210–211, 214–215, 228–229, 231, 236–237, 248, 250–251, 258, 264, 266, 297, 404–405, 414, 419, 427
Schwartz, Glenn 29, 37, 158, 300
Schwartz, Rick 218, 243, 247–248, 257, 293, 402, 405–407, 412, 414, 417, 420, 426
Sciota 196
Scotrun 127, 196
Scranton 5, 48, 59, 73, 95, 99, 104, 106, 110, 112, 114–115, 118, 120–125, 127, 129, 132, 134–135, 138, 140–147, 151–155, 157–161, 166–167, 171, 191–197, 199, 201–205, 221, 230, 244, 247, 257, 261, 263–264, 271, 292–293, 305, 320, 331, 337, 349, 353–356, 366, 368, 370–371, 373, 375–378, 381–383, 385, 408, 419, 423, 430
Scranton, William 387
seafloor spreading 8, 10
sea surface temperatures 13, 18, 66, 68–69, 362, 369, 390, 393, 397, 433, 439, 446–447, 455–456
Seelyville Dam 240
Selinsgrove 156, 190, 368, 370, 372, 383, 409
Sellersville 264
Seven Springs 75, 119, 128, 162, 340
Sewickley 177; Township 327–328, 346

shale 7, 9, 460
Shamokin 97, 189, 305, 311, 341
Shapp, Milton 198, 249
Sharon 317, 326, 370, 374
Sharon Springs 136
Sharpsville 360
Shenandoah 78, 97, 124, 278, 311, 426
Shickshinny 306
Shippensburg 107, 120, 126, 134, 147, 231, 307, 334, 372, 380
Siberia 10, 15, 173, 442, 457
Silkworth 306
Sinnemahoning State Park 6
SKYWARN 44
Small, Alexander 31
Smethport 21, 157, 159, 190–191, 241, 312, 343
Smith, Joseph 149–150
Smith, Thomas 29
Smithfield Township 169, 193, 261, 264; Middle Smithfield 77
Smithsonian Institution 37–41, 93–94, 97, 149–150, 164, 178, 180, 215, 364, 405
Snyder County 332
snow 1, 11–13, 18–20, 25, 28, 33, 36, 42–45, 47, 52–56, 58–60, 63, 65–72, 74–170, 173–174, 177–180, 182–193, 195–198, 200, 202–206, 209, 211–214, 228, 230–231, 233, 237, 255, 302, 304, 319, 354, 402–403, 432, 437, 441–442, 455, 457, 459–460
Somerset County 6, 43, 75, 127–128, 137, 318, 322, 337–338, 340–341, 351, 417
Sonestown 261
South Fork Dam 222
Southern California 49, 52–53, 363, 454
Southern Hemisphere 14
Spartansburg 169, 227
Spring Grove 125, 134, 189, 318
Spring Lake Village, Mobile Home Park 329

Spring Mills 30, 335
Springville 291, 342
Sproul State Forest 327
Stark County 169
State College xvii, 43–44, 47, 101, 103–104, 106–108, 113–114, 116–118, 120, 122–128, 141–143, 147, 152, 155–156, 161–162, 166–168, 170, 190–192, 198, 201, 203, 205, 257, 262, 264, 329, 331, 333, 343, 346, 354, 358, 371, 384
Steubenville 109
St. Marys 151, 166, 197, 234, 337; County 289
Stoddartsville 290
Stokes, Anna Maria 40
Stokes, John N. 39, 183
Stokes, Samuel 40
Stokes, Stodgell, 39
Stonycreek River 222–223
Storm Prediction Center (SPC) 280, 336, 339
straight-line wind 286, 292, 301, 319, 332, 343, 350
Strongstown 75, 158
Stroudsburg 39–40, 43, 59, 72, 74, 77, 89–91, 93–94, 96, 99, 101, 103–117, 120, 127, 134, 139, 141–148, 150–157, 159–161, 163–171, 179–180, 182–194, 196–200, 202, 204, 207, 213, 216–220, 228–229, 232, 236, 239, 243–246, 249, 264, 270, 286, 296, 301, 304, 309–310, 314, 320, 344, 364–368, 372–375, 383–384, 404, 406, 409, 415, 419–420, 423, 431, 433
sun 1, 4, 13–14, 45–46, 49–50, 53, 55–57, 60, 66, 73, 173, 219, 267–268, 273, 282, 303–304, 385, 405, 436, 444, 449–450, 452; sunspots 450–451
sudden stratospheric warming 62, 457–458
Sugar Grove 69
Sullivan County 153, 261–262, 289, 348
Summerdale 43
Sunbury 96, 159, 210, 239, 250–251, 253, 295, 303
Susquehanna County 5, 305, 307, 341–342, 345, 348–349

Susquehanna River 6, 82, 84, 159, 176, 210–211, 215, 217, 221–222, 226, 228, 230–231, 236–237, 239, 249–253, 255–257, 259–263, 280, 290, 296, 308, 327, 329–330, 332, 355–356, 359, 386, 412, 417, 421, 427
Susquehanna Valley 20, 87, 93, 97, 120, 122, 135, 138–139, 180, 197, 237, 252, 263, 265, 370, 381, 421, 427
Swatara Creek 260, 263, 266; Township 263
Sweetwater 322
Swiftwater 44, 196
Syracuse 110, 120, 136, 143

T
Taconic Orogeny 9
Tamaqua 98, 104, 168, 215, 230, 244, 248, 258
Tambora 84–85, 452
Tanners Falls 240
Tannersville 118, 127, 132, 134, 145, 149, 160, 162, 190, 433
Taylorsville 78
Terrytown 341
thunderstorms 14, 19, 22, 33, 36, 48–50, 52–54, 56–58, 119, 128, 154, 208–209, 219, 241–242, 253, 256, 262, 266–268, 271–274, 276–277, 279, 281, 293, 302, 309–311, 313–314, 319, 323, 331–335, 338–339, 341, 343–344, 349–352, 354, 359, 365, 376, 390–391, 440, 445, 449, 453, 455, 459
Thornburgh, Dick 254
Tinicum Island 25
Tioga County 5, 19–20, 155, 166, 189–190, 194, 253, 321, 336, 348
Tippecanoe 9
Titusville 97, 167, 227, 359
Tobyhanna 19, 23, 105, 110, 115–117, 120, 124–129, 132, 134, 136, 140–141, 144, 147, 151, 154–155, 157, 159–162, 165, 167, 170–171, 185, 189–190, 194, 197–198, 200–205, 243, 260, 322, 329, 367, 375–376, 422, 425

tornadoes, classification of 284-286; storms 30, 36, 38, 47, 52, 56, 119, 207, 220–221, 229, 233, 271, 273–274, 267, 271, 274, 278–360, 389, 391, 399, 408, 410, 422, 420, 422–423, 427, 431–432

Totten Glacier 444

Towanda 106, 120, 135, 138–139, 145, 149, 152–153, 165–167, 190, 239, 371, 386

trade winds 14, 33, 46, 390, 422, 440, 451, 453–454

tropical cyclones (*see*: hurricanes) 19, 46, 58, 62, 389–391, 393–394, 396, 399–400, 436, 456

Trappe 29, 78–79

Trenton 7, 29–30, 78, 84, 94, 152, 160, 166, 170, 210, 308, 332

Triassic Period (251-191 million years ago) 10

Triassic-Jurassic extinction 10

Troy 100, 124, 138, 212, 308–309, 347

Trucksville 222

Truittsburg 334

Trumbull County 326

Turtle Creek 417

Tyrone 143, 225–226, 340, 354

U

Union City 325, 357; Filtration Plant 21, 204

Union County 183, 256, 329, 348, 416

Uniontown 74, 151, 316–317, 331, 355, 382–383, 409, 417

Unionville 301, 332

Upper Midwest 50, 52–54, 90, 200, 378, 445, 451, 458

Upper Perkiomen Valley 29

U.S. Weather Bureau 17, 42, 44, 98, 169, 242, 311, 355, 392, 410, 417

V

Valley Forge 29, 79, 288, 425

Vandergrift 371

Venango County 227, 305, 325, 327, 335, 348, 357

Vries, David Pieterszoon de 25

W

Wabush Creek 230

Walter,. Y. S. 41, 213, 294

Wapwallopen 215, 329

Warren 103, 140, 157, 159-160, 194, 199, 384

Warren (OH) 317

Warren Center 155

Warren County 206, 286, 325, 328, 348

Warren County (NJ) 70, 133, 169, 247–248

Washington 382, 409

Washington County 83, 121, 170, 254, 264, 291, 316–317, 334, 353, 357, 359, 368, 409

Washington Crossing 247, 259, 266, 351

Washington, D.C. 41, 82, 87–88, 90–92, 94, 96, 98–99, 101–102, 106–108, 110–111, 114, 118, 120, 122, 124, 130, 132–134, 141–144, 148, 150, 152, 159, 161, 163–164, 166, 168, 170, 176, 179, 181, 185–187, 190–191, 201–203, 205, 264, 276, 290, 298, 305, 309, 339, 354, 363, 365, 369, 375–376, 379–380, 403, 407, 410–411, 413–414, 416, 419, 425

Washington, George 27–28, 77–78, 80, 82, 163, 401–402; Bridge 331

Washington County 83, 170, 254, 291, 316–317, 353, 359, 368, 409

Waterford 27–28, 103, 107, 325

Watson, John 27, 33, 41, 76–77, 81, 86, 158, 223, 363, 403

Watsontown 329

Wayne County 40, 96, 99, 101, 109, 113, 123, 129, 141, 180, 186, 196, 221, 229–230, 236, 240–242, 247, 258–259, 266, 268, 294, 345–346

Waynesburg 317, 382

Waynesburg Hills 6

weathering, chemical and mechanical 3, 8–9, 11, 13, 438

Webster, Noah 35

Wegener, Alfred 8

Weissport 211, 216, 230–231, 236, 241

Wellsboro 92, 120, 122, 138, 142, 146, 153, 156, 161, 165–166, 168, 190, 194, 198, 226, 300
Wellsville 321, 372, 374
West Bingham 152, 157, 160, 171, 193, 369
West Chester 35, 43, 87, 110–112, 115, 141, 152, 160, 201, 221, 278, 319, 331, 357, 375, 381–382, 414
West Coast 8, 52, 454
West Mifflin 286, 319–320
West Nanticoke 231
West Newton 233, 417
Westmoreland County 36, 153, 156, 270, 321, 339, 346–348, 357–359, 430
Wharton 234
Wheatland 326
White Haven 180, 216, 241
Whitehall 182–183
whiteout 70, 115, 199
Wilkes-Barre 80, 86, 96, 99–100, 103, 120, 126, 145, 149, 158, 163, 165–166, 170, 189–190, 211–212, 217, 221–222, 228, 230–231, 239, 250–251, 255, 257, 259, 261, 289–290, 292–293, 301, 305–306, 310–311, 331, 347, 355, 366–367, 408, 411, 421
Wilkinsburg 287
Williamson, Hugh 33–34
Williamsport 6, 9, 95–96, 101, 104, 106, 112, 114, 116, 121–122, 124, 127, 135, 138, 146–147, 155, 161–162, 168, 185, 198–199, 201, 203, 205, 217, 225–226, 228, 230, 239, 250, 253, 255, 257, 263, 282, 292, 303, 305, 329, 337, 348, 355, 366, 368–371, 373, 376–378, 381–382, 384, 421, 424
Windber 143, 319–320
Wind Gap 7, 314
winter storms 55, 65, 68, 111, 178, 230, 237
Wisconsinan glaciation 12
Wissahickon Creek 228
Wolfe, Tom 136
Wooddale 106, 359
Wrightsville 116, 308, 318, 330, 355

Wyalusing 129, 211, 333, 355
Wyoming County 231, 335, 341–342
Wyomissing 208, 279

Y

Yardley 248, 255, 330, 333
York 30, 102, 111, 116, 125, 128, 134, 141–142, 148, 150, 189, 197, 231, 236, 249, 251–252, 254, 260–261, 264, 292–293, 295, 300, 303, 318 320–321, 330–331, 367, 369–372, 383–384, 411, 414
York County 131, 134, 137, 243, 261, 266, 280, 293, 308, 311, 314, 318, 320, 330, 340, 355, 360, 411, 414, 418
York Haven 130
York Springs 423
Youghiogheny River 82, 233, 321, 417
Younger Dryas 15
Yucatan Peninsula 10, 17, 249, 391, 396, 421, 432

Z

Zelienople 328, 349

About the Author

BEN GELBER is an Emmy Award-winning television meteorologist and a science reporter at WCMH-TV. He earned first place for Best Weathercast from the Ohio Associated Press Media Editors. He has authored several regional weather history books, including *The Pennsylvania Weather Book* and *Pocono Weather: A History of Eastern Pennsylvania, the Poconos, Northwestern New Jersey*. Ben's articles and interviews have appeared in the *New York Times* ("Ben Franklin on Global Warming") and *Weatherwise* ("Rising Water: The Northeast Floods of 1955"), and many other newspapers and journals. A lecturer in meteorology at The Ohio State University, he also visits Ohio schools and moderates weather and climate presentations. Ben Gelber was inducted into the Ohio Associated Press Media Editors Hall of Fame "for serving with exceptional distinction and honor and upholding the highest ideals of journalism."

www.ingramcontent.com/pod-product-compliance
Lightning Source LLC
Chambersburg PA
CBHW080721230426
43665CB00020B/2573